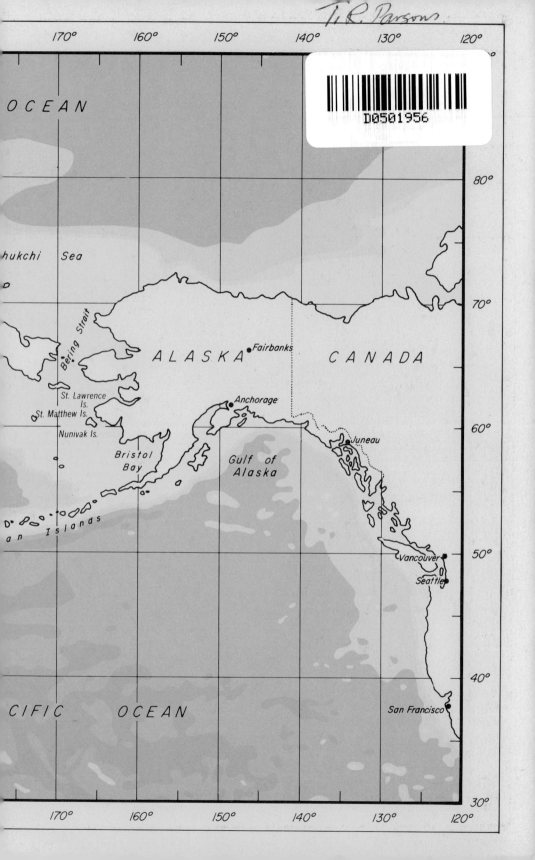

D0501956

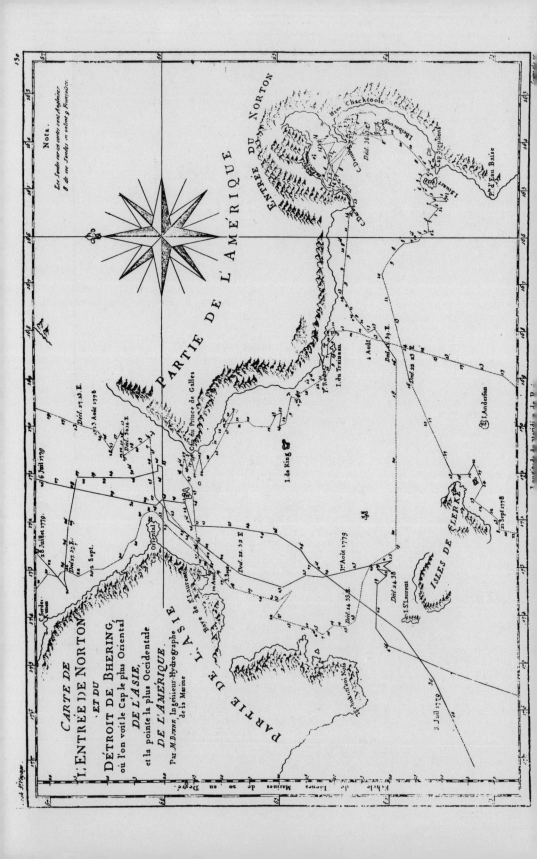

CARTE DE
L'ENTRÉE DE NORTON,
ET DU
DÉTROIT DE BHERING,
où l'on voit le Cap le plus Oriental
DE L'ASIE,
et la pointe la plus Occidentale
DE L'AMÉRIQUE.
Par M. Bonne, Ingénieur-Hydrographe
de la Marine.

PARTIE DE L'AMÉRIQUE

PARTIE DE L'ASIE

ENTRÉE DU NORTON

ISLES DE CLERKE

Nota.
Les Sondes sur cet ombre sont Anglaises
& de ces Sondes on relent 9 Fransaises.

OCEANOGRAPHY OF THE BERING SEA
with emphasis on renewable resources

MAP CREDITS:

French map (opposite) of Bering Strait and Norton Sound, used on Capt. Cook's 3rd Voyage, appeared in the *Atlas Encyclopédique de Padoue,* Paris, 1789. Copy provided by courtesy of F. F. Wright, Division of Statewide Services, University of Alaska, Marine Advisory Board: Anchorage.

Contemporary map shown in Mercator projection (inside front cover) rendered by S. A. Wilson, Institute of Marine Science, University of Alaska: Fairbanks.

OCEANOGRAPHY OF THE BERING SEA

with emphasis on renewable resources

Edited by D. W. HOOD AND E. J. KELLEY

Occasional Publication No. 2
Institute of Marine Science
University of Alaska, Fairbanks

International Symposium for Bering Sea Study

Hakodate, Japan : : 31 January–4 February 1972

cosponsored by

Faculty of Fisheries	Institute of Marine Science
Hokkaido University	University of Alaska
Hakodate, Japan	Fairbanks, Alaska
A. Y. Takenouti	D. W. Hood
Symposium Host-Convener	U. S. Chairman

under support of the U. S. Office for the International Decade of Oceanographic Exploration, National Science Foundation, and the Japan Polar Research Association

foreword

The Bering Sea is an uncommon ocean; neither arctic nor temperate, it has a unique combination of features. Encompassed by Alaska and Siberia, its waters are at times an extension of the North Pacific and yet in winter an icy cove of the Arctic Ocean. In winter the Bering Sea is subjected to limited light, the continental shelf is covered with ice, and the deep western basin undergoes tempestuous storms. During summer the waters are free of ice, but the surface temperature seldom exceeds 8 C.

Uniquely productive, the Bering Sea bountifully provides living resources to nations of North America and Asia. Its international stocks of whales, pinnipeds, and fishes are harvested annually by fishermen from the United States, Japan, the Soviet Union, Canada, and Korea. Such stocks are the foundation of culture and economy for Aleuts and Eskimos on the Alaskan shore. In prehistoric times a land bridge now covered by the Bering Sea was the migratory route of people into North America; the success of these migrations was substantially supported by the high productivity of these waters.

Research in the Bering Sea has been conducted mainly by the Russians, with contributions by Japanese and U. S. investigators. Although most of the work has been done in the calmer seas of the summer, U. S. Coast Guard icebreakers have supported limited work in the winter during recent years. Still, nearly all of the open sea winter studies were made by the Russians over ten years ago.

Today the Bering Sea remains largely a frontier to the marine scientist. The ocean system that sustains such high production on which several nations depend is obscure in spite of a number of excellent research studies. Herein lies the challenge to contemporary oceanographers.

preface

In response to U. S. observations that many unjoined energies of a parallel direction were being invested in study of the Bering Sea by separate institutions and individual investigators, an informal meeting was called on the occasion of the World Oceans Conference in Tokyo in September 1970. U. S. conferees S. Neshyba (Oregon State University) and D. W. Hood (University of Alaska) consulted with Prof. A. Y. Takenouti and other interested Japanese participants present on the possibility of organizing an international Bering Sea symposium and future cooperative program. As a result of this gathering, Professor Takenouti of Hokkaido University, Faculty of Fisheries, assumed leadership and arranged a subsequent meeting in Hakodate in January 1971 between Dr. Hood and Japanese scientists. At this later meeting, a symposium program was outlined, and it was agreed that emphasis should be placed on renewable resources of the Bering Sea. The Faculty of Fisheries of Hokkaido University agreed to host the symposium in Hakodate, and the Institute of Marine Science, University of Alaska, was committed to publish the proceedings in the form of a compendium volume of existing knowledge from which future work could be meaningfully developed.

The International Symposium for Bering Sea Study, co-sponsored by the Institute of Marine Science, University of Alaska, and the Faculty of Fisheries, Hokkaido University, was held in Hakodate from 31 January to 4 February The purpose of the Symposium was to review existing knowledge of the Bering Sea and to discuss informally the scientific problems to be proposed for study under long-range international cooperation. Fifty-one scientists participated (28 from Japan, 17 from the USA, 4 from the USSR and 2 from Canada; listed by roster at the back of the book), in addition to a scientific audience of about 30 members. The scientific portion was composed of eight sessions, in which 31 papers were presented and discussion invited. During two additional days, planning meetings were held, both in plenary assembly and by partitioned working groups, to initiate a comprehensive international program for Bering Sea oceanographic studies.

Support for the host institution was aided by the Japan Polar Research Association, and United States participation was supported by the Office of International Decade for Ocean Exploration under grant NSF GX 30197 ·of the National Science Foundation.

Institute of Marine Science,
University of Alaska, Fairbanks

D. W. Hood
Director

INTERNATIONAL SYMPOSIUM FOR BERING SEA STUDY

Arrangement Committee

Chairman	Dr. Tsuneyuki Saito
Vice Chairman	Dr. Naoichi Inoue
Secretary	Dr. A. Yositada Takenouti

MEMBERS

Mr. Yoshio Akiba	Dr. Masakichi Nishimura
Dr. Takeo Harada	Mr. Tsuneo Nishiyama
Dr. Mitsuo Konda	Dr. Kiyotaka Ohtani
Dr. Teruyoshi Kawamura	Dr. Kenji Shimazaki
Dr. Takashi Minoda	Dr. Shizuo Tsunogai
Mr. Tatsuaki Maeda	Dr. Tokimi Tsujita
Dr. Akira Taniguchi	Dr. Fumio Yamazaki
Dr. Satoshi Nishiyama	Mr. Seikichi Mishima

contents

Introduction

D. W. HOOD *and* E. J. KELLEY

Institute of Marine Science, University of Alaska, Fairbanks

In the geographic fold of two continents, the Bering Sea extends between 51–66°N and 157°W–163°E. Bounded by the land mass of Siberia on the west and that of Alaska on the east, its waters encompass a surface area of 2.3×10^6 km² and comprise a volume of 3.7×10^6 km³ with a mean depth of 1636 meters. As such, the Bering Sea is distinguished as second in size only to the Mediterranean among the relatively confined seas of the World Ocean.

In one sense, the Bering Sea is the northernmost extension of the Pacific Ocean. Through the Aleutian passes Pacific water exchanges relatively freely with the Bering Sea, and many characteristic parameters of North Pacific Ocean water are evident within the Bering Sea. From another standpoint, however, the Bering Sea is a very well-defined oceanic region. It is markedly separated from the Pacific Ocean by about 1200 miles of the Aleutian island arc, braided with limiting sills; its link with the Arctic Ocean is the narrow (85 km) and shallow (45 m) Bering Strait.

EARLY EXPLORATION

'Last of the unsailed seas'

Staging the last great voyages of discovery in the physical world, the Bering Sea was opened for general exploration less than 250 years ago, mainly through the two monumental expeditions of Vitus Bering in 1725–1743, followed 35 years later by the third and final voyage of Captain James Cook.

Although the North Pacific south of about 40° north latitude between Japan and the American coast had been fairly reliably described on maps dating from the late sixteenth century, abyssal ignorance prevailed with respect to the vast region north of that transect. It had not even been determined whether the Asian and American continents were separate or joined, and many notions supported fanciful concepts that assorted islands or another continent lurked in the dim, distant area to the north. Jesuit missionaries arriving in Japan in the mid-sixteenth century were told of an undefined land mass called Yezo and described as "some part of the continent Tartaria." Spiked with gossip from Spanish voyages, seventeenth-century rumors persisted throughout the commercial world that such a land, cultured in silver and gold, abounded in the foggy Pacific.

It was not until 1643, during a Dutch East India Company expedition commanded by Maerten Gerritszoon Vries in the vicinity of Japan, that Yezo became a reality—discovered among what are now known as the Kurile Islands. Many exotic interpretations of Yezo and the neighboring Gama Land found their way onto ensuing maps through the turn of the century, and imaginative speculations prevailed as to the great region beyond. In a memoir read before the Paris Academy of Sciences in 1720, Guillaume Delisle claimed Yezo to be a part of Asia with Japan attached as a peninsula.

Meanwhile, a bargeload of Cossacks had made their way down a Siberian river in 1639 to the Sea of Okhotsk for a first but uncomprehending view of the "Eastern Ocean." By the end of the seventeenth century it had become determined through Russian excursions that the Ob, Yenisey, and Lena rivers emptied into the "Icy Sea" to the north, while the Amur, Ud, and Okhota rivers flowed south to the "Eastern Sea." Still there was no knowledge of a passage between the two bodies of water.

The problem of whether America and Asia were united attracted the dynamic intrigue of Peter I, Czar of Russia from 1682–1725, who initiated his program of westernization by sending Swedish prisoners of war into Siberian ports as teachers of shipbuilding and navigational arts. In what was to become a deathbed effort continued on by successor Empress Catherine, Peter the Great commissioned the first expedition of Danish explorer Vitus Bering and Lt. Alexei Chirikov, who discovered the East Cape of Asia. The sister ships *St. Peter* and *St. Paul* sailed through the Bering Strait in 1728 to settle once and for all the question of two continents. Driven back by native hostilities and climatic severities during his first expedition, Bering embarked on a second voyage in 1733. Leaving in detachments from St. Petersburg, then capital of the Russian empire, a total of 570 men (including 30 or 40 "academists") departed for their destination nearly 5500 miles straightline distant, supplied with a chart from the Imperial Academy showing a total void of land east of Siberia. At the end of Bering's second expedition, costing the explorer's life due to scurvy in 1941, much of the arctic coast of Asia was charted, as well as the southwest Alaskan coast and the Aleutian arc.

In spite of Bering's colossal successes, his work was not altogether accepted as conclusive and was to some extent disregarded. The pressing question of a northern passage from the Pacific prompted the British Admiralty to issue instructions in 1778 to Scotch-born Captain James Cook to proceed on a third voyage "in further search of a North East, or North West passage, from the Pacific Ocean into the Atlantic Ocean, or the North Sea" of Europe, by way of Russia. Cook was known to be aware of recent Spanish expeditions and was likely ordered to follow Bering's route. At the helm of the *Resolution* and assisted by Lt. Charles Clarke in command of the *Discovery*, Cook passed through Bering Strait into the Arctic Ocean during the summer of 1778, working back and forth through the icepack off the American and Asian coasts. Penetrating north to over 70°, Cook viewed the sea called "Chuckchee" and the extremities of both continents—proving nonexistent the navigable northwest or northeast passage to India so long sought from the Atlantic side.

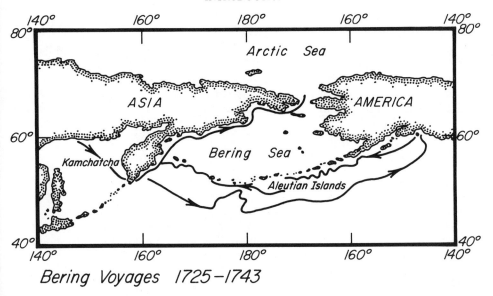

Bering Voyages 1725–1743

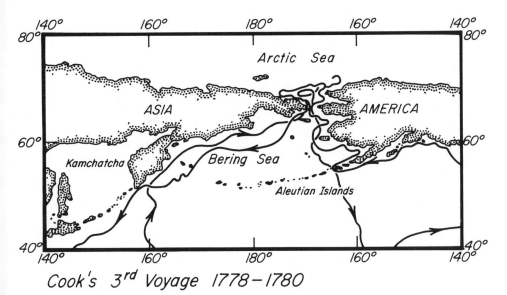

Cook's 3rd Voyage 1778–1780

"In justice to Behrings Memory, I must say he has deleneated this Coast very well and fixed the latitude and longitude of the points better than could be expected from the Methods he had to go by . . ."

Capt. James Cook
4 September 1778

Dramatically demonstrated was the remarkable accuracy of Bering's earlier cartography. The southbound expedition, enroute to warmer waters to wait out the arctic winter, stood into Norton Sound and the river mouths and bays of the southwestern Alaskan coast. The following summer's voyage resumed back through Bering Strait on a similar course but without Captain Cook, the world's most heavily acclaimed sailor, who had been murdered by Sandwich Island natives in 1779.

OCEANOGRAPHIC FEATURES

A productive sea, the Bering

To this day the borders and islands framing the Bering Sea are inhabited by Eskimos in the arctic and southwest coastal areas and by Aleuts along the Aleutian arc. As ancient cultures, these natives subsisted on a hunting and fishing economy for centuries in compatibility with an abundance of fish and animals in their area. Not until western man began exploiting the region to supply large populations of people with furs and fish was the natural living wealth of the area affected. As a result, the sea otter was exploited to the point of near-extinction, flounder stocks were depleted, and the herring ecological niche appears to have been displaced by pollock. Another abundant resource, the sockeye salmon, once endangered and still under pressure, can probably be restored but will require assistance at a level of international cooperation. Only during the last decade has the wide diversity of the living and non-living resource potential of the Bering Sea been fully appreciated. Large harvests of pollock, cod, ocean perch, black cod, halibut, and rattails are now being taken by Korea and the three largest world powers—the Soviet Union, Japan and the United States.

The Bering Sea has an unusually large expanse of continental shelf, which is by far the most productive area of the World Ocean. Of the total area of the Bering Sea (2.3×10^6 km²), 44 percent is shelf, 43 percent is abyssal, and 13 percent is continental slope.

In the world's atmospheric regime, the Bering Sea lies in one of the regions of maximum gradients and, hence, of maximum atmospheric vigor—comparable only to the "roaring forties" of the Norwegian Sea and "furious fifties" of the Antarctic. As a well-defined region, the Bering Sea represents a tractable area for study of oceanic response to intense atmospheric energy input.

Due to its latitude, the Bering Sea lies in a region with very large annual variations in certain properties. Incident radiation in the northern region varies annually from almost complete darkness to almost total light; the wind torque on the sea is an order of magnitude greater in winter than in summer, and an extensive ice cover in winter is totally absent in summer. Thus the Bering Sea appears to be an ideal place to examine certain basic questions such as how organisms maintain themselves during nonproductive periods and how the wind-driven ocean systems respond to changes in wind stress.

Despite what may appear to be a harsh climate for phytoplankton growth, the Bering Sea shelf region supports production rates that are surpassed only by upwelled regions of the world such as exist on the eastern borders of the

Pacific and Atlantic oceans. Oceanographic features of the area must provide the elements necessary for such high sustained productivity; however, the specific functions cannot yet be clearly defined.

The ice cover of the Bering Sea provides oceanographic conditions which support high primary productivity on the undersurface of the ice during periods of the year when the water column itself is not productive. This phenomenon can best be further studied in the Bering Sea, where background data are now available.

The relatively cold waters of the Bering Sea support a large biomass per unit area. One explanation may be that in colder water, the effect of temperature is much greater on respiration rates than it is on photosynthetic rates; if relatively less photosynthetic energy is consumed in respiration, the energy thus reserved can support a larger biomass.

Although warmer seas contain more diverse populations, the colder seas support much larger individual populations. The Bering Sea seems to excel in this respect, as exemplified by such features as one of the largest marine mammal populations in the world, possibly the largest clam population in the world, one of the world's largest salmon runs, one of the largest bird populations per unit area in the world, the world's largest eelgrass (*Zostera*) beds, and high potential annual yields of pelagic fish per unit area.

Some of the most promising oil and heavy-mineral provinces of the world, yet untapped, lie on certain continental shelves most prominently represented by portions of the Canadian Archipelago and the Bering Sea.

At this time there is no significant contamination of the Bering Sea by direct additions of man-made substances, although there is some contribution of airborne materials. The opportunity exists, then, to study the effect of windborne contaminants on the world ocean systems and at the same time to provide baseline data for determining the impact of man on this system as a result of the imminent industrialization of the area.

Political oceanography of the Bering Sea

A unique situation is posed by the geographic location of the Bering Sea between the United States and Russia and by the recent convergence of fisheries interests among these two countries, Japan, Korea and others. It is essential that the principles of national exploitation of the Bering Sea renewable resources be developed through close coordination of research and commercial efforts of all countries concerned in order to assure rational utilization of common properties.

The situation is further complicated by the potential for mineral development, particularly on the eastern Bering Sea shelf. Extensive known mineral deposits exist on the land regions, often accessible only by sea, and in submerged lands which will be developed through underwater placer mining techniques or by oil drilling. The extraction of these mineral resources is imminent and will impinge upon the habitats and ecosystems of the renewable resources. Only through an international understanding of and respect for the total environment as an oceanic system will it be possible to resolve the conflicts that will inevitably arise from exploitation of both renewable and non-renewable resources from the Bering Sea.

DALL, W. H.

1890   A critical review of Bering's First Expedition, 1725–30. *Nat. Geogr. Mag.* 2(2).

DAVIDSON, G.

1901   *The tracks and landfalls of Bering and Chirikov.* Press of John Partridge, San Francisco.

GOLDER, F. A.

1968   *Bering's voyages,* Vol. 1. Amer. Geogr. Soc., Research Series No. 1.

GOVERNMENT SERVICES ADMINISTRATION

1964   *Records relating to the U. S. surveying expedition to the North Pacific Ocean 1852–1863.* National Archives and Records Service, GSA, Washington, D. C. (microfilm avail. Univ. Alaska, Fairbanks).

GREELY, GEN. A. W.

1892   The cartography and observations of Bering's First Voyage. *Nat. Geogr. Mag.* 3: 205–230.

GULLAND, J. A., and A. R. TUSSING

1972   Fish stocks of Alaska and the Northeast Pacific Ocean. In *Alaska fisheries policy,* edited by A. R. Tussing, T. A. Morehouse and J. D. Babb, Jr. Inst. Soc., Econ., and Gov't. Res., Univ. Alaska, Fairbanks, pp. 75–116.

LAURIDSEN, P.

1889   *Vitus Bering: the discoverer of Bering Strait.* Trans., 1969, by J. E. Olson, Select Bibliographics, Reprint Series, Books for Libraries Press, Freeport, N. Y.

McROY, C. P.

1970   On the biology of eelgrass in Alaska. Ph.D. Thesis. Univ. Alaska, Fairbanks.

MENDENHALL, T. C.

1891   *Early expeditions to the region of Bering Sea and Strait.* U. S. Coast and Geodetic Survey, Appendix No. 19—Report for 1890. Government Printing Office, Washington, D. C.

MEYERHOFF, H. A., and MEYERHOFF, A. A.

1973   Arctic geopolitics. In *Arctic geology,* edited by M. G. Pitcher. AAPO Memoir 19: 646–670.

NELSON, H., and D. M. HOPKINS

1969   Sedimentary processes and distribution of particulate gold in the northern Bering Sea. *U. S. Geol. Survey Open File Report.*

PRICE, A. G. (ED).

1971   *The explorations of Captain James Cook in the Pacific* (as told by selections of his own journals 1768–1779). Dover Publications, Inc., New York.

SANGER, G. A.

1972 Preliminary standing stock and biomass estimates of sea birds in the sub-arctic Pacific region. In *Biological oceanography of the northern North Pacific Ocean* [Motoda commemorative volume], edited by A. Y. Takenouti et al. Idemitsu-shoten, Tokyo.

SHEFFER, V. B.

1958 *Seals, sea lions and walruses: A review of the Pinnipedia.* Stanford Univ. Press, Stanford, California.

SHIELS, W. E., and D. W. HOOD

1970 Artificial upwelling in Alaskan fjord estuaries. *The Northern Engineer* 2: 8–12.

VILLIERS, A.

1967 *Captain James Cook.* Charles Scribner's Sons, New York.

Part 1

PHYSICAL PROCESSES RELATED TO BIOLOGICAL PRODUCTIVITY

CHAPTER 1

Flow into the Bering Sea through Aleutian island passes

Felix Favorite

NOAA, National Marine Fisheries Service, Northwest Fisheries Center, Seattle, Washington

Abstract

Present knowledge concerning flow through Aleutian island passes has been accumulated from the records of historical oceanographic cruises and expeditions and through the results of oceanographic research conducted in this area by the Northwest Fisheries Center of the National Marine Fisheries Service since 1955. Flow through the various openings in the Aleutian-Commander island arc is shown to be quite variable in direction and magnitude. Westward volume transports in the Alaskan Stream south of the Aleutian Islands in summer vary more than 50 percent, but there is little evidence of the annual winter intensification suggested by wind-stress transports and reflected in sea level data in the Gulf of Alaska. Analyses of volume transport data indicate a mean northward flow of 14 Sv through openings in the Aleutian-Commander island arc east of the Commander Islands. This result falls between two Soviet estimates of 8 and 19.5 Sv, and the value is quite close to the estimate of 16 Sv obtained from wind-stress data. It is suggested that moored buoy arrays be established to obtain long-term measurements of actual flow and that a series of monitoring stations be designated and occupied on a cooperative basis.

INTRODUCTION

The Northwest Fisheries Center of the National Marine Fisheries Service (NMFS), formerly the Seattle Biological Laboratory of the Bureau of Commercial Fisheries, has conducted an oceanographic field program in the central North Pacific Ocean and Bering Sea since 1955. A substantial portion of the

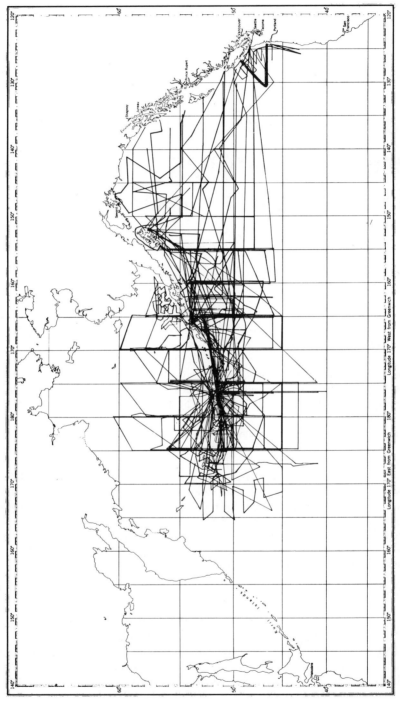

Fig. 1.1 Cruise tracks of Northwest Fisheries Center vessels along which oceanographic data were obtained, 1955–70.

oceanographic data-base in this subarctic region (Fig. 1.1) has been provided by cruises both of charter vessels (*Paragon, Mitkof, Celtic, Tordenskjold, Attu, Pioneer, Marine View* and *Bertha Ann*) and by government research vessels such as the *John N. Cobb, George B. Kelez,* and *Miller Freeman,* (Favorite 1969). These cruise data have been used in numerous monographs as well as two major oceanographic summaries (Dodimead et al. 1963; Arsen'ev 1967). The NMFS program was conducted to ascertain knowledge of conditions and processes in the environment of the Pacific salmon that would contribute to solutions of problems confronting the International North Pacific Fisheries Commission. Although there was little chance to conduct oceanographic investigations independent of fishing operations, there was an excellent opportunity to acquire extensive oceanographic observations in an area where knowledge was either non-existent or fragmentary. The large concentrations of salmon, particularly sockeye, in the Aleutian area during late spring and early summer, when the Japanese high seas salmon mothership fleet was operating, resulted in focusing much of the field work on this region. Because the major flow of water into the Bering Sea occurs through the passes of the Aleutian-Commander island arc, this work is particularly pertinent to the scope of the International Symposium for Bering Sea Study. Although accumulation of knowledge has been painfully slow and is still quite inadequate, a chronology of this accumulation should prove instructive in devising guidelines for further studies.

Our studies in the Aleutian area can be divided into the following phases: (1) *General exploration* (1955 to 1962): observations were obtained from aboard charter vessels over a wide area, and continuity of the Alaskan Stream as well as general environmental conditions during summer were defined. (2) *Vessel acquisition and gear development* (1963 to 1965): the R/V *George B. Kelez,* a vessel of sufficient size to permit year-round cruises in subarctic waters, was acquired and equipped with modern electronic oceanographic devices, and field work was limited to test cruises off the Washington coast. (3) *Monitoring* (1966 to 1968): conditions in the Alaskan Stream south of Adak Island were monitored to ascertain seasonal fluctuations. (4) *Special studies* (1969 to present): direct current measurements were made in the Alaskan Stream, and intensive investigations were conducted in the central and eastern Bering Sea.

BACKGROUND

Aleutian-Commander island arc

Favorite (1967) lists 39 openings through the Aleutian-Commander island arc, only 14 of which have an area greater than 1 km^2 or sill depth of at least 200 m. The Aleutian passes can be classified into three groups: eastern, central and western, in addition to the Commander-Near and Kamchatka straits (Table 1.1). These five major openings constitute 99.9 percent of the area penetrating the island arc (Fig. 1.2). Not until 170°W do depths exceed 400 m. Amchitka Pass, in the central Aleutian group, is the only pass east of 172°E that exceeds 1000 m, and only in Kamchatka Strait are depths in excess of

TABLE 1.1   Depth and area of the major openings in the Aleutian-Commander island
arc

General opening	Pass/Strait	Depth (m)	Area (km²)	
East Aleutian group	Samalga	200	3.9	
	Chuginadak	210	1.0	
	Herbert	275	4.8	
	Yunaska	457	6.6	
	Amukta	430	19.3	
	Seguam	165	2.1	37.7
Central Aleutian group	Tanaga	235	3.6	
	Amchitka	1155	45.7	49.3
West Aleutian group	Kiska	110	6.8	
	Buldir	640	28.0	
	Semichi	105	1.7	36.5
Commander-Near Strait	Near	2000	239.0	
	Commander	105	3.5	242.5
Kamchatka Strait		4420		335.3
			Total area	701.3

2000 m. Most of the early analyses of geostrophic currents and transports are
based upon reference levels of 1000 or 1500 db (approximately 1000 and
1500 m); when looking for continuity of flow, it is necessary to take into
account the depths of the individual passes in relation to this reference level.

Early oceanographic investigations
Although there is evidence that surface observations of temperature and even
density were made in the Aleutian area by early whalers, sealers, and cod
fishermen, these data are not documented. It was not until 1888 when the
U. S. Fish Commission Steamer *Albatross* commenced a series of cruises that
not only surface but also bottom temperatures became part of the oceanographic literature (Townsend 1901). In fact, most of the early information concerning flow in the eastern Aleutian passes probably stems from reports based
upon data from *Albatross* cruises. For example, while the vessel was in Unalga
Pass on 28 July 1888, it was reported that the tide rushed through the narrows
with great force, causing heavy rips and at times overfalls but was quite smooth
at the time of high water (Tanner et al. 1890). It was also noted that the only
excuse for describing the pass was that there was no published information
concerning it, although, particularly in thick weather, it was the best route to
or from Unalaska. Rathbun (1894), in a brief statement summarizing summer
conditions in the eastern Bering Sea near the Alaska Peninsula in 1890, noted
that tidal currents were strongest in the vicinity of Unimak Pass with the flood
setting northward and ebb to the southward, the former being invariably the
stronger. *Albatross* cruises in this area were terminated in 1906; although local

knowledge was undoubtedly obtained from reports of vessels in this area, there are no available records of any subsequent scientific studies until 1933. The following excerpt was attributed to the United States Coast Pilot for 1931 (U. S. Coast Guard 1936), however, and it has only been in the last decade that the validity of these generalities have been challenged.

"As far west as Attu Island water flows through the passes of the Aleutian Islands from the Pacific to the Bering Sea; a rising tide increases the current to the north. A falling tide reverses it to the south but at a smaller velocity; and immediately north of the Aleutian Islands, from Attu Island to Unalaska Island the currents set toward the east and are not affected by tides."

Although a series of observations well offshore along the north side of the Aleutian Islands was obtained during a cruise of the Soviet vessel *Krasnoarmeets* in 1932 (Lisitsyn 1966, p. 7), the first modern-day oceanographic investigation adjacent to the Aleutian Islands was conducted in the vicinity of Adak Island (176–178°W) as part of the Aleutian Islands Survey Expedition in 1933 aboard the USS *Gannet*. Results of this and a subsequent investigation aboard the USCGC *Chelan* northward of Unalaska Island (166–168°W) in 1934 were reported by Barnes and Thompson (1938). The lines of stations occupied by the USS *Gannet* were generally north-south, and east-west flow was deduced from dynamic typography referred to 1000 db. Surface waters at 65 km south of the island were reported to move north with velocities at times greater than 15 cm/sec. Closer inshore this flow was deflected to the west by the land, and the major flow was considered to be through Tanaga Pass with a sill depth of less than 300 m. Just north of the island the direction of flow changed markedly from northeasterly to easterly from June to August, and maximum velocities in excess of 30 cm/sec were reported. Water of abnormally low temperature in this area in August (1.74 C versus 3.14 C in June) was believed to have originated along the Siberian coast.

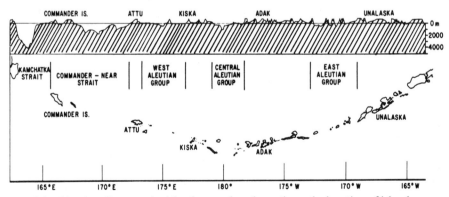

Fig. 1.2 Aleutian-Commander island arc and a schematic vertical section of island passes and straits.

Cruises of the Japanese warships *Komahashi* in 1934, 1935 and 1936 and *Itsukushima* in 1935 (Japan Agricultural Technical Association 1954) provided widely spaced station data (temperature and salinity to 4000 m) in the vicinity of Attu and Commander islands. Mishima and Nishizawa (1955), in a composite estimate of dynamic topography based upon data widely spaced in time (*Carnegie* 1929, *Gannet* 1933, *Itsukushima* 1935, *Oshoro Maru* 1953 and 1954), suggested that westerly flow along the south side of the Aleutian Islands was continuous to 165°E. Sugiura (1958) combined data from the 1934 to 1936 cruises of the *Komahashi* and derived a similar conclusion, as did Koto and Fujii (1958) in combining data from the 1936 cruise of the *Komahashi* and a 1956 cruise of the *Oshoro Maru*. Although it was difficult to assess the validity of the proposed continuity of flow based upon these data because of the separation in time and space of station data, the data from recent cruises of Japanese research vessels have provided sufficient information to support these conclusions.

The cruise tracks of the USS *Oglala* in the vicinity of the Rat Islands (179°W to 177°E) in 1935 as shown by Lisitsyn (1966, p. 7) suggest a survey as comprehensive as that of the USS *Gannet* in 1933, but any oceanographic data collected or records of analyses appear to be lost.

Goodman et al. (1942), reporting on field work in summer 1937 and 1938 aboard the USCGC *Northland*, noted a pronounced movement of water into the Bering Sea through all the Aleutian passes; in a schematic current diagram embracing all known investigations, including those of Ratmanoff (1937), however, an intense northward flow was shown through the Commander-Near Strait and extending in a northeasterly direction across the Bering Sea basin, together with a somewhat reduced northward flow at the extreme eastward end of the island arc near Unalga and Unimak passes (where sill depths are less than 100 m).

Some oceanographic stations were obtained in Unimak Pass in summer 1940 during a cruise of the USCGT *Redwing* (Favorite et al. 1961). Except for a report of an HMCS *Cedarwood* cruise in the northeastern Bering Sea in summer 1949 (Saur et al. 1952), in which the existence of a northeasterly surface current across the Bering Sea basin was challenged, no further oceanographic field work was reported in the Aleutian area for almost a decade.

Direct current measurements

Records of direct current measurements in the Aleutian passes by the U. S. Coast and Geodetic Survey commenced in 1934 with data collected near Unimak Pass. These data are of considerable value for ascertaining maximum tidal velocities and thereby denoting potential hazards to vessels or for assistance in navigation; however, non-tidal components of these measurements are certainly of limited value because of fluctuations in flow of the Alaskan Stream, which will be discussed in a subsequent section. Five of the 18 records[1] were of only one-day duration; the 13 other measurements shown, taken during the period 1934 to 1946, represent average non-tidal flows over periods of two to eight days (Fig. 1.3). They are shown here to demonstrate the near futility of

---

[1] On file at NOAA Headquarters, Rockville, Maryland 20952.

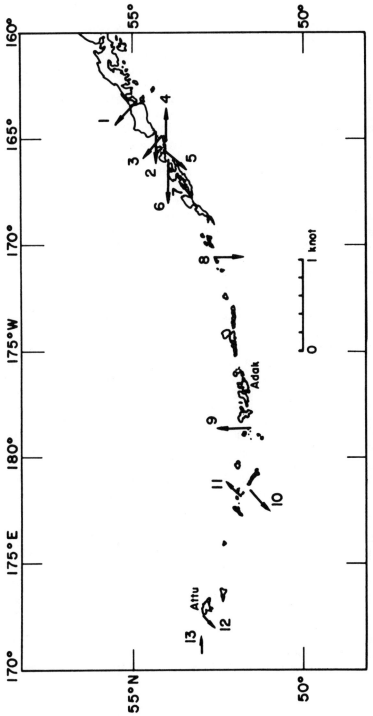

Fig. 1.3 Nontidal components of surface flow in passes, 1934–46 (from direct current measurements of duration in excess of 24 hours).

mounting a limited program of direct current measurements to define conditions through the passes.

Reed (1971) has analyzed 25 surface tidal flow records from radio current meters suspended from buoys anchored in various passes for intervals of over 4 days during the period 1949 to 1956. It is not known to what extent motion of the buoy contributed to these records, but northerly and southerly net flows were evident from Unimak to Oglala passes. Both north and south net flows were recorded at some locations, although at different times. Observations from, or telemetered to, an anchored vessel are particularly expensive and at times impractical, especially in the deeper areas, because the movement of the vessel or buoy can be greater than the flow. This was clearly demonstrated by observations obtained aboard the USCGS *Surveyor* at anchor in Amchitka Pass (51°35′N, 180°) in summer 1963.[2] Precise positioning at 15-min intervals indicated vessel velocities as great as 85 cm/sec.

Although these direct measurements challenged the concept that a net northward flow occurs through all passes, they provide little information on the origin or character of the water constituting the flow that is obtained from oceanographic station data. Although station data cannot be as easily interpreted with regard to flow as direct current observations because of various *in situ* processes, they do enhance the validity of short-term, direct measurements and provide some understanding of the forces operating in the system.

NORPAC Expeditions
It is clear that knowledge of oceanographic conditions in the Aleutian area prior to 1955 was extremely limited, as was information on conditions in the North Pacific Ocean (Fleming 1955). Two major events occurred in 1955, however, that resulted in a significant increase in oceanographic activities. The first was the NORPAC Expeditions, a major international oceanographic survey of the North Pacific Ocean (NORPAC Committee 1960). Data from the NORPAC Expeditions in the Aleutian area were limited, and an interpretation of circulation as deduced from geopotential topography (referred to 1000 db) suggested not only that almost all flow out of the Gulf of Alaska along the Alaska Peninsula was discharged northward into the Bering Sea through Unimak Pass but that any residual westward flow terminated near 175°W. The former was actually a physical impossibility because the sill depth of Unimak Pass (165°W) is only 60 m. The latter idea was equally perplexing; although it was generally in agreement with the work of Barnes and Thompson (1938), it differed from the conclusions of Mishima and Nishizawa (1955), Koto and Fujii (1958), and Sugiura (1958). There is little question that if these results were known prior to the NORPAC Expeditions, however, some attempt would have been made to obtain more data in the western Aleutian passes and Commander-Near Strait.

INPFC Investigations
The second event to occur in 1955 was the commencement of extensive field studies in the subarctic region by the International North Pacific Fisheries

---

[2] OPR 421 on file at NOAA Pacific Oceanographic Research Laboratory, Seattle, Washington 98102.

Commission (INPFC); some aspects of this program are still being conducted today. Although the grand and synoptic summer oceanographic picture of the North Pacific Ocean provided by NORPAC was of value to INPFC studies, there were other considerations of equal or more importance. There was a constant real-time urgency to the INPFC research because of the extensive commercial salmon fishing that was being conducted during spring and summer by literally hundreds of ocean-going vessels. A need existed for continuity in time and space of oceanographic observations throughout the year for investigation of the relations between salmon and their environment. Aspects of the early phases of these studies have been summarized by Uda (1963) and by Dodimead, Favorite et al. (1963). It is primarily to this Commission that one must look for the impetus for much of the environmental research conducted in the Bering Sea in recent years. Arsen'ev (1967) analyzed these data up to 1959, together with all historical data, and presented the most recent schematic of flow showing anticyclonic gyres around the central Aleutian Islands in summer. These flows are referenced to 3000 db; however, only Amchitka Pass has depths greater than 1000 m. He also shows an intense northward flow in the eastern Bering Sea during winter.

OCEANOGRAPHIC CONDITIONS

Surface conditions

General oceanographic surveys in the late 1950s showed clearly that surface salinity was an indicator of surface flow. Although great discretion should be exercised in using a single chemical parameter to deduce flow, there is no question that snowmelt and runoff from the mountains ringing the Gulf of Alaska result in a dilute coastal band of water that is carried westward along the south side of the Alaska Peninsula (Fig. 1.4). Normally this flow diverges near the eastern Aleutian passes, sending one branch southward to form part of the cyclonic circulation in the Gulf of Alaska and the other westward along the south side of the Aleutian arc as the Alaskan Stream; however, there may be one or more branches of this type. Although there is some flow northward into the Bering Sea through the more eastern passes of this group, there is a southward flow from the Bering Sea through the more western passes of this group as indicated by the occurrence of high salinity water (33.0‰) characteristic of the Bering Sea near 170°W. A northward flow is indicated through the central Aleutian passes.

During the monitoring phase of our studies, data from a constantly recording surface salinity and temperature device attached to the ship's seawater intake line permitted boundary observations of this band of dilute water (<32.8‰), which extends westward along the south side of the Aleutian Islands to south of Adak Island (176°20′W). In June 1969, both the north and south boundaries were clearly identified by marked changes in salinity (Fig. 1.5). Although there is little evidence of the southern boundary in temperature values and only a change of 1 C at the northern boundary, by August a change of over 6 C occurred at the northern boundary and is attributed to vertical mixing in the inshore waters, most likely in the passes. The most striking feature

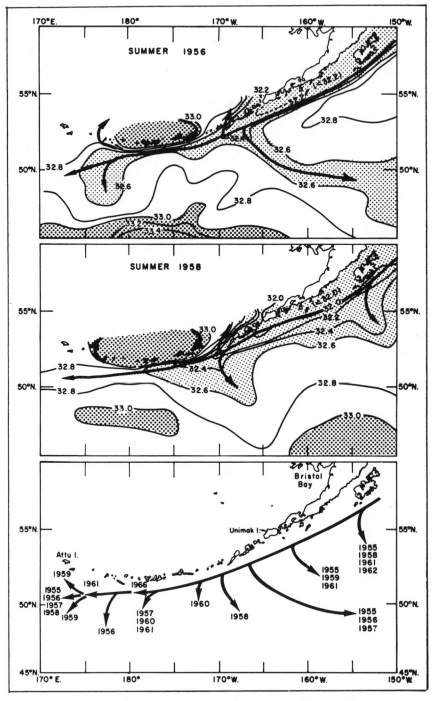

Fig. 1.4   Distribution of surface salinity and implied surface flow, 1955–61.

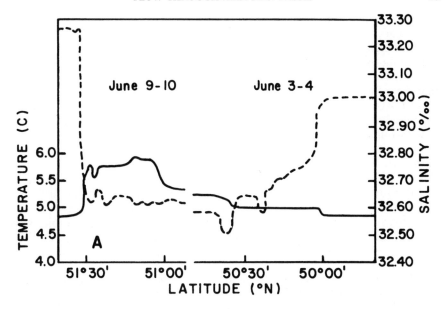

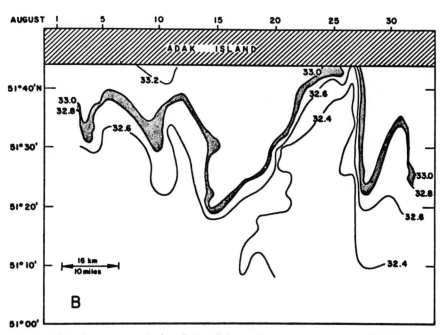

Fig. 1.5 Continuous record of surface salinity and temperature south of Adak Island during two periods in June 1969, showing (A) the sharp boundaries of dilute surface water flowing westward in the Alaskan Stream and (B) fluctuations in latitude of the salinity front at the north boundary in August 1969.

of the northern boundary, however, is its fluctuation with time. A composite picture of numerous crossings of the front over a 30-day period shows that its distance offshore fluctuated from 0– >40 km. The representation of the wave-like configuration of the front with time is of course influenced by the sampling periods, but there is no question concerning the amplitude and no question that these fluctuations are reflected in changes in flow through the eastern and central Aleutian passes. It would appear that direct current measurements over weeks, not days, would be necessary to ascertain any indication of net flow through the passes.

Drift bottle experiments

The limitations of such experiments are well known; discounting wind effects on the bottle, one may find in the results only a reflection of Ekman drift. It is further recognized that the relatively uninhabited, volcanic island arc presents an unusually severe challenge to glass bottles; yet, interesting results have emerged. Eight drift experiments were made from 1957 to the present (Favorite and Fisk 1971; Fisk 1971). Only some results of three of these experiments are presented here.

Drift bottles were released from the M/Vs *Attu* and *Pioneer* at 10 locations near the Aleutian Islands in late spring 1958 (Fig. 1.6). One bottle released south of Unimak Island indicated a westward flow south of the eastern Aleutian passes, and another released north of these passes was recovered far north on St. Lawrence Island. Similar drifts were reported from the studies of the 1930s, but new information concerning surface circulation in the Aleutian area was provided by recoveries on the Washington-Oregon coast of bottles released south of Adak Island and Amchitka Pass and by recoveries on Attu Island of bottles released east and south of the island. Most of these drifts were suggested by the surface salinity distribution (Fig. 1.4), but the apparent westward (rather than north or northeast) drift in the central Bering Sea was unexpected.

Releases of bottles well south of the Aleutian Islands from a merchant ship-of-opportunity in fall 1964 implied a northerly drift from five of seven release locations southward of Adak Island, several recoveries being made on Attu Island (Fig. 1.7). The affinity of Attu Island for drift bottles released from all directions is reflected in recoveries of bottles released from the R/V *George B. Kelez* during an oceanographic cruise in winter 1966 (Fig. 1.8). Oceanographic conditions observed at the times of release reflected an intense northward flow through the Commander-Near Strait (Favorite et al. 1967; McAlister et al. 1970; Ohtani 1970).

Subsurface conditions

Water structure all along the south side of the Aleutian Islands is typified by conditions along 175°E, which passes through Buldir Pass in the western Aleutian area (Fig. 1.9). The southern boundary of the warm coastal water (>4 C) is marked by the ridging of isopleths, not only of salinity (and therefore density, because the relatively uniform 3 to 4 C temperatures have little effect on density in this area) but also of dissolved oxygen. This feature serves to indicate the southern boundary to the westward flowing Alaskan Stream, and

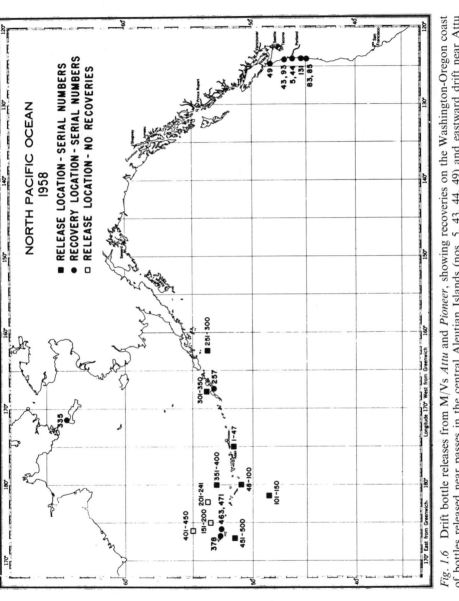

Fig. 1.6 Drift bottle releases from M/Vs *Attu* and *Pioneer*, showing recoveries on the Washington-Oregon coast of bottles released near passes in the central Aleutian Islands (nos. 5, 43, 44, 49) and eastward drift near Attu Island (no. 378).

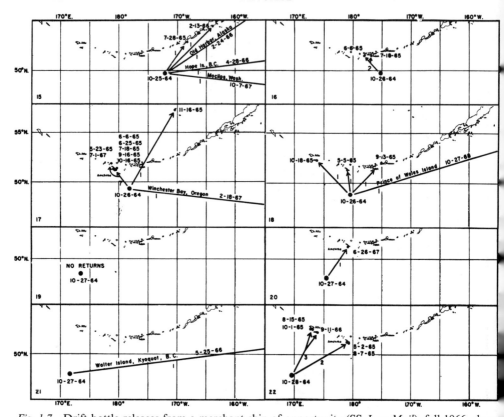

Fig. 1.7 Drift bottle releases from a merchant ship of opportunity (SS *Java Mail*), fall 1966, show northward drift south of the central Aleutian Islands (after Fisk 1971).

little north-south exchange is possible except in the very upper layers (< 100–200 m) without eliminating the ridge structure; although this occurs, it is the exception rather than the rule in this area.

In regard to flow through the Aleutian passes, the distribution of temperature at 100 m in 1958 (Fig. 1.10) indicates flow into the Bering Sea through the eastern, central and western Aleutian passes. This is a common occurrence in summer, but it is difficult to ascertain the precise nature of the temperature distribution or to estimate the relative volume of flow for any given time period because of the paucity of observations in time and space across or north of any of the passes.

A southward intrusion of water at the western side of the eastern Aleutian passes is evident when temperatures of <4 C occur at depths greater than 200 m. This has been observed only in late winter and early spring, and the intruded water is found only westward of passes in the eastern Aleutian group (e.g., south of Atka and Adak islands). Arsen'ev (1967) suggests that the southward intrusion occurs during all seasons and shows an intense anti-cyclonic circulation around the islands between the eastern Aleutian and central Aleutian passes. North and south of the island arc, this flow is evidently

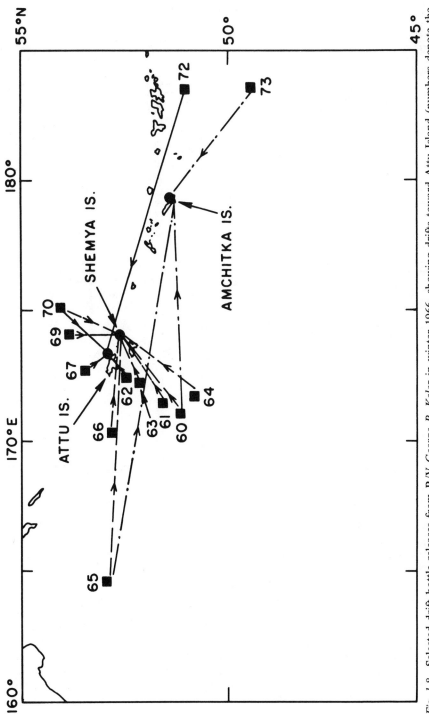

Fig. 1.8 Selected drift bottle releases from R/V *George B. Kelez* in winter 1966, showing drifts toward Attu Island (numbers denote the serial numbers of drift bottle cards released in quantities of 96 at the indicated locations).

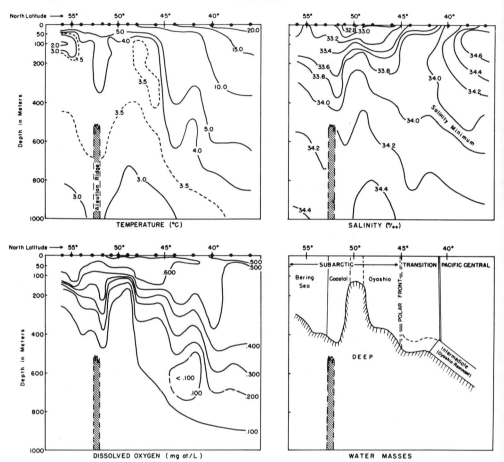

Fig. 1.9 Schematic of water masses north and south of the Aleutian Ridge at 175°E (Buldir Pass), showing ridge structure at southern boundary of the coastal water mass (Alaskan Stream).

relative to 3000 db; depths in the passes, however, are <500 and 1100 m, respectively. McAlister et al. (1969, 1970a) show evidence of an eastward countercurrent in a vertical section of temperature that was observed to extend in winter 1967 along the Alaska Peninsula (162°W) to a depth of 1000 m influence. This was attributed to a southward movement of Bering Sea water through the eastern Aleutian passes. As indicated above, however, the maximum depth of these passes is <500 m. The water structure, which in this case reflected an eastward transport of about 2 Sv,[3] is quite similar to that which occurred south of the Alaska Peninsula as far east as 155°W in winter 1962 (Favorite et al. 1964). Its cause requires further investigation. In both instances the main axis of the high velocity near-surface flow of the Alaskan Stream normally found adjacent to the continental shelf was shifted over 100 km

---

[3] Sv indicates Sverdrups = $10^6 m^3/sec$.

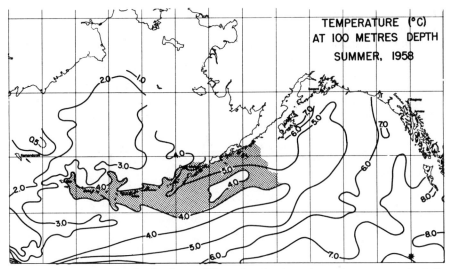

Fig. 1.10   Temperature distribution at 100 m, showing intrusions of warm water (<4 C) moving westward along the south side of the Aleutian Islands.

farther offshore. Thus, its axis has a variable position along the peninsula (Fig. 1.11).

Conditions south of Adak Island (176°20′W) were monitored during cruises of the R/V *George B. Kelez* from 1966 to 1969. Variability is evident in distributions of temperature and salinity and is reflected in geostrophic velocities from August 1967 to July 1968 (Fig. 1.12). At this location, it appears

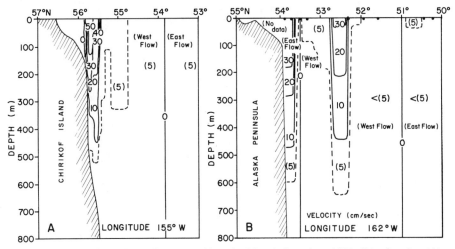

Fig. 1.11   Vertical profiles of geostrophic velocities (referred to 1000 db), showing (A) main axis of Alaskan Stream near the continental shelf at 155°W, but (B) well offshore at 162°W (R/V *George B. Kelez*, winter 1967).

that in late winter and early spring 1968 there was an interruption in the continuity of warm (>4 C) water in the Alaskan Stream. This is attributed to a southward intrusion of Bering Sea water through the eastern Aleutian passes and its subsequent advection westward. This is not an anomalous condition, having been reported at this location in 1957 (INPFC 1959, p. 80). There is evidence that during this period recirculation into the Gulf of Alaska may be intensified; however, as will be shown by volume transports, the westward flow in the Alaskan Stream was undiminished in winter or spring 1968. Anoma-

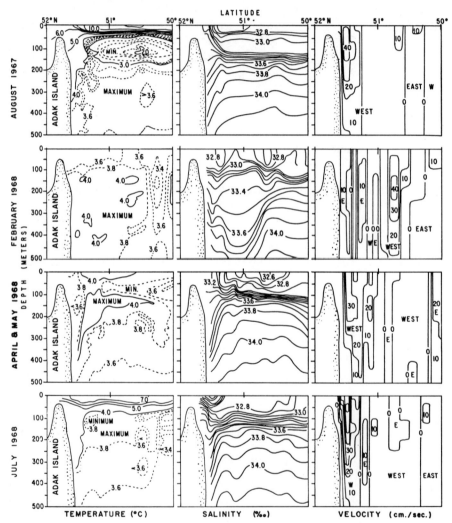

Fig. 1.12 Vertical profiles of temperature, salinity, and geostrophic velocities (referred to 1500 db), showing seasonal variability of conditions in the Alaskan Stream (R/V *George B. Kelez*, August 1967–July 1968).

lous conditions occurred in February 1968, and the main axis of the Alaskan Stream (as evidenced by velocities >40 cm/sec) was over 100 km farther offshore to the southward than during the other periods.

Subsurface temperatures in summer 1957 suggest a southward flow in the western Aleutian passes, contrasted to a northward flow in summer 1958 and 1959 (INPFC 1959, p. 80). Temperature data for other years are insufficient to indicate with any assurance whether or not a southward flow could occur on the eastern side and a northward flow on the western side, or vice versa. Kitano (1967) shows north and south flow in these passes and in the Commander-Near Strait in summer 1964. This dilemma can be avoided by considering net flow based upon volume transports.

VOLUME TRANSPORTS

Into the Bering Sea

Several conflicting opinions prevail in regard to recent interpretations of the amount and location of flow through the passes. Leonov (1960) reported that flow occurred through numerous passes but that it was not known through which passes or in what amounts; he presumed that undoubtedly inflow and outflow could occur at the same depth or at various depths in any of the passes. Flow through the Kamchatka and Commander-Near straits was considered of the greatest significance, but little exchange occurred in the eastern and central Aleutian passes. Leonov estimated an annual water budget in which 100,000 km³/yr (3.1 Sv) flowed in and 62,136 km³/yr (2.0 Sv) flowed out through the Aleutian-Commander island arc; maximum inflow occurred in September and minimum inflow in March. On the other hand, Natarov (1963) reported that in both winter and summer most of the flow into the Bering Sea moved through the eastern and central Aleutian passes; on the basis of 1958 and 1959 data, he concluded that the Commander-Near Strait was not a significant source of Bering Sea water. It is possible, however, that Natarov was referring to only surface waters rather than to total transport. Moiseev (1964) reported that surface waters flowed into the Bering Sea mainly through the eastern Aleutian passes but that this inflow increased in winter as a result of increased wind-stress (atmospheric circulation). Batalin (1964) reported an inflow of 100,000 km³/yr through the Kamchatka and Commander-Near straits and of 50,000 km³/yr through all Aleutian passes, a total of 150,000 km³/yr (4.8 Sv) based upon a heat budget study; but he suggests also a flow of 8 Sv based upon wind-stress calculations across the western straits, or 9.5 Sv if the local wind field over Bering Sea is considered.

It is to Arsen'ev's (1967) excellent summary of oceanographic conditions in the Bering Sea that one must look for resolution of these Soviet conclusions; after an exhaustive analysis of all station data prior to 1959, he clarifies many of the dilemmas (Table 1.2). The Commander-Near Strait is emphatically considered the most significant source of inflow (14.4 Sv), with a considerably reduced flow in the central Aleutian passes (4.4 Sv); no net exchange was considered to occur in the eastern Aleutian passes. Although a rather vague

TABLE 1.2   Flow (Sv) through Aleutian-Commander island arc according to Arsen'ev (1967)

Pass	Transport out	Transport in
Kamchatka	21.0[1]	2.6[2]
Commander-Near	—	14.4
Western Aleutian group	—	0.7
Central Aleutian group	—	4.4
Total	21.0	22.1[3]

[1] Above 3000 m
[2] Below 3000 m
[3] Loss through Bering Strait −1.1; total transport east of Commander Islands −19.5

method was used to ascertain a standard reference level, it appears that these transports are basically referred to 3000 db. It was also apparent from Arsen'ev's analysis that, except for the deep inflow in Kamchatka Strait, almost the entire flow into the Bering Sea is derived from the Alaskan Stream. Ohtani (1970) has also reported this to be the case in winter 1966.

In the Alaskan Stream
First estimates of volume transport in the Alaskan Stream along the south side of the Aleutian Islands were presented by Sugiura (1958). Since that time numerous data have been presented (Table 1.3), most of them obtained by the R/Vs *Oshoro Maru* and *George B. Kelez*. Favorite (1967) indicated continuity of flow from 165°W to 172°E of about 6 Sv during summer 1959; Ohtani (1970) has shown similar continuity of about 8 Sv during winter 1966. Both analyses suggest a southward flow through eastern Aleutian passes (0.2 and 0.7 Sv, respectively) and a northward flow through the central Aleutian passes (1.0 and 4.3 Sv, respectively). The summer data reflect a northward flow through western Aleutian passes (1.8 Sv), and the winter data show a southward flow (2.8 Sv). During both periods, a northward transport of 3 Sv occurred at the southern boundary of the Stream near 180° and compensated for some of the losses through the central Aleutian passes. These flows are relative to the 1000-db level and thus represent flow outside the bounds of all passes that are <1000 m deep.

In the Aleutian area, it is only in the central Aleutian passes (Amchitka Pass) that depths exceed 1000 m. Yet, Ohtani's estimate of northward flow into the Bering Sea through the central Aleutian pass group (4.3 Sv) is almost identical to that of Arsen'ev (4.4. Sv). Equally surprising are the values obtained from data on Amchitka Pass during two cruises of the R/V *George B. Kelez* (February 1968 and September 1970; Fig. 1.13), which reflect northward transports of 5 and 4 Sv, respectively.

In order to compare westward transports in the Alaskan Stream[4] in Table 1.3 with flow through the Commander-Near Strait given in Table 1.2, a factor

[4] Referred to 1000 db

TABLE 1.3 Relative westward transport (Sv) in the Alaskan Stream

Season	Year	Location							Reference Level (db)	Author
		170–172°E	175°E	180°	173–177°W	170–172°W	165°W	160°W		
Summer	1935	10.0	10.0	10.0	5.0	5.0	—	—	0/1500	Sugiura (1958)[1]
Summer	1956	—	5.8[2]	—	—	7.3[2]	—	—	0/600[2]	Koto and Fujii (1958)
Summer	1957	3.6	—	5.0	6.2	5.3	—	—	0/600/800	Ohtani (1970)
Summer	1959	4.9	6.0	4.8	—	4.2	5.9	8.2	0/1000	Favorite (1967)
Summer	1959	—	—	2.8	—	5.1	—	—	0/1000	Ohtani (1970)
Summer	1961	5.4	—	—	—	—	—	—	0/1000	Ohtani (1970)
Summer	1962	5.6	4.2	—	—	—	—	—	0/1000	Ohtani (1970)
Summer	1963	—	—	6.6	—	—	—	—	0/1000	Ohtani (1970)
Spring	1964	7.5	—	7.2	6.2	—	—	—	0/1000	Ingraham and Favorite (1968)
Fall	1965	—	—	—	—	—	8.0	—	0/1000	Ohtani (1970)
Winter	1966	10.0	—	7.0	9.6	—	—	—	0/1000	Ingraham and Favorite (1968)
Winter	1966	—	—	—	9.9	—	—	—	0/1000	Ingraham and Favorite (1968)
Summer	1966	—	—	—	5.2	—	—	—	0/1000	Ingraham and Favorite (1968)
Spring	1967	—	—	—	9.5	—	—	—	0/1500	McAlister, Ingraham et al. (1970)
Summer	1967	—	—	—	8.4	—	—	—	0/1500	McAlister, Ingraham et al. (1970)
Winter	1968	—	—	—	8.2	—	—	—	0/1500	McAlister, Ingraham et al. (1970)
Spring	1968	—	—	—	8.4	—	—	—	0/1500	McAlister, Ingraham et al. (1970)
Summer	1968	—	—	—	5.1	—	—	—	0/1500	McAlister, Ingraham et al. (1970)

[1] From schematic diagram.

[2] Correct values appear to be: 175°E −5.9, 172°W −4.7, 0/800 db.

must be applied which extends these data to 2000 m, the maximum depth of the Commander-Near Strait. According to Ingraham and Favorite (1968), this factor (across the main axis of the Alaskan Stream adjacent to the Aleutian Islands) is 2.2; thus, flow in the Alaskan Stream in summer and winter would be 13.2 to 17.6 Sv, respectively. If we assume no net exchange through either the eastern or western Aleutian pass groups, a loss of 4 Sv into the Bering Sea through the central Aleutian pass group, a gain of 3 Sv across the southern boundary of the stream, and that the residual westward flow west of the western Aleutian passes moves northward into the Bering Sea through the Commander-Near Strait—this results in a flow of 12.2 to 16.6 Sv, which compares favorably with the 15.5 Sv shown in Table 1.2. Data collected during our monitoring cruises south of Adak Island from 1966 to 1968 indicate, however, that although westward volume transport in the Alaskan Stream can be relatively constant over several seasons, it can also vary 50 percent or more from one summer to the next; and, although volume transports in winter 1966 were the highest in recent years, there is usually little change in volume transport between winter and summer.

Reference level
The inadequacies of geostrophic calculations are well known and will not be discussed here, but one cannot help but ponder the validity of volume transports in the Aleutian area in terms of questioning what constitutes an appropriate reference level. During the winter cruise of the R/V *George B. Kelez* in February 1966, which was conducted to ascertain the effect of the Commander Ridge on the westward flow of the Alaskan Stream and the nature of the flow through the Commander-Near Strait (McAlister et al. 1970), we obtained a maximum westward transport south of Adak Island when using a reference level near 4000 db. This implies that westward flow below 2000 m along the Aleutian-Commander island arc would exist as far west as Kamchatka Strait, where it would enter the Bering Sea.

Direct current measurements using drogued buoys south of the central Aleutian Islands in 1959 (Reed and Taylor 1965) indicated that velocities at 300 m (55 cm/sec) were over 70 percent of surface velocities (80 cm/sec). In 1969, we conducted similar observations to 1000 m south of Adak Island to test the validity of geostrophic transports referred to 1000 db. A surface geostrophic velocity of only 15 cm/sec and a maximum geostrophic velocity of 30 cm/sec at 200 m depth were calculated from a reference level of 1000 db (Fig. 1.14). A westward velocity of 28 cm/sec at 1000 m was observed, however, and if the geostrophic velocity profile is aligned to account for this, a surface velocity of 43 cm/sec and a maximum velocity of 60 cm/sec at 150 m are obtained. It is not unusual to find a maximum geostrophic velocity at depth in this area.

Although at this time the *Kelez* was not equipped for serial observations below 1500 m, data obtained in 1966 indicated that surface velocities computed from 4000 db were 1.5 times those computed from 1000 db (Ingraham and Favorite 1968). If this factor is applied, velocities of 22.5 cm/sec at the surface and 45 cm/sec at 200 m could be anticipated. These results suggest that geostrophic velocities computed from levels near 3000–4000 db are more repre-

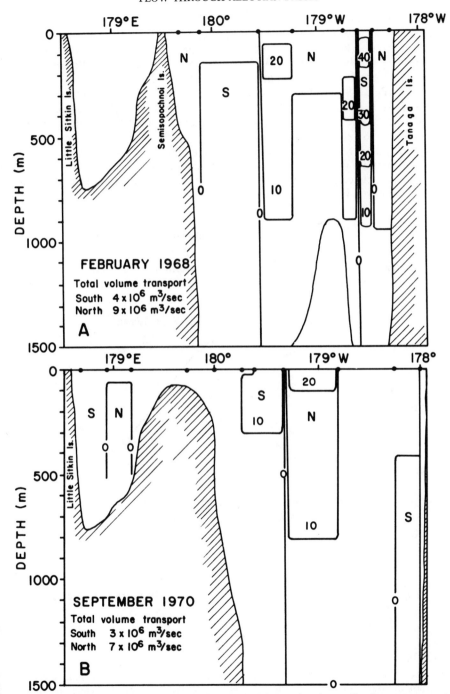

Fig. 1.13 Vertical profile of geostrophic velocities and values of volume transports (referred to 1000 db) across Amchitka Pass, showing north and south components occurring during (A) winter 1968 and (B) summer 1970 (R/V *George B. Kelez*).

sentative of actual flow than those computed from shallower levels. If 4000 db
is used as a level of no motion, mean annual transport in the Alaskan Stream
south of Adak Island (based upon data in Table 1.3) is estimated to be 12.5
Sv (11 Sv − 0/2000 db, 8 Sv − 0/1500 db, or 5 Sv − 0/1000 db). Because only
Kamchatka Strait is deeper than 2000 m, 1.5 Sv of the stream reaches this
strait. Of the 11 Sv above 2000 db, 4 Sv is lost through the central Aleutian
passes, 3 Sv is gained across the southern boundary (no net exchange occurs
in the western Aleutian passes) and 10 Sv flows northward through the Commander-Near Strait west of Attu Island. Thus, without considering exchanges
in Kamchatka Strait, (which is discussed by Hughes et al. in Chapter 3) a net
annual flow into Bering Sea of 14 Sv is obtained. This result is essentially a
mean value between the transports of 8.0 Sv proposed by Batalin (1964) and
19.5 Sv proposed by Arsen'ev (1967).

WIND-STRESS TRANSPORTS

Another question concerning the validity of volume transports in the Aleutian
area is why do not winter transports reflect the marked intensification of winds
which occur in winter.

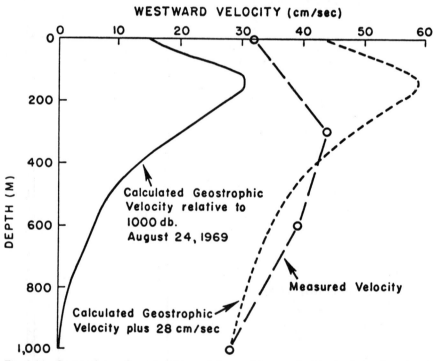

Fig. 1.14 Comparison of geostrophic and drogued buoy velocities in the main axis of
the Alaskan Stream south of Adak Island, summer 1969 (R/V *George B. Kelez*), showing
adjustment of former at 1000 db to observed velocity of latter at 1000 m.

TABLE 1.4   Monthly maximum, minimum, and mean total integrated wind-stress transport (Sv) in the Alaskan Stream at longitude 180°, 1950–59

Month	Maximum	Minimum	Mean
January	61	−16[1]	18
February	32	−17	12
March	46	−4	22
April	28	−4	15
May	42	3	21
June	20	−11	6
July	16	−2	6
August	8	−13	−4
September	20	−12	3
October	46	9	25
November	58	2	28
December	72	2	34

Decade mean   16

[1] Negative values indicate eastward transport.

The first adequate data to permit an accurate computation of volume transport in the Alaskan Stream south of Adak Island during winter were obtained aboard the R/V *George B. Kelez* in 1966 and reported by Ingraham and Favorite (1968). An increase in transport of 50 percent over the preceding fall and nearly 100 percent over the subsequent summer was apparent, providing good evidence of winter intensification of flow.

The concept of wind-stress transports derived by Sverdrup (1947), presented in the form of monthly charts by Fofonoff and Dobson (1963), provides a method of deriving theoretical transports without operations at sea. Agreement between wind-stress transports and volume transports in summer 1959 (Favorite 1967) led to further investigation of this technique. Values of total integrated west-east wind-stress transport in the Alaskan Stream at 180° (Table 1.4) are quite variable, but mean values for the decade 1950 to 1959 indicate a marked increase in transport in winter. This is attributed to the nature of the wind-stress on the sea surface from the high winds associated with the annual low pressure system in the Gulf of Alaska during this period. Transport data from our monitoring cruises from 1966 to 1968 (Table 1.3), however, indicated a gradual decrease from the maximum geostrophic transport in March 1966 throughout the entire period, and there was no evidence of markedly increased volume transports in winter.

Analysis of sea level measurements at Adak Island for 1947 to 1956[5] (Fig. 1.15) revealed considerable fluctuations in monthly mean sea level even after pressure effects were removed, but there was little suggestion of a significant increase in westward transport during winter. The fact that minimum monthly anomalies occurred in 1947–51 and maximum monthly anomalies were noted in 1951–56 suggests significant long-term variations in mean sea

[5] Analyses for the decade 1950 to 1959 would have been more consistent with other data, but the Adak Island reference level appeared to shift in 1957.

level. Of course, little coherence could be expected, since the surface exchange between the Bering Sea and the Pacific Ocean at this location is relatively free, and the Alaskan Stream is often found well offshore in this area.

Still, it seemed feasible that an increase in transport in winter should occur. The answer to this problem was sought in sea level data for the decade 1950 to 1959 in the Gulf of Alaska at Yakutat, where neither of these adverse conditions existed. If the effects of atmospheric pressure and of seasonal steric and precipitation changes are eliminated, an anomalous increase in mean sea level which occurs during winter is attributed to intensification of circulation (Fig. 1.16). The relation between the mean monthly increases in corrected sea level and modified wind-driven (Sverdrup) transport is nearly linear except during the summer, when low winds and considerable anomalous dilution, as a result of extensive runoff, occur in the harbor (Fig. 1.17). This strongly suggests that a significant increase in transport occurs during winter in the Gulf of Alaska; this could be imparted to the Alaskan Stream and thereby result in an increased inflow into the Bering Sea. A linear increase in slope from the central Gulf of Alaska to Yakutat of the magnitude indicated (~ 13 cm) would correspond to a substantial increase in transport of about 30 Sv, but it is assumed that the change in slope is more pronounced near the coast with a resulting decrease in transport. It appears, however, that the resulting

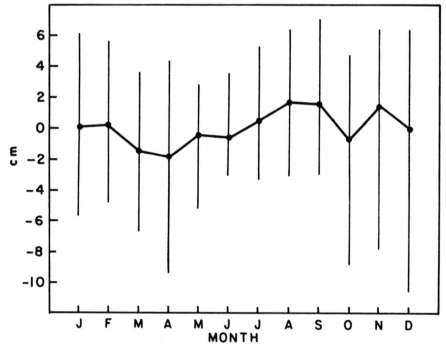

Fig. 1.15 Monthly mean sea level anomalies (cm) at Adak Island corrected for normal pressure, 1947–56, showing little monthly variation (vertical lines show range).

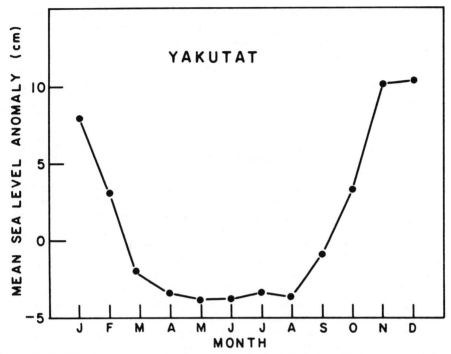

Fig. 1.16 Monthly mean sea level anomalies (cm) at Yakutat, Alaska, corrected for normal pressure, seasonal steric effects, and precipitation, showing anomalous increase in fall and winter.

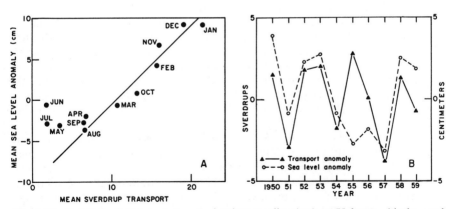

Fig. 1.17 Relations between mean sea level anomalies (cm) at Yakutat, Alaska, and modified, northward (A) mean monthly integrated wind-stress transports and (B) mean annual wind-stress transport anomalies, in the Gulf of Alaska, 1950–59.

sea surface slope is not imposed for a sufficiently long period to allow the distribution of density to adjust and reflect the increased volume transport before the summer period of low winds occurs. This would imply that there are significant changes in the barotropic mode and quite variable total transports. Geostrophic calculations are based on the baroclinic mode, however, which is a quasi-steady state condition that reflects mean values of a system forced in winter and relaxed in summer, with friction at the existing velocities preventing any build-up beyond present mean values. This would certainly account for the relatively constant values of volume transport found.

Although the validity of the geostrophic method as an indicator of flow in the Alaskan Stream (a boundary current) during winter has been challenged, the question of whether or not geostrophic flow in the Bering Sea in late spring and summer is representative of actual flow is still open to question. Favorite and Ingraham (1972) have shown an anomalous, southward flow along the eastern side of Bowers Ridge in the central Bering Sea in summer 1970. This was determined by geostrophic computations and actual drift of the R/V *George B. Kelez* while on station and is not shown by Arsen'ev (1967) or any other authors. If this discrepancy is perhaps due to inadequate spacing of previous station data it can be corrected by devising cooperative cruise plans in the future. Data from a cruise of the R/V *George B. Kelez* in the eastern Bering Sea in spring 1971, however, raises the question of the validity of the geostrophic method in another boundary current along the edge of the continental shelf in the eastern Bering Sea. At this location, where all previous reports have shown an intense northward flow, an extensive tongue of dilute water ($<32.6\%_{oo}$) extended from the continental shelf south of the Pribilof Islands over 150 km southward (Fig. 1.18). Isopleths of geostrophic velocities generally conform to the pattern of the isohalines but provide little indication of how such a feature could exist in the presence of the previously supposed intense northward flow in this area.

Finally to be considered is the fact that the estimate of mean annual volume transport into Bering Sea given here (14 Sv) corresponds closely to the mean annual wind-stress transport for the decade 1950–59 (Fig. 1.19). Assuming that the only northward flow through the Aleutian Island passes is 4 Sv through the central Aleutian passes (Amchitka Pass), a water budget for the wind-stress transport of 15 Sv in the Alaskan Stream west of 160°W is as follows: loss southward by recirculation into the Gulf of Alaska, 5 Sv; loss through the central Aleutian passes (Amchitka Pass), 4 Sv; and gain across the southern boundary, 3 Sv. This leaves a residual westward flow south of Attu Island of 12 Sv.

Wind-stress data across 55°N in the Bering Sea (i.e., between the Commander Islands and the Alaskan Peninsula) require a northward wind-driven transport of 16 Sv. This requirement can be satisfied if, in addition to the 4 Sv through the central Aleutian passes, the 12 Sv flow south of Attu Island moves northward into Bering Sea through the Commander-Near Strait. Assuming a loss of 1 Sv through Bering Strait, then 15 Sv must flow southward out of the Bering Sea as a boundary current through Kamchatka Strait.

Correspondence of wind-stress with volume transports may be merely

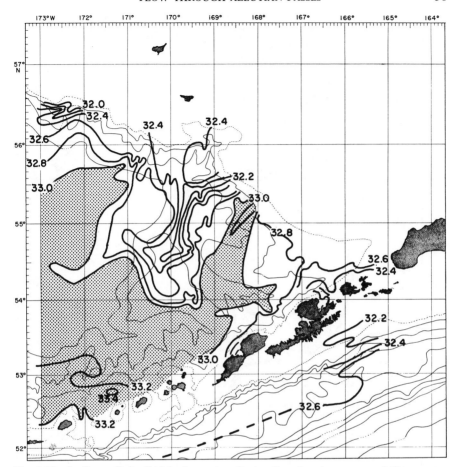

Fig. 1.18 Surface salinity (‰) in the eastern Bering Sea, showing tongue of dilute water extending southward from the edge of the continental shelf near the Pribilof Islands and little evidence of northward flow through the Aleutian passes, spring 1971 (R/V *George B. Kelez*).

fortuitous, and it is not exactly clear that either method is basically sound in this area. It can also be speculated, however, that if the mean annual wind-stress transport in Bering Sea is 16 Sv, then intensification of transport must occur in winter (as in the Gulf of Alaska) because of the winter wind field. This may in part justify the remarkably large southward volume transport of 24 Sv near Kamchatka Strait observed in winter 1966 by Reid (1966).

CONCLUSIONS

1. Volume transport in the Alaskan Stream, the major source of water flowing through the Aleutian-Commander island arc into the Bering Sea, is quite

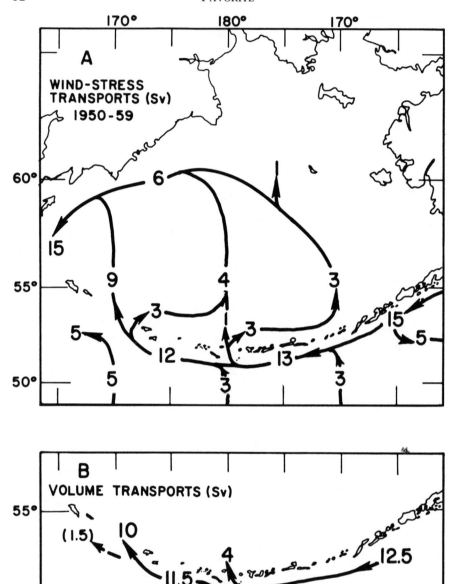

Fig. 1.19 Comparison of (A) mean integrated total wind-stress transports, 1950–59, and (B) estimated total geostrophic volume transports, showing flows of 16 and 14 Sv, respectively, through the Aleutian-Commander island arc east of Kamchatka Strait.

variable from year to year. The relatively equivalent values of volume transports in winter and summer, however, raises some question of the validity of this method of estimating flow because of the intensification of flow in the Gulf of Alaska in winter based on sea level data. Because the distribution of density cannot adjust to actual winter flow conditions before the less intense summer wind field is established, volume transports may represent a long-term steady state with actual winter transports significantly underestimated.

2. Flow through the various openings in the island arc is highly variable based upon any time scale, from days to years. North and south flows can occur simultaneously in all major Aleutian passes, and existing data are inadequate to ascertain if a net flow exists in either direction in any particular season except perhaps through Amchitka Pass. The presence of swift tidal currents and complex bottom topography in this area further complicate the estimation of volume transport.

3. If we accept volume transports as being a reliable estimate of flow in the Alaskan Stream, the following conclusions can be drawn:

(a) Net flow through the eastern Aleutian passes is poorly documented and appears to be greatly influenced by how far south of the passes the main axis of the Alaskan Stream occurs and at what longitude the main recirculation of coastal water into the Gulf of Alaska takes place. It is estimated that there is no net annual exchange.

(b) Using 4000 db as a level of no motion, mean annual volume transport in the Alaskan Stream south of Adak Island is estimated to be 12.5 Sv (11 Sv − 0/2000 db; 8 Sv − 0/1500 db; and 5 Sv − 0/1000 db).

(c) There is an estimated northward volume transport of 4 Sv through the central Aleutian passes (Amchitka Pass) and 10 Sv through the Commander-Near Strait, or a total transport of 14 Sv into the Bering Sea east of the Commander Islands. This may be compared to two Soviet estimates of 8.0 and 19.5 Sv and corresponds closely with a mean annual wind-stress transport requirement of 16 Sv.

4. The existence of a volume transport of 14 Sv in an area where about 20 years ago flow was assumed to be much less than 5 Sv, suggests the significant advances in oceanographic research in the Aleutian area in recent years. New advances and accurate estimates of flow into the Bering Sea will only come through long-term data from judiciously placed current meter arrays. Co-operative plans should be made to establish lines of monitoring stations to replace the sporadic observations obtained during the last few years.

5. The possibility of anticyclonic gyres around the eastern, central and western Aleutian Islands could have significant effects upon valuable populations of resident living marine resources having a planktonic egg or larval stage such as halibut and king crab and should be definitely investigated. Knowledge of real-time data concerning flow through all major passes would be of great value to high seas research and commercial vessels following the migrations of the Pacific salmon, whales and other inhabitants of the sea.

Discussion

UDA: Is the difference between your measured current and geostrophic velocity south of Adak Island in Summer 1969 due to wind current or gradient current?

FAVORITE: This difference is attributed to the fact that a depth of 1000 m is too shallow a zero reference level on which to compute geostrophic currents in the area of the Alaskan Stream; 3000–4000 m would be more appropriate.

Acknowledgments

The assistance of W. J. Ingraham, Jr., R. S. Pearson, V. Fiscus, and M. Morey in the preparation of this report is gratefully acknowledged.

REFERENCES

ARSEN'EV, V. S.
     1967   Currents and water masses of the Bering Sea [in Russian, English summary]. Izd. Nauka, Moscow. (Transl., 1968, Nat. Mar. Fish. Serv., Northwest Fish. Center, Seattle, Wash.). 135 pp.

BARNES, C. A., and T. G. THOMPSON
     1938   Physical and chemical investigations in the Bering Sea and portions of the North Pacific Ocean. *Univ. Wash. Publ. Oceanogr.* 3(2): 35–79 + Appendix pp. 1–164.

BATALIN, A. M.
     1964   On the water exchange between the Bering Sea and the Pacific Ocean [in Russian]. *Trudy VNIRO* 49: 7–16. (Transl., 1968, in *Soviet fisheries investigations in the northeastern Pacific*, Part 2, pp. 1–12, avail. Nat. Tech. Inf. Serv., Springfield, Va., TT 67-51024).

DODIMEAD, A. J., F. FAVORITE, and T. HIRANO
     1963   Salmon of the North Pacific Ocean. Part 2. Review of oceanography of the subarctic Pacific region. *Bull. Int. North Pac. Fish. Comm.* 13, 195 pp.

FAVORITE, F.
     1967   The Alaskan Stream. *Bull. Int. North Pac. Fish. Comm.* 21: 1–20.
     1969   A summary of BCF investigations of the physical-chemical oceanic environment of Pacific salmon, 1955–68. INPFC Doc. 1216, Bur. Com. Fish., Biol. Lab., Seattle, Wash., 38 pp.

FAVORITE, F., and D. M. FISK
     1971   Drift bottle experiments in the North Pacific Ocean and Bering Sea—1957–60, 1962, 1966, and 1970. U. S. Dep. Comm., Nat. Oceanic Atmos. Admin., Nat. Mar. Fish. Serv., Data Rep. 67, 20 pp. (on 1 microfiche).

FAVORITE, F., and W. J. INGRAHAM, JR.
  1972   Influence of Bowers Ridge on circulation in Bering Sea and influence of
         Amchitka Branch, Alaskan Stream, on migration paths of sockeye salmon.
         In *Biological oceanography of the northern North Pacific Ocean* [Motoda
         commemorative volume], edited by A. Y. Takenouti et al. Idemitsu Shoten,
         Tokyo, pp. 13–29.

FAVORITE, F., W. B. MCALISTER, W. J. INGRAHAM, JR., and D. DAY
  1967   Oceanography. Int. North Pac. Fish. Comm., Annual Rep. 1966, pp. 89–96.

FAVORITE, F., B. A. MORSE, A. H. HASELWOOD, and R. A. PRESTON, JR.
  1964   North Pacific oceanography, February–April 1962. U.S. Fish Wildl. Serv.,
         Spec. Sci. Rep. Fish. 477, 66 pp.

FAVORITE, F., J. W. SCHANTZ, and C. R. HEBARD
  1961   Oceanographic observations in Bristol Bay and the Bering Sea, 1939–41.
         (USCGT *Redwing*). U. S. Fish Wildl. Serv., Spec. Sci. Rep. Fish. 381, 323 pp.

FISK, D. M.
  1971   Recoveries from 1964 through 1968 of drift bottles released from a merchant
         vessel, S. S. *Java Mail*, en route Seattle to Yokohama, October 1964. *Pac.
         Sci.* 25: 171–177.

FLEMING, R. H.
  1955   Review of oceanography of the northern Pacific. *Bull. Int. North Pac. Fish.
         Comm.* 2: 1–43.

FOFONOFF, N. P., and F. W. DOBSON
  1963   Transport computations for the North Pacific Ocean 1950–1959. 10-year
         means and standard deviations by months. Wind-stress and vertical velocity
         annual means, 1955–1960. Fish. Res. Bd. Canada, MS Rep. Ser. No. 166,
         179 pp.

GOODMAN, J. R., J. H. LINCOLN, T. G. THOMPSON, and F. A. ZEUSLER
  1942   Physical and chemical investigations: Bering Sea, Bering Strait and Chukchi
         Sea during the summers of 1937 and 1938. *Univ. Wash., Publ. Oceanogr.*
         3(4): 105–169+Appendix pp. 1–117.

INGRAHAM, W. J., JR., and F. FAVORITE
  1968   The Alaskan Stream south of Adak Island. *Deep-Sea Res.* 15: 493–496.

INTERNATIONAL NORTH PACIFIC FISHERIES COMMISSION
  1959   Annual Rep. 1958, 119 pp.

JAPAN AGRICULTURAL TECHNICAL ASSOCIATION
  1954   Oceanographic data in the northern waters of the North Pacific. 552 pp.

KITANO, K.
  1967   Oceanographic structure near the western terminus of the Alaskan Stream.
         Hokkaido Reg. Fish. Res. Lab. 32: 23–40.

KOTO, H., and T. FUJII

1958 Structure of the waters in the Bering Sea and the Aleutian region. *Bull. Fac. Fish., Hokkaido Univ.* 9: 149–170.

LEONOV, A. K.

1960 Water balance of the Bering Sea [in Russian]. In his *Regional'naya okeanografiya*, pp. 79–103. Gidrometeorol. Izd., Leningrad. (Transl. avail. Nat. Tech. Inf. Serv., Springfield, Va., AD 627508, TT 66-60450).

LISITSYN, A. P.

1966 Recent sedimentation in the Bering Sea [in Russian]. Izd. Nauka, Moscow, 584 pp. (Transl., 1969, avail. Nat. Tech. Inf. Serv., Springfield, Va., TT 68-50315).

MCALISTER, W. B., F. FAVORITE, and W. J. INGRAHAM, JR.

1970 Influence of the Komandorskie Ridge on surface and deep circulation in the western North Pacific Ocean. In *Kuroshio—Symposium on the Japan Current*, edited by John C. Marr. East-West Center Press, Honolulu, pp. 85–96.

MCALISTER, W. B., W. J. INGRAHAM, JR., D. DAY, and J. LARRANCE

1969 Oceanography. Int. North Pac. Fish. Comm., Annual Rep. 1967, pp. 97–107.
1970 Oceanography. Int. North Pac. Fish. Comm., Annual Rep. 1968, pp. 90–101.

MISHIMA, S., and S. NISHIZAWA

1955 Report on hydrographic investigations in the Aleutian waters and the southern Bering Sea in the early summers of 1953 and 1954. *Bull. Fac. Fish., Hokkaido Univ.* 6(2): 85–124.

MOISEEV, P. A.

1964 Some results of investigations carried out by the Bering Sea research expedition [in Russian]. *Trudy VNIRO* 53: 7–29. (Transl., 1968, in *Soviet fisheries investigations in the northeastern Pacific*, Part 3, pp. 1–21, avail. Nat. Tech. Inf. Serv., Springfield, Va., TT 67-51205).

NATAROV, V. N.

1963 Water masses and currents of the Bering Sea [in Russian]. *Trudy VNIRO* 48: 111–133. (Transl., 1968, in *Soviet fisheries investigations in the northeastern Pacific*, Part 1, pp. 110–130, avail. Nat. Tech. Inf. Serv., Springfield, Va., TT 67-51203).

NORPAC COMMITTEE

1960 Oceanographic observations of the Pacific—1955. In *The NORPAC atlas*. Univ. Calif. (Berkeley and Los Angeles); Univ. Tokyo, 8 pp. + 123 plates.

OHTANI, K.

1970 Relative transport in the Alaskan Stream in winter. *J. Oceanogr. Soc. Japan* 26: 271–282.

RATHBUN, R.

1894 Summary of the fishing investigations conducted in the North Pacific Ocean and Bering Sea from July 1, 1888, to July 1, 1892, by the U. S. Fish Commission Steamer *Albatross. Bull. U. S. Fish. Comm.* 12: 127–201.

RATMANOFF, G. E.
1937   Explorations of the seas of Russia. *Publ. Hydrol. Inst.* 25, pp. 1–175.

REED, R. K.
1971   Nontidal flow in the Aleutian Island passes [letter to editor]. *Deep-Sea Res.* 18: 379–380.

REED, R. K., and N. E. TAYLOR
1965   Some measurements of the Alaskan Stream with parachute drogues. *Deep-Sea Res.* 12: 777–784.

REID, J. L.
1966   ZETES Expedition. *Trans. Amer. Geophys. Un.* 47: 555–561.

SAUR, J. F. T., R. M. LESSER, A. J. CARSOLA, and W. M. CAMERON
1952   Oceanographic cruise to the Bering and Chukchi Seas, summer 1949. Part 3. Physical observations and sound velocity in the deep Bering Sea. U. S. Navy Electronics Lab. Rep. 298, San Diego, Calif., 38 pp.

SUGIURA, J.
1958   Oceanographic conditions in the northwestern North Pacific based upon the data obtained on board the *Komahashi* from 1934 to 1936. *J. Oceanogr. Soc. Japan* 14(3): 1–5.

SVERDRUP, H. U.
1947   Wind-driven currents in a baroclinic ocean; with applications to the equatorial currents of the eastern Pacific. *Proc. Nat. Acad. Sci.* 33: 318–326.

TANNER, Z. L., et al.
1890   Explorations of the fishing grounds of Alaska, Washington Territory, and Oregon, during 1888, by the U. S. Fish Commission Steamer *Albatross*, Lieut. Comdr. Z. L. Tanner. U. S. Navy, commanding. *Bull. U. S. Fish. Comm.* 8: 1–92.

TOWNSEND, C. H.
1901   Dredging and other records of the United States Fish Commission Steamer *Albatross*, with bibliography relative to the work of the vessel. U. S. Comm. Fish and Fish., Part 26, Rep. Comm. 1900, pp. 387–560.

UDA, M.
1963   Oceanography of the subarctic Pacific Ocean. *J. Fish. Res. Bd. Can.* 20: 119–179.

U. S. COAST GUARD
1936   Report of the oceanographic cruise, United States Coast Guard Cutter *Chelan* Bering Sea and Bering Strait, 1934 (F. A. Zeusler). U. S. Coast Guard Headquarters, Washington, D.C., 72 pp.

CHAPTER 2

Currents and water masses in the Bering Sea: A review of Japanese work

A. Yositada Takenouti *and* Kiyotaka Ohtani

Faculty of Fisheries, Hokkaido University, Hakodate, Japan

Abstract

About half of the volume transport of the Alaskan Stream enters the Bering Sea through Aleutian island passes and the rest from west of Attu Island. The highly stratified Alaskan Stream water loses its characteristic structure upon entering the Bering Sea during its first step of transformation from Eastern to Western Subarctic water. A general counterclockwise circulation and small eddies prevail in the Bering Sea.

Dispersion of low-salinity shelf water in the surface layer and upwelling of deeper water associated with eddies reconstruct a stratified vertical pattern in the Bering Sea basin. This water then flows out as the East Kamchatka Current, mixing horizontally with cold low-salinity water from the Okhotsk Sea, and gradually becomes Western Subarctic in nature.

The continental shelf of the eastern Bering Sea is characterized by various types of vertical temperature and salinity structures. Freshwater dilution of the surface layer, the intrusion of warm saline water near the bottom, and strong vertical mixing associated with winter cooling cause formation of dichothermal water. In regions where a conspicuous halocline is present as a barrier to winter convection activity, cold bottom water is absent.

INTRODUCTION

Strata of minimum temperature in the western North Pacific were recognized as early as the end of the last century, but very few investigations on their distribution and movement had been made until the mid-1930s. Uda (1935) concluded that the dichothermal water was formed through cooling and vertical convection during winter. Although Uda's study was confined to the Okhotsk Sea and the North Pacific Ocean south of the Kurile Islands, his concepts have been extended by many authors to the Bering Sea.

Most oceanographic data in Japan have been obtained on board the *Oshoro Maru* and the *Hokusei Maru* of Hokkaido University. In 1955 Bering Sea observations were made from the *Oshoro Maru* as part of the NORPAC Expedition. The *Oshoro Maru* has visited the Bering Sea each year since, and the *Hokusei Maru* has made occasional visits. Because the primary purpose of these cruises was cadet training, the time available for oceanographic observations has been limited and has allowed only partial coverage of the Bering Sea each year. The area of observations has varied from year to year, however, and consequently the whole Bering Sea has eventually been covered by Hokkaido University vessels. Fisheries agency survey ships have also visited the Bering Sea but only during the warm season of the year. For data on other seasons, Japanese oceanographers have had to use the results obtained on ships of other nations, mainly those of Canada and the United States.

In this review paper, the authors have briefly summarized the results of investigations made by Japanese oceanographers in recent years without particular reference to those of foreign scientists.

ALASKAN STREAM

Volume transport

Since the historical work of Barnes and Thompson (1938), it has been known that Pacific water enters the Bering Sea. From data obtained by *Oshoro Maru* cruises in 1953 and 1954, Mishima and Nishizawa (1955) reported a westward flow along the southern side of the Aleutian Islands. Later Koto and Fujii (1958) computed the volume transport of this current across the meridians of 172°W and 175°E above 600 m and obtained values of 7.25 and 5.80 Sv, respectively. From the data of *Oshoro Maru* cruises in 1955 and 1956, as well as those obtained by the *Komahashi* in 1936, they concluded that the current passed west of Attu Island and turned northeast into the Bering Sea west of the island. From the decrease of transport observed between these two sections, it was inferred that a considerable amount of Pacific water entered the Bering Sea through Amchitka Pass.

The name *Alaskan Stream* has been generally accepted by Japanese oceanographers since Dodimead et al. (1963) published a hydrographic summary of the subarctic Pacific Ocean. It is understood that the Alaskan Stream represents (as elucidated by Favorite 1967) the extension of the Alaska Current westward along the Aleutian Islands, providing a source of dilute and relatively warm (4 C) surface water at a depth of 300 m to the northwestern North Pacific Ocean continuous as far westward as 170°E.

Ohtani (1965) computed the volume transport in summer of the Alaskan Stream from data of 1957–63 *Oshoro Maru* and *Hokusei Maru* cruises, obtaining values ranging from 2.8 to 6.5 Sv over 600 m. Ohtani (1970) calculated further the volume transport in winter over the 1000-db surface and concluded that half of the transport enters the Bering Sea through Amchitka Pass, while the rest enters through the pass west of Attu Island. A part of the Subarctic Current also joins the northward flow coming from the Alaskan Stream, causing a combined volume transport of northward flow estimated at 11 Sv. Water thus entering the Bering Sea flows out as the East Kamchatka Current except for a small portion that reaches the Arctic Ocean through the Bering Strait.

Structure

The Alaskan Stream has a complicated structure with varying vertical patterns of temperature and salinity across the current, as illustrated in Figure 2.1, which shows the distribution along longitude 175°E. Near the southern boundary of the current, a very sharp halocline is observed just below the 100-m level but disappears gradually towards the north until no appreciable halocline is observed near the northern Aleutian Ridge boundary of the current. Entering the Bering Sea, the water of the Alaskan Stream loses its stratified structure but still maintains the high-temperature mesothermal layer characteristic of the subarctic region.

BERING SEA BASIN

Current pattern

A general counterclockwise circulation in the Bering Sea has been reported by Koto and Fujii (1958), but there are considerable variations in current patterns presented by different authors. Ohtani et al. (1972) considered that in a subarctic region the upper layer is flowing uniformly and that the dynamic topography at the 100- to 1000-db surface represents the general current pattern of the upper layer (Fig. 2.2). Although stations were occupied in different years, the computed values of dynamic heights generally corresponded closely, suggesting that the current pattern is stable from year to year. The data show continuous isohypses ranging from 0.85 to 0.95 dyn. m which describe a continuous counterclockwise circulation in the deep Bering Sea basin.

Between these continuous isohypses are many small eddies, including clockwise eddies near Adak, Attu and two other islands near the continental shelf that are represented by isohypses over 1.0 dyn. m. These are thought to be of a permanent nature, although the positions of their centers and scopes vary from year to year. Small eddies north of Bowers Bank represent cut-offs of warm Alaskan Stream water forming isolated clockwise eddies that move towards Cape Olyutorski; these eddies, however, appear unstable.

The current speed in the Bering Sea is generally a few centimeters per second except along the continental slope, the coast of Kamchatka and in certain eddies, where values of 10–15 cm/sec have been obtained by dynamic calculation. The Alaskan Stream water which enters the Bering Sea remains there for over a year before flowing out into the Pacific Ocean, and thus the Bering Sea fulfills its definition as a region of water mass formation (Tully 1964).

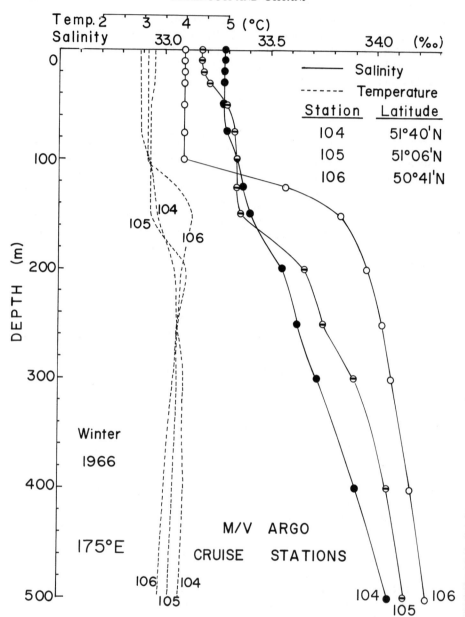

Fig. 2.1 Vertical structures of salinity and temperature of the Alaskan Stream along longitude 175°E in winter.

Water masses

In the Bering Sea and adjacent region, three basic water masses are distinguished; namely, Western Subarctic, Bering Sea, and the Alaskan Stream. These waters are characterized in common by dichothermal strata in the upper layer and a conspicuous halocline.

Koto and Fujii (1958) studied the characteristics of dichothermal strata in the Bering Sea region and found that the sigma-t of the minimum-temperature water decreased towards the south along the section parallel to the 180° meridian. They concluded that although the dichothermal water *seemed* to spread southward according to the vertical temperature and salinity distributions observed, this was not actually the case, thus supporting Uda's (1935) theory of dichothermal water formation. They speculated further, however, that dichothermal strata in the Subarctic Current are not formed in the same

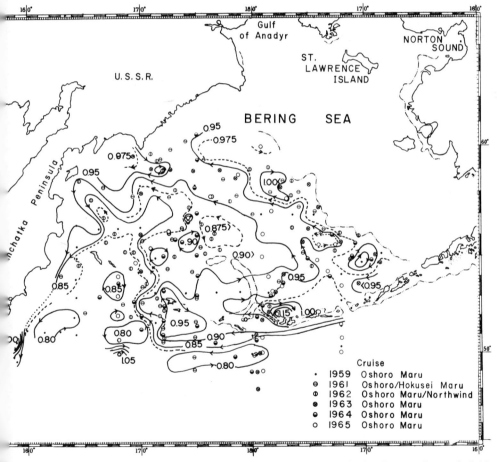

2.2   Dynamic topography (dyn. m) of 100 to 1000 db in Bering Sea basin. Contour intervals 0.05 m (——) and 0.025 dyn. m (. . . .).

way as in the Bering Sea basin but are caused rather by horizontal mixing with the adjacent waters. In the northern basin there are two temperature minima in the dichothermal strata: based on salinity values the lower temperature minimum was interpreted as the bottom of vertical mixing in winter, while the upper one is formed by lateral mixing with cold and less saline water on the shallow shelf region.

Kitano (1970a, b) has studied the distributions of temperature and salinity in the Bering Sea and has introduced data indicating several water masses in the region. Ohtani et al. (1972) have distinguished five types of vertical structures in the Bering Sea (Table 2.1), which are suggestive of the underlying transformation process.

TABLE 2.1 Characteristics of various vertical structures in the Bering Sea

Type of vertical structure	A	B	C	D	E
Depth (m) of upper layer	100–140	140–300	150–200	300 or more	***
Gradient ($^o/_{oo}$ m^{-1}) of halocline	0.01	0.006	0.004	0.002	***
Salinity ($^o/_{oo}$) at bottom of halocline	33.8	33.7–33.8	33.7–33.8	***	***
Temp. (°C) of dichothermal layer	1.0–2.0	1.0–2.0	2.0–3.0	2.5–3.5	***
Temp. (°C) of mesothermal layer	3.6–3.7	3.7–3.8	3.8–3.9	3.9 or more	***

During summer the surface layer is under strong influence of precipitation and heating; consequently, the vertical distribution of various parameters exhibits very complicated features. In winter, however, strong mixing occurs in the surface layer due to cooling and atmospheric disturbances, and the characteristics are relatively simplified and clear. Examples of summer patterns are shown in Figures 2.3–2.6; Figure 2.7 illustrates those in winter, indicating that the temperature of the surface layer in the Bering Sea basin is coldest in the order of types B, A, C, and D. In types A and C the surface layers are completely homogenous with respect to temperature and salinity. In type D the surface water is relatively warm, only slightly colder than the mesothermal layer and characteristic of Eastern Subarctic water. The upper layer temperature and salinity of type B with small positive gradients are lower than those of types A or C, while the salinity at the bottom of the halocline coincides with that of type C.

Type E, found only in limited localities near the Kamchatka Peninsula and the Aleutian Islands, is characterized by a deep upper layer having small gradients both in temperature and in salinity. Plots of these five types are superimposed in Figure 2.8 against the depth of the 33.8‰ isohaline surface (bottom of the halocline).

Formation of Western Subarctic water

Ohtani et al. (1972) discussed the processes of formation of different types of vertical structures in the Bering Sea water column and concluded that

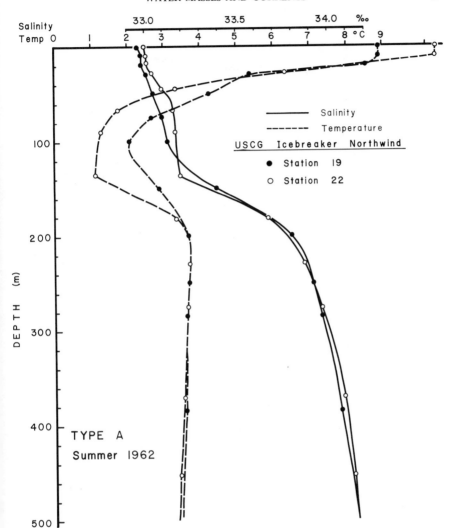

Fig. 2.3 Type A vertical structures of salinity and temperature in the Bering Sea.

Western Subarctic water is formed in the Bering Sea. Figure 2.8 shows that boundaries between the different types correspond with isobaths. For instance, the 250-m line corresponds to the boundary of type A, and the 350-m line is the boundary between C and D near the continental shelf. Comparison of Figures 2.8 and 2.2 indicates that isobaths of the bottom of the halocline are generally parallel to isohypses. The relationship of the type distribution both with the depth of the bottom of the halocline and with the stream contour lines suggests that a transformation process occurs among the various types.

Type A is a structure typical of Pacific Subarctic water. It flows westward along the Aleutian Ridge as the Alaskan Stream or enters the Bering Sea

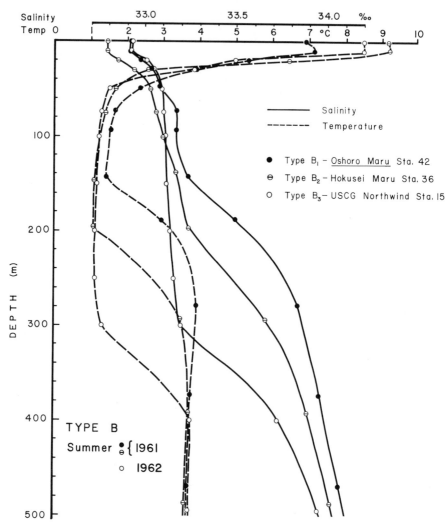

Fig. 2.4 Type B vertical structures of salinity and temperature in the Bering Sea.

through shallow and narrow passes between the islands. Strong turbulence, both vertical and horizontal, is generated by topographical effects of the complicated bottom configuration of the island chain and causes very strong mixing in this region, a phenomenon not known to occur in the mid-ocean. The highly stratified salinity pattern observed in type A is thus destroyed and a homogenous salinity gradient is formed. Type D, near the Aleutian Islands and at the northern boundary of the Alaskan Stream, is formed by this mechanism.

To cause vertical mixing across a stable pycnocline, a supply of energy is necessary because of the increase of potential energy associated with mixing. The amount of work necessary to transform a stratified water mass such as type A into a homogenous type D is much greater than the kinetic energy of the Alaskan Stream; however, the dissipation of tidal energy is sufficient for this transformation. The Alaskan Stream water, having the vertical structure of type A, enters the Bering Sea through passes between the Aleutian Islands while it is being mixed by tidal energy and wind to form type D water.

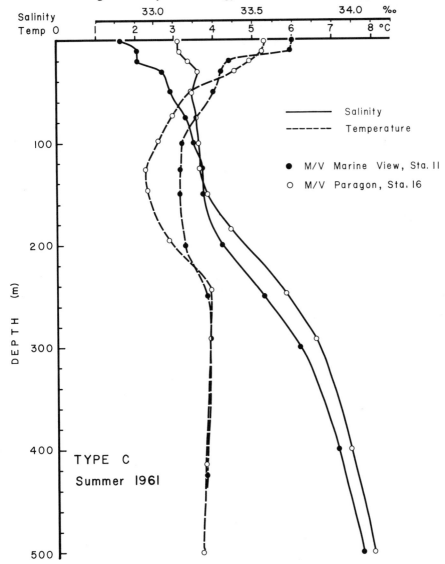

Fig. 2.5 Type C vertical structures of salinity and temperature in the Bering Sea.

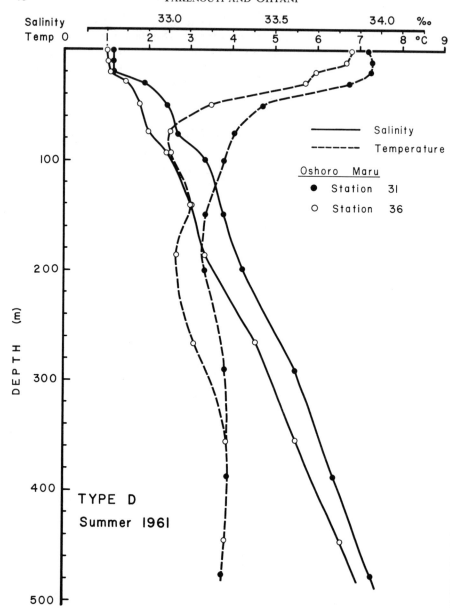

Fig. 2.6 Type D vertical structures of salinity and temperature in the Bering Sea.

Approximately half of the original volume of the Alaskan Stream is converted to type D water by this process.

Types B and C are formed from type D by upwelling of the deep layer associated with counterclockwise circulation or divergent flow in the surface layer. In the case of type B, there is less intrusion of low-saline shelf water than with type C water; yet, stronger winter cooling occurs. Some type B water has a thick surface layer due to depression of surface water by the clockwise eddies. In regions of extensive sinking due to strong currents, the thickness of the upper layer exceeds 500 m with only slight temperature and salinity gradients, thus forming type E water. This type of vertical structure was observed in 1936 and 1959 in areas off the Kamchatka Peninsula, Kamchatkii Bay, and southeast of Cape Lopatka (Koto and Fujii 1958, Ohtani 1965).

Type B water flows southwest from the Bering Sea along Kamchatka Peninsula as the East Kamchatka Current, which passes along the Kurile Islands with a volume transport of about 9 Sv. The cold and less saline Okhotsk

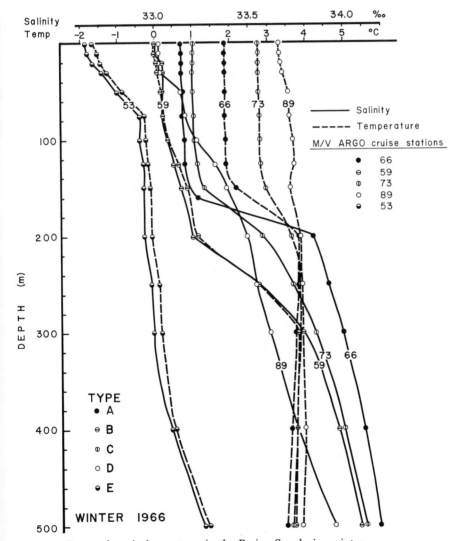

Fig. 2.7 Types of vertical structures in the Bering Sea during winter.

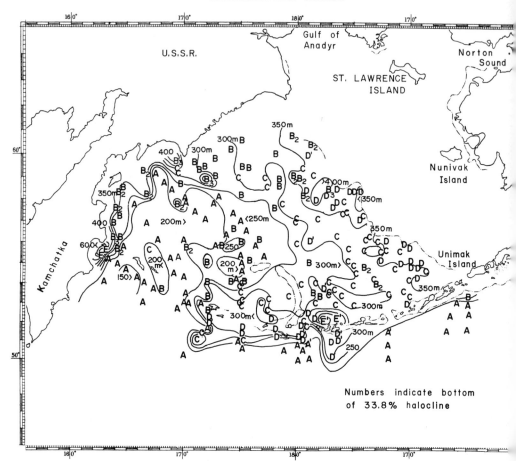

Fig. 2.8   Distribution of five types of vertical structures plotted against depth of isohaline surface (bottom of halocline) in the Bering Sea.

water mixes laterally with waters of the East Kamchatka Current near the Kurile Islands, forming the typical Western Subarctic water. A part of it turns eastward and flows parallel to the west wind drift as the Subarctic Current, while the other part flows further southwest as the Oyashio Current.

CONTINENTAL SHELF REGION

Temperature distribution in bottom water
Most of the continental shelf waters in the eastern Bering Sea are shallower than 100 m. This results in the formation of strata in the deep Bering Sea basin which appear on the continental shelf as cold bottom water. Results of investigations on dichothermal strata and on cold bottom water formation are interapplicable.

Koto and Maeda (1965) analyzed the results of oceanographic observations carried out on board the *Oshoro Maru* and the *Kin'yo Maru* in 1956. Strong interface was evident on the continental shelf of warm water from the deep basin against cold water coming from the north. During the period from late April to late June, despite a general increase in bottom water temperature, the interface was found to have shifted northward. An increase in water column temperature was noted at stations occupied twice every two weeks and appeared to coincide closely with the surface heat influx.

Maeda et al. (1967, 1968) reported that the bottom water temperature on the trawl fishing grounds in the eastern Bering Sea was warmer in 1963 than in 1960; that is, the core of cold bottom water was 1 C higher in 1963 and the 0°C isotherm was found only near St. Matthew Island. This was thought to be due to the warmer winter in 1963, during which time the sea ice remained further north than in 1960. Year-to-year variations of the temperature and salinity distributions of bottom water on the continental shelf in July in the eastern Bering Sea showed a center of cold water on the sea bottom in the

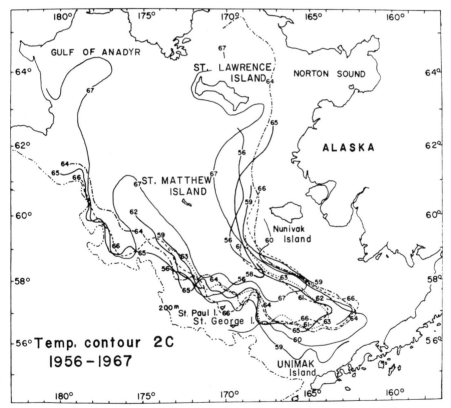

Fig. 2.9a Horizontal distribution of bottom seawater 2 C temperature contour in eastern Bering Sea during summer 1956–67.

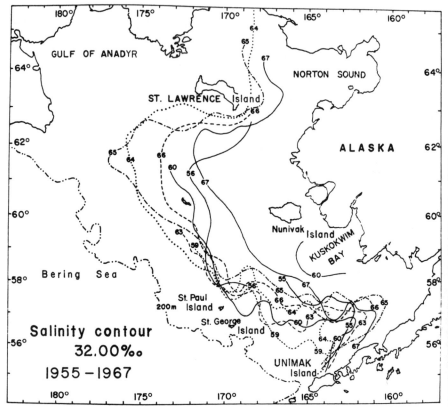

Fig. 2.9b Horizontal distribution of bottom seawater 32.00% salinity contour in eastern
Bering Sea during summer 1955–67 (years indicated by last two digits).

area southwest of St. Lawrence Island, extending southeast toward Bristol
Bay. A positive correlation was observed between the salinity and temperature
of the bottom cold water. The 2 C isotherm and the 32.00‰ isohaline contour
were considered the boundary of bottom water formed by winter cooling and
the intrusion of warm saline water from the open ocean (Fig. 2.9a, b).

Kihara and Uda (1969) extended the above findings by plotting the
salinities and temperatures of bottom water in the eastern Bering Sea and
suggested three water masses as the source of eastern Bering Sea water: Alaskan
coastal, the Alaskan Stream extension, and Bering boreal waters. It was not
clearly indicated how and where these basic water masses were formed.

Water masses on the continental shelf

Vertical structure of waters on the continental shelf of the eastern Bering Sea
was studied by Koto and Maeda (1965) in connection with their studies on the
variation of bottom water temperature. Ohtani (1969) discussed the distributions of water temperature, salinity, and density from the results of summer
observations made aboard the *Oshoro Maru* from 1963 to 1966. Along the

edge of the shelf there is a discontinuous zone of temperature and salinity between the oceanic water and the shelf water. In summer the surface water temperature rises to 6 to 8 C in the southern part, or to 10 to 12 C in the northern part; cold water ranging from −1.7 to 2 C remains under the surface water. The isohalines are nearly parallel to the isobaths and decrease gradually toward the coast. A distinct pycnocline appears between the surface and lower layers at the depth of 20–30 m in the region above 60°N. This results in very low heat transfer from the surface layer to the lower layer during the summer.

Extremely cold bottom water is found southwest of St. Lawrence Island. In the Gulf of Anadyr, oceanic water having high salinity and temperature penetrates to the bottom, and oceanographic conditions in the continental shelf region are governed by the mixing of oceanic water and freshwater derived from precipitation, melting of sea ice and river discharge. From the patterns of vertical distribution of temperature and salinity both in summer and winter, as well as the condition of sea ice, Ohtani (1969) classified seven water types: two oceanic, three in the continental shelf region, and two in areas under strong influence of freshwater runoff from the land.

The same data obtained by the *Oshoro Maru* and used by Ohtani (1969) were analyzed also by Kitano (1970a, b), who considered several water masses in determining the source of the cold bottom water southwest of St. Lawrence Island. His conclusion that its origin was in the Gulf of Anadyr conflicted with that of Ohtani. (This difference may have arisen from Ohtani's use of Gulf of Anadyr data obtained in 1966 but not used by Kitano.)

Ohtani (1969) further computed the penetration depth of convection from winter cooling. He assumed salinities at the surface and the lower layers, as well as the thickness of the surface layer and the dichothermal strata. By introducing the assumed numerical values, he concluded that in the regions of oceanic waters over the continental shelf and the southern part of the continental shelf, the halocline is weak and thermal convection in the winter can reach the bottom. Near the coast and the northern part of the continental shelf, as well as areas such as the Gulf of Anadyr and Norton Sound under strong influence of freshwater runoff from land, it was found that a very strong halocline develops between the surface and lower layers; consequently the winter cooling cannot reach the bottom in these regions.

The ice in the Okhotsk Sea was found to be about one meter thick and the vertical convection due to sea ice formation was as much as 100 m (Tabata 1960). Ohtani (1969) studied the Bering Sea by a similar method and concluded that, because of low salinity in the surface layer on the continental shelf, the depth of vertical convection due to ice formation was only about 50 m. Particularly in the Gulf of Anadyr, where the oceanic water enters at a lower level, convection is restricted to the upper 30 m, and no cold high-salinity water was found at the bottom. In the northern part of the continental shelf, the halocline is moderate and the effect of winter cooling sufficient to form sea ice; cold and high-salinity water reaches the bottom and remains there until the following summer. In the shallow region near the coast, strong vertical mixing due to tidal currents occurs, and the cold high-salinity water disappears in summer. Characteristics of various waters on the continental shelf in the eastern Bering Sea are shown in Table 2.2 and Figure 2.10.

TABLE 2.2   Characteristic temperature and salinity structures of continental shelf waters

Water system	Region	Layer	Vertical distributions of temperature and salinity		Sea ice
			summer	winter	
Oceanic water	AS	Upper	0–20 m 8 C, 32.0–32.6$^o/_{oo}$	deepening 2 to 3 C	None
		Thermocline	20–50 m, 0.2/m	vanish	
		Halocline	20/30–100 m, 0.01/m	vanish on shelf	
		Lower	4 C, 33.0–33.2$^o/_{oo}$		
	BSW	Upper	0–20 m 7–8 C, 32.6–32.9$^o/_{oo}$		None
		Thermocline	20–30/50 m, <0.3/m	isothermohaline	
		Halocline	20–30 m, 0.02/m	1 to 2 C	
		Lower	3–2 C, 32.7–33.2$^o/_{oo}$		
	CA	Upper	0–20/30 m 7 C, 31.0–32.5$^o/_{oo}$		Drift ice
		Thermocline	10/20–30 m, >0.3/m	isothermohaline	
		Halocline	20–30 m, <0.2$^o/_{oo}$	0 to 2 C	
		Lower	0.3 C,		
	CW	Surface	0–10 m, 5/6–3/4 C	isothermohaline	Drift ice and
		Bottom	3–4 C, <31.6$^o/_{oo}$	−1.7 to 0°C	freezing
Shelf water	IFA (deep)	Upper	0–20 m 8–9 C, <31.6$^o/_{oo}$	freezing point	Freezing
		Thermocline	20–30 m, 0.7–0.9/m		
		Halocline	10/20–20/30 m, <0.4$^o/_{oo}$		
		Lower	−1.7–0 C, 31.8–32.2$^o/_{oo}$	vanish	
		Bottom	−0.5 to 0.5 C	freezing point	
	IFA (shallow)	Upper	0–10 m, 5/7 C	freezing point	Freezing
		Thermocline	10–20 m, 0.5/m		
		Halocline 1	10–20/30 m, 0.03/m	vanish	
		Halocline 2	30/50–70 m, 0.03/m		
		Lower	0.5–1.5 C, 31.5–32.3$^o/_{oo}$	freezing point	
Land water	NS	Surface	0–10/20 m, 9–11 C	freezing point	Freezing
		Thermocline	1–0.6/m		
		Halocline	0/10–20/30 m, 0.1/m	vanish	
		Bottom	1–3 C, 32$^o/_{oo}$	freezing point	
	AG	Surface	0–10/20 m 10–12 C, 32$^o/_{oo}$	freezing point	Freezing
		Thermocline	1/m	inversion	
		Halocline	0.1/m		
		Lower	1–2 C, 33$^o/_{oo}$	freezing point	

Remarks: AS—Alaskan Stream Water   BSW—Bering Sea Water
              CA—Convective Area   CW—Coastal Water
              IFA—Ice Forming Area
              NS—Norton Sound   AG—Gulf of Anadyr Water

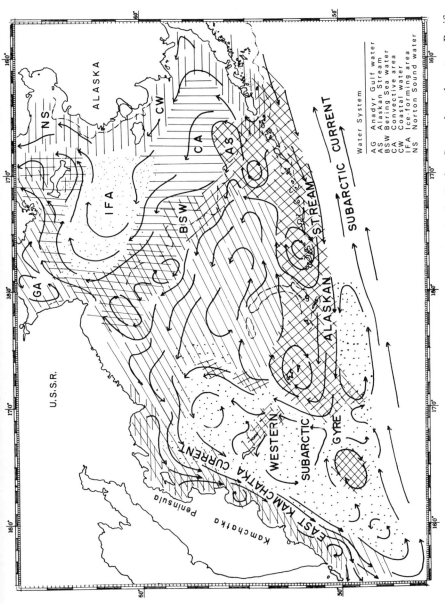

Water System

AG Anadyr Gulf water
AS Alaskan Stream
BSW Bering Sea water
CA Convective area
CW Coastal water
IFA Ice-forming area
NS Norton Sound water

Fig. 2.10 Schematic diagram of circulation and extent of water masses in the Bering Sea and northwestern Pacific Ocean.

Discussion

COACHMAN: It is becoming quite clear that there is a strong time-dependency in terms of the location of the water masses, the strength of the circulation, and hence the quantitative interactions prevailing in the area.

REFERENCES

Barnes, C. A., and T. G. Thompson
> 1938   Physical and chemical investigations in the Bering Sea and portions of the North Pacific Ocean. *Univ. Wash. Publ. Oceanogr.* 3(2): 35–79 + Appendix pp. 1–164.

Dodimead, A. J., F. Favorite, and T. Hirano
> 1963   Salmon of the North Pacific Ocean, Part 2. Review of oceanography of the subarctic Pacific region. *Bull. Int. North Pac. Fish. Comm.* 13:195.

Favorite, F.
> 1967   The Alaskan Stream. *Bull. Int. North Pac. Fish. Comm.* 21: 1–20.

Kihara, K., and M. Uda
> 1969   Analytical studies on the mechanism concerning the formation of demersal fishing grounds in relation to the bottom water masses in the eastern Bering Sea. Part 1. Studies on the formation of demersal grounds. *Tokyo Univ. Fish.* 55(2): 83–90.

Kitano, K.
> 1970a   A note on the thermal structure of the eastern Bering Sea. *J. Geophys. Res.* 75(6): 1110–1115.
> 1970b   A note on the salinity structure of the eastern Bering Sea [in Japanese]. *Bull. Tohoku Reg. Fish. Lab.* 30: 79–85.

Koto, H., and T. Fujii
> 1958   Structure of the waters in the Bering Sea and the Aleutian region. *Bull. Fac. Fish., Hokkaido Univ.* 9(3): 149–170.

Koto, H., and T. Maeda
> 1965   On the movement of fish shoals and the change of bottom temperature on the trawl-fishing ground of the eastern Bering Sea [in Japanese]. *Bull. Jap. Soc. Scient. Fish.* 31(4): 263–268.

Maeda, T., T. Fujii, and K. Masuda
> 1967   On the oceanographic condition and distribution of fish shoals in 1963 [in Japanese]. Part 1. Studies on the trawl fishing grounds of the eastern Bering Sea. *Bull. Jap. Soc. Scient. Fish.* 33(8): 713–720.
> 1968   On the annual fluctuation of oceanographical conditions in summer season [in Japanese]. Part 2. Studies on the trawl fishing grounds of the eastern Bering Sea. *Bull. Jap. Soc. Scient. Fish.* 34(7): 586–593.

MISHIMA, S., and S. NISHIZAWA

1955 Report on the hydrographic investigations in Aleutian waters and the south-
      ern Bering Sea in early summers of 1953 and 1954. *Bull. Fac. Fish., Hokkaido
      Univ.* 6(2): 85–124.

OHTANI, K.

1965 On the Alaskan Stream in Summer [in Japanese]. *Bull. Fac. Fish., Hokkaido
      Univ.* 15(4): 260–273.

1969 On the oceanographic structure and the ice formation on the continental
      shelf in the eastern Bering Sea [in Japanese]. *Bull. Fac. Fish., Hokkaido Univ.*
      20(2): 94–117.

1970 Relative transport in the Alaskan Stream in winter. *J. Oceanogr. Soc. Japan*
      26(5): 271–282.

OHTANI, K., Y. AKIBI, and Y. TAKENOUTI

1972 Formation of Western Subarctic water in the Bering Sea. In *Biological
      oceanography of the northern North Pacific Ocean* [Motoda commemorative
      volume], edited by A. Y. Takenouti et al. Idemitsu-shoten, Tokyo, pp.
      31–44.

TULLY, J. P.

1964 Oceanographic regions and processes in the seasonal zone of the North
      Pacific Ocean. In *Studies on oceanography*, edited by K. Yoshida. Univ.
      Tokyo Press, Tokyo, pp. 68–84.

UDA, M.

1935 On the distribution, formation and movement of the dichothermal water in
      the North-Western Pacific. *Umi to Sora* 15(2): 445–452.

CHAPTER 3

Circulation, transport and water exchange in the western Bering Sea

F. W. HUGHES, L. K. COACHMAN, *and* K. AAGAARD

Department of Oceanography, University of Washington, Seattle, Washington

Abstract

Drogue measurements in the deep straits of the southwestern Bering Sea indicate a volume transport of $6-8 \times 10^6$ m^3/sec, or three times the computed geostrophic transport relative to 1000 db. Similar differences between measured and computed values were found for southward flow through the western third of Near Strait in summer.

The primary driving mechanism for the Bering Sea circulation is concluded to lie in the field of wind stress. Results suggest that the available wind torque is more than adequate for driving the observed transports and there is an annual variation in the wind torque, the winter values being close to an order of magnitude greater than in summer.

INTRODUCTION

Bases for assessments of the circulation

Previous assessments of the circulation of the Bering Sea have been based predominantly on three techniques: the dynamic method for computing the geostrophic circulation, water mass analysis, and wind-driven circulation models. Very few direct flow measurements are available from the deep basin area. One major program of direct measurements was conducted by Taguchi (1959), who released drift floats with hooks attached in a limited region of the central and southwest Bering Sea and adjacent regions of the northwest Pacific Ocean in May and June 1959. These floats were recovered in gill nets by the Japanese high seas fishing fleet. Based on the release and recovery points of the floats, the surface circulation was deduced. Interpretation of such measurements is subjected to a large amount of artistic license because of the lack of intermediate positions.

Limited drift bottle studies were carried out in 1957, 1958, and 1960 by the Biological Laboratory, Bureau of Commercial Fisheries (now National Marine Fisheries Service, Northwest Fisheries Center), Seattle, in the same region (Dodimead et al. 1963). More recent bottle studies have been discussed by Favorite (Chapter 1). Also, some direct measurements using a geomagnetic electrokinetograph (GEK) have been reported (Faculty of Fisheries, Hokkaido University 1959). The latter measurements, which were taken while the ship was steaming between hydrographic stations, are not well suited for interpretation of the circulation as they are instantaneous values which may contain sizeable contributions from periodic components such as tides. In addition, the classic interpretation of GEK measurements is problematic (Sanford and Schmitz 1971).

The indirect techniques for assessing circulation also have shortcomings. Very few quasi-synoptic data are available on which to base dynamic method calculations. Also, the relative importance of baroclinic and barotropic components in the circulation has not as yet been established.

Assessment of the circulation from water properties tends to be ambiguous. At least five different water mass classifications have been proposed for the Bering Sea (Leonov 1947, 1960; Smetanin 1958; Dobrovol'skii and Arsen'ev 1961; Uda 1963; Kitano 1970). Qualitative studies of the wind-driven circulation have been made by Federov (1956) and discussed by Dobrovol'skii and Arsen'ev (1959), but there has been only one quantitative study (Gurikova et al. 1964). The calculations must be done with considerable care, because the results depend strongly on the details of the atmospheric pressure maps and their averaging (Aagaard 1970; also Chapter 4 this volume).

The present study is based on direct current measurements and on hydrographic data encompassing the entire central and western Bering Sea at least once within a single season. This has made possible compilation of a current chart for the summer season using quasi-synoptic data. Measurements in selected areas were repeated in other years to permit conclusions concerning the permanency of observed features. We have also qualitatively considered the effects of bathymetry on the flow.

The evolution of circulatory schemes
The geographic nomenclature for the Bering Sea is shown in Figure 3.1. The first circulation scheme was proposed by Dall (1881), who also published the first chart of the surface currents (1882). The first oceanographic cruises which incorporated current measurements and modern temperature and salinity measurements were reported by Ratmanov (1937). His chart (Fig. 3.2) based on dynamic calculations became generally accepted as best representing the summer circulation. He indicated correctly the importance of the influx of Pacific water and the insignificant role of exchange with the Arctic Ocean. Ratmanov concluded that one of the main causes of the Bering Sea circulation is the exchange of water through the passes of the Komandorski-Aleutian island arc.

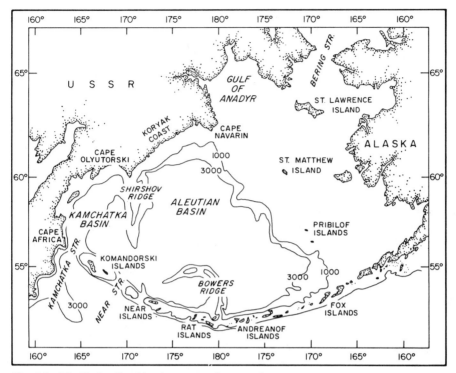

Fig. 3.1 The Bering Sea (depth contours in meters).

Goodman et al. (1942) presented a chart of summer surface currents (Fig. 3.3), essentially like Ratmanov's scheme except for the northern portion of the sea. Circulation deduced in the latter region was based on new observations of currents over the northern and eastern parts of the shelf and in the vicinity of Bering Strait.

In 1958 the U. S. Navy Hydrographic Office published a chart of the prevailing summer surface currents (Fig. 3.4). The overall scheme was similar to that of Ratmanov, with additional details provided for the Bristol Bay area where neither Ratmanov nor Goodman et al. had data. No cruises or other data sources for this chart were cited, nor were the analytical techniques discussed. The chart was the first to show current speeds, although again no observational basis was documented.

A significant departure from the earlier circulation schemes was published by Dobrovol'skii and Arsen'ev (1959). Their chart (Fig. 3.5) incorporated the results of cruises to the western, central, and southern portions of the sea and combined them with the work of the previous investigators, principally Ratmanov, whose scheme was adopted in regions where more recent data were unavailable. Their findings included a major outflow through the western portion of Near Strait; a major current flowing northwest from the eastern Aleutian passes, parallel with the continental shelf edge, and extending nearly to Cape Navarin; and the presence of numerous gyres in the deep basin.

The next major contribution to knowledge of the circulation was made by Natarov (1963). His observations encompassed nearly the entire area within a single season, obviating the need for data compiled from a variety of sources. For the first time, fairly extensive data were also available for a winter season (cruise of the *Pervenets*, 1961). The observed winter and summer circulation patterns are shown in Figure 3.6.

Arsen'ev (1967), using a large compilation of data, presented charts of the geostrophic currents, wind-driven currents, and the combined resultant surface currents for both summer and winter. The methods by which the charts were prepared were not specified; he was most confident of his summer results (Fig. 3.7). His charts depict the current systems within both the Bering Sea and the portion of the Pacific Ocean lying immediately south of the Komandorski-Aleutian island arc.

Evaluation and comparison of previous circulation schemes
The foregoing circulation schemes are all subject to one or more of the following major inadequacies: they are derived from an inadequate data base; they fail to consider adequately the effects of bathymetry on the flow; they give no consideration to the deep circulation; and they provide little or no information on temporal variability.

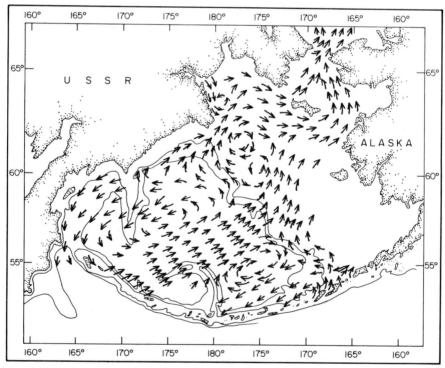

Fig. 3.2 Bering Sea circulation according to Ratmanov (1937). Contours are 1000 m and 3000 m depths.

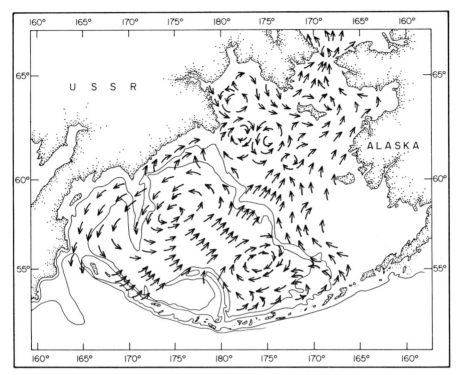

Fig. 3.3 Bering Sea circulation according to Goodman *et al.* (1942).

Only Dall's (1882) chart used direct current measurements as the primary basis for the deduced circulation pattern. His conclusions about surface currents were based on the sets of numerous ships, but there was no attempt to correlate differences between individual current measurements with seasonal variations. The nature of the measurements precluded the elimination of effects due to tidal motions. The U. S. Navy Hydrographic Office (1958) chart, which indicated current speeds, made no mention of the basis for their determination. Presumably a mixture of direct and indirect techniques were employed. All the remaining circulation schemes were based primarily on geostrophic calculations because of the general lack of direct current measurements. Where such measurements were available, they were said to compare favorably with the indirect schemes (Ratmanov 1937; Dobrovol'skii and Arsen'ev 1959). Wind-driven currents were calculated (method unknown) by Arsen'ev (1967) and vectorially added to the geostrophic currents. In spring and summer, the addition of wind currents was felt to simply reinforce the geostrophic flow.

The reference level employed for the dynamic height computations has varied. Ratmanov (1937) used a very shallow level (100 db) except for small areas near Cape Olyutorski and Kamchatka Strait where 500 db was used. Goodman et al. (1942), considering shallow water over the continental shelf, were precluded from using deep reference levels. Dobrovol'skii and Arsen'ev

(1959) and Natarov (1963) used 1000 db, while Arsen'ev (1967) used several levels from 1000 to 3000 db.

Some consideration of the qualitative effects of bathymetry on the circulation was included in the schemes of Ratmanov (1937), Natarov (1963), and Arsen'ev (1967). For example, Natarov felt that the locations of permanent gyres as well as some current speeds were partially controlled by bathymetry.

The nature of the deep circulation was discussed by Leonov (1960) and Arsen'ev (1967), but Arsen'ev was the only investigator to have sufficient data to consider the problem in any detail.

As to temporal variability, the chart published by the U. S. Navy Hydrographic Office (1958) was the first published to mention that changing wind conditions and tidal currents could produce large day-to-day variations in the general flow. Dobrovol'skii and Arsen'ev stated that their data were insufficient to resolve the origin or duration of small gyres which they felt existed within the circulation pattern. The problem of small gyres was raised again by Natarov (1963), who discussed their local significance, although his data also were

Fig. 3.4   Bering Sea circulation according to U. S. Navy Hydrographic Office (1958).

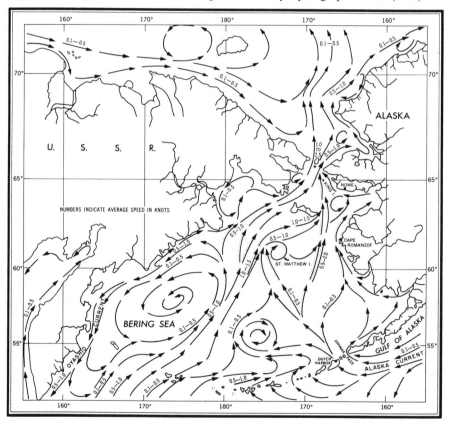

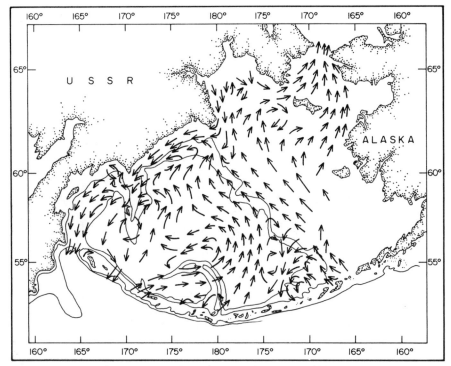

Fig. 3.5 Bering Sea circulation according to Dobrovol'skii and Arsen'ev (1959).

insufficient to define the gyres either spatially or temporally. Having data from both winter and summer, however, Natarov was able to provide a more complete discussion of seasonal variations than had been possible previously. On the other hand, Arsen'ev (1967) provided little or no information on seasonal variations, winter conditions, possible secular changes, or the validity of the long-term averages which he used in his analyses, although he had a large compilation of data covering all seasons.

Figures 3.2–3.7 allow comparison of the major features of the circulation as determined by each of the previous investigators. Features to be compared are the locations of the major inflows and outflows, the location and definition of gyres and delineation of the main currents within the sea. Brief consideration will be given to the differences between winter and summer circulations.

Prior to the work of Dobrovol'skii and Arsen'ev (1959), the main inflow was considered to occur across the whole width of Near Strait, while the major outflow was through Kamchatka Strait (Ratmanov 1937). Ratmanov postulated that the inflow through Near Strait was a branch of the Kuroshio, but such a hypothesis was never widely accepted. According to Ratmanov, other large inflows were through passes near Unalaska Island (Fig. 3.2). Dobrovol'skii and Arsen'ev (1959) showed that there was significant outflow through western Near Strait as well as through Kamchatka Strait. In addition they

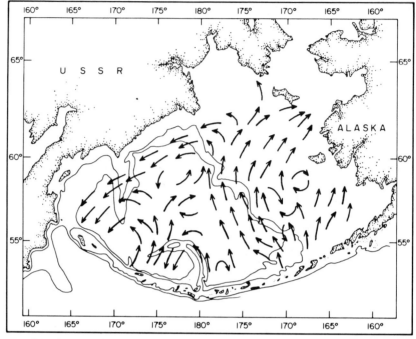

Fig. 3.6   Summer (above) and winter (below) circulation in the Bering Sea according to Natarov (1963).

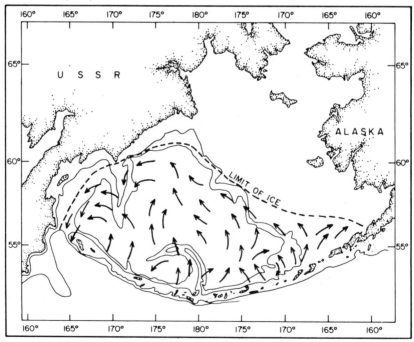

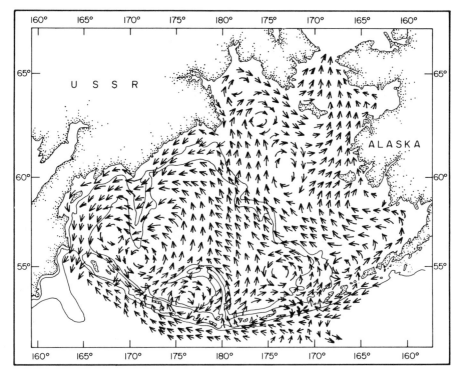

Fig. 3.7   Bering Sea circulation according to Arsen'ev (1967).

showed that part of the water flowing in through the eastern Aleutians returns to the Pacific, mainly through Amukta Pass and other straits of the Islands of Four Mountains. According to their scheme, water enters the Bering Sea primarily in three areas: eastern Near Strait, Amchitka and Tanaga passes, and the area of the Fox Islands (Fig. 3.5). They also indicated some inflow through eastern Kamchatka Strait; this was attributed to a local anticyclonic circulation around the Komandorskii Islands.

Natarov (1963) regarded as erroneous the opinion that Near Strait admits the bulk of Pacific Ocean water. He claimed that the main inflow takes place through the numerous straits in the eastern part of the Aleutian arc (Fig. 3.6); however, a nearly constant transport in through the eastern portion of Near Strait in the vicinity of Stalemate Bank was noted. Again, the main outflow was presumed to be through Kamchatka and western Near straits.

Arsen'ev (1967) reiterated the idea that the main inflow was through eastern Near Strait (Fig. 3.7). His views about outflow were conventional, although his chart did introduce a number of permanent anticyclonic gyres around groups of the Aleutian Islands. (It should be noted that all the circulation diagrams discussed in this paper show flow northward through the Bering Strait into the Arctic Ocean.)

All of the schemes showed at least one major cyclonic gyre in the deep western part of the sea. In all other respects there are numerous differences between the various schemes in the number, location, size, and even the direction of rotation of gyres depicted. This prompted Natarov (1963) to conclude that the description and origin of the smaller gyres was the single most complicated problem in the study of the current dynamics, especially in the deep water regime.

As in the matter of inflow and outflow through Near Strait, the discussion by Dobrovol'skii and Arsen'ev (1959) of the location of the main currents within the sea represented a change from the earlier ideas which had originated with Ratmanov (1937). They showed that the main flow, rather than crossing the deep basin from Near Strait to St. Matthew Island, was easterly from Near Strait along the northern side of the Aleutian arc. North of the Adreanof Islands the flow turned north and then northwest, becoming part of the general northwesterly flow which extends from the eastern Aleutian passes to Cape Navarin and runs parallel to the continental slope (Fig. 3.5). A significant portion of this flow was shown to turn southwest, proceeding along the Koryak coast and eventually reaching the Kamchatka Strait. The southwestward flow along the Koryak coast represented another major departure from the basic scheme of Ratmanov (Fig. 3.2). Because of his conclusions concerning the relatively small magnitude of the inflow through Near Strait, Natarov (1963) did not indicate an eastward flow north of the Aleutians, but showed only that portion of the northwesterly flow along the continental slope which had originated in the eastern Aleutian passes (Fig. 3.6).

All of the circulation schemes discussed so far are representative of the summer season. Winter observations reported by Natarov (1963) showed a general cyclonic circulation which appeared to have been strengthened relative to the summer circulation; the strengthening was presumably due to intensification of the atmospheric circulation during winter, which was in accord with a hypothesis of Drogaitsev advanced by Dobrovol'skii and Arsen'ev (1959). Both Natarov (1963) and Arsen'ev (1967) associated the intensified flow during winter with a simplified general circulation in which there were fewer gyres and less branching of the currents.

PRESENT OBSERVATIONS

Three summer cruises of approximately one month each have been made aboard the R/V *Thomas G. Thompson* of the University of Washington during the years 1969–71. Cruise designations and dates are given in Table 3.1 below; station locations and cruise tracks are shown in Figures 3.8–3.10.

TABLE 3.1   Summer cruises of the R/V *Thomas G. Thompson* in the western Bering Sea, 1969–71

Cruise	Dates
TT–041	13 July–15 August 1969
TT–050	16 July–11 August 1970
TT–062	30 July–23 August 1971

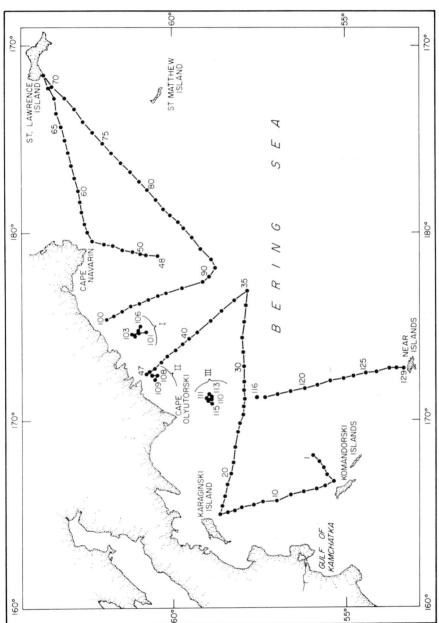

Fig. 3.8 Station locations, *T. G. Thompson* cruise TT-041, 1969.

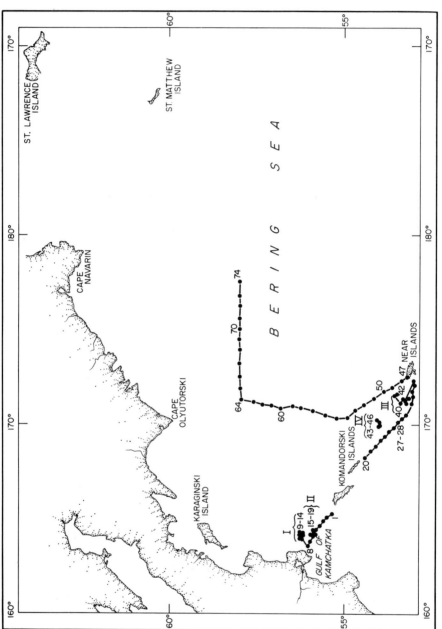

Fig. 3.9 Station locations, *T. G. Thompson* cruise TT-050, 1970.

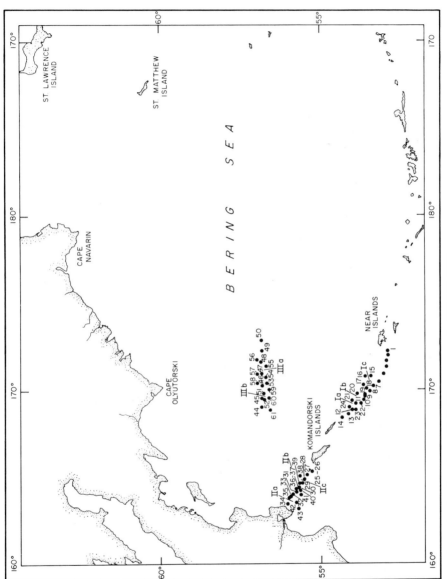

Fig. 3.10 Station locations, *T. G. Thompson* cruise TT-062, 1971.

These cruises were to the western Bering Sea from where data have been sparse and where the major exchange with the Pacific Ocean occurs. Both direct current measurements using parachute drogues and indirect measurements using STD and standard Nansen bottle hydrographic casts were made. Locations of the drogue measurements are shown by the roman numerals in Figures 3.8–3.10.

STD casts were taken on most oceanographic stations to a depth of 1500 m, or in shallower waters to within 10 to 25 m of the bottom. A grid of STD stations was taken, encompassing the drogue tracks, on each release of drogues. Digital recordings of the temperature, salinity, and depth parameters were made for direct input to a shipboard IBM 1130 computer. For calibration purposes a water bottle was attached 5 m above the STD sensors on each STD cast.

Hydrographic casts were taken at selected stations for routine STD calibration, for sampling the deep water, and while the STD was inoperative. These casts were made to within 25 m of the bottom at the deeper stations and to within 5 m of the bottom at the shallower stations. Nansen bottles were used to obtain water samples for salinity and dissolved oxygen analyses. The salinity determinations and the dissolved oxygen analyses were all performed aboard ship using an inductive salinometer and a modified Winkler method, respectively.

Direct measurement of the currents was accomplished during all three cruises by means of parachute drogues. The drogue system (Fig. 3.11) was modeled after those described previously by Volkmann et al. (1956) and Gerard and Salkind (1965). The surface float was a 2-m Roberts current meter buoy (Roberts 1947) modified to support a 3-m aluminum mast and a 30-kg keel. Atop the mast was mounted an active radar transponder, rather than the more conventional passive radar reflector. Use of transponders was a major innovation and resulted in the consistent achievement of radar ranges substantially greater than those using systems with reflectors. Typical maximum ranges obtained were about 10 n. miles, compared with a reflector range of 2–3 n. miles. The maximum theoretical range for the transponders, based on a 3-m transponder antenna height and a 15.3-m height of the radar antenna aboard the *Thompson*, is 12.5 n. miles. The achievable range using reflectors depends much more critically on sea state than does the range using transponders, due to the difficulty in differentiating the reflector image from the sea return on the radar scope. With a transponder system, the radar set aboard ship can be tuned to maximize the transponder signal while minimizing the sea return. Additionally, the transponder can be coded to transmit a multiple delayed signal which is easily distinguishable from the sea return.

The transponders operate in the X-band and are compatible with the Decca RM429 radar aboard the *Thompson*. During all cruises, Model SST 119X-A transponders manufactured by Motorola were used. These have 50-125 watt peak power and employ a 24-30 VDC power supply. During cruise TT-062, Model 430 MA transponders manufactured by Alpine Pioneer were used together with the Motorola packages. The Alpine units have 100-watt peak power and employ a 12 VDC power supply. Both types of transponders return to a stand-by condition if the units are not interrogated within about one minute, thus conserving the power supply. The power was provided by four heavy-duty

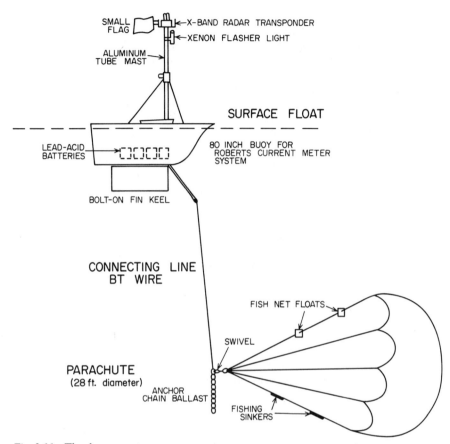

Fig. 3.11 The drogue system.

12-volt automobile batteries mounted in racks installed low within the floats to thus also serve as ballast.

The parachutes (approximately 9 m in diameter) were suspended using galvanized bathythermograph (BT) wire of 2.4 mm diameter. Wire of 4.0 mm diameter was used when the supply of BT wire was exhausted. Weights of 70–90 kg were used to keep the wire nearly vertical. To assist in keeping the parachute shrouds from entangling, as well as in maintaining parachute orientation, small fishing sinkers were attached to the lower shrouds, styrofoam fish-net floats were attached to the upper shrouds, and the entire parachute assembly was attached to the connecting line with a swivel. No attempts were made to recover the parachute assemblies or connecting line; they were simply cut loose at the time of retrieval of the surface floats.

Positioning was most frequently based on satellite fixes using the Navy Navigation Satellite System. Such systems were on board the *Thompson* during both cruise TT-050 and TT-062. During cruise TT-050, a Magnavox MX/702/hp navigation system was in use and during TT-062 a Magnavox MX/706/CA. There was an average of 24 usable fixes per day. The ship was equipped with a Kelvin-Hughes LORAN A/C set during TT-041 and TT-050 but only with LORAN A during TT-062. There was also an RCA CRM-N2A-30 radar available for navigational use in addition to the X-band Decca RM429 radar.

The accuracy of the satellite navigation system is about 55 m RMS for a stationary platform, while the overall fix error due to a one-knot velocity estimate error is 360 m (Stansell 1969). When the ship drifts slowly, as was frequently the case during tracking of the drogues, the estimated fix error is about 120 m RMS. Buoy tracking inaccuracies due to the radar bearing were repeatable by different observers to within 0.5° and .05 n. miles, respectively, so that only minor "noise" was believed to be introduced into the drogue tracks through the radar readings themselves. LORAN C has an accuracy of about 0.25 n. mile at 1000 n. miles range and LORAN A about 5 n. miles at 1500 n. mile range.

Accuracy of the drogue measurements is influenced not only by uncertainties in position but also by the nature of the drogue system itself, principally by the total drag being due in part to the surface float and connecting wire rather than just to the parachute. The most complete analytical study of the drag forces acting on parachute drogue systems was made by Terhune (1968). Based on his results, the error in the present drogue system is from 12 to 20 percent of the mean integrated velocity shear, where the latter is defined as the absolute value of the vector difference between the current velocity at the surface and at drogue depth.

Drogues were deployed in pairs, with one drogue being set at a relatively shallow depth (65–200 m) and the other at a greater depth (600–1300 m). Such placement provided a measure of the mean vertical shear and allowed a comparison with the geostrophic shear calculated from concurrent synoptic hydrographic information. During TT-041 and TT-050 only one pair of drogues was deployed at a time, so that experience and confidence could be gained in the method itself and to obtain a detailed record of each track (Figs. 3.12–3.17). (The track of the first drogue set made during TT-041 is not shown due to positioning uncertainties. Unusual atmospheric layering existed, which caused strong ducting of the radar signals and resulted in unknown range errors). The locations of the drogue sets can be seen in Figures 3.8–3.10 and Table 3.2.

During TT-062 a maximum of three pairs of drogues were employed simultaneously, which more efficiently utilized the great ranging capability of the system and (combined with the excellent positioning ability of the satellite navigator) allowed the direct measurement of horizontal shears over great distances. Tracks of these drogues are shown in Figures 3.12–3.20.

During drogue tracking in Kamchatka Strait (Fig. 3.19), a topographic feature resembling a guyot was noted and was not indicated on any bathymetric charts of the region. The boundary of the top of the feature was crossed in the vicinity of 55°49′ and 163°54′ while heading southwest and again in the vicinity of 55°43′ and 163°58′ while heading east into deeper water. The depth atop the feature was uniform and about 1720 m, while the surrounding, relatively flat bottom was about 4240 m. The 2500-m high sides were extremely steep. The

Drogue track figure	Cruise	Figure/Area	Depth (m)	Launch location	Launch time (local)	Zone time used	Recover location	Recover time	Track duration hours
—	TT–041	8/I	180	60°54'N, 174°40'E	1510, 3 Aug 69	+11	61°01'N, 174°52'E	1120, 5 Aug 69	41.5
			850	60°51'N, 174°39'E	1800, 3 Aug 69	+11	60°58'N, 174°50'E	1145, 5 Aug 69	41.5
12		8/II	200	60°32'N, 172°24'E	2055, 5 Aug 69	+11	60°28'N, 172°20'E	1345, 7 Aug 69	39.5
			100	60°32'N, 172°24'E	2210, 5 Aug 69	+11	60°30'N, 172°15'E	1315, 7 Aug 69	37
13		8/III	160	59°01'N, 171°13'E	0620, 8 Aug 69	+11	58°52'N, 170°42'E	1330, 11 Aug 69	51
			850	59°01'N, 171°17'E	0925, 8 Aug 69	+11	58°56'N, 171°00'E	0930, 11 Aug 69	65.5
			Reference*	58°58'N, 171°16'E	1540, 8 Aug 69	+11		0830, 11 Aug 69	
14	TT–050	9/I	65	56°16'N, 164°19'E	1720, 22 Jul 70	–11	56°02'N, 163°50'E	1302, 25 Jul 70	63
			1020	56°16'N, 164°19'E	1827, 22 Jul 70	–11	56°17'N, 164°07'E	1035, 25 Jul 70	65.5
15		9/II	200	55°53'N, 164°22'E	1700, 25 Jul 70	–11	55°44'N, 164°14'E	0940, 28 Jul 70	63
			1000	55°53'N, 164°22'E	1630, 25 Jul 70	–11	55°50'N, 164°20'E	0845, 27 Jul 70	62
16		9/III†	125*	53°05'N, 171°17'E	1713, 30 Jul 70	–12	53°32'N, 171°36'E	1850, 1 Aug 70	72.5
			700	53°05'N, 171°17'E	1650, 30 Jul 70	–12	53°28'N, 171°36'E	1934, 1 Aug 70	72.5
17		9/IV	100*	53°53'N, 170°02'E	0700, 2 Aug 70	+12	53°58'N, 170°03'E	1650, 4 Aug 70	57
			600*	53°53'N, 170°02'E	0632, 2 Aug 70	+12	54°00'N, 170°01'E	1630, 4 Aug 70	57
18	TT–062	10/Ia	125	54°03'N, 169°12'E	1615, 3 Aug 71	+11	53°58'N, 168°53'E	1926, 7 Aug 71	99
			750	54°03'N, 169°12'E	1410, 3 Aug 71	+11	53°59'N, 169°05'E	1835, 7 Aug 71	98
18		10/Ib	125	53°46'N, 169°52'E	1020, 4 Aug 71	+11	53°51'N, 169°47'E	1652, 7 Aug 71	76.5
			750	53°45'N, 169°53'E	1110, 4 Aug 71	+11	53°48'N, 169°49'E	1615, 7 Aug 71	74
18		10/Ic	125	53°32'N, 170°21'E	2018, 4 Aug 71	+11	53°43'N, 170°49'E	1045, 7 Aug 71	61
			750	53°32'N, 170°24'E	2103, 4 Aug 71	+11	53°43'N, 170°38'E	1147, 7 Aug 71	60.5
19		10/IIa	150	55°58'N, 163°54'E	1302, 9 Aug 71	+11	55°28'N, 163°11'E	1045, 12 Aug 71	68.5
			1300	55°58'N, 163°54'E	1347, 9 Aug 71	+11	55°50'N, 163°11'E	0830, 12 Aug 71	14
19		10/IIb	150	55°48'N, 164°19'E	0850, 10 Aug 71	+11	55°46'N, 163°45'E	2030, 12 Aug 71	58
			1300	55°48'N, 164°19'E	0937, 10 Aug 71	+11	55°52'N, 164°07'E	1500, 12 Aug 71	51.5
19		10/IIc	150	55°38'N, 164°47'E	2030, 10 Aug 71	+11	55°45'N, 164°19'E	0847, 13 Aug 71	59
			1300	55°38'N, 164°47'E	2105, 10 Aug 71	+11	55°42'N, 164°27'E	1006, 13 Aug 71	59
20		10/IIIa	150	56°49'N, 170°54'E	0947, 15 Aug 71	+11	56°55'N, 171°09'E	1829, 18 Aug 71	79
			1000	56°48'N, 170°55'E	1021, 15 Aug 71	+11	56°54'N, 171°10'E	1811, 18 Aug 71	77.5
20		10/IIIb	150	56°49'N, 169°38'E	2050, 15 Aug 71	+11	56°48'N, 169°16'E	0900, 18 Aug 71	59
			1000	56°49'N, 169°38'E	2134, 15 Aug 71	+11	56°46'N, 169°36'E	1020, 18 Aug 71	59

*Denotes use of 4.0 mm wire.
†Time zone changed to +12 during tracking period.

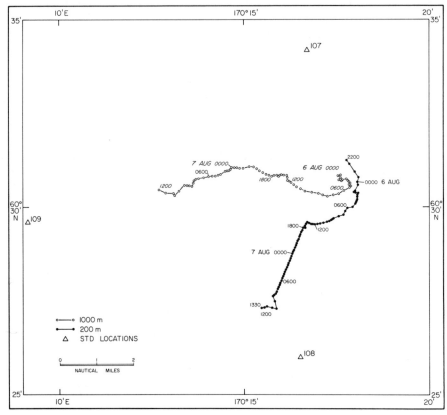

Fig. 3.12   Drogue tracks, TT-041 set II, 1969.

number of additional such bathymetric features and their effect on the flow in a strait previously considered to be deep and unobstructed is not known at present. The movement of the deep drogues in the strait can be interpreted as being influenced by the seamount (Figure 3.19).

ACCURACY OF GEOSTROPHIC CALCULATIONS IN QUANTITATIVE DEFINITION OF FLOW FIELD

The accuracy of geostrophic calculations in defining the flow field was assessed by comparing the motion measured by each pair of drogues to geostrophic flow calculated from pairs of STD stations bracketing the drogue track. Thirty-four comparisons associated with the 15 drogue pairs from the three cruises are summarized in Table 3.3.

The time interval used for calculating the speed and direction of drogue movement varied with the position data available. Cruise TT-050 provided the greatest number of STD station pairs as well as frequent fixes over relatively long drogue tracks, and therefore certain of these cases were examined for the

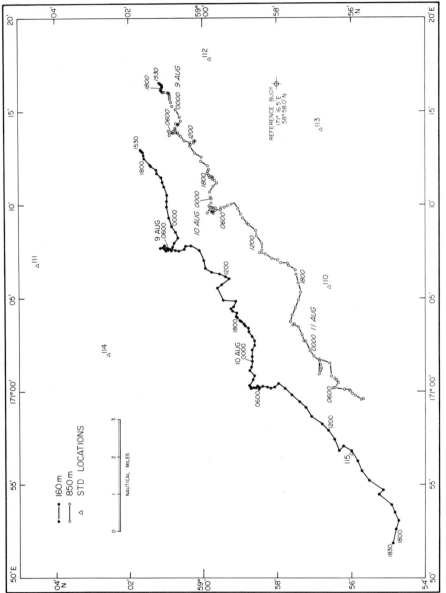

Fig. 3.13 Drogue tracks, TT-041, set III, 1969.

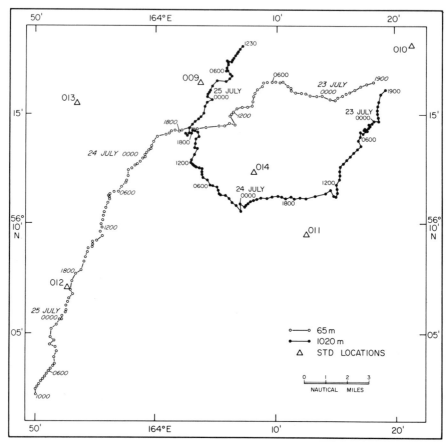

Fig. 3.14   Drogue tracks, TT-050, set I, 1970.

effects of using different averaging intervals. Selecting intervals 12 hours before
the station occupations, 12 hours after the stations, or 24 hours spanning the
stations made little difference in the results. In two cases (TT-050 set I, TT-062
set IIa), distinctly better agreement between the geostrophic and calculated
shears was obtained by averaging the flow over three-day intervals. This was not
true for other cases, although in no instance did the longer time interval lessen
the agreement. Table 3.3 is based on intervals of one day or longer.

    In general, the geostrophic and measured shears agreed well; the values
in more than half the cases were within 2×10^{-3} cm/sec-m, which can be
considered excellent agreement. In only 15 percent of the cases did the shears
disagree in direction; one such case was off the Koryak Coast, another in
eastern Near Strait, and three instances occurred in eastern and mid-Kamchatka
Strait. We conclude that geostrophic equilibrium was approached in those
regions where the measurements were made. The cases of opposite shear could
in each instance be attributed to a hydrographic station which was anomalous

with respect to other stations in the area. These are probably transient situations and should be anticipated to occur about 15 percent of the time.

In Table 3.3 the dynamic calculations are based on 1500 db whenever possible, otherwise on 1000 db; these have been the reference levels used by the majority of previous investigators. Agreement between measured and calculated currents is rather poor. For both shallow and deep levels the geostrophic calculation gave a velocity opposite in direction to the measured flow in fully one-fourth of the instances. In only one-fifth of the instances did the velocities agree within 20 percent.

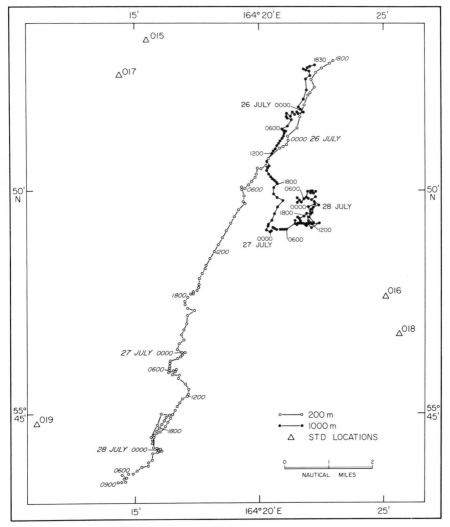

Fig. 3.15   Drogue tracks, TT-050, set II, 1970.

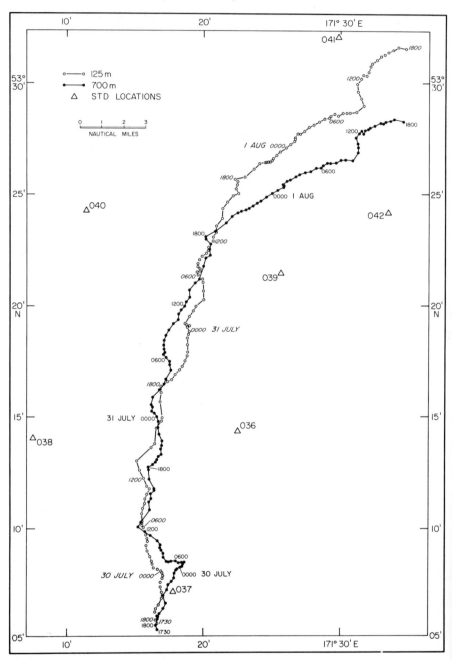

Fig. 3.16   Drogue tracks, TT-050, set III, 1970.

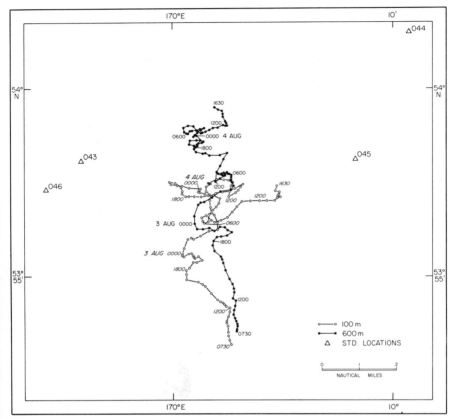

Fig. 3.17 Drogue tracks, TT-050, set IV, 1970.

There are two reasons for the poor agreement. One is the transient occurrence of anomalous water at an occasional station, causing marked departures from geostrophic balance; as noted, however, this occurred in only 15 percent of our cases. The second and more important factor is the use of an inappropriate reference level. By fitting the calculated geostrophic profile to the measured currents, we have estimated both the appropriate reference level in each case and the surface geostrophic velocity component (Table 3.3).

The appropriate reference level for use in the western Bering Sea varies markedly, and hence circulation patterns based on geostrophic calculations using a fixed reference level can even qualitatively contain major errors. With this major reservation, the following fixed reference levels are indicated by our data as being most nearly applicable: 1500 m for the western part of the Aleutian Basin; about 1000 m near the Shirshov Ridge; no reference level established across Near Strait, flow existing at all depths; and about 1500 m seems indicated in the eastern and middle parts of Kamchatka Strait, deepening to more than 2000 m in the western part.

The geostrophic velocity component at the surface appeared to show some systematic variation. Away from Near and Kamchatka straits the values were in the range of 5–18 cm/sec. In Near Strait the values were relatively large (11–23 cm/sec) and northerly in the east, smaller in the middle (5–10 cm/sec) and variable in direction, and of the same magnitude but southerly (6 cm/sec) in the western part. In Kamchatka Strait the component was small (3 cm/sec) and northerly in the east, changing to southerly in mid-strait (4–14 cm/sec) and increasing to very large southerly values (14–35 cm/sec) in the west.

The cyclonic horizontal shear across the major straits appears in both the measured and calculated flow fields and is indicated by the average surface geostrophic velocity components given in Table 3.4. The corresponding Rossby numbers (ratio of relative to planetary vorticity) are of order 10^{-2}.

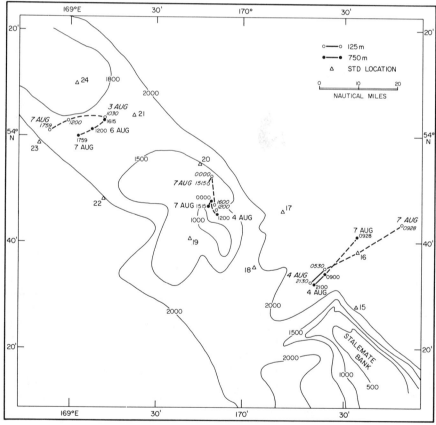

Fig. 3.18 Drogue tracks, TT-062, sets Ia, Ib, Ic, 1971. Depths in meters.

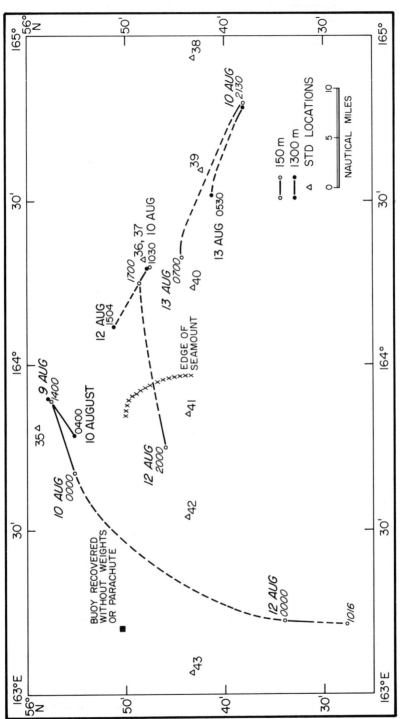

Fig. 3.19 Drogue tracks, TT-062, sets IIa, IIb, IIc, 1971.

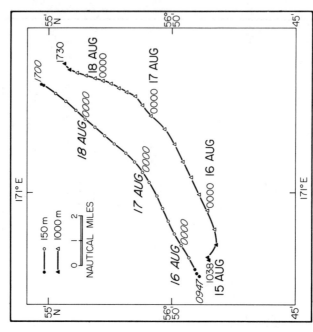

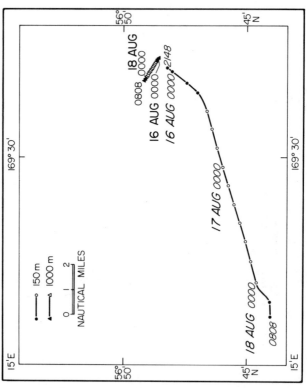

Fig. 3.20 Drogue tracks, TT-062, set IIIa (right) and IIIb (left), 1971.

TABLE 3.3 Calculated geostrophic—vs—measured flow

Location	Cruise, Sec. no.	Date	Speed (cm/sec) and direction				Qualitative Assessment				Estimated Surface Geostrophic Velocity Component		Estimated Reference Level (m)
			Shallow		Deep		Shear (a) Agreement	Opposite Shear	Flow Opposite in direction				
			Geostrophic	Measured	Geostrophic	Measured			Shallow	Deep	Speed	Direction	
Koryak coast	041	I	8-04-69	4 NE	16 NE	<1 NE	12 NE	E	X		X	18	NE	1700
	041	II	8-07-69	5 W	3 W	3 E	6 W	E		X	X	3	SW	none
Shirshov Ridge (NE)	041	III	8-07-69	3 NE	6 SW	6 NE	3 SW	E				11	SW	1300
	041		8-09-69	4 W	5 W	3 W	3 W	E				11	SW	1300
			8-09-69	6 SW	10 SW	2 SW	5 W	G				5	SW	none
	062		8-10-69	<1 SW	14 SW	1 NE	6 NE	E				6	NE	none
(east side)	062	IIIa	8-17-71	<1 SW	6 NE	0	6 NE	G		X	X	10	SW	1000
(west side)	062	IIIb	8-17-71	1 SW	10 SW	0	0	E						none
Near Strait (east)	050	III	7-30-70	14 N	22 N	8 N	15 N	E				18	N	none
			7-30-70	18 N	22 N	7 N	15 N	E				11	N	none
			7-31-70	5 N	22 N	11 N	24 N	G				18	N	none
			8-01-70	18 NE	20 NE	5 NE	18 NE	E				23	NE	none
(east)	062	Ic	8-05-71	7 W	21 E	1 W	10 E	E	X	X	X	19	NE	none
			8-05-71	6 NE	21 NE	0	13 NE	E				10	N	none
(mid)	050	IV	8-03-70	14 S	8 N	10 S	9 N	E		X	X	5	S	2000
			8-04-70	6 S	<1 S	1 S	3 N	E			X	6	N	2000
(mid)	062	Ib	8-06-71	4 N	6 N	<1 N	2 N	E				6	N	2000
			8-06-71	4 N	5 N	2 N	2 N	E				6	SW	none
(west)	062	Ia	8-07-71	<1 S	4 S	1 N	3 S	E		X	X	6	SW	none
			8-07-71	<1 S	5 S	<1 S	3 S	E				6	SW	none
			8-07-71	6 W	7 W	1 W	2 W	E				6	SW	none

(a)
E—excellent agreement, values differ by $<2 \times 10^{-3}$ cm/sec-m
G—sign of shear agrees, values differ by $>2 \times 10^{-3}$ cm/sec-m

TABLE 3.3 Calculated geostrophic—vs—measured flow (continued)

Location	Cruise, Sec. no.	Date	Speed (cm/sec) and direction Shallow Geostrophic	Shallow Measured	Deep Geostrophic	Deep Measured	Qualitative Assessment Shear (a) Agreement	Opposite Shear	Flow Opposite in direction Shallow	Deep	Estimated Surface Geostrophic Velocity Component Speed	Direction	Estimated Reference Level (m)
Kamchatka Strait (east)	062 IIc	8–10–71	2 S	6 N	0	3 N	G	X	X		3	N	1400
(mid)	050 II	8–11–71	6 N	2 N	<1 N	1 N	E				14	SW	1800
		7–26–70	8 SW	14 SW	0	6 SW		X	X	X			
		7–27–70	2 NE	6 SW	0	<1 NE		X	X	X			
		7–27–70	4 NE	6 SW	0	<1 NE							
(mid)	062 IIb	8–10–71	21 S	11 S	<1 S	0	G				14	S	1300
		8–11–71	5 S	3 S	0	3 N	E				4	S	800
(west)	050 I	7–23–70	20 W	18 SW	2 SW	14 SW	G				35	SW	2000
		7–23–70	17 NW	15 NW	3 NW	8 NW	G				35	SW	2000
		7–24–70	6 S	19 S	5 N	12 N	G				14	S	700,1400
		7–24–70	1 SE	7 SE	4 NW	8 NW	G				motion cross-channel		
(west)	062 IIa	8–10–71	22 SW	30 SW	<1 SW	15 SW	G				30	SW	3000
		8–11–71	3 S	22 S	0	9 S	G				24	S	2000

(a)
E—excellent agreement, values differ by <2 × 10⁻³ cm/sec-m
G—sign of shear agrees, values differ by >2 × 10⁻³ cm/sec-m

TABLE 3.4   Average surface geostrophic velocity components (cm/sec)

Strait	West	Mid	East
Near	6 S	4 N	18 N
Kamchatka	25 S	11 S	3 N

PROPOSED CIRCULATION SCHEME

A new surface circulation scheme for the western Bering Sea during summer is proposed based on the results of our three cruises. The following factors are taken into account: the direct current measurements made with the drogues; the bathymetry of the region (qualitatively); and geostrophic calculations based on our own observations as well as those of other recent investigators. The deduced surface circulation scheme is shown in Figure 3.21.

The inferred circulation pattern in the vicinity of Bowers Ridge is based principally on the dynamic computations of Favorite and Ingraham (1971).

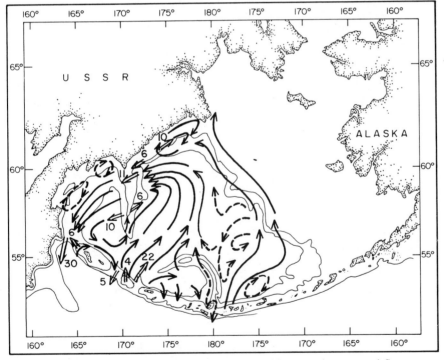

Fig. 3.21 Proposed surface circulation scheme. Double arrow is measured flow vector, with speed in cm/sec. Qualitative key:

        ⟶ certain

        ---→ less certain

        ⇒ direct measurements (cm/sec)

Gyres east of Karaginsky Island and in Olyutorski Bay are inferred from the observations of Arsen'ev (1967), of Dobrovol'skii and Arsen'ev (1959) and from our own data. The gyre north of the Andreanof Islands first appeared in Ratmanov's (1937) interpretation of the circulation and has taken various forms subsequently (Dobrovol'skii and Arsen'ev 1959; Natarov 1963; Arsen'ev 1967). The circulation pattern northeast of Bowers Ridge is inferred from the charts of these same investigators and has been made to comform with the recent results of Favorite and Ingraham (1971) in the vicinity of the ridge itself.

The overall circulation in the deep basin is cyclonic in nature. The effect of the bathymetry on the flow is particularly evident in the region of Bowers Ridge and at the southern end of Shirshov Ridge. The latter feature seems to cause a divergence in the flow. Farther north, in the region southeast of Cape Olyutorski, the flow is indicated to be convergent before passing over Shirshov Ridge. The minimum cross section over the ridge axis north of 57°05′ is about 2.3×10^8 m². If we assume an average current over this cross section of 8 cm/sec, which is reasonable in view of the directly measured values, then the volume transport of this flow is 18.4 Sv. This is in reasonable agreement with the estimated summer outflow through Kamchatka Strait (see Table 3.7). On a smaller scale, there is undoubtedly some bending of the streamlines as the flow crosses the ridge, first bending northward and then westward. Akiba (personal communication) has found evidence of this.

Further intensification of the flow is shown north of Kamchatka Strait, in conformance with the observed outflow which takes place mainly through the western half of the strait. Evidence of westward intensification of a southerly flow was not, however, found along the entire western boundary of the two deep basins. Along the northern part of the Koryak coast, in fact, a northeastward flow was observed. Such a flow was noted by Ratmanov (1937) but has not appeared in circulation schemes subsequent to and including that of Dobrovol'skii and Arsen'ev (1959).

Outflow is shown through the passes southwest of Bowers Ridge and east of the Near Islands (Buldir and Kiska passes). This outflow is inferred from continuity requirements and will be discussed subsequently.

The water between 1000 and 1500 m appears to follow the overall circulation scheme shown by the surface layers, although outflow at these levels is, of course, no longer possible through Buldir and Kiska passes as shown for the surface layers. Sufficient data are not available from our three cruises to define the circulation below 1000–1500 m. Direct current measurements at 1000 m at the southern end of Shirshov Ridge suggest, however, that below these depths the bathymetry forces the development of two separate cyclonic gyres, one in the Kamchatka Basin and one in the Aleutian Basin.

TRANSPORTS IN THE BERING SEA

Wind-driven transport calculations
The primary source of relative vorticity for the ocean is the torque generated by the wind. This relationship was first explored in a classic paper by Sverdrup

(1947), who described the vorticity balance between the curl of the wind-stress and the advection of planetary vorticity. This model has later been shown by numerous investigators to be a realistic one in the eastern and central portions of the ocean, as long as the effects of variable depth are not important.

We have calculated the steady wind-driven circulation associated with the field of wind-stress over the Bering Sea. This has been done in the form of Sverdrup transport (the ratio of wind-stress curl to the variation with latitude of the Coriolis parameter) integrated westward from the eastern margin of the sea. Such calculations have been successful in quantitatively explaining the mean circulation of another subpolar gyre, that of the Greenland-Norwegian Sea (Aagaard 1970). The calculations are based on Fofonoff's (1962) method and utilize the U. S. Weather Bureau climatological monthly surface pressure charts (U. S. Weather Bureau 1961).

Table 3.5a gives the resultant Kamchatka Current transport that would be necessary to close the Sverdrup circulation. Aagaard (1970) has shown that the use of monthly average pressure charts can lead both to an underestimate of the actual wind-stress field by a half order of magnitude and likewise to a similarly large underestimate of the western boundary current transport. In light of this, it is clear from Table 3.5 that the available wind torque is more than adequate for forcing return transports as large as any that have been estimated for the Kamchatka Current. An annual variation in wind torque is also apparent, with the winter values being close to an order of magnitude greater than summer values. This is in accord with results cited by Arsen'ev (1967) and also with the fact that dynamic height calculations indicate that the transport of the Kamchatka Current during summer is one-half to one order of magnitude less than during winter; e.g., contrast the present calculated geostrophic summer transports (Table 3.6) with those of Reid (1966) for winter, which exceeded 20 Sv based on the 1500 db surface.

We have also calculated the integrated Sverdrup transports for June-August 1969 based on four-times daily pressure maps. These results show a predominant anticyclonic wind torque over the Bering Sea, whereas our oceanographic observations point to a weak cyclonic circulation. This appears to be in accord with the conclusions of Aagaard (1970) concerning the wind-driven circulation of the Greenland Sea, where even prolonged (five months) reversals of the wind-stress curl did not force reversal of the western boundary current.

TABLE 3.5   Kamchatka current transports necessary to close the Sverdrup circulation (Sv) based on climatological data

January	13	July	3
February	13	August	3
March	11	September	4
April	3	October	5
May	2	November	6
June	<1	December	9

Transport and water balance

The water exchange between the Bering Sea and the North Pacific Ocean is very poorly known, although the exchange with the Arctic Ocean through Bering Strait (Coachman and Aagaard 1966) is reasonably well understood. As part of the water transport computations, the cross-sectional area of the passes must be determined. The first diagram showing the depths of the major passes through the Komandorski-Aleutian island arc was presented by Fleming (1955) based largely on data from the U. S. Navy Hydrographic Office. A similar diagram was presented by Udintsev et al. (1959), who also provided a tabular listing of the names, depths, and cross-sectional areas for the 47 passes shown. The only three passes deeper than 1000 m are Kamchatka, Near, and Amchitka.

Initial transport computations for the Komandorski-Aleutian passes were made by Leonov (1960), but no supporting data were supplied and the arguments lacked credibility. Transports through Kamchatka and Near straits were computed by Batalin (1964) using both heat balance data and a method based on the inhomogeneity of the zonal wind field. The most detailed water balance published was computed by Arsen'ev (1967), who assumed that the bulk of the water exchange occurred through the five major passes with the predominantly largest cross sections (Near, Kamchatka, Amchitka, Buldir, and Amukta).

Transports through major straits

Using both the direct current measurements and the hydrographic information obtained during the 1970 and 1971 cruises, the volume transports through Near and Kamchatka straits were computed. The components of the direct current measurements normal to the hydrographic sections were calculated, and the geostrophic profiles were adjusted to conform to these measured values. The adjusted velocity profiles across the straits were then contoured and planimetered to arrive at the measured volume transports.

In addition, geostrophic transports based solely on dynamic heights were calculated. These were referenced to 1500 db in Near Strait and 3000 db in Kamchatka Strait. The selected reference levels resulted in nearly equal transport cross sections for the two straits, and they corresponded to the deepest levels of no motion in common use. The results are shown in Table 3.6. The net transport through Near Strait is into the Bering Sea while that through Kamchatka Strait is out to the Pacific.

The dynamic topography in Near Strait showed little relief during both cruises, and there was generally no reference level. As a result, the calculated geostrophic flow was much less than the measured flow. The measured net flow through Near Strait was nearly the same for both cruises and occurred predominantly through the eastern part on either side of Stalemate Bank. This agrees with the findings of Natarov (1963), although the magnitude of the flow through Near Strait is certainly not what Natarov termed an insignificant amount. Since his results were based on geostrophic calculations, however, it is understandable that he substantially underestimated the inflow (see Tables 3.3 and 3.6). Our measured inflow through the eastern two-thirds of Near Strait in excess of 30 Sv contrasts with the net inflow of 15 Sv obtained by

TABLE 3.6   Volume transports (Sv) through Near and Kamchatka straits during 1970 and 1971

Strait	Cruise	Area of hydrographic section $(10^6 \ m^2)$	Geostrophic			Measured		
			Inflow +	Outflow −	Net	Inflow +	Outflow −	Net
Near	TT–050	358	3.9	0.2	+3.7	30.4	4.5	+25.9
	TT–062	319	5.5	0	+5.5	33.6	7.9	+25.7
Kamchatka	TT–050	305	0.1	11.6	−11.5	2.2	17.0	−14.8
	TT–062	353	2.7	13.8	−11.1	3.4	22.9	−19.5

Arsen'ev (1967) using the dynamic method. Our measurements show a southward transport through the western third of Near Strait of about 6 Sv in summer. The measured values are about a half order of magnitude greater than those obtained solely from our dynamic height calculations. No outflow was even indicated in Near Strait by the hydrographic data taken during TT-062. This puts in a new perspective Ratmanov's (1937) conclusion, based on geostrophic calculations, that inflow occurred across the entire width of Near Strait. There is thus reason to question the validity of dynamic calculations for even a qualitative expression of the flow field and water exchange.

In Kamchatka Strait a reference level can generally be found, but it varies across the strait, shoaling from west to east. A small inflow in the upper layers in the vicinity of the Komandorski Islands is indicated in both the geostrophic and measured profiles. This is in agreement with the conclusions of Batalin (1964) and Arsen'ev (1967), that a wind-driven anticyclonic circulation exists around these islands. A deep inflow below 1300 m was observed in the eastern portion of Kamchatka Strait during both cruises. A counter current below 1000 m in the western portion of the strait was observed during TT-050 (Figs. 3.14 and 3.15). Such a current was neither measured by the drogues nor indicated by the hydrographic data, however, during TT-062 (Figure 3.19).

There was much larger variation in the outflow through Kamchatka Strait than in the inflow through Near Strait between the two cruises. This is due probably to short-term variations in the flows.

Proposed seasonal water budget
Based on our own observations during summer, the winter measurements of other investigators and the known intensification of the circulation in winter (Dobrovol'skii and Arsen'ev 1959; Natarov 1963; Arsen'ev 1967), an approximate water budget for the Bering Sea in both summer and winter is proposed. This budget excludes the exchange of water with the Arctic Ocean through Bering Strait, on the assumption that it is balanced by inflow through the eastern passes of the Aleutians. The proposed budget is given in Table 3.7 below:

Considering the summer values first, outflow through Buldir and Kiska passes is consistent with the proposed circulation scheme (Fig. 3.21). The total cross-sectional area of these passes is given by Udintsev et al. (1959) as

40.3×10^6 m² and by Favorite (1967) as 34.8×10^6 m². Using the smallest of these estimates, a mean current of 14.4 cm/sec would provide the needed transport. An overall average current in Buldir Pass of approximately 8 cm/sec, based on geostrophic calculations, was estimated by Arsen'ev (1967); this was inflow, however. Southward flow through these passes was shown by Natarov (1963), and direct observations in summer also showed only southerly flow (Reed 1971). Thus, outflow does not appear to be an unrealistic assumption for this pass.

Little seasonal variation has been found in the surface flow through Near Strait (Arsen'ev 1967; Natarov 1963), and we have therefore assumed the same inflow in winter as was measured in summer. This same volume transport was computed geostrophically (reference level unspecified) using data collected in March 1966 by Favorite et al. (1967). The proposed seasonal variations in Kamchatka Strait are founded principally on the measurements of Reid (1966) and the long-term computations of Arsen'ev (1967). Arsen'ev (1967) based all of his geostrophic calculations for Kamchatka Strait on a level of no motion at 3000 m and determined that the average annual surface velocity through the strait was 17 cm/sec. Average surface velocities for the months of January, July and August were 20, 13 and 10 cm/sec, respectively. Based on an annual average velocity of 7 cm/sec above 3000 m, Arsen'ev computed a discharge of 21.0 Sv. Assuming that the average velocity and hence the volume transport through the cross section varies in direct proportion to the average surface velocity, transports for January, July, and August would be 24.7, 16.1, and 12.4 Sv, respectively. These summer transports compare very favorably (Table 3.6) with those we have measured. Reid (1966) reported that winter measurements showed the transport of the Kamchatka Current relative to 1500 db to be 23 Sv. This was about three times the value he obtained from summer measurements (8 Sv). The indicated winter surface speed of the Kamchatka Current near Cape Africa was 60 cm/sec (Reid 1966). Assuming that the shallow reference level used by Reid (1966) underestimated the winter flow in the same proportion it underestimated the summer flow computed by Arsen'ev (1967), an outflow through Kamchatka Strait above 3000 m of about 35 Sv in winter is implied.

The difference between the estimated winter inflow and outflow through the deep western straits must be compensated for by exchange through the shallower Aleutian passes. Considering only the passes with sill depths greater than 300 m (Yunaska, Amukta, Amchitka, and Buldir), they constitute a total cross-sectional area of 99.6 km² out of the 130 km² east of Attu Island (Favorite

TABLE 3.7   Proposed approximate water balance (Sv) for the Bering Sea

	Summer	Winter
Near Strait	+25	+25
Kamchatka Strait	−20	−35
Buldir and Kiska passes	−5	
Aleutian passes		+10 (including Buldir and Kiska)

1967). An average speed of 10 cm/sec through this cross section is needed to satisfy continuity and appears to be a reasonable value when compared with either the winter values in the vicinity of these passes computed geostrophically by Natarov (1963) or similar summer values obtained by Arsen'ev (1967). Over any significant period of time, the water balance of the Bering Sea must be satisfied very accurately. An imbalance of only 0.1 Sv persisting for a period of 1 month would change the water level of the Bering Sea by about 10 cm.

Implications of seasonal variations
The seasonal variations in the proposed water budget imply changes in the internal circulation of the sea. The Kamchatka Current flowing southward along the western boundary must intensify in winter, and the proposed summer circulation scheme (Fig. 3.21) must be modified to account for the large inflow needed through the Aleutian passes. Thus, the cyclonic gyre which dominates the deep basin in summer must encompass nearly the entire sea in winter. Such a simplified winter circulation scheme (Fig. 3.6) was proposed by Natarov (1963), and similar patterns had been hypothesized by numerous previous investigators (Schulz 1911; Federov 1956; Dobrovol'skii and Arsen'ev 1959). The intensification during winter was related to the intensified atmospheric circulation over the Bering Sea by Dobrovol'skii and Arsen'ev (1959). The wind-curl calculations presented in this paper provide a quantitative physical basis for this intensification and its relationship to the wind field.

STATUS OF KNOWLEDGE

A number of problems exist which have still not achieved satisfactory resolution. Among the most important of these are the deep water circulation, quantitative effects of bathymetry on the circulation, secular, and seasonal variations in the flow, the exact role of the wind-driven circulation, and quantitative estimates of the exchange with the North Pacific Ocean and between the deep basin and the shelf.

Studies at the University of Washington are aimed at obtaining solutions to the above problems. Direct measurements to define the flow field are planned over the continental slope west of St. Matthew Island and the Pribilofs, as well as north of the Andreanof Islands. The role of the wind as a driving force is under study, and hopefully a model with which to forecast current patterns from routine atmospheric parameters can be developed. No winter cruises have as yet been possible, but it is hoped to be able to conduct cruises covering the ice-free area of the sea, with sections penetrating the ice wherever feasible. The determination of an appropriate reference level and its variations throughout the deep basin is being attempted.

Other problems which need to be investigated concern the location and permanence of small gyres, and detailed studies of the current in each of the passes. Both of these are significant to the fisheries. Using presently available data, it is possible to describe the circulation pattern and water masses of the Bering Sea better than has so far been accomplished, e.g., by Arsen'ev (1967), who based his study entirely on data from 1959 and earlier. About 35 cruises were made in 1960–66 alone by Japanese, Russian, and U. S. agencies. We anticipate a thorough isentropic, volumetric, and core analysis to identify the source, distribution, and variability of the water masses. Sufficient data are also available to attempt a salt balance for the Bering Sea. Such a balance has not yet appeared in literature available to us.

Discussion

UDA: Is the circulation you propose in the Bering Sea the dominant one? I think that the circulation varies year by year due to fluctuations in the atmospheric or wind circulation, because especially in winter and spring the energy supply comes mainly from the wind-stress. Have you made reasonable, theoretical computations for the actual measured results?

HUGHES: We believe that our proposed scheme represents the overall tendency of the circulation. The existence of small gyrals and temporal variations of the flow are problems which are as yet unresolved. Some annual variation does seem to exist, however, as you point out. Our wind-driven transport calculations do confirm a winter intensification of the transport. These computations support both the values we have measured for the summer and those we have proposed for winter. The latter values are also based on results of other investigators such as Reid (1966).

UDA: The cyclonic circulation causes an upwelling of deep water. Concerning the calculation of water transport, I believe that this supplied water quantity should be taken into account. Have your computations included this phenomenon?

HUGHES: We hope to make such calculations in the near future as our investigations continue. The results will be of importance also to studies of the deep circulation.

TAKENOUTI: What correction is necessary for using a relatively large float with active radar transponder for a parachute drogue?

HUGHES: The surface floats are not really as large as it might seem, since the transponders being supported are quite small. Our floats are only 80 inches long, weigh about 600 pounds when fully outfitted (exclusive of connecting wire, parachutes, and sinking weights), and have a streamlined design.

A detailed discussion of the drag forces acting on parachute drogues has been given by Terhune (1968). Based on his results, the error in the present drogue system ranges from 12 to 20 percent of the velocity shear, which is defined as the absolute value of the vector differences between water velocities at the surface and at drogue depths. This was very small in most cases. In all cases, the force acting on the parachute was 85 percent or more of the total of the forces acting on the system.

FAVORITE: I must question your evidence for a Bering Sea winter circulation approximating the transports of major systems, the Gulf Stream and Kuroshio

Current; I can see no real way for wind-stress to establish such a magnitude of flow.

COACHMAN: The wind-stress *is* there: the curl calculations from the wind-stress field show that the torque on the system is an order of magnitude larger in the winter than in the summer, and it apparently drives the flow regime which is two or three times greater.

FAVORITE: It must be assumed that there is sufficient time for this flow to be reflected in the distribution of mass, and there is not any more time for this to occur in the Bering Sea than in the Alaskan Stream. What is the total source of this flow? If a 40 Sv flow occurs in the winter through a 2000 m pass, then the westward flow in the Alaskan Stream—a major component of flow into the Bering Sea—would have to be tremendous, and I see no evidence of this.

Acknowledgments

This study was supported by a grant from the National Science Foundation. Cdr. Celia Barteau was instrumental in computer processing of the atmospheric pressure data. The research was performed while the senior author was on active duty with the U. S. Navy as a participant in the Junior Line Officer Advanced Scientific Education Program (Burke Program) administered by the Bureau of Naval Personnel.

REFERENCES

AAGAARD, K.
> 1970   Wind-driven transports in the Greenland and Norwegian seas, *Deep-Sea Res.* 17: 281–291.
> 1972   On the drift of the Greenland Pack ice. In *Sea ice*, edited by T. Karlsson. Proc. int. conf. held in Reykjavik, Iceland, 10–13 May 1971. Nat. Res. Counc. Iceland, pp. 17–21.

ARSEN'EV, V. S.
> 1967   Currents and water masses of the Bering Sea [in Russian, English summary]. Izd. Nauka, Moscow. (Transl., 1968, Nat. Mar. Fish. Serv., Northwest Fish. Center, Seattle, Wash.), 135 pp.

BATALIN, A. M.
> 1964   On the water exchange between the Bering Sea and the Pacific Ocean [in Russian]. *Trudy VNIRO* 49: 7–16. (Transl., 1968, in *Soviet fisheries investigations in the northeastern Pacific*, Part 2., pp. 1–12, avail. Nat. Tech. Inf. Serv., Springfield, Va., TT 67-51024).

COACHMAN, L. M., and K. AAGAARD
> 1966   On the water exchange through Bering Strait. *Limnol. Oceanogr.* 11: 44–59.

DALL, N. H.
> 1881   Hydrologie des Bering-Meeres und der benachbarten Gewässer. *Pet. Geog. Mitt.* 27(10): 361–380 and 27(11): 443–448.

DALL, N. H.

1882   Report on the currents and temperatures of Bering Sea and adjacent waters. Appendix 16—report for 1880. U. S. Coast and Geodetic Survey, U. S. Government Printing Office, Washington, D.C., 46 pp.

DOBROVOL'SKII, A. D., and V. S. ARSEN'EV

1959   On the question of the currents of the Bering Sea [in Russian]. *Probl. Severa* (3): 3–9. (Transl., Nat. Res. Coun. Can., Ottawa).

1961   Hydrological character of the Bering Sea [in Russian]. *Tr. Inst. Okeanol. Akad. Nauk SSSR* 38: 64–96.

DODIMEAD, A. J., F. FAVORITE, and T. HIRANO

1963   Salmon of the North Pacific Ocean, Part 2. Review of oceanography of the subarctic Pacific region. *Bull. Int. North Pac. Fish. Comm.* 13: 195 pp.

FAVORITE, F.

1967   The Alaskan Stream. *Bull. Int. North Pac. Fish. Comm.*, 21: 1–20 pp.

FACULTY OF FISHERIES, HOKKAIDO UNIVERSITY

1959   Data record of oceanographic observations and exploratory fishing, No. 3, 296 pp.

FAVORITE, F., W. B. MCALISTER, W. J. INGRAHAM, JR., and D. DAY

1967   Oceanography. In *Investigations by the United States for the International North Pacific Fisheries Commission—1966*. Int. North Pac. Fish. Comm. Annual Rep., 1966, pp. 89–96.

FAVORITE, F., and W. J. INGRAHAM, JR.

1972   Influence of Bowers Ridge on circulation in Bering Sea and influence of Amchitka Branch, Alaskan Stream, on migration paths of sockeye salmon. In *Biological oceanography of the northern North Pacific Ocean* [Motoda commemorative volume], edited by A. Y. Takenouti et al. Idemitsu Shoten, Tokyo, pp. 13–29.

FEDEROV, K. N.

1956   Results of modeling of the absolute currents excited by the wind in the sea [in Russian]. *Inst. Okeanol. Akad. Nauk SSSR* 19: 83–97.

FLEMING, R. H.

1955   Review of the oceanography of the northern Pacific. *Bull. Int. North Pac. Fish. Comm.*, 2: 43 pp.

FOFONOFF, N. P.

1962   Machine computations of mass transport in the North Pacific Ocean. *J. Fish. Res. Bd. Can.*, 19(6): 1121–1141.

GERARD, R., and M. SALKIND

1965   A note on the depth stability of deep parachute drogues. *Deep-Sea Res.* 12: 377–379.

GOODMAN, J. R., J. A. LINCOLN, T. G. THOMPSON, and F. A. ZEUSLER

1942   Physical and chemical investigations: Bering Sea, Bering Strait and Chukchi

Sea during summers of 1937 and 1938. *Univ. Wash. Publ. Oceanogr.* 3(4): 105–169.

GURIKOVA, K. F., T. T. VINOKUROVA, and V. V. NATAROV

1964 A model of the wind-driven currents in the Bering Sea in August 1959 and 1960 [in Russian]. *Trudy VNIRO* 49. (Transl., 1968, in *Soviet fisheries investigations in the northeastern Pacific*, Part 2, pp. 48–77, avail. Nat. Tech. Inf. Serv., Springfield, Va., TT 67-51204).

KITANO, K.

1970 A note on the thermal structure of the eastern Bering Sea. *J. Geophys. Res.* 75: 1110–1115.

LEONOV, A. K.

1947 Water masses of the Bering Sea and its surface currents [in Russian]. *Meteorol. Gidrol.*, No. 2.

1960 Regional oceanography, Part 1 [in Russian]. *Gidrometeoizdat* (Leningrad), 765 pp. (Transl. avail. Nat. Tech. Inf. Serv., Springfield, Va., AD 627508 and AD 689680).

NATAROV, V. V.

1963 Water masses and currents of the Bering Sea [in Russian]. *Trudy VNIRO* 48: 111–133. (Transl., 1968, in *Soviet fisheries investigations in the northeastern Pacific*, Part 2, pp. 110–130, avail. Nat. Tech. Inf. Serv., Springfield, Va., TT 67-51204).

RATMANOV, G. E.

1937 On the hydrology of the Bering and Chukchi seas [in Russian]. *Issled. Morei SSSR. Gidrometeoizdat* (Leningrad) 25: 10–118. (Transl., avail. Univ. Wash. Lib., Seattle).

REED, R. K.

1971 Nontidal flow in the Aleutian Island passes. *Deep-Sea Res.* 18: 379–380.

REID, J. L.

1966 ZETES Expedition. *Trans. Am. Geophys. Un.* 47: 555–561.

ROBERTS, E. B.

1947 Roberts radio current meter operating manual. U. S. Dep. Commerce (Coast and Geodetic Survey), Washington, D.C., 32 pp.

SANFORD, T. B., and W. J. SCHMITZ, JR.

1971 A comparison of direct measurements and GEK observations in the Florida Current off Miami. *J. Mar. Res.* 29(3): 347–359.

SCHULZ, B.

1911 Die Strömungen und die Temperaturvehältnisse des Stillen Ozeans nördlich von 40°N-Br. einschlie Blich des Bering-Meeres. *Annalen der Hydrographie und Maritimen Meteorologie* 39: 4(1, 2): 171–190; 5(3, 4): 242–264.

SMETANIN, D. A.

1958 Hydrochemistry of the area of the Kurile-Kamchatka deep water trench. First report. Some questions about the hydrology and chemistry of the lower subarctic waters in the area of the Kurile-Kamchatka trench [in Russian]. *Inst. Okeanol., Akad. Nauk SSSR* 27: 22–54.

Stansell, T. A., Jr.

1969   An integrated geophysical navigation system using satellite-derived position fixes. Offshore Technology Conference, 1969. Preprints. Vol. 2: 227–244.

Sverdrup, H. U.

1947   Wind-driven currents in a baroclinic ocean; with application to the equatorial currents of the eastern Pacific. *Proc. Nat. Acad. Sci.* 33: 318–326.

Taguchi, K.

1959   On the surface currents in the mother ship fishing ground based on the recovery of drifting floats [in Japanese]. Japan. Fish. Ag. (Int. North Pac. Fish. Comm. Doc. 323), 70 pp.

Terhune, L. D. B.

1968   Free-floating current followers. Fish. Res. Bd. Can., Tech. Rep. No. 85, 34 pp.

Uda, M.

1963   Oceanography of the Subarctic Pacific Ocean. *J. Fish. Res. Bd. Can.* 20: 119–179.

Udintsev, G. B., I. G. Boichenko, and V. F. Kanaev

1959   The bottom relief of the Bering Sea [in Russian]. *Inst. Okeanol., Akad. Nauk SSSR* 29: 17–64. (Transl. avail. Nat. Tech. Inf. Serv., Springfield, Va., TT-64-11837).

U. S. Navy Hydrographic Office

1958   Oceanographic Survey results, Bering Sea area, winter and spring, 1955. Tech. Rep. No. 46, 95 pp.

U. S. Weather Bureau

1961   Climatological and oceanographic atlas for mariners. Vol. 2. North Pacific Ocean. U. S. Government Printing Office, Washington, D.C.

Volkmann, G., J. Knauss, and A. Vine

1956   The use of parachute drogues in the measurement of subsurface ocean currents. *Trans. Am. Geophys. Un.* 37: 573–577.

CHAPTER 4

Some dynamic and budgetary implications of circulation schemes for the Bering Sea

WILLIAM M. CAMERON

Marine Science Directorate, West Vancouver, British Columbia, Canada

Having occasion on this short notice to give some thought to the Bering Sea again after some 20 years of administrative digression, I have been intrigued to realize, perhaps more than before, what a unique region this sea is. I can think of no other sea that presents so many interesting challenges—not only to the physical oceanographer, but to the chemical and biological oceanographers as well. It is a northern, subarctic sea of tremendous expanse. The only other seas of this magnitude in the same latitude are the Norwegian and Greenland seas, but they differ markedly from this very interesting body of water. Whereas they are open to the south and north to the Atlantic and Arctic oceans, the Bering Sea on the other hand is barred for much of its southern expanse from the Pacific and is connected to the Arctic Ocean only by a very narrow and shallow pass. Furthermore, the southern passes—though relatively narrow—are of oceanic depths that permit the easy flow of deep water, unlike the case in the Atlantic seas.

Thus to a physical oceanographer, the Bering Sea presents an opportunity to study a miniature ocean with boundaries which are amenable both to physical study and to simplistic modeling. This fact, together with recent technological advances, makes it possible for the first time to make boundary measurements that will go a long way toward easing an understanding of the Bering Sea itself.

There are more specific features which deserve review. The Bering Sea is characterized by a strong wind-stress curl that prevails over its waters, an amazingly high productivity, a sharp division into a deep oceanic region and an extensive shelf, the observable effect of river discharge, and by a dramatic change in conditions in winter when the ice covers half the sea and in summer when the ice has disappeared.

What I think is really important about the Bering Sea, however, is that it presents the challenging possibility of developing interdisciplinary programs that will move toward a solution of basic questions that have not been amenable to attack elsewhere. Because the Bering Sea is so sharply bounded, the physical oceanographers can, within reasonable limits of expense, provide the biologist with accurate figures of the net inflow or outflow of biologically important elements. With this as a base, the biologist can look for possible sinks or sources within the sea and, defining their magnitude, provide the physicist with checks on his own measurements.

What has prevented the proper integration of physical and biological studies heretofore and elsewhere has been that the biologist has not been sure what he wanted from the physicist, and if he was, the physicist was not able to provide it to him. I think that now the physicist can provide the information asked. Even more intriguing is that here the one discipline assists the other and in doing so is itself aided by reciprocal reward.

I consider that budgetary studies of conservative and non-conservative properties in the Bering Sea are both possible and potentially highly productive.

Prompted by the papers presented in this session and by what I have recently read, I offer two brief comments:

There has been sharp difference of opinion on the magnitude of transport into and out of the Bering Sea. There has been reference to baroclinic and barotropic flow and to Sverdrup transport as though they were separate or rather separable phenomena. I suggest this overlooks the implications of Sverdrup's integrated transport equations. You will recall that Sverdrup integrated vertically the internal friction from an infinitely deep or frictionless boundary to the surface and equated that integral to the surface stress. In taking the curl he removed the internal pressure field and by implication the variation in the mass field. But this manipulation did not presume an absence of either baroclinic or barotropic flow. Naturally the bottom boundary conditions implicitly limited the nature of both these flows, but this is not the time to press into this more deeply.

Perhaps what makes one uneasy is the large magnitudes of transport that derive from the wind-stress curl, especially if the interval of summation of the wind vectors is reduced. I suggest that the discrepancy between what might seem reasonable and what is calculated may be due to the inadequate knowledge that still prevails of the relation of stress to mean velocity especially at high wind speeds. In other words, if there is a real and observable difference between the baroclinic and barotropic transports and the Sverdrup transports, I would be more inclined to fault our wind-stress calculations than the applicability of the Sverdrup treatment.

In the papers of this session and in those to which they made reference, there has been no mention of the salt balance in the Bering Sea. This omission is particularly interesting in that the data necessary for geostrophic calculations provide also the data sufficient for salt transport. Had these calculations been made, I am sure that certain estimates would have been rendered uncomfortably dubious.

I grant that when one is effecting a volume transport balance by the summation of several separably definable flows in the upper levels, the calculation of the salt balance may be unproductive inasmuch as the small differences obtained might be within the accuracy of the volume transport estimates. I suggest, though, that the calculation should at least be made.

On the other hand, when one suggests that a strong volume transport inward is balanced by an excess of outward shallow flow, he must look at the implications for the salt balance very critically. The constant long-term sea level in the Bering Sea demands that the following well-known equation apply:

$$V_i + R = V_u \qquad (1)$$

where V_i is the inward transport
V_u is the outward transport
R is the excess of precipitation and river run-off over evaporation

If the salt content of the Bering Sea is to remain constant over the long term, the following equation must hold:

$$S_i V_i = S_u V_u \qquad (2)$$

where S_i and S_u are the *effective* salinities of the inflowing and outflowing waters, respectively. Effective salinity is that salinity which, when multiplied by the volume transport, expresses the same salt transport as is more properly described by the integral of the local salt transport, integrated over the section.

Combining equations (1) and (2) derives:

$$S_i = S_u \left(1 + \frac{R}{V_i}\right) \qquad (3)$$

In the case of the Bering Sea, the ratio $\frac{R}{V_i}$ is small. Arsen'ev (1967) presents $R = 0.036$ and $V_i = 21.0$ expressed in Sverdrups (1 Sv = 1×10^6 m³/sec). A quick calculation suggests that the difference in effective salinities should be of the order of 0.06 percent. Recognizing that there is a net outward flow of low-salinity water through the Bering Strait, it is apparent that the opposing streams in the southern passes would differ by a smaller effective salinity and the difference, though of necessity real, might be indistinguishable within the accuracy of the calculations.

A much different situation applies, however, when a deep flow is involved. Using the subfix "d" to apply to the deep current and neglecting the small freshwater effect discussed above, we have:

$$V_i S_i + V_{id} S_d = (V_i + V_{id}) S_u \qquad (4)$$

where V_i and S_i represent the shallow inflowing transport and its effective salinity
where V_{id} and S_d represent the deep inflow transport and its effective salinity
and where S_u is the effective salinity of the outflowing transport $V_u = V_i + V_{id}$

From the above equation can be derived:

$$\frac{S_u - S_i}{S_d - S_u} = \frac{V_{id}}{V_i} \tag{5}$$

Without inserting any quantities, it can be seen that since the right-hand side of equation (5) is positive and $(S_d - S_i)$ is certain to be positive in the Pacific Ocean, the outflowing shallow water must be of a higher effective salinity than the shallow inflowing water. One must be careful, however, not to confuse *effective* salinity with *mean* salinity and be cautioned that this relationship might pertain if the outflowing water were transported through a significantly greater depth than the inflowing stream.

My remarks are not intended to be critical of the concept of deep currents flowing into the Bering Sea. Indeed it would be absurd to dismiss them on the basis of the foregoing argument alone. I have tried to illustrate the power of the budgetary consideration and suggest again that the Bering Sea lends itself admirably to this type of analysis.

Discussion

FAVORITE: Possibly you are criticizing our application of wind stress in considering volume transport, but in any case you do not seem to be correctly taking into account the Bering Strait discharge of 1.1 Sv. It is such problems that make oceanography interesting.

CAMERON: I was not proposing to assign the flows to specific areas. Arsen'ev's figures were used only as an example, since it was he who suggested that deep water flow is the key to the volume balance. I am not fully in agreement with Arsen'ev's magnitudes; I am only pointing out that no matter what magnitudes are used, one must provide for a salt balance.

UDA: In my opinion, salinities in the North Pacific and in the Bering Sea are characterized mainly by freshwater dilution from river discharge and deep water advection from particularly the Kuroshio Current. Dr. Tully (Tully and Barber 1961) has suggested a two-layered ocean in the subarctic and the importance of the entrainment of the deeper water. Upwelling should be considered in this case.

CAMERON: I agree. If in fact there is a deep flow in, it has to be upwardly entrained into the upper water in order to get out. Although I must confess that I disagree with Dr. Tully's emphasis on the estuarine character of the North Pacific Ocean, the fundamental difference in our nomenclature attitudes probably stems merely from the differing perspectives of our studies: Pritchard and I had been concerned with the *dynamics* of estuaries; Tully, on the other hand, was looking at their *kinematics*. Thus, to Tully, estuary entrainment and open ocean entrainment were similar.* To Pritchard and me, the dynamics of the two cases deserve distinction: in the two-layered estuary, the energy for entrainment derives from the solenoidal field generated by freshwater input. As river water entrains saltwater in its flow to the sea, the downward pressure gradient maintained by this dilution is sufficient to accelerate the increased volume seaward and to overcome friction. In the open ocean, however, the solenoidal field alone is insufficient to drive such an entrainment, and it is on this basis that Pritchard and I would decline an estuarine analogy.

UDA: Do you agree that cyclonic circulation in the upper levels acts as a source of entrainment energy?

CAMERON: Yes. There does not appear to be enough energy derived from cross isobaric flow in the open ocean to provide for entrainment, and I suggest the required energy must come from the wind.

* NOTE: Abstract of "An Estuarine Analogy in the Subarctic Pacific Ocean," submitted by J. P. Tully, March 1972:

An estuarine analogy occurs in any region of the Subarctic Pacific Ocean or Bering Sea in which freshwater input exceeds loss by evaporation. Definite pycnoclines are formed between the upper (low density) brackish or fresh water and the underlying, denser zones of higher salinity. A turbulent velocity gradient is established in the pycnocline as the lighter water in the upper zone persistently moves out of the region in a net amount equal to its inflow plus the seawater it entrains enroute. Conversely, as deep water is entrained upward, it follows that there must be a compensating net inflow of dense water to the region. If this were not so, one component or the other would accumulate and sea level would change.

The phenomenon of entrainment is not concerned with the various causes of the velocity gradient. Along the shores northward of about 48°N, the coastal currents driven by freshwater runoff are observed to flow in the same direction as the wind-driven Subarctic Gyre. It is notable that in these regions the depth of the permanent halocline coincides closely with Ekman's depth (D) of wind influence. Evidently there is a gradient of velocity between the upper brackish and lower saline zones due primarily to wind but also to runoff and local tides. In the ultimate course of Subarctic Pacific circulation, the California Current returns flow to the subtropic region, where evaporation now exceeds freshwater input. As the waters move enroute, the salinity of the upper zone increases until the halocline finally vanishes at about 40°N on the eastern side of the ocean. Thus the conditions for an estuarine analogy are fulfilled.

REFERENCES

ARSEN'EV, V. S.

    1967   Currents and water masses of the Bering Sea [in Russian, English summary]. Izd. Nauka, Moscow. (Transl., 1968, avail. Nat. Mar. Fish. Serv., Northwest Fish. Center, Seattle, Wash.). 135 pp.

CAMERON, W. M., and D. W. PRITCHARD

    1963   Estuaries. In The sea, Vol. 2, edited by M. N. Hill. Interscience Publishers, New York.

TULLY, J. P., and F. G. BARBER

    1961   An estuarine model of the subarctic Pacific Ocean. In Oceanography, edited by M. Sears. Amer. Ass. Adv. Sci., New York, N. Y.

Part 2

DYNAMICS OF
CHEMICAL PARAMETERS

CHAPTER 5

The carbon dioxide system
of the Bering Sea

P. Kilho Park, Louis I. Gordon, *and* Saul Alvarez-Borrego

School of Oceanography, Oregon State University, Corvallis, Oregon

Abstract

Existing CO_2 system data show that the surface carbon dioxide concentration varies between 1–4×10^{-4} atmospheres. It is affected by photosynthesis of marine plants, changes in water temperatures, biochemical oxidation, upwelling, upward divergence of deep water by cyclonic gyrals, changes in the depth of the surface mixed layer, and by CO_2-rich river runoff. In addition, the impact of open leads and polynyi on the dynamics of CO_2 air-sea exchange is emphasized by the non-equilibrium nature of the open water with respect to the atmospheric CO_2 and by rapid heat flux between the air and the water. Subsurface water has a high CO_2 concentration, about 13×10^{-4} atmospheres at 800 m depth.

Supersaturation by carbonate minerals, calcite and aragonite exists near the surface, and undersaturation occurs in the deep waters. Most undersaturation occurs at 1000 m depth with 55 percent saturation for calcite and 35 percent saturation for aragonite.

A linear relationship exists between the AOU (apparent oxygen utilization) and phosphate concentrations below the euphotic zone. The AOU/PO_4 slope is twice that of the Redfield biochemical oxidation model of 138. Excluding the AOU minimum zone at 1000 m, there exists an inverse relationship between AOU and pH. There is also a positive, almost-linear relationship between AOU and total CO_2 in the deep Bering Sea, represented by a $AOU/\Sigma CO_2$ similar to the Redfield biochemical oxidation model of 138/106.

INTRODUCTION

A unique feature of the Bering Sea CO_2 system is that a large portion of the sea surface is covered by ice during winter months (Fig. 5.1). This ice cover does not necessarily mean, however, that the air-sea exchange of CO_2 ceases completely during these months. The ever-shifting ice cover produces open leads and polynyi, at the surface of which the CO_2 concentration in the seawater is found to be supersaturated in CO_2 with respect to air by about 14 ppm (Kelley and Hood 1971a). In addition, the turbulent heat flux to the atmosphere can increase about 100 times during the winter over an artificial lead (Badgley 1966). Therefore, if the ice were removed for a significant period of time, CO_2 could be released significantly from the seawater to the atmospheric reservoir. In this review article, attention is focused on what is known of the CO_2 dynamics of the Bering Sea, a system interwoven with air-sea interaction, biogeochemical cycles, and the hydrodynamic pattern of the sea.

DATA SOURCES

Ivanenkov's (1964) book, *Hydrochemistry of the Bering Sea*, essentially summarizes the hydrochemical study of the Bering Sea for the period 1932 to 1960. The author's assessment of existing data shows tabulation of about 6000 alkalinity measurements and about 10,000 pH determinations. Based on these available data, he prepared a comprehensive review of the CO_2 system of the Bering Sea.

In 1966, Park (Barstow et al. 1968) led a hydrochemical oceanographic cruise aboard the R/V *Yaquina* (Oregon State University) in the Northeastern Pacific Ocean. During this cruise seven hydrographic stations were occupied in the Bering Sea near the Aleutian Islands (Fig. 5.2) and resulted in hydrochemical data, from the sea surface to the bottom, which proved useful in

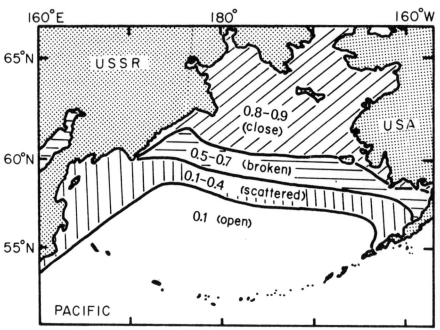

Fig. 5.1 Ice cover in the Bering Sea during February (U. S. Government, 1961).

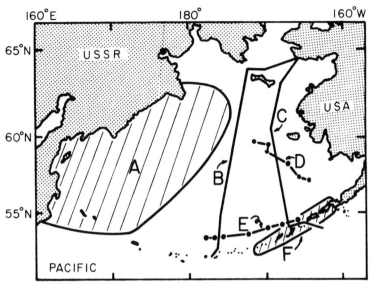

Fig. 5.2 Regions in the Bering Sea studied for CO_2 dynamics. (A) Seasonal surface P_{CO_2} study (Ivanenkov 1964); (B) Surface P_{CO_2} study (Gordon et al. 1973); (C) Surface P_{CO_2} study (Kelley and Hood 1971b); (D) Ice-covered sea-surface P_{CO_2} study (Kelley and Hood 1971a); (E) Vertical CO_2 system study (Alvarez-Borrego et al. 1972—positions of the seven hydrographic stations are shown by dots); (F) Surface P_{CO_2} study (Kelley et al. 1971).

later chemical oceanological studies (Park 1966ab, 1967ab, 1968ab; Park et al. 1967; Culberson and Pytkowicz 1968; Hawley and Pytkowicz 1969). Alvarez-Borrego et al. (1972) discuss its scientific details elsewhere.

In 1968, infrared CO_2 analyzers were employed by two groups of scientists, from the University of Alaska and from Oregon State University, to map the surface distribution of P_{CO_2} (partial pressure of CO_2) in the eastern half of the Bering Sea (Kelley and Hood 1971b; Gordon et al. 1973) (Fig. 5.3).

Recently, University of Alaska scientists have begun a systematic and comprehensive study of the Bering Sea CO_2 system aboard the R/V *Acona* of the University of Alaska (Kelley et al. 1971), and aboard the U. S. Coast Guard ice breaker *Northwind* (Kelley and Hood 1971a). This study is still in progress.

ANALYTICAL METHODS

With the advent of measurements of ΣCO_2 and P_{CO_2}, in addition to the classical pH and alkalinity measurements, four parameters are now available for investigation of the CO_2 system in the ocean. Two of these four parameters, in conjunction with temperature and salinity, are all that are needed to estimate the concentrations of all the components of the CO_2 system (Park 1969).

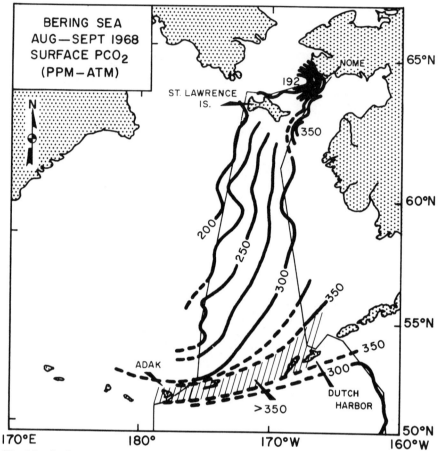

Fig. 5.3 Surface P_{CO_2} distribution in the eastern Bering Sea (Gordon et al. 1973).

Recent advancement in chemical instrumentation now enables measurement of any of these parameters with precision (Takahashi et al. 1970).

pH

Generally, pH is measured on board with a glass-calomel electrode pair. The method of Culberson and Pytkowicz (1970) yields a standard deviation of ± 0.003 pH. In their method an Orion (USA) Model 801 digital pH meter is used. The pH samples are drawn into 10-ml plastic syringes, which are placed into a 25 C water bath for 10 min. The pH sample is then injected into a cell based upon a Beckman (USA) 46850 micro blood pH electrode, and the cell potential is measured.

Earlier studies, for precision comparison, yielded an average standard deviation of ± 0.01 pH unit (Park 1966b) or greater.

Alkalinity

Two methods of alkalinity determination are in use at present. The potentio-metric acid titration method (Edmond 1970) yields a precision of 0.3 percent for alkalinity. The improved single acid addition method of Anderson and Robinson (1946), carried out by Culberson et al. (1970), shows a standard deviation of $\pm$ 0.003 meq/kg for individual replicates and $\pm$ 0.6 percent for measurements along the alkalinity-depth profile.

Edmond (1970) method: A 150-ml, flat-bottomed, four-neck flask is the titration chamber. A glass-calomel electrode pair and a 2.5-ml micrometer-screw burette are fitted into three of the necks. In the fourth neck, a vertical capillary is fitted to equalize the pressures inside and outside the titration chamber. The titration is performed in a constant-temperature (25 C) bath with a 0.3 N HCl solution. An Orion Model 801 digital pH meter is used to measure the electrode potential during the titration. The results of the titration are treated by the method of Dyrssen and Sillén (1967) to give the two carbonic acid endpoints from which alkalinity is calculated.

Culberson et al. (1970) method: Six ml of 0.01 N HCl is added to a 20-ml seawater sample. After driving off carbon dioxide from the solution by purging for 5 min with water saturated air, the pH of the solution is measured at 25.0 ± 0.1 C. The alkalinity is calculated from the following equation:

$$\text{Alk} = \left[\frac{1000}{d \cdot V_s} \right] \left[V \cdot N - (V_s + V) \frac{a_{H+}}{f_{H+}} \right] \tag{1}$$

where Alk is total alkalinity in milliequivalent per kilogram, f_{H+} is 0.741, V_s is the volume of seawater, V is the volume of HCl, N is the normality of HCl, d is the density of seawater at 25 C, and a_{H+} is 10^{-pH}.

ΣCO_2

A gas chromatographic method (Weiss 1970) and a gas extraction-infrared determination method (Wong 1970) have been developed recently. The Weiss (1970) gas-chromatography method yields a precision of $\pm$ 0.3 percent, while Wong's (1970) extraction method has precision capability of $\pm$ 0.2 percent.

Weiss (1970) method: A 2-ml aliquot of seawater is injected into a gas chromatograph and is stripped on line with hydrogen or helium carrier gas in a way similar to that described by Swinnerton et al. (1962a, b) and by Park et al. (1964). By adding a small amount of phosphoric acid to the stripping chamber before each analysis, bicarbonate and carbonate ions are converted into carbonic acid and hence to carbon dioxide. The evolved gases are separated on a silica-gel column and detected by thermal conductivity. One analysis takes approximately 5–10 minutes. This method is easily used at sea.

Wong (1970) method: A special pipette of known volume, 30–40 ml, is filled with the sample, which is then transferred to a 1-liter flat-bottom flask containing CO_2-free phosphoric acid. The flask is pumped under vacuum by means of a dry-ice trap to remove water vapor and then to a double-walled liquid nitrogen trap to collect CO_2. After about one hour of pumping, the condensed CO_2 is purified by warming the spherical trap and recondensing the CO_2 in a U-trap at liquid nitrogen temperature. This U-trap is warmed to dry ice temperature, and the CO_2 is condensed in a second U-trap in order to remove any residual water vapor. The second U-trap is then warmed and the CO_2 transferred in a stream of CO_2-free nitrogen into a previously evacuated 5-liter flask at room temperature and ambient pressure. The final concentration of CO_2 in the flask is determined by infrared analysis. This method is usually carried out on suitably preserved water samples at a shore laboratory.

P_{CO_2}

By means of infrared-absorption measurements of CO_2, values of P_{CO_2} in the surface water have been frequently obtained since the IGY (International Geophysical Year) of 1957. The P_{CO_2} measuring system was pioneered by Keeling et al. (1965), Hood et al. (1963), Kanwisher (1960), Takahashi (1961), Teal and Kanwisher (1966) and others.

The infrared methods used in the Bering Sea are described by Kelley and Hood (1971a), Kelley et al. (1971), and Gordon (1973). All of these methods rely on infrared absorption measurements; only the seawater equilibrator systems differ among them. Described below are the methods of Kelley et al. (1971) and Gordon (1973).

Kelley et al. (1971) method: Measurements of CO_2 concentration are accomplished with a six-stage water-aspirator equilibrator (Fig. 5.4). The instrument consists of six individual phase separation chambers into which six water-driven aspirators, in parallel arrangement, discharge a stream of gas. After the gas is then fed in series from chamber to chamber, it is finally passed through the infrared analyzer (Ibert and Hood 1963).

Gordon (1973) method: A single-stage equilibrator is used by Gordon and his co-workers (Gordon et al. 1971, Gordon et al. 1973, Park et al. 1969). This technique is a modification of the methods of Takahashi (1961) and Kanwisher (1960). In conjunction with CO_2 measurements, the temperature and pH of seawater are measured concurrently (Fig. 5.5). The 99 percent response time for the Gordon equilibrator-gas analyzer system to a change in P_{CO_2} in seawater is about 5 min. During the First Canadian Trans-Pacific Oceanographic Cruise in 1969, this method yielded a continuous $P_{CO_2} - pH$ profile for 15-day and 20-day intervals without interruption.

SAMPLE WATER STREAM & BYPASS

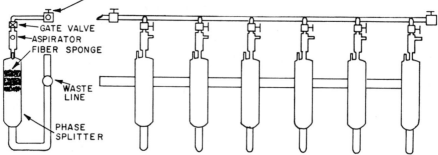

Fig. 5.4 University of Alaska CO_2 equilibrator system (Kelley et al. 1971).

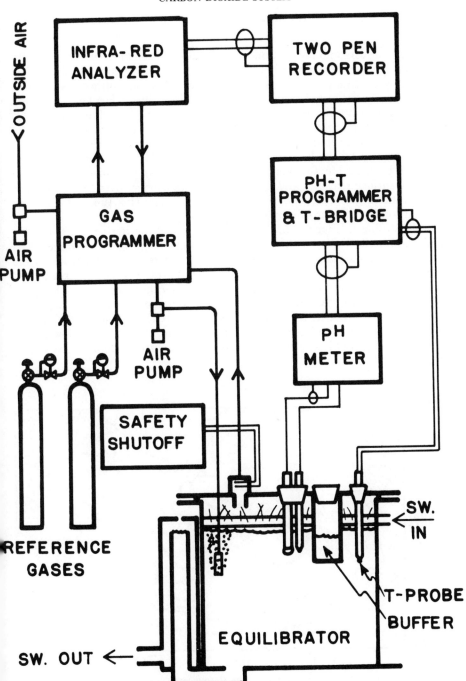

Fig. 5.5 Oregon State University CO_2 equilibrator system (Gordon 1973).

P_{CO_2} SURFACE DISTRIBUTION

In order to understand the dynamics of CO_2 air-sea exchange, surface mapping of the P_{CO_2} has been extensively carried out by Keeling (1968) and others. Due to scanty CO_2 data in the North Pacific above 40°N, the entire subarctic and Bering Sea regions were left untouched by Keeling (1968) when he constructed his global distribution map. To augment his mapping, Gordon et al. (1973) traced the distribution of P_{CO_2} in the Bering Sea and in the subarctic region of the Pacific for late summer (Fig. 5.6). The P_{CO_2} values in the central and eastern regions (shown by the large shaded area in Fig. 5.2) of the Bering Sea were obtained by direct measurement of the concentration of CO_2 in air equilibrated with the surface seawater and of CO_2 in the atmosphere over the sea surface using infrared gas analyzers. The western region data (Ivanenkov 1964) were derived from pH and alkalinity measurements.

According to Figure 5.6, the major portion of the Bering Sea surface is undersaturated in CO_2 with respect to the atmosphere in late summer. The only probable mechanism to induce the surface water to undersaturate is the utilization of CO_2 by phytoplankton during photosynthesis. The supersaturated regions, with respect to CO_2, are near the westernmost part and along the Alaskan coast and Aleutian Islands. These observed conditions of supersaturation can occur as a result of upwelling of CO_2-rich water, by the introduction of CO_2-rich freshwater (such as from the Yukon River), and by the upward divergence of deep water through cyclonic current action. These effects are discussed later.

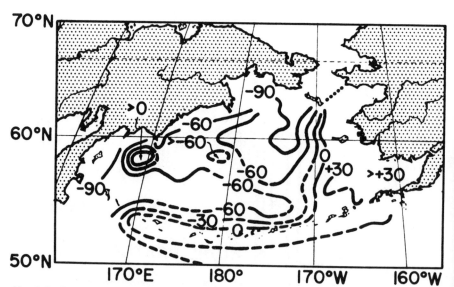

Fig. 5.6 Saturation of surface waters of the Bering Sea with respect to the atmospheric CO_2 in late summer. The numerals are $\Delta P_{CO_2} = $ (sea P_{CO_2} − air P_{CO_2}) in ppm (Gordon et al. 1973).

Seasonal distribution of surface P_{CO_2}

Ivanenkov (1964) gives seasonal distributions of surface P_{CO_2} in the western Bering Sea (Fig. 5.7–5.10). The basic equation he used to obtain P_{CO_2} is as follows:

$$P_{CO_2} = \frac{1}{\alpha_s} \frac{(a_{H+})^2 \, (C.A.)}{K'_1 \, (A_{H+} + 2K'_2)} \qquad (2)$$

where (C.A.) is carbonate alkalinity, K'_1 and K'_2 are the first and second apparent dissociation constants of carbonic acid in seawater, α_s is the solubility coefficient of carbon dioxide in seawater at one atmosphere. The validity of equation (2) has been tested by Gordon et al. (1971) and found quite good in the subarctic region of the Pacific (Fig. 5.11).

Since no atmospheric CO_2 was determined, Ivanenkov (1964) assumes it to be 320 ppm.

Winter distribution
Nearly all of the area covered is supersaturated, with P_{CO_2} values of 320–420 ppm (Fig. 5.7). Therefore, the sea must be a source of CO_2 into the atmosphere. The large P_{CO_2} values are attributed to the cessation of any appreciable photosynthesis, while biochemical oxidation continues during the winter months. In addition, the mixing of CO_2-rich subsurface water with the surface water would also increase the surface P_{CO_2}.

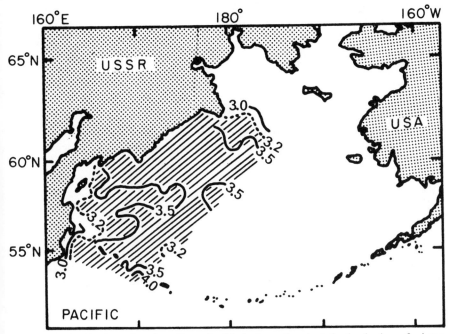

Fig. 5.7 Distribution of surface P_{CO_2} (in 10^{-4} atmosphere) at the beginning of winter 1952–53 (Ivanenkov 1964).

Spring distribution

Approximately half of the area that showed supersaturation during winter months is undersaturated with values as low as 100 ppm in the shallow water, whereas the remaining half over deep water still shows supersaturation with a value up to 410 ppm (Fig. 5.8). Therefore, the sea over the shallow depth should absorb CO_2 from the atmosphere while over the deep areas it should expel the excess CO_2 into the atmosphere. Ivanenkov (1964) attributes the emergence of the low P_{CO_2} area to the vigorous development of neritic species of phytoplankton.

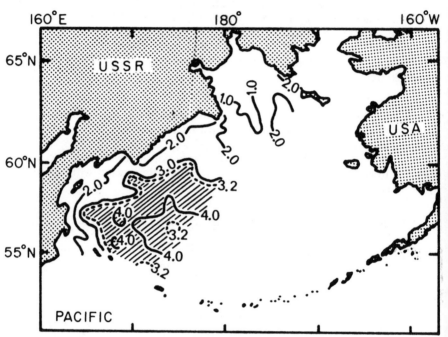

Fig. 5.8 Distribution of surface P_{CO_2} (in 10^{-4} atmosphere) in the spring of 1952 (Ivanenkov 1964).

Summer distribution

Since pH and alkalinity data are abundant for the summer months, the P_{CO_2} distribution over the entire Bering Sea can be estimated (Fig. 5.9). It is shown that the central part of the sea is undersaturated with values lower than 250 ppm, while supersaturation occurs only in a limited area over the deep western basin, in the Yukon River plume area and the region around the eastern Aleutian Islands. Photosynthesis probably would be the cause for the low P_{CO_2} area in the central part; mixing with either CO_2-rich deep water or CO_2-rich river water, coupled with *in situ* biochemical oxidation, would have caused these surface waters to possess high P_{CO_2} values.

There is a good semblance between the summer pattern (Fig. 5.9) and the late summer pattern prepared by Gordon et al. (1973) (Fig. 5.6).

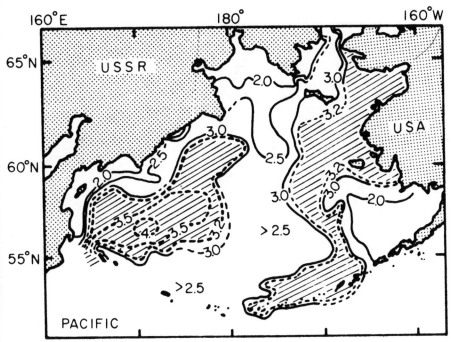

Fig. 5.9 Distribution of P_{CO_2} (in 10^{-4} atmosphere) in summer 1932–60 (Ivanenkov 1964).

Autumn distribution

Based on 1951 data Ivanenkov (1964) prepared an autumn P_{CO_2} distribution pattern as shown in Figure 5.10. Ivanenkov had stated, however, that the year 1951 was unusually warm. Therefore, Figure 5.10 may not represent a typical autumn P_{CO_2} distribution pattern for the western Bering Sea. It shows that undersaturation, with values as low as 250 ppm, covers almost all the area. We are of the opinion that photosynthesis was still strongly going on during the autumn months in 1951.

Probably the late summer distribution shown by Gordon et al. (1973) (Fig. 5.6) is a more accurate representation of the autumn season.

Near the eastern Aleutian Islands, studies were carried out aboard the R/V *Acona* for direct determination of CO_2 in surface waters during the summer and early autumn of 1970 (Kelley et al. 1971). Relative CO_2 concentrations between the atmosphere and the surface seawater in the vicinity of the Aleutian island passes are shown in Figure 5.12, which indicates complex distribution with seasonal variations observable near Unimak Pass even within the summer months. Among the observations that Kelley et al. (1971) made is the change in atmospheric concentration from 327 ppm in June 1970, which represents a value close to the winter and spring CO_2 level in the arctic and subarctic atmosphere, to 314 ppm in September 1971. Kelley (1968) attributed this decrease to uptake of CO_2 from the atmosphere during the summer autotrophic plant bloom.

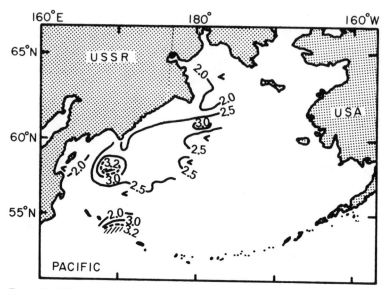

Fig. 5.10 Distribution of P_{CO_2} (in 10^{-4} atmosphere) at the beginning of autumn 1951 (Ivanenkov 1964).

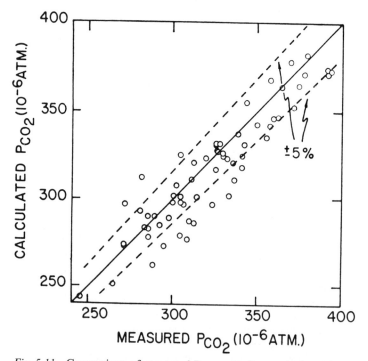

Fig. 5.11 Comparison of measured P_{CO_2} with P_{CO_2} calculated from pH and alkalinity (Gordon et al. 1971).

A further observation on the seasonal CO_2 change that clearly can be attributed to marine plant growth was seen in the area near Izembek Lagoon (see Fig. 5.12 for location). This lagoon is the largest and most important eelgrass (*Zostera marina L.*) area on the coast of Alaska (McRoy 1968). The seasonal range of values for CO_2 in the surface waters both inside and out of the lagoon (Fig. 5.13) shows undersaturation in June, near equilibrium in the beginning of September and supersaturation, with respect to the atmospheric CO_2, in October.

Additional evidence of seasonal variance of surface P_{CO_2} in the subarctic region, near the Aleutians, is reported by Gordon et al. (1971). The surface P_{CO_2} and pH between Canada and Japan were measured in March and April 1969 (Figs. 5.14–5.15). Comparison of the 1969 March data with the 1968 October data, shows about 60 ppm P_{CO_2} variance (Fig. 5.16).

All of the foregoing seasonal P_{CO_2} variance studies clearly show that seasonal studies of surface P_{CO_2} distribution are essential if we are to understand the dynamics of the CO_2 exchange across the air-sea interface. Future Bering Sea studies should include seasonal observations.

Processes affecting surface P_{CO_2} distribution

Processes causing undersaturation
The major process that reduces the surface CO_2 concentration is photosynthesis by marine plants. An inverse relationship between phytoplankton production, as approximated by measured chlorophyll *a* concentrations, and the P_{CO_2} in the subarctic water in the North Pacific near the Aleutian Islands was shown by Gordon et al. (1971) (Fig. 5.17).

Additional evidence that correlates photosynthetic activities and P_{CO_2} is the relation of nutrient uptake by phytoplankton to P_{CO_2} decrease. Kelley et al. (1971) showed that nitrate concentration decreased with decreasing CO_2 concentration in the surface water (Fig. 5.18) of their sampling area, located in the North Pacific Ocean near the southwestern tip of Umnak Island to the area north of Sequam Island. Figure 5.18 also can be used to show the trend of biochemical oxidation, since oxidation increases both P_{CO_2} and nitrate values.

Cooling of seawater produces a decrease in P_{CO_2} if exchange with the atmosphere is restricted (Eriksson 1963; Takahashi 1961); conversely, warming of seawater produces a rise in P_{CO_2}. Estimates of the magnitude of this effect range from 1 percent/degree C (Harvey 1955) to 7 percent/degree C (Revelle and Suess 1957). A rough calculation by Gordon et al. (1971), based on cruise data in the Pacific near Canada, seems to indicate a magnitude of 2–3 percent/degree C effect (Fig. 5.19).

Processes causing supersaturation
Apart from the temperature effect, there seem to be at least five identifiable processes that favor supersaturation of CO_2 in surface water, with respect to atmospheric CO_2: biochemical oxidation, addition of CO_2-rich river waters, increase in the depth of the surface-mixed layer, upwelling of CO_2-rich subsurface water, and upward divergence of deep water via cyclonic gyres.

The biochemical oxidation effect has already been shown by the positive correlation between nitrate concentration and P_{CO_2} (Fig. 5.18). It will be further discussed in the vertical P_{CO_2} distribution section.

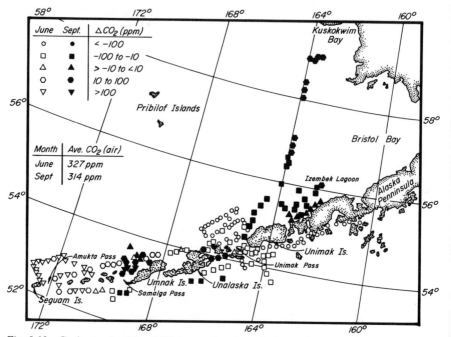

Fig. 5.12 Cruise track of the R/V *Acona*, showing relative carbon dioxide concentrations between air and surface seawater in the vicinity of the Aleutian island passes (Kelley et al. 1971).

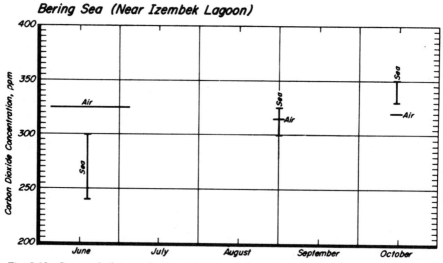

Fig. 5.13 Seasonal (late spring and fall) variation of carbon dioxide in the surface seawater (Kelley et al. 1971).

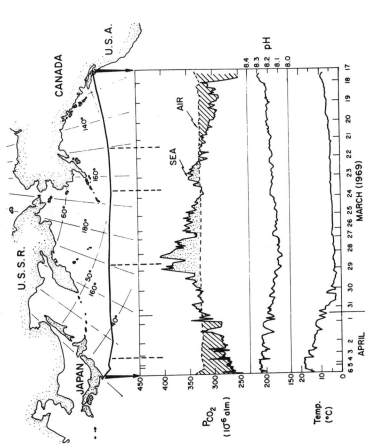

Fig. 5.14 Distribution of P_{CO_2} in the water and atmosphere (broken line), pH and temperature on the westbound portion of the *Endeavour* cruise (17 March–6 April 1969). The dates and values of the observations are projected vertically downward from the track which is drawn at the top (Gordon et al. 1971).

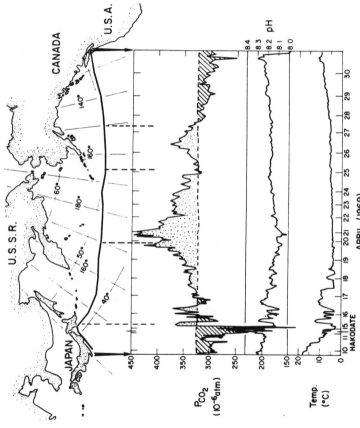

Fig. 5.15 Distribution of P_{CO_2} in the water and atmosphere (broken line), pH and temperature on the Tokyo to Hakodate (10–11 April 1969) and eastbound (15–30 April 1969) portions of the *Endeavour* cruise. The lack of near-continent rises in atmospheric CO_2 could be due to air circulation at the time of measurement (Gordon et al. 1971).

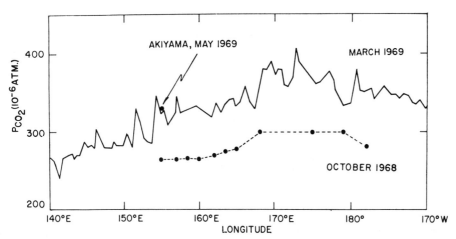

Fig. 5.16 Seasonal variation in P_{CO_2} as indicated by comparison of *Endeavour* data (solid line) and data of Akiyama (1969) (one point) with *Surveyor* data (broken line). The seasonal difference was about 60 ppm (Gordon et al. 1971).

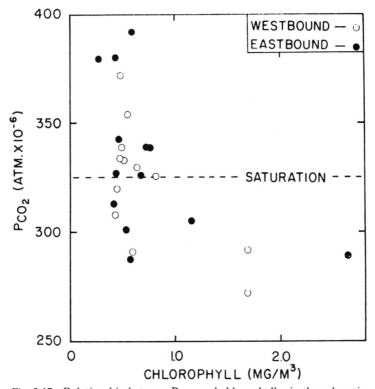

Fig. 5.17 Relationship between P_{CO_2} and chlorophyll *a* in the subarctic region of the Pacific (Gordon et al. 1971).

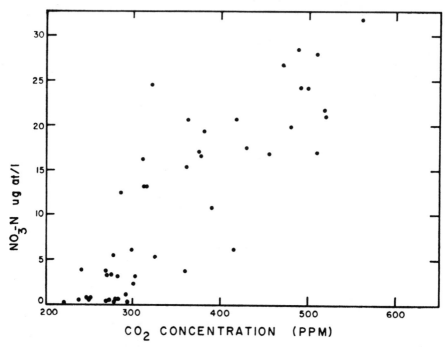

Fig. 5.18 Carbon dioxide and nitrate-N concentrations in the surface seawater in the vicinity of Amukta Pass (Kelley et al. 1971).

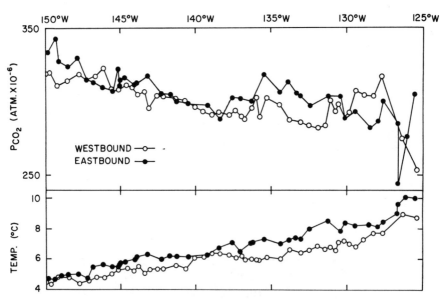

Fig. 5.19 Comparison of P_{CO_2} and temperatures from the westbound and eastbound legs of the *Endeavour* cruise. Only a small portion of the track (150°W to 125°W) is shown (Gordon et al. 1971).

River water often contains high P_{CO_2}. Near the mouth of the Columbia River in the northwestern United States, the observed P_{CO_2} values are 600–1000 ppm (Park et al. 1969). Since four large rivers (the Yukon, Kuskokwim, Anadyr, and Kamchatka) enter the Bering Sea with their major discharges occurring during summer months, the coastal region where salinity is diluted by the river waters may have high P_{CO_2} values. The summer P_{CO_2} surface distribution (Fig. 5.9), and late summer distribution (Fig. 5.6) both show a large area of CO_2 supersaturation near the Alaskan coast where salinity is also low ($<32\%_{oo}$). It is likely, therefore, that the contributions of the Yukon and other rivers that contain high P_{CO_2} help maintain the high P_{CO_2} values in the coastal region.

A positive correlation between depth of the surface-mixed layer and high P_{CO_2} was reported by Miyake and Sugimura (1969). For the subarctic region of the Pacific, Gordon et al. (1971) plotted the surface CO_2 concentration versus the depth of the top of the thermocline (Fig. 5.20). Although quite weak, the correlation is sufficient to indicate a possible influence of vertical mixing upon the P_{CO_2} distribution.

Upwelling of CO_2-rich deep water is a common mechanism to increase the surface CO_2 values. Along the Oregon coast, the wind-induced upwelling increases the surface P_{CO_2} to over 400 ppm. Kelley et al. (1971) do not report

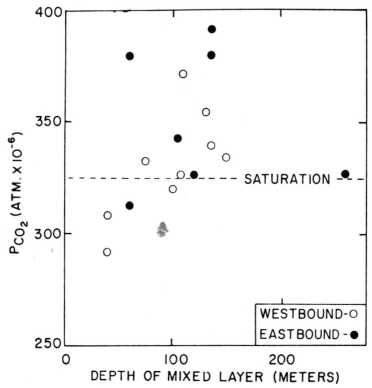

Fig. 5.20   Relationship between P_{CO_2} and depth of mixed layer in the subarctic region of the Pacific (Gordon et al. 1971).

any apparent wind-induced upwelling of CO_2-rich deep water to the surface in the Bering Sea. Their findings of upwelled regions near the Aleutian Islands, evidenced by high CO_2 and nutrient concentrations, are probably caused by inertial type upwelling as a result of deep easterly currents sliding up the continental slope with sufficient speed to cause surface outcropping.

The persistent CO_2-rich surface area over the western basin of the Bering Sea (Figs. 5.6–5.9) and near Amukta Pass in the Aleutians (Fig. 5.12) may be caused by the upward divergence of CO_2-rich deep water by cyclonic gyres. According to Arsen'ev (1967), the deep waters of the Bering Sea, similar to the overlying intermediate and subsurface waters, rise gradually to the surface in the cyclonic gyrals and are returned to the North Pacific as surface water.

All of these processes affect the observed surface P_{CO_2} distributions. In a particular region, one process may predominate over others; in other areas, the combined effects of these processes may control the surface P_{CO_2} distribution.

Open leads and polynyi

The importance of open leads and polynyi to the transport of CO_2 across the sea surface is emphasized by Kelley and Hood (1971a), who showed that supersaturation of CO_2 (with respect to the atmosphere) exists in open leads and polynyi during winter months. Their further investigations during the summer indicate that the seawater in leads and polynyi is undersaturated (Kelley and Hood 1971c), a condition attributed to utilization of CO_2 by phytoplankton based on measurements of chlorophyll a. When one considers that approximately 11 percent of the ice-covered areas are open at any one time (Fig. 5.1), the ice-covered sea cannot be ignored in estimating its CO_2 source/sink relationship on a seasonal basis (Kelley and Hood, 1971a). Further work is needed in the ice-covered regions to evaluate the importance of ice cover, open leads, and ice freezing and melting processes on the distribution of CO_2 between the sea and the air.

P_{CO_2} VERTICAL DISTRIBUTION

In the previous section we alluded often to the existence of CO_2-rich deep water without showing any CO_2 vertical distribution pattern. We present here a calculated P_{CO_2} vertical profile in the Bering Sea near the Aleutian Islands.

The application of infrared gas analyses to determine CO_2 in the depths of the ocean has not been widespread. Several groups of scientists at the University of Alaska, Oregon State University, Texas A&M University and others have used direct pumping of seawater to measure P_{CO_2} concentration in the upper 200 m with success.

For direct measurements of deep-sea P_{CO_2}, the method of Li (1967) and Li et al. (1969) is available. This method is designed for P_{CO_2} measurements in 20-liter water samples. The water sample is first transferred to a 20-liter glass bottle, and a carrier gas is recirculated through a gas dispersion tube immersed in the water sample at a rate of about 2 liter/min. Equilibration is attained primarily by gas exchange between the fine gas bubbles produced and dispersed in the water. The CO_2 concentration in the recirculating carrier gas is continuously monitored by an infrared gas analyzer. About 10 min are required to reach equilibrium. The GEOSECS project has used this method to determine deep-sea CO_2 concentrations in the Atlantic and Pacific oceans since July 1972. During the summer of 1973, the GEOSECS scientists occupied a station in the eastern basin of the Bering Sea to initiate the north-south GEOSECS transect of the Pacific Ocean.

There are no directly measured values of the vertical distribution of CO_2 in the Bering Sea. In this review, we use the calculated P_{CO_2} vertical profile of Alvarez-Borrego et al. (1972) to show a representative pattern. This work is based on the R/V *Yaquina* cruise data of summer, 24 June–4 July, 1966 (Barstow et al. 1968). The locations of the seven hydrochemical stations are shown in Figure 5.21.

During the 1966-*Yaquina* cruise, Park (1966b, 1967b) measured both pH and alkalinity at sea (Figs. 5.22–5.23). The pH was measured at 25 C and at one atmosphere. To obtain the *in situ* pH, the raw pH data are corrected for *in situ* temperature and pressure by the method of Culberson and Pytkowicz (1968), employing the following equation:

$$\frac{\Sigma CO_2}{\Sigma B} = \left[\frac{Alk}{\Sigma B} - \frac{K'_{B_p}}{A_{H_p} + K'_{B_p}}\right]\left[\frac{(A_{H_p})^2 + A_{H_p}K'_{1_p} + K'_{1_p}K'_{2_p}}{A_{H_p}K'_{1_p} + 2K'_{1_p}K'_{2_p}}\right] \quad (3)$$

where ΣCO_2 is the total concentration of carbon dioxide components, ΣB is the total concentration of boron, A_{H_p} is the hydrogen-ion activity at *in situ* pressure, K'_{B_p}, K'_{1_p}, and K'_{2_p} are the apparent boric acid dissociation constant, the first and second apparent carbonic acid dissociation constants, respectively, at *in situ* pressure. Note that ΣCO_2, ΣB, and alkalinity are not functions of temperature and pressure. ΣCO_2 in equation (3) can be calculated from alkalinity and raw pH values measured at 25 C and one atmosphere pressure (Park 1965). K'_{B_p}, K'_{1_p}, and K'_{2_p} are obtainable for any given temperature, salinity, and pressure using Lyman's (1956) apparent dissociation constants and Culberson's (1968) pressure coefficients. To obtain these constants, temperature (Fig. 5.24), salinity (Fig. 5.25) and pressure (depth) data are necessary. From equation (3), Alvarez-Borrego et al. (1972) used an iteration procedure suggested by Ben-Yaakov (1970) to solve for A_{H_p}, and thus for *in situ* pH values (Fig. 5.25).

In situ partial pressure of CO_2 is defined by Alvarez-Borrego et al. (1972) as follows:

$$P_{CO_2} = \frac{(A_{H_p})^2}{K'_{1_p} \cdot \alpha_s \cdot (A_{H_p} + 2K'_{2_p})}\left[Alk - \frac{K'_{B_p} \cdot \Sigma B}{K'_{B_p} + A_{H_p}}\right] \quad (4)$$

These authors contend that P_{CO_2} expressed by equation (4) is the partial pressure that carbon dioxide dissolved in seawater at a given depth would exert if that parcel of seawater were brought to the sea surface without changing the *in situ* equilibria of carbonic and boric acids. Figure 5.26 shows the results of computations (Alvarez-Borrego et al. 1972) of the vertical distribution of P_{CO_2} at stations in the Bering Sea.

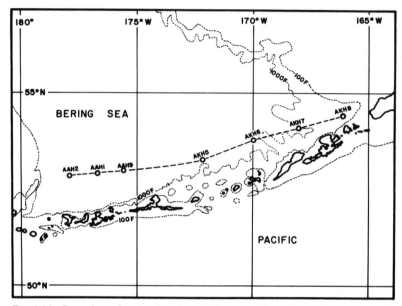

Fig. 5.21   Location of hydrochemical stations used by Alvarez-Borrego et al. (1972).

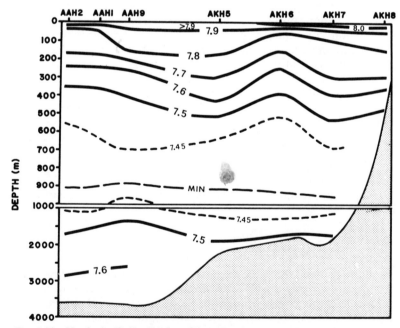

Fig. 5.22   Vertical pH distribution. The pH values are at 25 C and one atmosphere pressure (Alvarez-Borrego et al. 1972).

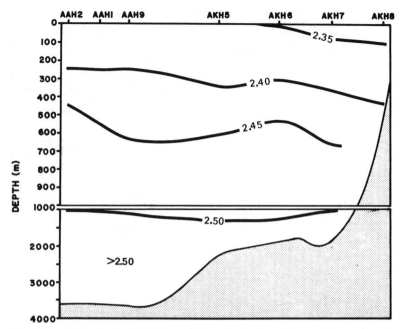

Fig. 5.23 Vertical alkalinity, in meq/liter, distribution (Alvarez-Borrego et al. 1972).

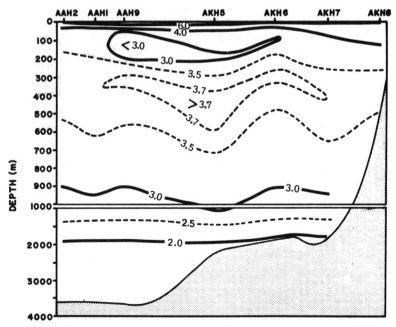

Fig. 5.24 Vertical temperature distribution (C) (Alvarez-Borrego et al. 1972).

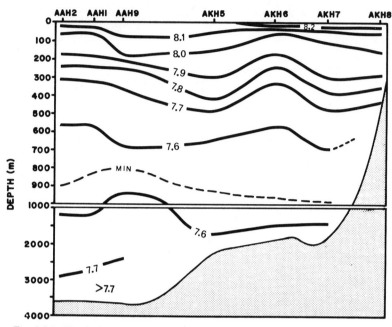

Fig. 5.25 Vertical *in situ* pH distribution (Alvarez-Borrego et al. 1972).

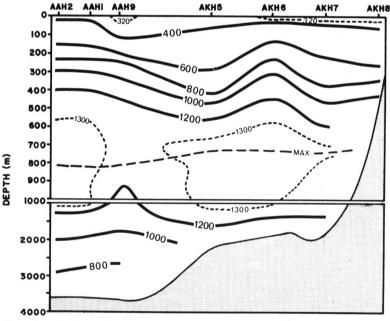

Fig. 5.26 Vertical P_{CO_2} distribution (ppm) (Alvarez-Borrego et al. 1972).

Figure 5.26 shows that at the sea surface the calculated P_{CO_2} deviates only slightly from the assumed atmospheric P_{CO_2} of 320 ppm. The 320 ppm isogram shows that about half of the cruise track is undersaturated with respect to atmospheric CO_2, while the remainder is supersaturated. This estimation agrees fairly well with the late summer P_{CO_2} distribution map, based on measurements made directly by use of the infrared CO_2 analyzer (Gordon et al. 1973; Fig. 5.6).

Below the surface, the CO_2 concentration increases with depth. A maximum is shown near 800 m with an average P_{CO_2} value of 1300 ppm, which is four times greater than the surface value. Below 800 m the P_{CO_2} gradually decreases and is less than 800 ppm at depths greater than 3000 m. Its vertical distribution resembles inversely the *in situ* pH (Fig. 5.25) and oxygen (Fig. 5.27) profiles, while it correlates positively with phosphate (Fig. 5.28) and nitrate (Fig. 5.29). All of these chemical parameters are greatly affected by the biological activity that occurs in the ocean; therefore, their resemblance (Figs. 5.25–5.29) is expected. A departure from this trend is found in the silicate distribution (Fig. 5.30), however, which does not exhibit a mid-depth maximum as do the other nutrient parameters. The difference is thought to be caused by the slow decomposition of diatom and other silica-rich organism tests—which does not depend totally on biochemical oxidations. Some inorganic precipitation has been found (Park 1967a) to occur in the deep ocean.

The horizontal P_{CO_2} distribution at any given depth (Fig. 5.26) is similar to the density distribution pattern (Fig. 5.31). At a given density surface, therefore, P_{CO_2} tends to remain constant.

The factors affecting the P_{CO_2} surface distribution discussed in the previous section, were upwelling, winter mixing, and upward divergence of deep water by cyclonic gyrals. From Figure 5.26 it is seen that CO_2 concentrations of 400 ppm or greater exist at 100 m depth, as much as 600 ppm is found at 200 m and 1000 ppm at 400 m. Therefore, any appreciable vertical turbulence or vertical transport of deep water to the surface can affect the surface P_{CO_2} distribution significantly. The *Acona* (Fig. 5.12) and the *Endeavour* cruise data (Figs. 5.14–5.15) show surface P_{CO_2} values greater than 400 ppm in the Pacific near the Aleutian Islands which can be explained by such vertical processes (Gordon et al. 1973; Kelley and Hood 1971b).

In the Bering Sea, the difference in seawater density between the surface and 200-m depth observed from the R/V *Yaquina* (Fig. 5.31) is in the order of 0.0004 g/cm³, or $\Delta\sigma_t$ of 0.4. At a salinity of 33.3‰ (Fig. 5.32), an increase in the surface seawater density of 0.0004 g/cm³ can be obtained by cooling surface water by about 5 C. Arsen'ev (1967) shows that in the deep parts of the Bering Sea, the annual surface temperature fluctuation is about 5 C; winter temperatures are about 2 C or less and summer temperatures about 7 C. Annual temperature fluctuation alone, therefore, can result in mixing of surface water with the 200-m deep waters—thus the surface P_{CO_2} values can be significantly increased.

CARBONATE SATURATION

The level of calcium carbonate and the processes controlling its saturation in the ocean are under question, as indicated by the results of several investigators (Ben-Yaakov and Kaplan 1971; Hawley and Pytkowicz 1969; Li et al.

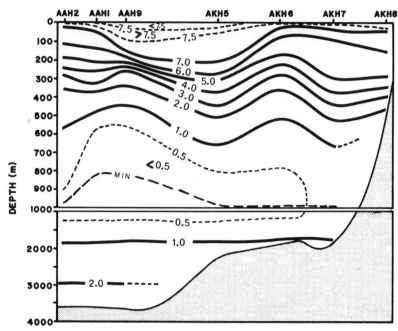

Fig. 5.27   Vertical oxygen distribution (ml/liter) (Alvarez-Borrego et al. 1972).

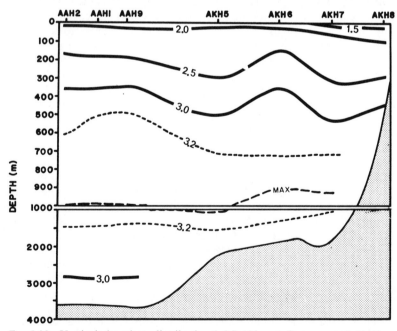

Fig. 5.28   Vertical phosphate distribution (μM) (Alvarez-Borrego et al. 1972).

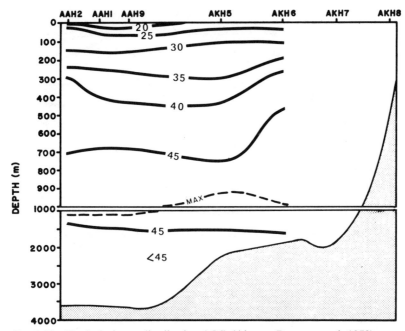

Fig. 5.29   Vertical nitrate distribution (μM) (Alvarez-Borrego et al. 1972).

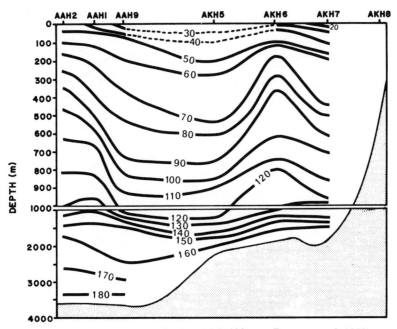

Fig. 5.30   Vertical silicate distribution (μM) (Alvarez-Borrego et al. 1972).

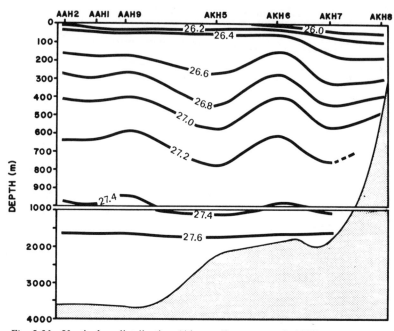

Fig. 5.31   Vertical σ_t distribution (Alvarez-Borrego et al. 1972).

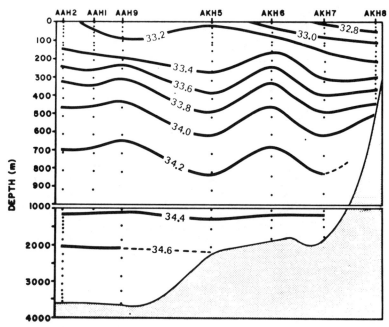

Fig. 5.32   Vertical salinity distribution (‰). Dots indicate sampling depths (Alvarez-Borrego et al. 1972).

1969; Lyakhin 1968). Oceanic carbonate saturation chemistry, therefore, is one of the current problems in chemical oceanography that indicates the need for further definitive work.

Data for the Bering Sea were obtained by Alvarez-Borrego et al. (1972) using the method of Hawley and Pytkowicz (1969). The *in situ* percentage saturation of calcium carbonate, both for calcite and aragonite, was calculated from pH, alkalinity, temperature, salinity and pressure (depth) data. The percent carbonate saturation is expressed as follows:

$$(\% \text{ Saturation}) = 100 \frac{\Sigma Ca(\Sigma CO_3)_p}{K'_{SP_p}} \tag{5}$$

where ΣCa and ΣCO_3 are the total stoichiometric concentrations of dissolved calcium and carbonate that include both free ions and ion-pairs (Garrels and Thompson 1962). The term K'_{SP_p} is the apparent solubility product of calcium carbonate at pressure p.

To calculate ΣCa, Alvarez-Borrego et al. (1972) used Wattenberg's (1936) relationship,

$$\Sigma Ca = \tfrac{1}{2}(C.A.) + 477 \times 10^{-16}(Cl\%_{oo}) \quad (\text{mole/liter}) \tag{6}$$

where $Cl\%$ is chlorinity in parts per thousand.

The term $(\Sigma CO_3)_p$ is calculated by the following equation:

$$(\Sigma CO_3)_p = (C.A.) \left[K'_{2_p}/(A_{H_p} + 2K'_{2_p}) \right] \tag{7}$$

To obtain the value for the apparent solubility product at *in situ* pressure, K'_{SP_p}, it is necessary to know the apparent solubility product at one atmosphere, K'_{SP}. Li et al. (1969) provide the following two empirical equations for K'_{SP}:

$$K'_{SP} \text{ (calcite)} = (0.69 - 0.0063T_p) \cdot 10^{-6} \cdot (Cl\%_{oo}/19) \tag{8}$$

$$K'_{SP} \text{ (aragonite)} = (1.09 - 0.0078T_p) \cdot 10^{-6} \cdot (Cl\%_{oo}/19) \tag{9}$$

The units for these two K'_{SP} are $mole^2/liter^2$. The term T_p is the *in situ* temperature (C). The apparent solubility product at *in situ* pressure is then obtained by the following equation:

$$K'_{SP_p} = (10^{-\Delta\bar{V} \cdot Z/23RT_k}) K'_{SP} \tag{10}$$

where $\Delta\bar{V}$ is the change in partial molar volume for calcium carbonate, Z is depth in meters, R is the gas constant and T_k is the *in situ* temperature in degrees Kelvin. By the use of the pressure coefficient data at 2 C obtained by Hawley and Pytkowicz (1969), $\Delta\bar{V}$ can be calculated as -35.8 cm^3/mole for calcite and -34.5 cm^3/mole for aragonite. These partial molar volumes are used to calculate K'_{SP_p} for both calcite and aragonite. By substituting the values obtained by equations (6), (7), and (10) into equation (5), Alvarez-Borrego et al. (1972) obtained vertical percentage saturation profiles for calcite (Fig. 5.33) and for aragonite (Fig. 5.34).

In the percentage saturation patterns for both calcite and aragonite, the surface layer is supersaturated, although the seawater temperature was colder than 7 C everywhere. Supersaturation is found down to about 250 m for calcite

(Fig. 5.33) and to about 100 m for aragonite (Fig. 5.34). Below these depths, the seawater is undersaturated for the respective carbonate minerals, with a minimum layer at 1000 m. At the minimum, the percentage saturation for calcite is about 55 percent (Fig. 5.33) and that for aragonite is about 35 percent (Fig. 5.34). Below the minimum depth, the percentages increase slightly, to about 60 percent for calcite and 40 percent for aragonite near the sea-floor.

There is a similarity between the vertical distributions of percentage carbonate saturation (Figs. 5.33–5.34) and apparent oxygen utilization (AOU) (Fig. 5.35); the depth of the carbonate saturation minimum (1000 m) is also the AOU maximum layer. This can be partially explained by the strong relationship between AOU and pH (Park 1968a, b), together with the concurrent pressure dependence of pH (Hawley and Pytkowicz 1969). Marine organisms by their AOU affect the carbonate saturation by changing the pH of seawater. Since the pressure effect on the carbonate solubility product increases monotonically with depth, therefore, the minimum observed in the percent saturation of carbonate at 1000 m depth is produced mainly by AOU. In addition, the dissolved calcium increase deduced from the alkalinity/chlorinity ratio or specific alkalinity increase with depth (Fig. 5.36), coupled with the vertical salinity increase (Fig. 5.32), is also monotonic without any major variance at 1000 m depth.

The alkalinity/chlorinity ratio is used to study carbonate chemistry in the ocean because it increases when carbonate minerals are dissolved and decreases when carbonate minerals are precipitated or when marine organisms assimilate calcium carbonate. Since biological assimilation takes place in the upper layers of the ocean and dissolution of calcareous skeletons takes place in the deep areas of the ocean (the latter process being facilitated by the carbonate undersaturation shown in Figs. 5.33–5.34), the alkalinity/chlorinity ratio generally increases with depth (Fig. 5.36). In the work of Alvarez-Borrego et al. (1972) the ratio is about 0.128 near the sea surface and increases to about 0.133 meq/liter/Cl‰ near the sea-floor. The net difference is 0.005 meq/liter/Cl‰. A similar change was observed by Ivanenkov (1964) for the western part of the Bering Sea; his values were 0.126 for the surface and 0.131 for deep waters. Hypothetically, this 0.005 increase can be produced by dissolving 0.05 mM of $CaCO_3$ in surface seawater.

When it is considered that the increase in the alkalinity/chlorinity ratio expresses the differential carbonate mineral dissolution with depth, the change in ratio can be used to assess its effect on the vertical pH distribution (Park 1968a, b). Carbonate dissolution increases pH by release of the carbonate ion into seawater. Therefore, all else being equal, an increase in the alkalinity/chlorinity ratio would elevate the pH of seawater (Park 1968a). In the Bering Sea work of Alvarez-Borrego et al., the alkalinity/chlorinity ratio was found to increase monotonically with depth and the AOU showed its maximum at 1000 m depth. The combined effect would be that the pH minimum should be shallower than the depth of the AOU maximum. Indeed, such is the case, for both the pH values measured at 25 C and at one atmosphere (Fig. 5.22) and the calculated *in situ* pH values (Fig. 5.25) show their minima at 900 m. Correspondingly, since the P_{CO_2} profile is strongly affected by pH (equation 4),

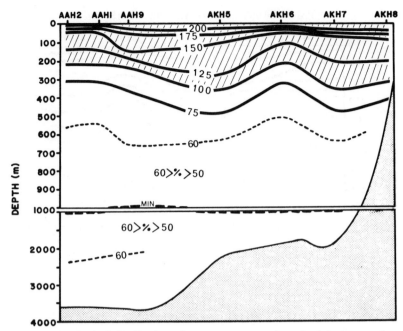

Fig. 5.33 Vertical distribution of percentage saturation of calcite in a section of the Bering Sea (Alvarez-Borrego et al. 1972).

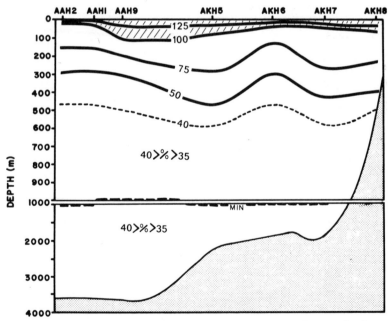

Fig. 5.34 Vertical distribution of percentage saturation of aragonite in a section of the Bering Sea (Alvarez-Borrego et al. 1972).

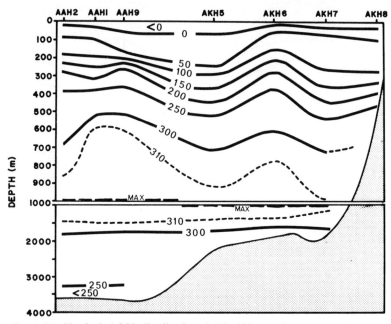

Fig. 5.35   Vertical AOU distribution (μM) (Alvarez-Borrego et al. 1972).

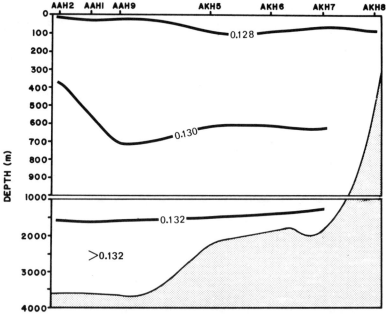

Fig. 5.36   Vertical distribution of alkalinity/chlorinity ratio (meq/liter-Cl‰) (Alvarez-Borrego et al. 1972).

the depth of its maximum also should be shallower than the depth of the AOU maximum. Alvarez-Borrego et al. (1972) found the calculated *in situ* P_{CO_2} maximum to exist at 800 m depth.

CO_2-O_2-NUTRIENT RELATIONSHIP

The data of the previous section show that negative AOU values occur consistently at the sea surface and are slightly over $+310$ μM at the 1000-m level. A similar vertical distribution of AOU was obtained from R/V *Yaquina* in the Pacific near the Aleutian Islands, between $170-180°$W (Park 1967b).

Based on the AOU (Fig. 5.35) and nutrient data (Figs. 5.28–5.30), Alvarez-Borrego et al. (1972) prepared Figure 5.37, which shows the relationship between AOU and nutrients. A linear relationship is drawn between the AOU and phosphate and between AOU and nitrate where AOU is greater than 100 μM, corresponding to depths below 250 m. A positive correlation, almost linear, also exists on the AOU-silicate graph between the AOU values of 100 to 300 μM which correspond to the upper 1000 m. Below 1000 m, AOU decreases and silicate continues to increase.

The slopes observable on the AOU-phosphate and AOU-nitrate plots for the deep water, where AOU is greater than 100 μM, is approximately twice the slope obtainable from the biochemical oxidation model of Redfield et al. (1963). This implies that the preformed nutrient distributions must vary at different depths. Since the AOU-phosphate diagram shows less scatter than the AOU-nitrate diagram, Alvarez-Borrego et al. (1972) chose to examine the preformed phosphate distribution only. Its distribution (Fig. 5.38) shows preformed phosphate values of approximately 2 μM at the surface and 1 μM for the deep waters. The AOU-preformed phosphate data are also plotted in Figure 5.37. Since the AOU-phosphate relationship is linear, the resulting AOU-preformed phosphate relationship also is linear, where AOU is greater than 100 μM. Although not plotted, the AOU-preformed nitrate relationship is expected to be linear, with surface values of between $20-30$ μM decreasing to less than 10 μM at the AOU maximum zone of 1000 m depth. The deep preformed phosphate value of about 1.0 μM is similar to those found in the subarctic region (Park 1967b) and off the Oregon coast (Park 1967a).

The CO_2, O_2 and nutrient cycles in the ocean are interwoven. Craig (1971) examined the question of a significant rate of *in situ* oxygen consumption in deep water by his vertical-diffusion and vertical-advection model. His work is applied elsewhere, and Culberson (personal communication) of Oregon State University developed a method to obtain the Redfield et al. (1963) ratio by plotting oxygen and phosphate against potential temperature for the vertical region where the potential temperature-salinity diagram gives a straight line. Hopefully, the Culberson method can be applied to total CO_2 data for the Bering Sea in the future.

By use of the Culberson method, Alvarez-Borrego et al. (1972) examined the ratio between O_2 consumption and phosphate release in the sea at $1300-3600$ m depth where the potential temperature-salinity diagram is linear. They

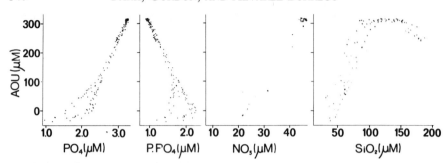

Fig. 5.37 AOU-nutrient relationship in the Bering Sea. P.PO$_4$ denotes preformed phosphate (Alvarez-Borrego et al. 1972).

obtained a ΔAOU:ΔP value of -3.4 ml:1 μM, which is close to the Redfield et al. (1963) value of -3.1 ml:1 μM.

Having established an empirical relation between AOU and nutrients (Fig. 5.37), we will now examine the relation between AOU and pH, pH and alkalinity, and AOU and ΣCO$_2$ (expressed as TCO$_2$ in Fig. 5.39). There is an inverse relationship between AOU and pH measured at 25 C and at one atmosphere (Fig. 5.39a). The pH values decrease as the AOU increases in the

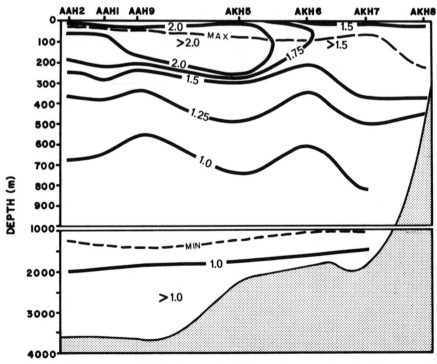

Fig. 5.38 Vertical preformed phosphate distribution (Alvarez-Borrego et al. 1972).

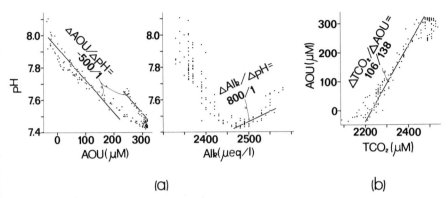

Fig. 5.39   The relationships between pH (measured at 25 C and one atmosphere pressure) and AOU, between pH and alkalinity, and between AOU and ΣCO_2, expressed as TCO_2, in the Bering Sea (Alvarez-Borrego et al. 1972).

upper 1000 m and decrease below 1000 m. Based on 1966 R/V *Yaquina* cruise data for the Gulf of Alaska (Barstow et al. 1968), Park (1968a, b) examined the two major processes that affect the vertical distribution of pH and estimated that the AOU effect on the pH change is several times greater than the carbonate dissolution effect. Alvarez-Borrego et al. (1972) employed Park's (1968a, b) empirical correlation slope ($\Delta AOU/\Delta pH$ equal to -500 $\mu M/pH$ unit) and found that Park's (1968a, b) value agreed fairly well with the observed pH-AOU relation, except for the region near 1000 m where AOU shows a maximum.

Park (1968a, b) also gives the carbonate dissolution effect on pH by an empirical number of $\Delta Alk/\Delta pH$ equal to 800 $\mu eq/pH$ unit as shown on the pH-alkalinity plot in Figure 5.39. The only region where the slope appears to agree is near 1000 m, or the AOU maximum. When the AOU effect is combined with the carbonate dissolution effect implied by the alkalinity/chlorinity ratio change, the vertical pH variations observed in the Bering Sea appear to be explainable by these two effects.

Alvarez-Borrego et al. (1972) further examined the relationship between AOU and ΣCO_2. The total CO_2 concentration was calculated from pH and alkalinity (Fig. 5.39). The Redfield et al. (1963) ratio for $\Delta CO_2/\Delta AOU$ of 106/138 holds in first approximation. Culberson and Pytkowicz (1970) have shown that a linear correlation between AOU and ΣCO_2, which satisfies the biochemical oxidation model of Redfield et al. (1963), is found when changes in ΣCO_2 concentration due to all the processes (such as carbonate dissolution and preformed CO_2 change) have been compensated. The conclusion of Culberson and Pytkowicz (1970) appears to agree with the Bering Sea data examined by Alvarez-Borrego et al. (1972; Fig. 5.39).

CONCLUDING REMARKS

Although available data are scanty, we have presented a detailed scientific scrutiny of the carbon dioxide system in the Bering Sea. The very dynamic nature of the CO_2 system is apparent. For instance, the surface P_{CO_2} distribution has a range from 100 to >400 ppm, and its value changes by many identifiable processes. We have also become aware of the fact that even though a good portion of the Bering Sea is covered by ice during the winter, we cannot ignore the important influence of open leads and polynyi on the dynamics of CO_2 air-sea exchange. If we are to understand the CO_2 system correctly, we need to execute a series of synoptic studies, in time and in space, for the entire sea. Without such studies we cannot intelligently assess the role of the Bering Sea as a source or a sink for the atmospheric CO_2.

Winter data is needed and is obtainable only under adverse sea conditions. A well equipped research platform or ship which uses the most advanced scientific instrumentation is essential for this study. As shown in this review, sufficiently sophisticated instrumentation is already available; competent and eager oceanographers are also available. It appears that the present stumbling block is of a political rather than scientific nature. We would like to urge that the scientists concerned with future Bering Sea study convince their respective governments as well as scientific research funding agencies of the importance of the intended scientific endeavor as well as the high probability of achieving the desired scientific success.

Discussion

HOLT: What location is planned for the 1973 GEOSECS station in the Bering Sea?

PARK: According to present plans the station will be located in the deepest part of the eastern basin, representing the northern terminus of the Pacific Ocean.

Acknowledgements

We acknowledge the aid of the U. S. National Science Foundation through grants GA-1281, GA-12113, and GX-28167. The Office for the International Decade of Ocean Exploration (IDOE), National Science Foundation, contributed to U. S. participation in the Hakodate symposium through a grant to the University of Alaska.

We acknowledge scientific research support by the U. S. Office of Naval Research through contract N00014-67-A-0369-0007 under project NR 083-102, and support by the Consejo Nacional de Ciencia y Tecnologia of Mexico in granting a graduate fellowship for Saúl Alvarez-Borrego. This review paper comprises part of two dissertations submitted by Louis I. Gordon and Saúl Alvarez-Borrego in partial fulfillment of their Ph.D. degrees at Oregon State University.

We thank John J. Kelley of the Institute of Marine Science, University of Alaska, for his scientific criticism.

REFERENCES

AKIYAMA, T.

1969 Carbon dioxide in the atmosphere and in seawater in the Pacific Ocean east of Japan. *Oceanogr. Mag.* 21: 129–135.

ALVAREZ-BORREGO, S., L. I. GORDON, L. B. JONES, P. K. PARK, and R. M. PYTKOWICZ

1972 Oxygen-carbon dioxide-nutrients relationships in the southeastern region of the Bering Sea. *J. Oceanogr. Soc. Japan* 28: 71–93.

ANDERSON, D. H., and R. J. ROBINSON

1946 Rapid electrometric determination of the alkalinity of seawater using a glass electrode. *Ind. Eng. Chem., Anal. Ed.* 18: 767–773.

ARSEN'EV, V. S.

1967 The currents and water masses of the Bering Sea [in Russian, English summary]. Izd. Nauka, Moscow. (Transl., 1968, Nat. Mar. Fish. Serv., Northwest Fish. Center, Seattle, Wash.). 135 pp.

BADGLEY, F. I.

1966 Proc., *Symposium on the arctic heat budget and atmospheric circulation.* Memorandum RM 5233-NSF, 267, Nat. Sci. Found., Washington, D.C.

BARSTOW, D., W. GILBERT, K. PARK, R. STILL, and B. WYATT

1968 Hydrographic data from Oregon waters, 1966. Data Rep. 33, Dep. Oceanogr., Oregon State Univ., Corvallis, 109 pp.

BEN-YAAKOV, S.

1970 A method for calculating the *in situ* pH of seawater. *Limnol. Oceanogr.* 15: 326–328.

BEN-YAAKOV, S., and I. R. KAPLAN

1971 Deep-sea *in situ* calcium carbonate saturometry. *J. Geophys. Res.* 76: 722–731.

CRAIG, H.

1971 The deep metabolism: oxygen consumption in abyssal ocean water. *J. Geophys. Res.* 76: 5078–5086.

CULBERSON, C.

1968 Pressure dependence of the apparent dissociation constants of carbonic and boric acid in seawater. M. S. Thesis, Oregon State Univ., Corvallis, 85 pp.

CULBERSON, C., and R. M. PYTKOWICZ

1968 Effect of pressure on carbonic acid, boric acid, and the pH in seawater. *Limnol. Oceanogr.* 13: 403–417.

1970 Oxygen-total carbon dioxide correlation in the Eastern Pacific Ocean. *J. Ocean. Soc. Japan* 26(2): 95–100.

CULBERSON, C., R. M. PYTKOWICZ, and J. E. HAWLEY

1970 Seawater alkalinity determination by the pH method. *J. Mar. Res.* 28: 15–21.

DYRSSEN, D., and L. G. SILLEN

1967   Alkalinity and total carbonate in seawater, a plea for p-T-independent data. *Tellus* 19: 113–121.

EDMOND, J. M.

1970   High precision determination of titration alkalinity and total carbon dioxide content of seawater by potentiometric titration. *Deep-Sea Res.* 17: 737–750.

ERIKSSON, E.

1963   Possible fluctuations in atmospheric carbon dioxide due to changes in the properties of the sea. *J. Geophys. Res.* 68: 3871–3876.

GARRELS, R. M., and M. E. THOMPSON

1962   A chemical model for seawater at 25 C and one atmosphere total pressure. *Am. J. Sci.* 260: 57–66.

GORDON, L. I.

1973   A study of carbon dioxide partial pressures in surface waters of the Pacific Ocean. Ph.D. Thesis. Oregon State Univ., Corvallis, 216 pp.

GORDON, L. I., P. K. PARK, S. W. HAGER, and T. R. PARSONS

1971   Carbon dioxide partial pressures in North Pacific surface waters—time variations. *J. Oceanogr. Soc. Japan* 27(3): 81–90.

GORDON, L. L., J. J. KELLEY, D. W. HOOD, and P. K. PARK

1973   Carbon dioxide partial pressures in the North Pacific surface waters. 2: General late-summer distribution. *Mar. Chem.* 1: 191–198.

HARVEY, H. W.

1955   The chemistry and fertility of seawaters. Cambridge Univ. Press, 224 pp.

HAWLEY, J., and R. M. PYTKOWICZ

1969   Solubility of calcium carbonate in seawater at high pressure and 2 C. *Geochim. Cosmochim. Acta* 33: 1557–1561.

HOOD, D. W., D. BERKSHIRE, I. SUPERNAW, and R. ADAMS

1963   Calcium carbonate saturation level of the ocean from latitudes of North American to Antarctica and other chemical oceanographic studies during cruise III of the USNS *Eltanin*. Data rep., Nat. Sci. Found., Texas A&M Univ., College Station.

IBERT, E. R., and D. W. HOOD

1963   The distribution of carbon dioxide between the atmosphere and the sea. Tech. Rep. 63–9–7, Texas A&M Univ., College Station, 131 pp.

IVANENKOV, V. I.

1964   *Hydrochemistry of the Bering Sea.* Acad. Sci. USSR, Izd. Nauka, Moscow, 137 pp.

KANWISHER, J.

1960 pCO_2 in seawater and its effect on the movement of CO_2 in nature. *Tellus* 12: 209–215.

KEELING, C. D.
    1968   Carbon dioxide in surface ocean waters. 4. Global distribution. *J. Geophys. Res.* 73: 4543–4544.

KEELING, C. D., N. W. RAKESTRAW, and L. S. WATERMAN
    1965   Carbon dioxide in surface waters of the Pacific Ocean. 1. Measurements of the distribution. *J. Geophys. Res.* 70: 6087–6097.

KELLEY, J. J.
    1968   Carbon dioxide and ozone studies in the arctic atmosphere. In *Arctic drifting stations*, edited by J. E. Sater. Arctic Inst. North Amer., Washington, D.C., pp. 155–166.

KELLEY, J. J., and D. W. HOOD
    1971a   Carbon dioxide in the surface water of the ice-covered Bering Sea. *Nature* 229: 37–39.
    1971b   Carbon dioxide in the Pacific Ocean and Bering Sea: upwelling and mixing. *J. Geophys. Res.* 76: 745–752.
    1971c   Carbon dioxide in the surface waters of the Barents Sea. *Southwind* Expedition—1970. Rep. 71–20, Inst. Mar. Sci., Univ. Alaska, Fairbanks, 9 pp.

KELLEY, J. J., L. L. Longerich, and D. W. HOOD
    1971   Measurements of carbon dioxide in northern seas. [Progress report to Office of Naval Research, Wash., D.C.] Inst. Mar. Sci., Univ. Alaska, Fairbanks, 36 pp.

LI, Y-H.
    1967   The degree of saturation of $CaCO_3$ in the oceans. Ph.D. Thesis. Columbia Univ., New York, 179 pp.

LI, Y-H., T. TAKAHASHI, and W. S. BROECKER
    1969   Degree of saturation of $CaCO_3$ in the oceans. *J. Geophys. Res.* 74: 5507–5525.

LYAKHIN, Y. I.
    1968   Calcium carbonate saturation of Pacific water. *Oceanology* 8: 58–68.

LYMAN, J.
    1956   Buffer mechanism of seawater. Ph.D. Thesis. Univ. Calif. Los Angeles, 196 pp.

McROY, C. P.
    1968   The distribution and biogeography of *Zostera marina* (eelgrass) in Alaska. *Pac. Sci.* 22: 507–513.

MIYAKE, Y., and Y. Sugimura
    1969   Carbon dioxide in the surface water and the atmosphere in the Pacific, the Indian and the Antarctic Ocean areas. *Rec. Oceanographic Works Japan* 10: 23–28.

PARK, K.
    1965   Total carbon dioxide in seawater. *J. Ocean. Soc. Japan* 21(2): 54–59.
    1966a   Deep-sea pH. *Science* 154: 1540–1542.

1966b Surface pH of the northeastern Pacific Ocean. *J. Oceanol. Soc. Korea* 1: 1–6.

1967a Nutrient regeneration and preformed nutrients off Oregon. *Limnol. Oceanogr.* 12: 353–357.

1967b Chemical features of the Subarctic boundary near 170°W. *J. Fish. Res. Bd. Can.* 24: 899–908.

1968a The processes contributing to the vertical distribution of apparent pH in the northeastern Pacific Ocean. *J. Oceanol. Soc. Korea* 3: 1–7.

1968b Seawater hydrogen-ion concentration: vertical distribution. *Science* 162: 357–358.

1969   Oceanic CO_2 system: an evaluation of ten methods of investigation. *Limnol. Oceanogr.* 14: 179–186.

PARK, K., H. C. CURL, JR., and W. A. GLOOSCHENKO

1967   Large surface carbon dioxide anomalies in the North Pacific Ocean. *Nature* 215: 380–381.

PARK, P. K., L. I. GORDON, S. W. HAGER, and M. C. CISSELL

1969   Carbon dioxide partial pressure in the Columbia River. *Science* 166: 867–868.

PARK, K., G. H. KENNEDY, and H. H. DOBSON

1964   Comparison of gas chromatographic method and pH-alkalinity method for determination of total carbon dioxide in seawater. *Anal. Chem.* 36: 1686.

REDFIELD, A. C., B. H. KETCHUM, and F. A. RICHARDS

1963   The influence of organisms on the composition of seawater. In *The sea*, Vol. 2, edited by M. N. Hill. Interscience Publishers, New York.

REVELLE, R., and H. E. SUESS

1957   Carbon dioxide exchange between atmosphere and ocean and the question of an increase of atmospheric CO_2 during the past decades. *Tellus* 9: 18–27.

SWINNERTON, J. W., V. J. LINNENBOM, and C. H. CHEEK

1962a Determination of dissolved gases in aqueous solutions by gas chromatography. *Anal. Chem.* 34: 483–485.

1962b Revised sampling procedure for determination of dissolved gases in solution by gas chromatography. *Anal. Chem.* 34: 1509.

TAKAHASHI, T.

1961   Carbon dioxide in the atmosphere and in the Atlantic Ocean water. *J. Geophys. Res.* 66: 477–494.

TAKAHASHI, T., R. F. WEISS, C. H. CULBERSON, J. M. EDMOND, D. E. HAMMOND, C. S. WONG, Y-H. LI, and A. E. BAINBRIDGE

1970   A carbonate chemistry profile at the 1969 GEOSECS intercalibration section in the Eastern Pacific Ocean. *J. Geophys. Res.* 75: 7648–7666.

TEAL, J. M., and J. KANWISHER

1966   The use of pCO_2 for the calculation of biological production, with examples from waters off Massachusetts. *J. Mar. Res.* 24: 4–14.

U. S. WEATHER BUREAU
1961   Climatological and oceanographic atlas for mariners. Vol. 2. North Pacific Ocean. 7 pp + 159 charts.

WATTENBERG, H.
1936   Kohlensäure und Kalziumkenbonat in Meere. *Fortschr. Mineral.* 20: 168–195.

WEISS, R. F.
1970   Dissolved gases and total inorganic carbon in seawater: distribution, solubilities, and shipboard gas chromatography. Ph.D. Thesis. Univ. Calif. San Diego, 130 pp.

WONG, C. S.
1970   Quantitative analysis of total carbon dioxide in seawater: a new extraction method. *Deep-Sea Res.* 17: 9–17.

CHAPTER 6

Assimilation and oxidation-reduction of inorganic nitrogen in the North Pacific Ocean

AKIHIKO HATTORI *and* EITARO WADA

Ocean Research Institute, University of Tokyo, Tokyo, Japan

Abstract

Available data are summarized on assimilation and oxidation-reduction of inorganic nitrogen over extended areas of the North Pacific Ocean, including coastal regions. The capacities for assimilation and oxidation-reduction of ammonia and nitrate vary to a considerable extent both regionally and seasonally, but some general trends can be seen. The most significant trend noted is the increased potential of the phytoplankton to utilize *in situ* concentrations of nitrate. Nitrate uptake in the offshore waters of the Bering Sea appears to range between 40 to 50 percent of total nitrogen uptake. Two simplistic models for nitrate balance are constructed in discussing the nitrogen budget in the epicontinental Eastern Bering Sea.

INTRODUCTION

The subarctic areas of the North Pacific Ocean are highly productive. The abundant supply of nutrient salts such as inorganic nitrogen and phosphorus that supports this high primary production is provided by extensive vertical mixing during winter. The distribution of inorganic nitrogen and phosphorus indicates that *nitrogen*, not phosphorus, is generally the principal factor limiting the growth of marine phytoplankton. Surveys of primary production, however, have traditionally been carried out only in terms of carbon assimilation. Dugdale and Goering (1967) first measured the primary production in the open ocean in terms of nitrogen, a fundamental component of living matter. In pointing out the importance of distinguishing between ammonia and nitrate as sources of nitrogen for natural phytoplankton populations, they proposed the term *regenerated* production for that associated with ammonia and *new* production for the case of nitrate and nitrite. Since ammonia is produced in the short-term regeneration process of organic nitrogen, only external sources such as deep water advection contribute sufficient amounts to increase the population size.

In this paper available data are summarized on assimilation and oxidation-reduction of inorganic nitrogen over extended areas of the North Pacific, including highly productive coastal and upwelling regions. The type and extent of nitrate metabolism are especially emphasized. Although capacities for assimilation of inorganic nitrogen vary to a considerable extent both regionally and seasonally, certain general trends appear.

Assimilation of inorganic nitrogen

Capacities for ammonia and nitrate assimilation (Table 6.1) were determined by means of ^{15}N technique with water samples collected from euphotic layers of oceanic and coastal regions of the North Pacific and the Bering Sea (MacIsaac and Dugdale 1969; McRoy et al. 1972; Hattori and Wada 1972; Miyazaki et al. 1973). In many cases the assimilation of both nitrate and ammonia was accelerated by the presence of light, suggesting that phytoplankton is responsible for these processes. A similarity between depth profiles of carbon and nitrogen assimilation (Dugdale and Goering 1967; Hattori and Wada 1972) can be considered evidence for this view. The ratio of carbon-nitrogen assimilation in various North Pacific and Bering Sea areas in summer varies widely with an average of 9.5 (Table 6.2), which would be expected from the elementary composition of marine phytoplankton (Fleming 1940).

Data obtained by Dugdale and Goering (1967) in the northeastern Pacific and in the northwestern Atlantic indicate that the ratios of nitrate uptake to total nitrogen uptake (nitrate plus ammonia uptake) vary to a great extent but tend to decrease directly with latitude (Table 6.3). In southern areas ammonia appears to be a more important source of nitrogen than is nitrate. Data collected in the central North Pacific along 155°W (Hattori and Wada 1972) and in the western North Pacific (T. Miyazaki, E. Wada, and A. Hattori, unpublished) show a similar trend. The highest ratio ever recorded was at a station on the western side of the Bering Strait, where four times more nitrate than ammonia was utilized preferentially by phytoplankton (McRoy et al. 1972). A high utilization of nitrate is found also in upwelling regions of the eastern equatorial Pacific (MacIsaac and Dugdale 1969).

In situ concentrations of nitrate in the euphotic layer in tropical areas and in temperate areas during summer are very low, usually less than 0.1 μg-atom N/liter, except in regions of significant deep-water upwelling. Nitrate concentrations in subarctic and temperate regions in winter are about 10 μg-atoms N/liter or higher. Ammonia concentrations range from 0 to 1 μg-atom N/liter, irrespective of depth, location and season. Taking into account the known concentration dependence of assimilation rates, the *in situ* capacities

TABLE 6.1 Ammonia and nitrate assimilation in the euphotic layer of the North Pacific Ocean and Bering Sea

| | | Assimilation rates (ng-atoms N/liter-hr) | | | | | |
| | | Potential activity* | | In situ capacity | | |
Area	Dates	NH_4^+	NO_3^-	NH_4^+	NO_3^-	References
Central North Pacific 50°N 155°W	August 1969	1.1–3.2	0.3–1.3	0.4	0.7	Hattori and Wada 1972
Central North Pacific 40°N 155°W	September 1969	11.0–16.0	1.0–11.0	3.3	0.1	Hattori and Wada 1972
Central North Pacific 10°N 155°W	September 1969	363.0	207.0	91.0	0.1	Hattori and Wada 1972
East Pacific off Mexico	Jan–Feb 1968	3.3–10.5	0.4–2.2			MacIsaac and Dugdale 1969
East Pacific 10°N–3°N	February 1967	6.8	0.4–3.9			MacIsaac and Dugdale 1969
Juneau	April 1965		77.0–94.0			MacIsaac and Dugdale 1969
Alaskan Coast	July 1965	19.1		24.1–34.8	4.6–18.8	MacIsaac and Dugdale 1969
Aleutian Islands	June 1968			4.9–24.9	83.6–118.2	McRoy et al. 1972
Bering Strait	June 1969					McRoy et al. 1972
Sagami Bay	October 1970	0.2–9.0	0.2–1.5	0.05–2.3	0.2–1.5	Miyazaki et al. 1973

*Assuming ammonia or nitrate saturation

TABLE 6.2   Relationship between carbon and nitrogen assimilation in the euphotic layer of the North Pacific Ocean and Bering Sea

Area	Dates	C/N Range	(Average)	References
Aleutian Islands	Summer	5.7–8.7	(6.8)	McRoy et al. 1972
Eastern Bering Sea	Winter	1.2–36.7	(15.4)	McRoy et al. 1972
Bering Strait	Summer	11.2–14.3	(12.6)	McRoy et al. 1972
Central North Pacific (50°N)	August 1969	4.1–27.0	(11.4)	Hattori and Wada 1972
Central North Pacific (40°N)	September 1969	0.8–2.3	(1.5)	Hattori and Wada 1972

for nitrogen assimilation can be calculated from the measured assimilation rates. In the central Pacific, the *in situ* capacities for nitrate assimilation are generally higher in northern areas than in southern areas. The high capacity is found also in nutrient-rich waters of coastal and upwelling regions.

High capacities for ammonia assimilation are sometimes observed in surface and subsurface waters of the tropical central Pacific that are poor in nutrients (Table 6.1). In this case, however, the light suppresses assimilation, and the C/N assimilation ratios are very low (A. Hattori and E. Wada, unpublished). It can be speculated that various biological agents, probably heterotrophic bacteria, are mainly responsible for this phenomenon.

Oxidation-reduction of inorganic nitrogen
Nitrite is of importance as a natural tracer in studying the nitrogen cycle in the sea because of its intermediate position in oxidation-reduction processes

TABLE 6.3   Summary of percentage nitrate uptake

Area	Dates	Percentage nitrate uptake	References
NE Pacific coast Juneau-Cape Spencer	Nov. 1964	27.6	Dugdale and Goering 1967
NE Pacific coast Seattle-Juneau	Sept. 1964	20.3	Dugdale and Goering 1967
North Pacific Vancouver-Honolulu	Feb. 1965	8.7	Dugdale and Goering 1967
Central Pacific at 155°W 50°N	Aug. 1969	28–50	Hattori and Wada 1972
40°N	Sept. 1969	0–44	
10°N	Sept. 1969	0	
West Pacific			T. Miyazaki, E. Wada, and A. Hattori (unpublished)
44°N 154°E	July 1971	40–50	
28°30′N 145°E	June 1971	0–14	
Bering Strait	Summer	77–89	McRoy et al. 1972
Aleutian Islands	Summer	13–40	McRoy et al. 1972
Eastern Bering Sea	Winter	22–91	McRoy et al. 1972

TABLE 6.4 Summary of data on oxidation and reduction of inorganic nitrogen to nitrate in shallow layers

Area	Date	Depth (m)	Potential activity (ng-atoms N/liter-hr) measured by ^{15}N* (or by colorimetric method‡)	
			NH_4^+ oxidation	NO_3^- reduction
Central Pacific	Sept. 1969	0	0‡	2.4‡
40°N, 155°W		10	0‡	2.0‡
		30	0‡	0.14‡
		50	1.34‡	0‡
		80	3.3‡	0‡
10°N, 155°W	Sept. 1969	0	0.4‡	4.0‡
		30	0.2‡	2.3‡
		45	3.0‡	2.9‡
		75	1.9‡	1.0‡
West Pacific	July 1971	10	1.2* (1.4)‡	5.1*(0.7)‡
44°N, 154°E		60	1.2* (2.8)‡	
28°30'N	June 1971	20	0.21*(0.44)‡	1.4*(1.0)‡
145°E		125	0.30*(0.52)‡	1.4*(1.1)‡
Sagami Bay	June 1970	0–50	0.2–5.0	1.1–3.0*
	Oct. 1970	5–90	0.7–2.2*	0.3–1.1*

between ammonia and nitrate (Rakestraw 1936; Hattori and Wada 1973). Localized zones rich in nitrite are considered the most active sites for biological oxidation-reduction reactions of nitrogen. In some locations the reaction rates are even so high that they can be estimated by following directly the change with time in nitrite concentrations in water samples (Table 6.4). The capacity of the surface waters in tropical and subtropical regions for nitrite production is several orders of magnitude lower than that for assimilation. In the boreal areas indicated, where approximately 10 μg-atoms N/liter or more of nitrate is present year round, the reduction of nitrate and nitrite are appreciable. This is also the case with coastal waters relatively rich in nutrients.

Application of the ^{15}N tracer technique in combination with other chemical methods allows us further to distinguish the respective contributions of ammonia oxidation and nitrate reduction to nitrite production (Table 6.4). In surface waters shallower than 100 m in Sagami Bay, ^{15}N was converted to nitrite at a rate comparable to that from either ^{15}N-ammonia or ^{15}N-nitrate. Since the rate of ^{15}N conversion is accelerated by light, the phytoplankton activity must be considered a contributing factor in addition to bacterial nitrification and nitrate reduction. The capacities for ammonia oxidation are consistent with nitrite distribution (Fig. 6.1).

The relative contribution of nitrate reduction to nitrite production is large in northern areas and less in southern areas, as is the case with nitrate assimilation.

The occurrence of denitrification, or N_2 production from nitrate, in layers of the secondary nitrite maximum was suggested earlier by Brandhorst (1959). Estimates of the denitrification rate are tabulated in Table 6.5. Denitrification is not commonly found, however, except in extremely oxygen-depleted zones

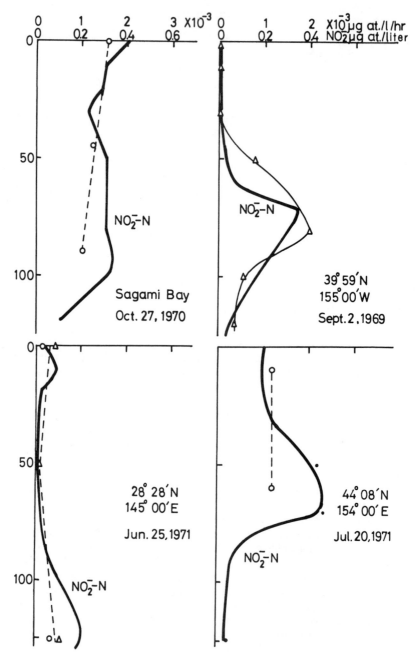

Fig. 6.1 Vertical profiles of nitrite and capacity for oxidation of ammonia in the central and western Pacific Ocean and Sagami Bay, Japan. (Open circles indicate capacities for ammonia oxidation measured by ^{15}N technique; triangles indicate measurement by colorimetric method).

TABLE 6.5   Summary of denitrification and nitrite production in deep waters (ug-atoms
N/liter-day)

Area	Denitrification	Nitrite production	References	
Darwin Bay,	1.0		Goering and Dugdale	1966
Galapagos		0.2	Richards and Broenkow	1971
Eastern Pacific				
off Mexico	18		Goering	1968
Central Pacific	0.002	0.015	Wada and Hattori	1971
Sagami Bay		0.024–0.141	Wada and Hattori	1971

as are common in the eastern tropical Pacific. Table 6.5 also includes data on
nitrate reduction, or nitrite production from nitrate at depth.

Distribution and metabolism of nitrate

Figure 6.2, based on the data described above, illustrates the variations of
nitrate concentrations in the surface layer, capacities for nitrate and ammonia
assimilation, primary production (Koblenz-Mishke 1965) and the percentage
of nitrate uptake as a function of latitude. A close correlation is seen between
the capacity for nitrate assimilation and primary production but not between
the capacity for ammonia assimilation and primary production. The high
capacity of nitrate utilization by the phytoplankton population is observed
consistently in nitrate-rich waters (Fig. 6.2D). Extensive literature reports
evidence that the nitrate-reducing enzyme can be induced by the presence of
nitrate (Kessler 1964). This may serve to explain regional variations in the
percentage of nitrate uptake. Unfortunately, however, no extensive investigation has been carried out with representative species of marine phytoplankton in this respect.

The in situ capacity for ammonia oxidation in shallow waters of the central
North Pacific is estimated to be about 10^{-5} μg-atoms N/liter-hr (Wada and
Hattori 1971), an order of magnitude lower than that for nitrate assimilation.
Besides its assimilation by phytoplankton, nitrate is also consumed by heterotrophic bacteria; therefore, ammonia oxidation must be extensive to maintain
a dynamic equilibrium with respect to nitrate. This requirement is probably
fulfilled by a ubiquitous occurrence of such activity in the massive volume of
deep water below the euphotic zone.

The estimation of the nitrate and ammonia assimilation capacities in the
Bering Sea is restricted almost to coastal areas. Certain aspects of nitrogen
metabolism can be applied to this area, however, by using the general rule
found over extended areas of the North Pacific Ocean. Between 40 and 50
percent nitrate uptake is estimated in offshore waters of the Bering Sea (Fig.
6.2D).

Nitrogen budget in the epicontinental Eastern Bering Sea

The deep Kamchatka and Aleutian basins of the Bering Sea are connected to
the North Pacific Ocean through numerous deep passes. The overlying water
has the oceanographic character of the North Pacific and can be distinguished

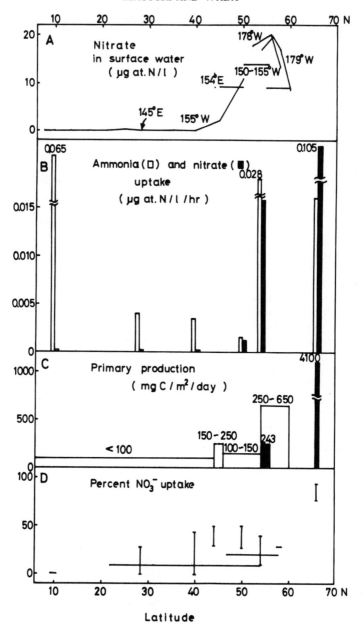

Fig. 6.2 Geographical variations of nitrate concentrations, *in situ* capacities for nitrate and ammonia assimilation, primary production and percentage nitrate uptake. Data sources: Faculty of Fisheries, Hokkaido Univ. 1968; Marumo 1970; Hattori and Wada 1972; Anderson et al. 1969; Koblentz-Mishke 1965; McRoy et al. 1972; and T. Miyazaki, E. Wada, and A. Hattori, (unpublished).

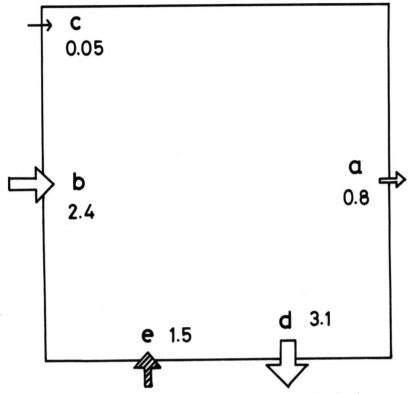

Fig. 6.3 Nitrogen budget in the epicontinental Eastern Bering Sea in summer (Model 1).

Northward water transport through Bering Strait in summer:

$$1.4 \times 10^6 \text{ m}^3/\text{sec or } 1.2 \times 10^{11} \text{ m}^3/\text{day}$$

Nitrate output (a):

$$7 \text{ } \mu\text{g-atoms N/liter} \times 1.2 \times 10^{11} \text{ m}^3/\text{day}$$
$$= 0.8 \times 10^9 \text{ g-atoms N/day}$$

Nitrate input:

Advection (b)

$$20 \text{ } \mu\text{g-atoms N/liter} \times 1.2 \times 10^{11} \text{ m}^3/\text{day}$$
$$= 2.4 \times 10^9 \text{ g-atoms N/day}$$

River runoff (c) 5×10^7 g-atoms N/day

Difference (input − output): 1.6×10^9 g-atoms N/day

Primary production: 300 mg C/m²/day
or 0.0031 g-atoms N/m²-day

Sea area: 10×10^{12} m²

Removal of dissolved inorganic nitrogen by primary production (d):

$$0.0031 \text{ g-atoms N/m}^2\text{-day} \times 1.0 \times 10^{12} \text{ m}^2$$
$$= 3.1 \times 10^9 \text{ g-atoms N/day}$$

Nitrate uptake: 51 percent

Nitrogen recycling (e): 1.5×10^9 g-atoms N/day

from the water over the epicontinental areas of the northeastern Bering Sea. The epicontinental Eastern Bering Sea covers about 1×10^6 km², or 44 percent of the total Bering Sea area, and is noted as one of the most productive areas in the world. Discussion is therefore focused on this area, and two simplistic models are considered.

The quantitative aspect of water exchange between the Bering Sea and the

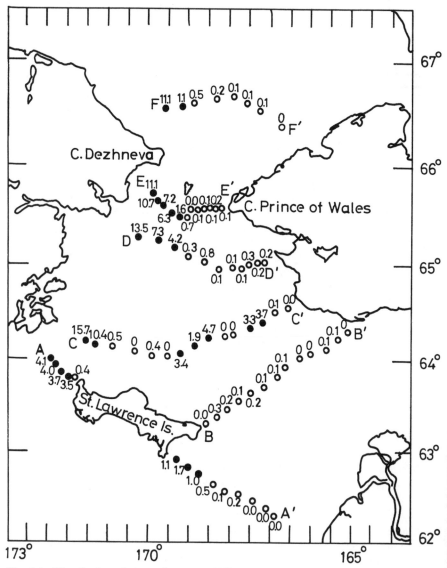

Fig. 6.4 Distribution of nitrate (μg-atoms N/liter) in surface layers of the northern Bering Sea and Bering Strait in summer 1970 (Husby and Hufford 1971).

North Pacific is not well known, but there are several estimates regarding flow to the Chukchi Sea through Bering Strait as summarized in Figure 6.3. A value of 1.4 Sv has been computed for the volume transport in summer (Coachman and Aagaard 1966). The amount of nitrate removed daily by the northbound water is calculated then to be 0.8×10^9 g-atoms N.

Nitrate is added by inflow to this area from sources in the outer ocean areas; the deep regions of the Bering Sea, Aleutian Stream, or both; and, river runoff. The daily addition in summer from outer seas and rivers is estimated to be 2.4×10^9 and 5×10^7 g-atoms N/day, respectively, assuming their nitrate content to be 20 μg-atoms N/liter. The net daily supply of nitrate thus amounts to 1.6×10^9 g-atoms N.

Mean daily primary production in the Aleutian Islands area is 243 mg C/m^2 (McRoy et al. 1972), and that in the Bering Sea Gyre region is 340 mg C/m^2 (Taniguchi 1969). A value of 300 mg C/m^2-day is assigned as the rate of primary production. Since C and N are taken up in a ratio close to 7:1 in the Bering Sea in summer (Table 6.3), the daily rate of N consumption is 3.1 mg-atoms N/m^2-day. From the sea area 3.1×10^9 g-atoms N of inorganic combined nitrogen is withdrawn daily by the phytoplankton. There is a deficit of 1.5×10^9 g-atoms N/day, which is probably furnished by the recycling process through ammonia. The nitrate uptake computed from these figures is 51 percent, a value consistent with observations made.

From hydrographic observations (Husby and Hufford 1971), at least two types of water masses are distinguished in the northern Bering Sea and Bering Strait; one is a warm low-salinity, nitrate-poor water mass found on the eastern side, and the other a cold, saline, nitrate-rich water mass on the western side (Fig. 6.4). Because of these marked differences in water characteristics a second model was constructed in which the epicontinental Eastern Bering Sea is divided into two portions which do not substantially interact (Fig. 6.5). It can be further postulated that the two masses with equal surface area are fed by water from different sources: the eastern portion is sustained by the Aleutian Stream and the western portion from the deep Bering Sea. Water flows out at an equal rate from both portions. Considering the high primary productivity (about 1 g C/m^2-day) observed on the western side of Bering Strait (McRoy et al. 1972) a value of 600 mg C/m^2-day is assigned. The calculated nitrate uptake is 45 percent in the eastern portion and 42 percent in the western part. In this model the high productivity in the western portion is balanced by an increased contribution of deeper source water rich in nitrate. Again, the calculation is consistent with the observations. Both of the models, however, fail to provide a clue as to the *extent* of nitrate regeneration by ammonia oxidation in the Bering Sea nitrogen cycle.

Discussion

McROY: From the limited data available, this is a very good summary of the nitrogen budget for the eastern Bering Sea. There is some evidence now that Yukon Delta sediment may act as a sink for nitrate in the winter, but more observations are needed.

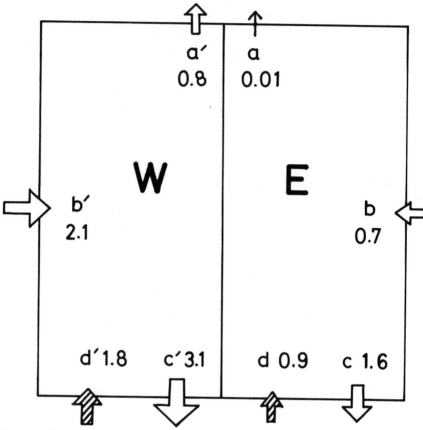

Fig. 6.5   Nitrogen budget in the epicontinental Eastern Bering Sea in summer (Model 2).

Western half

Northward transport: 6×10^{10} m³/day

Nitrate output (a'):

    13 μg-atoms N/liter $\times$ 6×10^{10} m³/day
            $= 0.8 \times 10^{9}$ g-atoms N/day

Nitrate input (b'):

    35 μg-atoms N/liter $\times$ 6×10^{10} m³/day
            $= 2.1 \times 10^{9}$ g-atoms N/day

Difference: 1.3×10^{9} g-atoms N/day

Loss by primary production (c'):

           3.1×10^{9} g-atoms N/day

Nitrate uptake: 42 percent

Nitrogen recycling (d'):

           1.8×10^{9} g-atoms N/day

Eastern half

Northward transport: 6×10^{10} m³/day

Nitrate output (a):

    0.2 μg-atoms N/liter $\times$ 6×10^{10} m³/day
            $= 0.1 \times 10^{8}$ g-atoms N/day

Nitrate input (b):

    12 μg-atoms N/liter $\times$ 6×10^{10} m³/day
            $= 7.2 \times 10^{8}$ g-atoms N/day

Difference: 7.1×10^{8} g-atoms N/day

Loss by primary production (c):

           1.6×10^{9} g-atoms N/day

Nitrate uptake: 45 percent

Nitrogen recycling (d):

           0.9×10^{9} g-atoms N/day

HOOD: What is the basis of your figure for nitrate advection in the Bering Sea?

HATTORI: The calculation was made by assuming that nitrate is supplied to the epicontinental Bering Sea by an inflow of nitrate-rich waters in the outer ocean that is balanced by outflow through the Bering Strait. Nitrate advection estimated in this way was compared to nitrate consumption by phytoplankton, calculated from primary production, using a 7:1 C/N assimilation ratio.

TAGUCHI: According to your calculation, there is a very large difference between the output-input of nitrate and the nitrate uptake by phytoplankton. Do you think this indicates that nutrient deficiency might be occurring in the Bering Sea?

HATTORI: I attributed this difference simply to nitrogen recycling.

REFERENCES

ANDERSON, G. C., T. R. PARSONS, and K. STEPHENS
  1969 Nitrate distribution in the subarctic northeast Pacific Ocean. *Deep-Sea Res.* 16: 329–334.

BRANDHORST, W.
  1959 Nitrification and denitrification in the eastern tropical North Pacific. *Conseil Int. Explor. Mer.* 25: 3–20.

COACHMAN, L. K., and K. AAGAARD
  1966 On the water exchange through Bering Strait. *Limnol. Oceanogr.* 11: 44–59.

DUGDALE, R. C., and J. J. GOERING
  1967 Uptake of new and regenerated forms of nitrogen in primary production. *Limnol. Oceanogr.* 12: 196–206.

FACULTY of FISHERIES, HOKKAIDO UNIVERSITY
  1968 Data record of oceanographic observations and exploratory fishing. *Bull. Fac. Fish. Hokkaido Univ.* 12: 291–420.

FLEMING, R. H.
  1940 Composition of plankton and units for reporting populations and production. *Proc., Sixth Pacific Sci. Congr.* (California 1939) 3: 535–540.

GOERING, J. J.
  1968 Denitrification in oxygen minimum layer of eastern tropical Pacific Ocean. *Deep Sea. Res.* 15: 157–169.

GOERING, J. J., and R. C. DUGDALE
  1966 Denitrification rates in an island bay in the equatorial Pacific Ocean. *Science* 154: 505–506.

HATTORI, A., and E. WADA

1972   Assimilation of inorganic nitrogen in the euphotic layer of the North Pacific Ocean. In *Biological oceanography in the northern North Pacific Ocean* [Motoda commemorative volume], edited by A. Y. Takenouti et al. Idemitsu Shoten, Tokyo, pp. 279–287.

1973   Biogeochemical cycle of inorganic nitrogen in marine environments with special reference to nitrite metabolism. In *Proceedings of symposium on hydrogeochemistry and biogeochemistry*, Vol. 2, edited by E. Ingerson. The Clarke Co., Wash., D.C., pp. 28–39.

HUSBY, D. M., and G. L. HUFFORD

1971   Oceanographic investigation of the northern Bering Sea and Bering Strait, 8–21 June 1969. U. S. Coast Guard Oceanographic Rep. 42, CG-373-42.

KESSLER, E.

1964   Nitrate assimilation by plants. *Ann. Rev. Plant Physiol.*, 15: 57–72.

KOBLENTZ-MISHKE, O. L.

1965   Primary production in the Pacific. *Oceanology* 5: 104–116.

MacISAAC, J. J., and R. C. DUGDALE

1969   The kinetics of nitrate and ammonia uptake by natural populations of marine phytoplankton. *Deep-Sea Res.* 16: 45–57.

MARUMO, R. (ed.).

1970   Preliminary report of the *Hakuho Maru* Cruise KH-69-4. Ocean Res. Inst., Univ. Tokyo, Nakano, Tokyo.

McROY, C. P., J. J. GOERING, and W. E. SHIELS

1972   Studies of primary production in the Eastern Bering Sea. In *Biological oceanography in the northern North Pacific Ocean* [Motoda commemorative volume], edited by A. Y. Takenouti et al. Idemitsu Shoten, Tokyo, pp. 199–216.

MIYAZAKI, T., E. WADA, and A. HATTORI

1973   Capacities of shallow waters of Sagami Bay for oxidation and reduction of inorganic nitrogen. *Deep-Sea Res.* 20: 571–577.

RAKESTRAW, N. W.

1936   The occurrence and significance of nitrite in the sea. *Biol. Bull. Mar. Biol. Lab., Woods Hole* 71: 131–167.

RICHARDS, F. A., and W. W. BROENKOW

1971   Chemical changes, including nitrate reduction, in Darwin Bay, Galapagos Archipelago, over a one month period, 1969. *Limnol. Oceanogr.* 16: 758–765.

TANIGUCHI, A.

1969   Regional variations of surface primary productivity in the Bering Sea in summer and the vertical stability of water affecting production. *Bull. Fac. Fish. Hokkaido Univ.* 20: 169–179.

WADA, E., and A. HATTORI

1971   Nitrite metabolism in the euphotic layer of the central North Pacific Ocean. *Limnol. Oceanogr.* 16: 766–772.

1972   Nitrite distribution and nitrate reduction in deep sea waters. *Deep-Sea Res.* 19: 123–132.

CHAPTER 7

Phosphate and oxygen problems with reference to water circulation in the subarctic Pacific region

YOSHIO SUGIURA

Geochemical Laboratory, Meteorological Research Institute, Tokyo, Japan

Abstract

Vertical mixing is suggested as a vital mechanism in maintaining high concentrations of conservative phosphate in the upper layer of the subarctic Pacific Ocean, particularly around the Aleutian Islands. The transformation of non-conservative (AOU-dependent) phosphate to the conservative form occurs when water is brought into the upper layer.

The horizontal distribution of conservative phosphate may be speculated from the current system and the location of intense vertical mixing. Also, the depth of the permanent pycnocline relative to the compensation depth appears to be an indicator for the amount of conservative phosphate in the intervening water layer.

Vertical distribution of dissolved oxygen in the subarctic Pacific is characterized by the presence of a thick (about 1000 m) deoxygenated zone beneath the permanent pycnocline. The appearance of the lower part of the deoxygenated zone is suggestive of limited local circulation with the more productive upper layer waters or perhaps intense mixing in the intermediate layer.

PHOSPHATE CLASSIFICATION

The relationship between phosphate and oxygen in seawater was studied from the western side of the Bering Sea (54°N) to the Kuroshio region (30°N) (Sugiura 1964). Figure 7.1 shows the relationship observed between phosphate and AOU (apparent oxygen utilization) in water samples classified by sigma-t

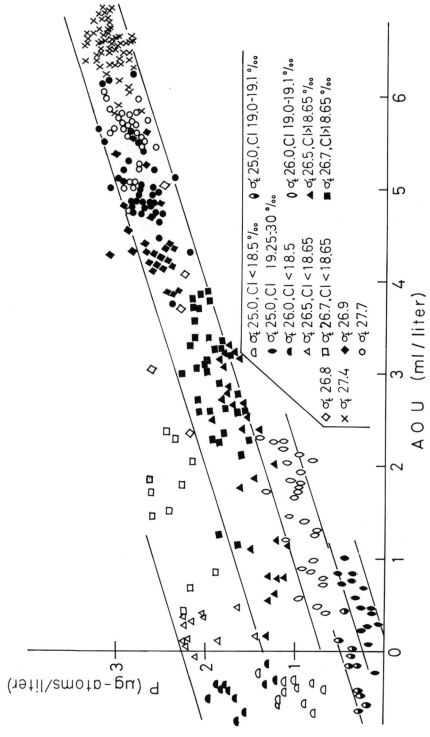

Fig. 7.1 Chlorinity and density relationships between phosphate and AOU (Apparent Oxygen Utilization) in the Kuroshio, Oyashio, and western Bering Sea waters (Sugiura and Yoshimura 1964).

and chlorinity into 14 groups. In Figure 7.1 each group of plotted points forms a straight interparallel band at a slope of about 1/270 based on the atomic ratio of phosphate to AOU. This result suggests that the phosphate concentration consists of two parts, one an AOU-dependent part and the other dependent on water temperature and chlorinity. The former is designated *non-conservative* phosphate (NC-P) and the latter form *conservative* phosphate (C-P). The phosphate concentration (P) can be expressed as follows:

$$P = NC\text{-}P + C\text{-}P$$

and $$NC\text{-}P \ (\mu g\text{-atoms/liter}) = (1/3) \ AOU \ (ml/liter)$$

because 270 μg-atoms of oxygen is approximately equivalent to 3 ml of oxygen at NTP.

Horizontal distribution of conservative phosphate

The distribution of conservative phosphate in the upper layer of the subarctic Pacific region (Fig. 7.2) was clarified (Sugiura 1965) by noting a distinct geographical pattern: concentrations higher than 1.8 μg-atoms/liter are confined north of approximately 47°N, decreasing somewhat along the Alaskan Stream and off the American continent.

The mechanism by which the highest concentration is maintained in the upper layer was then explored. Vertical mixing appeared to be the transport agent bringing phosphate from the lower layer and accounting for its transformation into the conservative type. The sea around the central Aleutian Islands is recognized as the source of phosphate supply. Based on the horizontal distribution of a degree of saturation of dissolved oxygen at the bottom of the upper layer (Fig. 7.3), vertical mixing between surface and subsurface waters is particularly active around the central Aleutian Islands, where a portion of phosphate-rich water is brought to the surface. In the subsurface waters where oxygen is undersaturated, phosphate occurs both in the conservative and nonconservative forms. Once subsurface water has been brought to the surface, its oxygen content is enhanced by the atmospheric contribution without change in phosphate concentration. This increase of oxygen content increases the conservative phosphate concentration and decreases the non-conservative phosphate concentration when the phosphate concentration is kept constant. As a net result, the conservative phosphate concentration increases in the surface waters.

The general pattern of the current system in the subarctic Pacific region (Dodimead et al. 1963) is presented in Figure 7.4. From this it is speculated that conservative phosphate-rich water entering the sea near the central Aleutian Islands circulates accordingly and obtains a horizontal distribution as shown in Figure 7.2. The decrease of conservative phosphate concentration off the American continent and along the Alaskan Stream is ascribed to dilution by water from the south that is low in conservative phosphate.

There is possibly another explanation for phosphate distribution. In general, two kinds of pycnoclines are known—seasonal and permanent. The seasonal pycnocline is produced by generation of the surface layer due to thaw dilution and solar heating of the water. The water which lies between the

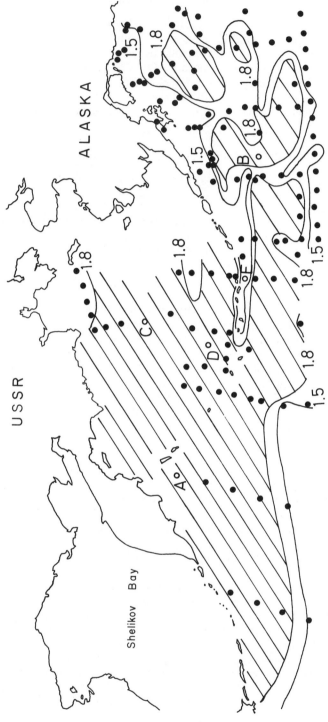

Fig. 7.2 Distribution of conservative phosphate in the upper layer of the Subarctic Pacific Region (Sugiura 1965).

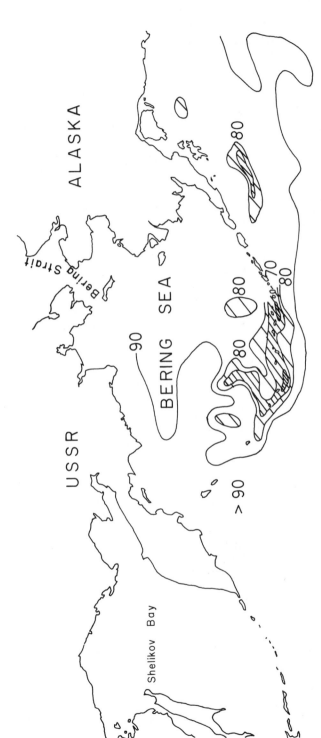

Fig. 7.3 Distribution of the degree of saturation of dissolved oxygen at the bottom of the upper water layer (Sugiura 1965).

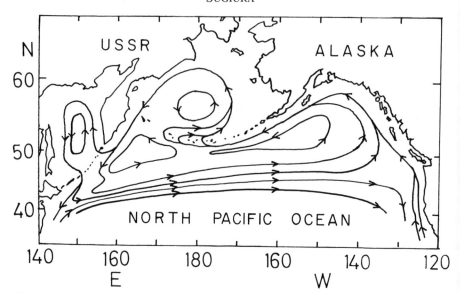

Fig. 7.4 General pattern of current system in upper layer of the subarctic Pacific waters (Dodimead et al. 1963).

seasonal and permanent pycnoclines is the residue left in place after winter convection. The highest seasonal variation of nutrient concentration occurs in winter, attributed to the predominance of oxidative decomposition taking place during this season. The upper layer above the depth of the winter convection cell abounds with oxygen from the atmosphere and with nutrient material from the decomposition of organic material. If regenerated nutrients enter the euphotic zone, they are re-utilized by phytoplankton; in the case of the aphotic zone, however, the nutrients will remain in the water. Accordingly, when the depth of the permanent pycnocline exceeds the compensation depth, the water in the layer between the compensation depth and the depth of permanent pycnocline abounds with oxygen and nutrients. Under such conditions the conservative phosphate concentration likewise increases. Conversely, when the compensation depth exceeds the depth of the permanent pycnocline, the amount of conservative phosphate is reduced.

The thickness of the upper layer is then the determining factor for the content of conservative phosphate: the vertical distribution of salinity and phosphate (Fig. 7.5) of water samples collected at stations shown in Figure 7.2 reveals that the thickness of the upper layer is over 100 m in the case of stations in the area of conservative phosphate concentrations of 1.8 μg-atoms/liter or more, as shown at stations A, B, C, and D, but it is thin in the case of stations in the area of the lower conservative phosphate concentrations shown at stations E and F.

Vertical distribution of dissolved oxygen
The characteristic features of the vertical distribution of dissolved oxygen in the subarctic Pacific region is described in Figure 7.6 (1969). Of note is the

existence of a thick (about 1000 m) zone of oxygen-poor (about 0.5 ml/liter) water, which is here designated the *deoxygenated zone* starting beneath the permanent pycnocline. This characteristic pattern is known to appear also in several other regions, including the sea off Central America and Peru in the Pacific and in the Arabian Sea and Bay of Bengal in the Indian Ocean. These areas are characterized by high primary productivity and a well-defined permanent pycnocline.

The upper layer in the Antarctic circumpolar region is recognized for its abundance of nutrients and high productivity, but a deoxygenated zone cannot be detected. On the contrary, an oxygen-rich condition prevails, because the Antarctic circumpolar water is heavily aerated by the intense convection due to cold weather.

A large amount of planktonic detritus derived from high productivity in the upper layer falls through the pycnocline under gravitational force, and oxygen is consumed in the process. The depletion of oxygen is offset somewhat by mixing in the oxygen-rich upper layer water, but the pycnocline prevents mixing between the upper and lower layers, and oxygen depletion persists below the pycnocline.

In the vertical distribution of organic carbon and nitrogen, a remarkable gradient can be seen above the depth of about 400 m (Menzel and Ryther 1970). This fact suggests that most oxidative decomposition of organic matter takes place above the depth of about 400 m; therefore, when the depth of the permanent pycnocline exceeds this depth, a clearly deoxygenated zone will not appear. In the case of the regions mentioned above, the permanent pycnocline is located at about 150 m, however, and the deoxygenated zone starts from a depth of about 200 m. This explains why dissolved oxygen is depleted around

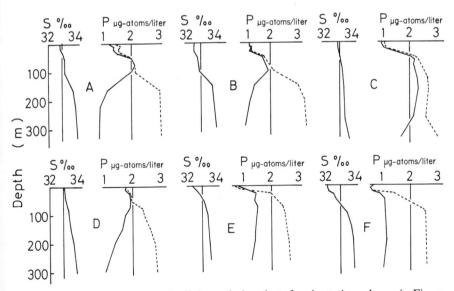

Fig. 7.5 Vertical distribution of salinity and phosphate for six stations shown in Figure 7.2. Station F (not shown) is located at 130°00′W, 51°30′N. (Sugiura 1965).

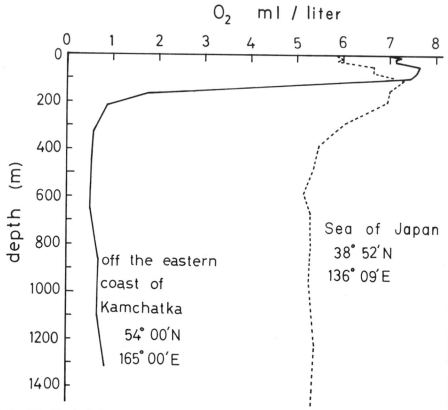

Fig. 7.6 Vertical distributions of dissolved oxygen in the Subarctic Pacific Region and Sea of Japan (Sugiura 1969).

the depth of 300 m, beneath the permanent pycnocline. Unanswered, however, is why the concentration of dissolved oxygen is still as low as 0.5 ml/liter even at a depth of 1000 m or more.

Considering the minute size and light density of planktonic detritus, a long time (to the order of 100 years) is thought to be required for such material to reach the bottom of the deoxygenated zone, during which process the lateral displacement of water may reach 10^4 to 10^5 km. In order for water in the intermediate layer to remain in a region of high productivity in the upper layer despite such a lengthy transport, the water must circulate within a certain limited domain. Intense mixing in the intermediate layer would be suggested as an answer to this problem if approached without knowledge of the local circulation.

The two established minima of dissolved oxygen concentrations in the Indian Ocean can be explained by the intrusion of oxygen-rich Antarctic water into the deoxygenated zone originating from the Arabian Sea or Bay of Bengal. As in this case, most vertical distribution patterns of dissolved oxygen might be explained by some modification to the deoxygenated zone by oxygenated water. The vertical distribution of dissolved oxygen in the Sea of Japan was compared (Fig. 7.6) with that in the subarctic Pacific region. Although a similarity can be seen between the two patterns, concentration levels below the pycnocline do differ, which seems to be due to the difference not only in the depth of the upper layer but also in the primary productivity between those two regions. The Sea of Japan has a closed basin with a sill-depth of about 200 m; the water below the sill-depth circulates counterclockwise as evidenced by the distribution of iron and aluminum (Sugiura and Yamamoto 1968). The case of the subarctic Pacific region appears to hold as well for the circulation of water in the Sea of Japan.

Discussion

PARK: I agree with Dr. Sugiura that physical processes, including air-sea dynamics, are vital to an understanding of the nutrient chemistry of the Bering Sea.

REFERENCES

DODIMEAD, A. J., F. FAVORITE, and T. HIRANO

    1963   Review of oceanography of the subarctic Pacific region. *Bull. Int. North Pac. Fish. Comm.* 13, 195 pp.

MENZEL, D. W., and J. H. RYTHER

    1970   Distribution and cycling of organic matter in the oceans. In *Organic matter in natural waters*, edited by D. W. Hood. Occas. Publ. No. 1, Inst. Mar. Sci., Univ. Alaska: Fairbanks, pp. 31–54.

SUGIURA, Y.

    1965   Distribution of reserved (preformed) phosphate in the subarctic Pacific region. *Papers Met. Geophys.* 15: 208–215.

    1969   On the oxygen-minimum layer in the oceans. *La Mer* 7: 161–167.

SUGIURA, Y., and K. YAMAMOTO

    1968   Distribution of iron and aluminum in seawater of the Japan Sea and its oceanographical significance. *La Mer* 6: 177–189.

SUGIURA, Y., and H. YOSHIMURA

    1964   Distribution and mutual relation of dissolved oxygen and phosphate in the Oyashio and northern Kuroshio regions. *J. Oceanogr. Soc. Japan* 20: 14–23.

CHAPTER 8

Dynamics of particulate material in the ocean

Part 1. Production and decomposition
of particulate organic carbon in the
northern North Pacific Ocean and Bering Sea

SATOSHI NISHIZAWA *and* SHIZUO TSUNOGAI

Faculty of Fisheries, Hokkaido University, Hakodate, Japan

Abstract

The mean concentration of particulate organic carbon in the upper 50-m layer of the Bering Sea ranges from 65 to 300 μg C/liter and is highest along the Aleutian island arc and the continental rise. This range is much higher than is typically found in southern areas, reflecting the high level of primary production in the Bering Sea. The concentration of carbon nearly always increases exponentially towards the sea surface, with a marked maximum usually at the air-sea interface; the highest single value reported at the sea surface was 530 μg C/liter. The phenomenal increase of particulate carbon towards the sea surface occurs concomitantly with a similarly distinct decrease in the concentration of chlorophyll, a situation generally common to most oceans and seas.

Minimum concentrations of particulate carbon in the range of 10 to 25 μg C/liter are found close to the center of the dichothermal layer. Often the lowest absolute concentration within the entire water column so far observed has occurred in this range. Below the dichothermal layer, the level of particulate organic material rapidly increases with depth and reaches an average value that is two to five times higher than in the dichothermal layer. The concentration in deeper layers is remarkably variable with depth, however, and is characterized by marked maxima and minima.

Recent extensive examinations of data collected from various areas in the North Pacific and adjacent seas revealed that the average particle concentration in deeper layers is closely correlated to the average particle concentration in the upper 50-m layer. This seems to imply that a major fraction of particulate material in deeper layers is derived directly from the surface layer of that area and the time rate of material transfer from the surface layer down to deeper layers is much quicker than previously estimated from simple sinking experiments.

There is a good possibility that insoluble radioactive nuclides such as ^{210}Pb, ^{210}Po and ^{234}Th can be used as tracers of particulate matter; laboratory experiments and observations at sea show that these nuclides do not enter a pure solid phase but move together with seston. Calcium carbonate tests are not expected to dissolve in the surface water, and thus high ratios of calcium to total particulate matter were found in the 100- to 150-m layer of the Bering Sea.

Discussion

CAMERON: I commend Dr. Nishizawa on his very detailed presentation. I would like only to remark that there are many properties of the ocean that vary exponentially with depth; it is encouraging to see an attempt to explain the vertical distribution of one of these properties on physical principles.

PARK: What depths do you consider for "deep" particulate organic carbon?

NISHIZAWA: The average concentration of POC is in the 200–500 m depth range.

PARK: Based on your calculations, how much time do you estimate it would take for a single particle of organic carbon to drop from the surface to the bottom of the typical ocean (5000 m)?

NISHIZAWA: That would have to be investigated, of course, but I would guess simply one week or less.

PARK: By Dr. Sugiura's method, this process might take a number of years.

NISHIZAWA: My reason for supposing relatively rapid transport of the particles to the deeper layer is based on our very good correlation between the surface and deep water content of POC which marks the seasonal variation.

PARK: Where does decomposition of POC take place?

NISHIZAWA: About two-thirds of the particulate matter is decomposed in the surface layer, and the remaining portion presumably decomposes as it drops to deeper water.

PARK: What is the constituency of particulate organic carbon?

NISHIZAWA: POC is generally comprised of a variety of living and non-living material, particularly including organic aggregates of special significance in marine ecology. Probably more than 50 percent of particulate carbon is organic in nature.

Dynamics of particulate material in the ocean

Part 2. Behavior of particulate material in the ocean studied by inorganic and radioactive tracers

S. Tsunogai, Y. Nozaki, M. Minagawa, *and* S. Yamamoto

Faculty of Fisheries, Hokkaido University, Hakodate, Japan

Abstract

The production and decomposition of particulate material in the ocean have been studied by inorganic and radioactive tracers. The high ratio of calcium in seston to total particulate matter was found in the upper 150-m layer in the Bering Sea. This layer is expected to contain actively decomposing particulate material, because calcium in the surface water is super-saturated with respect to calcium carbonate. Insoluble radioactive nuclide, lead-210 and thorium-234 in seawater were used as tracers or clocks of particulate material in the ocean. Laboratory experiments and observations at sea show that these nuclides do not attain their respective pure solid phases but probably aggregate with seston—both living and non-living organic matter, which comprises the largest part of particulate material in the ocean. The mean falling velocity of particulate material has been estimated to be 10^{-4} to 10^{-3} cm/sec in the surface water. The usual filtration method is apt to give unreliable data, because it does not consider the contribution of particles smaller than the pore size of the filters used.

INTRODUCTION

In this Symposium, Dr. Nishizawa has discussed the production and decomposition of seston in the ocean in terms of organic material in seawater. There are two additional approaches to seston evaluation. The first method, studying the change of dissolved chemical constituents related to biological activity, has been used by Tsunogai (1972). More particulate material is transported to the deep water of the northern North Pacific Ocean than in the tropical

175

Pacific. The second method is to introduce appropriate inorganic and radio-active tracers that aggregate with particulate organic matter. Reported here are some results obtained by the latter method.

Calcium was chosen as the inorganic tracer; a schematic diagram of the behavior of calcium responsible for biological activity is shown in Figure 8.1. The calcium concentration is not expected to increase in the surface water, where calcium is already supersaturated with respect to calcite or aragonite.

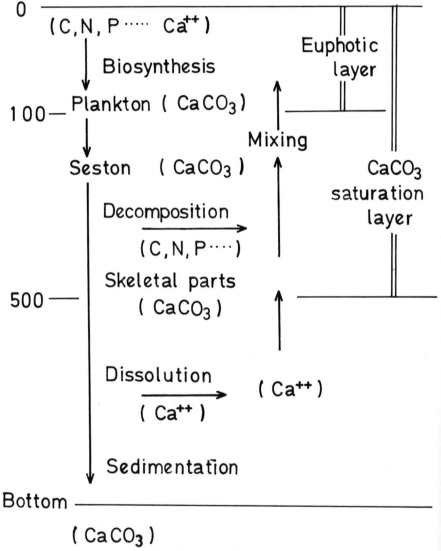

Fig. 8.1 Schematic diagram of biochemical cycle of calcium in the ocean.

Thus, calcium carbonate tests may persist in seston even when the decomposition of seston occurs in the surface water. Our objective was to detect the layer containing much particulate calcium.

As the radioactive tracers, we used thorium-234 and lead-210. Thorium-234 (half-life, 24 days) is a decay product (daughter) of uranium-238 in seawater. Although uranium is soluble in seawater by carbonate complexation, thorium is insoluble and forms a hydroxide in seawater. Lead-210 (half-life, 21.4 yrs) is produced from airborne radon-222 or radium-226 in seawater. According to Tsunogai and Nozaki (1971), lead behaves as an insoluble particle in seawater.

These nuclides exist in the ocean at first as atomic or very small particles and then may attach themselves to large particles (seston), thus rendering themselves useful as tracers or clocks to study the fate of particulate material in seawater.

Organic portion of particulate matter in seawater

The amount of particulate organic matter was determined by measuring ignition loss of particulate matter in seawater. Samples of particulate matter were collected in the Bering Sea during the cruise of the T/S *Oshoro Maru*, Hokkaido University, by Nakajima and Nishizawa. Locations of sampling stations are shown in Figure 8.2. The samples were treated according to the procedure shown in Figure 8.3. Particulate matter on the filter paper was washed with isotonic ammonium formate solution (8%) to remove sea salts, and then the ammonium formate was removed by drying at 60 C. Chemical constituents were determined by the atomic absorption method for sodium, magnesium, potassium and calcium, and by the colorimetric method of Iwasaki et al. (1956) for chloride. Results are given in Table 8.1.

TABLE 8.1   Dry weight, ash weight and calcium content of particulate material in the Bering Sea. Mean values with or without standard deviations are shown.

Depth m	No. of samples	Dry weight mg/liter	Ignition loss mg/liter	Ash/Dry %	Acid-soluble Ca μg/liter	Ca/Dry %o
		Stations in the Bering Sea				
0–100	40	0.69 ± 0.43	0.45 ± 0.25	33 ± 11	1.2 ± 1.0	1.8 ± 1.8
100–200	16	0.24 ± 0.08	0.16 ± 0.06	35 ± 11	2.5 ± 4.7	10 ± 19
200–500	17	0.26 ± 0.14	0.18 ± 0.13	33 ± 17	1.2 ± 0.8	5 ± 4
500–1500	24	0.23 ± 0.18	0.17 ± 0.14	31 ± 13	1.2 ± 1.0	5 ± 6
		Station, Os-2				
0–100	7	0.28	0.22	22	1.3	5
100–200	2	0.27	0.22	18	1.3	5
200–500	4	0.19	0.16	23	0.8	4
500–1500	4	0.39	0.29	23	1.4	4

Although the dry amount of particulate material varies widely from sample to sample, a fairly large mean value is found in the surface water of the Bering Sea. The highest value (1.7 mg/liter) was observed at 20- to 30-m depth at stations Os-5 and Os-6. This seems to be due to the high productivity

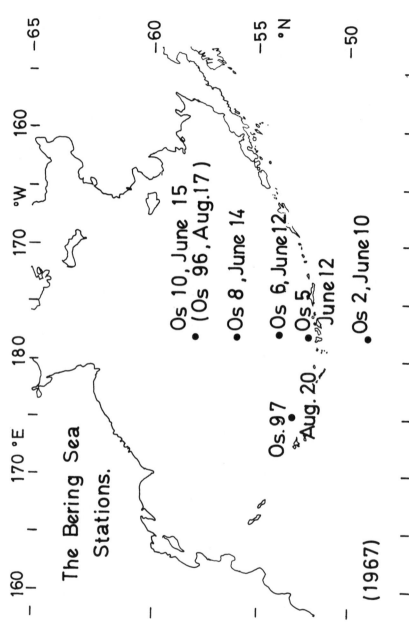

Fig. 8.2 Locations of sampling stations during cruise 24 of T/S *Oshoro Maru* of Hokkaido University.

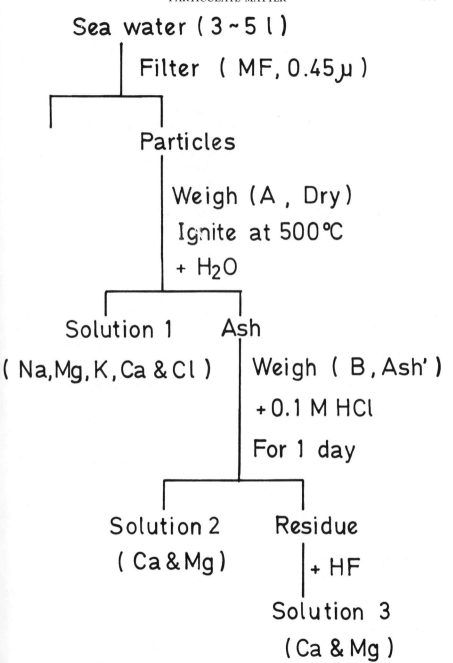

Fig. 8.3 Analytical procedure for filtered samples.

and the slow decomposition of organic matter in the water. The ash (inorganic) fraction was about one-third for samples from the Bering Sea and about one-fifth for samples outside the Aleutian Ridge in the northern North Pacific. Thus, the major part of particulate material in seawater is found to be seston. The larger ash content in the Bering Sea water may be caused by insoluble matter of terrestrial origin. The content of organic particulate matter in the water below 100 m indicated virtually no consistency.

Nakajima (1969) measured twice the amount of particulate organic matter as found in our own studies. He found the highest concentration value in the surface water, although the organic part comprised less than 50 percent of total particulate matter in the surface layer and consisted of 20 to 40 percent in the water below 100 m. A chief cause of the discrepancy between these results and ours seems to be due to accounting for smaller size particles that pass a 0.45-μ membrane filter. In Nakajima's determination of total particulate organic carbon, he used a 0.8-μ Whatman glass-fiber filter. It is necessary for the discussion of this problem to elucidate the size distribution of particulate material in seawater.

Size distribution of particulate material

Using a radioactive tracer, we studied the size distribution of particulate material. About 2×10^5 dpm (disintegrations per minute) of thorium-234 in 1 ml of 0.01 M hydrochloric acid solution were added to 1 liter of seawater. The seawater was stirred throughout the experiment by a magnetic stirrer, and occasionally 10-ml samples were drawn. (Small sample volume is desirable to avoid changes in the filtration condition). The radioactivity of particulate material was counted after samples were placed on filters of various pore sizes (Fig. 8.4). The solution did not attain an equilibrium state; with increase in time, the size of the radioactive particles increased due to the aggregation of the particulate material by stirring. Three days from the onset of the experiment, part of the particulate material was still visible. Although we could not reconcile the substantial difference in results between filtration procedures using 0.8-μ and 0.45-μ filters, a significant amount of particulate residue was found in the <0.45-μ pore size fraction. This observation suggests that the meaningfulness of our conventional definition for particulate material based on the matter retained on filter papers having pore sizes of either 0.45-μ or 0.8-μ is doubtful. Another problem is whether these results can be reproduced.

Reliability of the data obtained for particulate matter by the filtration method.

Certain factors apparently responsible for the scatter of data obtained by the filtration method include errors in measuring the amount of particulate matter, poor reproducibility of the filtration procedure, and the heterogeneity of particulate matter in seawater.

As shown in Figure 8.5, the vertical distribution of particulate matter was delineated. All samples were collected at the same station (28.5°N and 145.0°E) in the North Pacific during the same cruise (KH 71-3 or R/V *Hakuho Maru*). The seawater was sampled twice from 21 layers at the station, and all

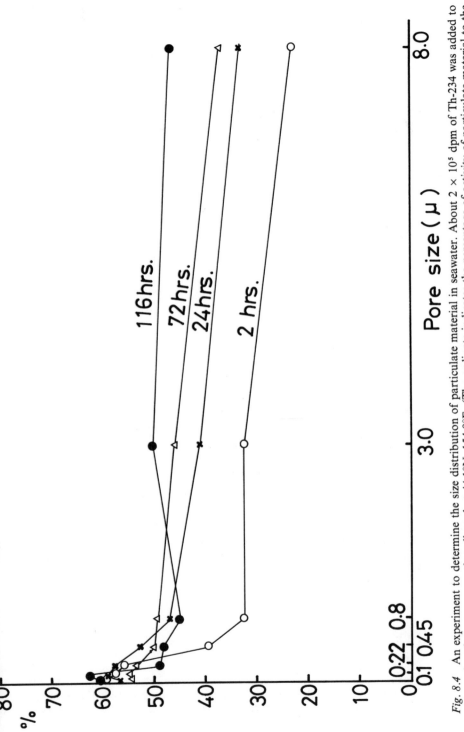

Fig. 8.4 An experiment to determine the size distribution of particulate material in seawater. About 2×10^5 dpm of Th-234 was added to 1 liter of surface-water sample collected at 44.1°N, 154.0°E. (The ordinate indicates the percentage of activity of particulate material to the total added activity; the abscissa indicates the Millipore filter size used).

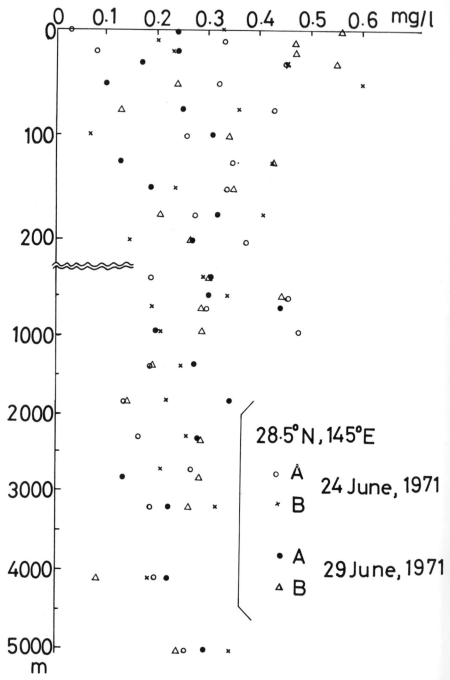

Fig. 8.5 Dry weight of particulate matter in seawater collected during cruise KH 71-3 of R/V *Hakuho Maru* of Ocean Research Institute, University of Tokyo. The samples were collected twice from 21 layers, and each sample was duplicately filtered with 0.45-μ HA Millipore filter.

duplicate samples were filtered with a 0.45-μ (HA) Millipore filter. The water was filtered in 10-liter volumes, or the amount filterable in two hours. In this manner we obtained four sets of data from the station. The results showed a wide variation from sample to sample and no definite tendency is shown.

There is no systematic error found among four sets of data, and the random error at the measurement of weight is much smaller than the range of the variation. Since large zooplankton were excluded from the filtered samples, the heterogeneous distribution of particulate matter is expected to be insufficient to cause the large variation. The variation must therefore be due to the filtration procedure. It is necessary to improve the filtration method or to develop a new technique for the study of particulate matter.

Calcium in seston

The calcium content was about an order of magnitude smaller than that found by Wangersky and Gordon (1965), who measured the amount of carbonate with an infrared carbon dioxide analyzer. Our results for calcium in the minor ash fraction seem to be reliable. The ratios of acid-soluble calcium to the dry amount of seston are shown in Table 8.1 and Figure 8.6. The effect of variation due to the filtration process tends to be cancelled out by using the ratio. The high ratios are found at depths near 158 m (station Os-10) and 150 m (stations

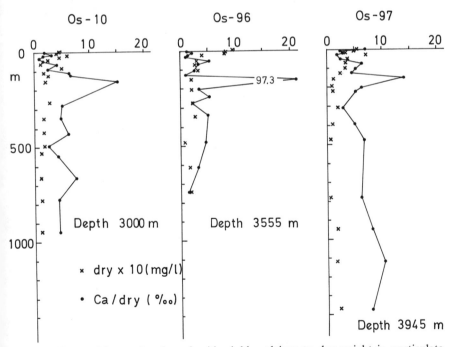

Fig. 8.6 Dry weights and ratios of acid-soluble calcium to dry weight in particulate matter. The samples were collected in the Bering Sea at stations Os-10 (58°03′N, 178°04′W) on 15 June 1971, Os-96 (58°00′N, 178°00′W) on 17 August, and Os-97 (53°08′N, 175°20′E) on 20 August.

Os-96 and 97). The peaks are probably related to the decomposition process of particulate organic matter, but we cannot presently estimate the rate because of lack of knowledge on the mixing rate of water.

Falling velocity of particulate matter

Thorium-234 and lead were used as clocks of particulate matter. Their falling velocities were determined as described below:

A particle bearing a radioactive nuclide moves in seawater together with the water and sinks gravitationally. The situation is shown schematically in Figure 8.7. For simplicity, it can be assumed that on the vertical water movement the vertical diffusion and advection constants do not vary with depth, and we can neglect the horizontal gradient of the concentration of the nuclide. The change of the concentration of the nuclide (C) with time (t) is given as follows:

$$\frac{\partial C}{\partial t} = D \frac{\partial^2 C}{\partial Z^2} - A \frac{\partial C}{\partial Z} + R - \lambda C \tag{1}$$

where D = vertical eddy diffusion constant; A = falling velocity, including the vertical advection velocity; R = production rate of the nuclide from its progenitor; λ = decay constant of the nuclide; and Z = depth.

The production rate R and the boundary condition are peculiar for each nuclide. The equation (1) is then solved separately for thorium-234 and lead-210.

a) Thorium-234

Since the concentration of uranium in seawater is fairly constant, because it forms a soluble complex with carbonate, the production rate of thorium-234 can be taken as a constant. By assuming a steady state, equation (1) is solved as follows:

$$C = \left(C_o - \frac{R}{\lambda}\right) \exp\left(\frac{A \pm \sqrt{A^2 + 4\lambda D_z}}{2D}\right) + \frac{R}{\lambda} \tag{2}$$

where C_o is the concentration of thorium at the surface.

In equation (2), we have used the boundary conditions that the thorium-234 concentration converges to a constant ($C_\infty = R/\lambda$) when the depth approaches infinity. Since R is identical to the activity of uranium-238, we used the activity ratio of thorium-234 to uranium-238, $X = \lambda C/R$. By substituting C and R with X, we obtain:

$$(1 - X) = (1 - X_o) \exp\left(\frac{A - \sqrt{A^2 + 4\lambda D_z}}{2D}\right) \tag{2'}$$

where X_o is the ratio at the surface.

When the upward flux of thorium-234 is balanced by the downward flux at the surface, we then obtain:

$$D\left(\frac{\partial C}{\partial Z}\right)_{Z=0} - AC_o = 0 \tag{3}$$

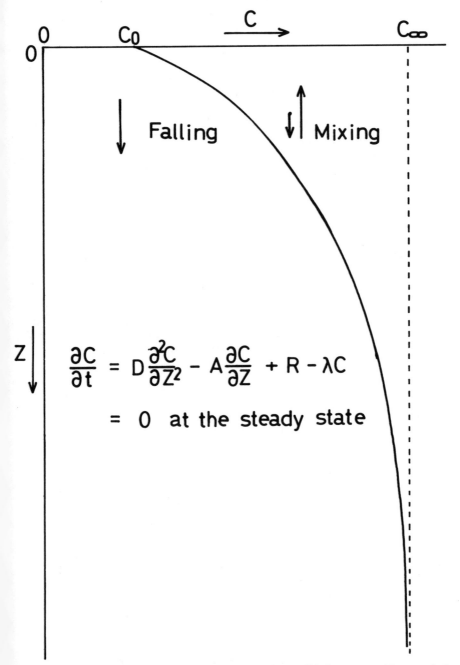

Fig. 8.7 Schematic diagram of an insoluble radioactive nuclide in seawater. D = vertical eddy diffusion constant; A = falling velocity, including vertical advection velocity; and R and λ = production rate and decay constant of the nuclide, respectively.

The above relationship can be expressed by using the activity ratio X as follows:

$$D = \frac{X_o}{\lambda}\left(\frac{A}{1 - X_o}\right)^2 \tag{3'}$$

By introducing the relationship (3)' into solution (2)', we obtain:

$$\ln(1 - X) = \ln(1 - X_o) - \frac{\lambda(1 - X_o)}{A} Z \tag{4}$$

$$= a - bZ \tag{4'}$$

where a and b are constants.

Thus, a logarithmic plot of $(1 - X)$ vs. depths must give a straight line, and the eddy diffusion constant and falling velocity of particulate matter are obtained by calculation using its tangent and its intercept at $Z = 0$. An example of our method is shown in Figure 8.8, where samples were collected at 42°02′N and 141°18′E. We used 1.72×10^{-7} as the ratio of uranium content and chlorinity (Miyake et al. 1970). Our result in computing the falling velocity of particulate matter is 1.1×10^{-3} cm/sec. Nearly the same values for water samples from various parts of the Pacific Ocean were obtained by Y. Miyake, Y. Sugimura, and E. Matsumoto (unpublished, 1970).

b) Lead-210

The production rate of lead-210 from its progenitor, radon-222 or radium-226, is not constant with depth. We assume the following relationship for the production rate on the basis of results obtained by Broecker et al. (1967):

$$R = R_1 + R_2 \exp(-\mu Z) \tag{5}$$

$$= 6.2 \times 10^{-6} - 5.5 \times 10^{-6} \exp(-5.6 \times 10^{-6} Z) \tag{5'}$$

where the units of R and Z are atoms/cm³-sec and -cm, respectively. The solution is obtained analogously to equation (2) as follows:

$$C = C_o \exp(-\sigma Z) + \frac{R_1}{\lambda}[1 - \exp(-\sigma Z)]$$

$$+ \frac{R_2}{D\mu^2 + A - \lambda}[\exp(-\mu Z) - \exp(-\sigma Z)] \tag{6}$$

where

$$\sigma = \frac{A}{2D}\left(\sqrt{1 + \frac{4\lambda D}{A^2}} - 1\right)$$

The boundary condition at the surface is written as follows:

$$I = -D\left(\frac{\partial C}{\partial Z}\right)_{Z=0} + AC_o$$

$$= AC_o + D\sigma\left(C_o - \frac{R_1}{\lambda}\right) + \frac{DR_2(\mu - \sigma)}{D\mu^2 + A\mu - \lambda} \tag{7}$$

where I is the input rate of lead-210 (atoms/cm²-sec) from the atmosphere as estimated by Lambert and Nezami (1965).

Log(1 - X)

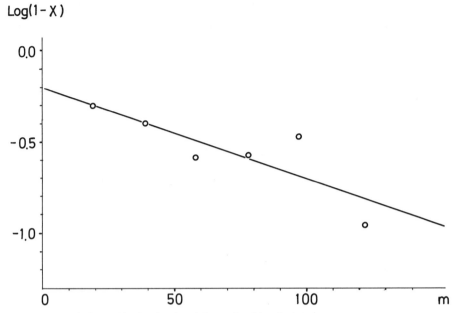

Fig. 8.8 Variation with depth of activity ratio (X) of Th-234 to U-238. Samples were collected off Usujiri at 42°02′N, 141°18′E, on 10 November 1971.

When the vertical distribution of lead-210 has been determined, it is possible to estimate the falling velocity and the eddy diffusion constant by solving two simultaneous equations (3 and 4). Since the half-life of lead-210 is 21.4 years, the axis of the water depth should be extended to the deep water. Consequently, the assumption of the vertically constant diffusion is not justified. Fortunately, however, the variation of the eddy diffusion constant is not sensitive to the solution of falling velocity. This implies that we can obtain the falling velocity of particulate matter from the concentration of lead-210 alone in the surface water by assuming some range of the vertical eddy diffusion constant. Tsunogai and Nozaki (1971) estimated the falling velocity of particulate matter labeled with lead-210 to be $(1 - 5) \times 10^{-4}$ cm/sec in the Pacific Ocean.

The falling velocities obtained by the thorium-234 method are several times larger than those estimated by the lead-210 method. One of the reasons for this discrepancy is due apparently to the decrease of velocity with increasing depth, because the falling velocity estimated by long-lived lead-210 is more influenced by the velocity in deep water below the thermocline than that by short-lived thorium-234. At any rate, the falling velocities estimated by radio-active nuclides are smaller than the values estimated by visible organic par-ticulates (Kajihara 1971). Thus, we must consider the effect of a large number of small-size particles on the transport process of carrying particulate matter from the surface to the deep areas. The falling velocity of lead is an order of magnitude larger than 1×10^{-5} cm/sec (mean falling velocity of lead deduced from its residence time in the ocean). This suggests that the fraction of larger particulate material decreases rapidly as decomposition with depth increases, and the falling velocity gradually decreases with depth.

Although fundamental research is still necessary to apply these methods to the actual ocean, inorganic and radioactive tracers can be used for the preliminary investigation of certain problems in the Bering Sea.

Acknowledgment

We are deeply appreciative to Professor M. Nishimura for valuable discussions, and we are grateful also to Professor P. K. Park for the manuscript review.

REFERENCES

BROECKER, W. S., Y. H. LI, and J. CROMWELL

1967   Radium-226 and radon-222: concentration in Atlantic and Pacific oceans. *Science* 158: 1307–1310.

IWASAKI, I., S. UTSUMI, K. HAGINO, and T. OZAWA

1956   A new spectrophotometric method for the determination of small amounts of chlorine using the mercuric thiocyanate method. *Bull. Chem. Soc. Japan* 29: 860–864.

KAJIHARA, M.

1971   Settling velocity and porosity of large suspended particle. *J. Oceanogr. Soc. Japan* 27: 158–162.

LAMBERT, G., and M. NEZAMI

1965   Importance des retombées sèches dans le bilan du plomb 210. *Ann. Geophysique* 21: 245–251.

MIYAKE, Y., Y. SUGIMURA, and M. MAEDA

1970   The uranium content and the activity ratio $^{234}U/^{238}U$ in marine organisms and seawater in the western North Pacific. *J. Oceanogr. Soc. Japan* 26: 123–129.

NAKAJIMA, K.

1969   Suspended particulate matter in the waters on the both sides of the Aleutian Ridge. *J. Oceanogr. Soc. Japan* 25: 239–248.

TSUNOGAI, S.

1972   An estimate of the rate of decomposition of organic matter in the deep water of the Pacific Ocean. In *Biological oceanography of the northern North Pacific Ocean* [Motoda commemorative volume], edited by A. Y. Takenouti et al. Idemitsu Shoten, Tokyo, pp. 517–533.

TSUNOGAI, S., and Y. NOZAKI

    1971   Lead-210 and polonium-210 in the surface water of the Pacific. *Geochem J.* 5: 165–173.

WANGERSKY, P. J., and D. C. GORDON, JR.

    1965   Particulate carbonate, organic carbon and Mn^{++} in the open ocean. *Limnol. Oceanogr.* 10: 544–550.

CHAPTER 9

Chemistry of the Bering Sea: An overview

DONALD W. HOOD *and* WILLIAM S. REEBURGH

Institute of Marine Science, University of Alaska, Fairbanks, Alaska

The preceding papers on nitrogen, phosphorus, the carbon dioxide system, and particulate and organic carbon in the Bering Sea illustrate that much of the chemical interest the Bering Sea has attracted is due to its high biological productivity and wide ranges of climatic conditions. Although we are generally aware of which factors are important in maintaining this high productivity, we do not know *how* the factors interact seasonally. Seasonal variations can be applied to advantage in future approaches to study of the Bering Sea.

Chemical inputs

Since most of the sources of water and chemical substances added to the Bering Sea can be identified, it is appropriate to view the Bering Sea from the standpoint of chemical inputs. Each input consists of two factors—the amount or rate of supply of water and the concentration of any given chemical element or species it contains. Both factors must be known for a wide variety of sources. The chemical inputs considered here are rivers, coastal lagoons, circulation and transport of deep water to the surface in the Aleutian passes, and contributions from sediment-water mixing on the eastern shelf.

Rivers

Four major rivers enter the Bering Sea. The Anadyr and Kamchatka rivers enter the western Bering Sea from the Soviet Union, and the Yukon and Kuskokwim rivers enter on the extensive eastern shelf from Alaska. Data on the amounts and seasonal distribution of freshwater added by these rivers have been assembled and reviewed by Roden (1967). Estimates of Yukon River flow, based on the sum of data from inland gauging stations on tributaries, have been made for about 20 years and have resulted in some information on the major element composition of river water. A comprehensive study of the composition of river water and suspended sediments has been performed (R. Gibbs, unpublished).

Since these rivers are fed largely by snow and ice melt, their freshwater arrives essentially as one large seasonal contribution each year. In terms of mean annual discharge, the Yukon River, the largest entering the Bering Sea, is about the same size as the Columbia River. During its peak discharge in June and July, however, the size of the Yukon approaches that of the Mississippi River. The months of June, July, and August account for almost 60 percent of the Yukon River's average annual flow and 90 percent of its flow between May and October. The areal extent of the effect of the freshwater contribution of these rivers is largely unknown.

Two methods of delineating plumes of other rivers would be useful here. Measurements of P_{CO_2} have yielded high values in areas influenced by rivers and continental drainage. Since these measurements are usually made continuously while a ship is underway, they lend themselves well to delineating river plumes. The Columbia River has also been studied on the basis of specific alkalinity—the ratio of titration alkalinity to the chlorinity (Park 1966)—and employing radioisotopes introduced by the Hanford Atomic Works (Gross et al. 1965; Gross 1966).

The process of sea ice formation poses a possible complication to the use of specific alkalinity as a river plume tracer in the Bering Sea. The segregation of salts during the formation of sea ice has been heavily studied; yet, there appears to be little agreement as to exactly what happens during this complex process. Laktionov (1931) indicates that alkalinity is high in old sea ice. Assur (1958) gives data indicating that $CaCO_3 \cdot 6H_2O$ begins to form in the ice phase at a temperature just below the freezing point of sea ice. If carbonate precipitation occurs in the ice phase on a large scale, high specific alkalinity meltwater might result and be confused with river water.

Lagoons

A series of lagoons, separated from the open sea by barrier islands and spits, are found along the coast of the eastern Bering Sea. These lagoons exchange much of their volume daily with the Bering Sea through tidal action and contribute water that has been modified by residence. The coastal lagoons harbor dense stands of eelgrass, which attracts many species of marine organisms seeking food and shelter. Izembek Lagoon, on the Alaska Peninsula, has received particular study (Barsdate et al., chapter 28, this volume; McRoy 1966, 1970; McRoy and Barsdate 1970; McRoy et al. 1972). Results indicate that the lagoons contribute phosphorus and large amounts of organic detritus and that they remove copper and probably other heavy metals from the adjacent shelf waters. Unlike the rivers, the lagoons contribute water and detritus throughout the year.

Sediment-water interaction

Inputs from the sediments may be considered from two standpoints; reaction of the sediments themselves with the seawater and also reactions taking place within the rather restricted interstitial environment. Reactions of the first sort are relatively unimportant biologically and usually involve release of silica from diatom frustules. The second type of reactions is usually biochemically mediated; due to restricted circulation in the interstitial water, these processes often lead to accumulation of large quantities of nutrients that are released by regeneration or decomposition reactions. No direct measurements of diffusion or transport of nutrients from sediments are available, but measurements of depth distributions of gases in sediments (Reeburgh 1969) indicate that in some environments as much as 30 cm of the sediments may be worked and mixed by physical processes or burrowing organisms. Although unimportant in the deep ocean, this interstitial water-nutrient reservoir could be quite important in the shallow eastern Bering Sea, and more knowledge should be acquired of the rates of mixing or interaction. Sediments in this area are generally sandy, so conventional coring and interstitial water sampling techniques are difficult to apply. In terms of a chemical input, diffusion or additions from the sediments would probably continue at about the same rate year-round.

Aleutian island passes
Estimates have been made of water transport through the Aleutian passes into the Bering Sea (Dodimead et al. 1963) and from the Bering Sea through the Bering Strait (Coachman and Aagaard 1966; Coachman and Tripp 1970; Favorite, Takenouti and Ohtani, Hughes et al.—chapters 1 to 3, this volume). Although these measurements are important in understanding oceanography as a whole, the only time that transport measurements influence chemical inputs as considered in the current context is when the water and nutrients are actually supplied to the euphotic zone. This is a form of upwelling, a condition which occurs in the Aleutian passes as a result of turbulent mixing due to converging currents and interaction with bottom topography. In recent years studies have been conducted aboard the R/V *Acona* (Institute of Marine Science, University of Alaska) to investigate the areal extent to upwelled water in Unimak and Samalga passes. The mixed water is easily identified by its low temperature and high concentrations of nitrate, phosphate, and P_{CO_2}. To assess turbulent tidal mixing as a chemical input, a means of estimating the vertical advective velocities is needed. J. J. Kelley and D. W. Hood (1973, unpublished) have suggested measurements which require determination of the evasion rate of CO_2 as a means of estimating the amount of water brought to the surface from depth. Mass transfer measurements of CO_2 have been made (Sugiura et al. 1963; Hood et al. 1963) using an inverted capsule of known area and volume to enclose a gas phase whose P_{CO_2} was continuously monitored. By removing CO_2 from the capsule or adding CO_2 to the system, evasion and invasion rates of CO_2 can be measured. These measurements probably yield a minimum rate due to the possibility of the capsule interfering with turbulence in the gas phase and affecting the turbulence of the gas-liquid interface layer of the surface.

Sea ice chemistry
Much of our knowledge of sea ice chemistry results from the laboratory studies of T. G. Thompson and his coworkers. Lewis and Thompson (1950) experimentally determined the effect of freezing on the sulfate-chlorinity ratio of sea-

water, and Nelson and Thompson (1954) studied the order of formation of solid phases in sea ice as seawater freezes. Nelson and Thompson's work largely confirmed the work of Ringer (1906) and extended it to lower temperatures.

The sea ice studies showed that large quantities of sulfate precipitate as sodium sulfate decahydrate between -8.2 and -22.9 C. Since Ringer's time, sulfate has been considered a prime cause of major variations in the chlorinity-density relationships of seawater in areas of sea ice formation. During the *Maud* expedition, Sverdrup (1929) and Malmgren (1927) considered that sulfate might be responsible for the differences in the measured density of sea ice meltwater and the calculated density obtained by measuring chlorinity and applying Knudsen's (1901) tables. No sulfate measurements were available, however, and the matter was put aside as an instance in which Knudsen's tables were not applicable. Almost all subsequent work has involved laboratory freezing experiments (Nelson and Thompson 1954; Lewis and Thompson 1950; Burkhalter 1967). Wiese (1930) and Bennington (1962) made measurements on natural samples of sea ice. Bennington's differences for natural samples are between 10 and 100 times larger than the laboratory samples of Nelson and Thompson (1954). Bennington concluded that sea ice showed no variations. Perhaps the only instance where sea ice formation has been directly implicated as a mechanism for the changing of seawater composition is that of Fukai and Shiokawa (1955). They reported variations in the sulfate-chlorinity ratio in samples from a temperature-minimum layer in the North Pacific. Since it was necessary for them to pool samples to perform a complete analysis, however, the variations reported were probably masked by the standard deviation.

Based on laboratory work, it appears unlikely that sulfate could remain in sea ice in significant concentrations for more than one season. As soon as the ice temperature approaches -8.2 C, the sodium sulfate decahydrate is observed to dissolve and form brine pockets that can drain from the ice.

Although not reported by Ringer or Thompson, it appears that calcium carbonate begins to precipitate just below the freezing point of seawater (Assur 1958). Since calcium is lost as a carbonate, the loss can be expressed as a change in the carbonate alkalinity of either the ice formed or of the seawater. In this frame of reference, the 40 mg/liter of calcium lost between -2 and -10 C, expressed as alkalinity, equals 2 meq/liter, or almost all of the initial seawater alkalinity. Since calcium carbonate begins to precipitate at -2 C, which is near the freezing point of seawater, it might be expected to be the last constituent to redissolve and drain from the ice as brine. Further, effects such as organic masking of the particles (Chave 1970) and the well-known kinetic problems involving calcium carbonate could be expected to keep the carbonate from redissolving even above the precipitation temperature. While the amounts of calcium or alkalinity cannot increase or undergo concentration, the ratio of calcium or alkalinity can be expected to increase with time or increased temperature because of the relative ease with which chloride can drain from the ice. An increase with time in the ratio of alkalinity to chlorosity (specific alkalinity) was reported by Laktionov (1931), who showed that the "A:Cl" factor varied between 1 for new ice and 105 for multi-year ice, although no units are given.

There is no evidence of alkalinity measurements performed on sea ice by other investigators.

One technique that may be useful in studying mixing and exchange processes in the Bering Sea involves measurements of radon-222 and its parent, radium-226. A method for the shipboard extraction and analysis of radon-222 has been developed and applied at Lamont-Doherty Geological Observatory (Broecker 1964; Broecker et al. 1967, 1968; Broecker and Peng 1971). Such measurements are currently being made routinely on the GEOSECS cruises (Broecker and Kaufman 1970).

Radon-222 (half-life, 3.86 days) is an alpha-emitting inert gas, the decay product of radium-226 (half-life, 1622 yrs). In Broecker's shipboard technique for radon-222, the gas is stripped from large (30-liter) samples of seawater using a circulating supply of helium. The radon is captured in traps cooled with liquid nitrogen. Following the removal of carbon dioxide and water by circulation through ascarite, the radon is quantitatively transferred to an alpha-counting scintillation counter. By sealing the seawater sample and allowing the radium present to decay for 7 to 10 days, the water sample can be stripped again and the radon obtained used as a measure of the radium in the sample.

In parts of the ocean well removed from the sea floor and sea surface, the decay rate of radon-222 equals that of its parent, and a measurement of radon-222 is also an indicator of radium-226. Broecker et al. (1967) report concentrations in the water column ranging from $5-8 \times 10^{-14}$ equivalent grams of Ra-226/liter. Broecker predicted (1964) and has demonstrated large concentration gradients in radon-222 near the sea floor (Broecker et al. 1968) and the sea surface (Broecker et al. 1971). Water in the pores of deep-sea sediments contains 10^4 to 10^5 times more radon-222 than the overlying seawater. Excess radon-222 diffuses into the overlying water, allowing the distribution of unsupported radon-222 with height above the bottom to be used for estimates of vertical eddy diffusivities. Standing crops of radon-222 in sediments increase with increased sedimentation rates. The radon-222 content in the atmosphere is about an order of magnitude lower than that of seawater (0.5×10^{-14} equivalent grams of Ra-226/liter); the atmosphere thus acts as a sink, and radon continually escapes from the sea to deplete the surface layers. The rate of escape depends on the rates of vertical mixing and gas exchange, which can be determined by analyzing the vertical distribution of radon depletion.

Few measurements give as much information of such a varied nature as those of radon and radium. Broecker's examples of depletion near the surface and accumulation near the bottom appear to have been obtained in sites as far removed from terrigenous contamination as possible. In an area such as the Bering Sea, however, contamination might be used to advantage as a qualitative tracer of river and lagoon contributions. Distributions of radon and radium near the surface could be used to estimate vertical advective velocities in the central Bering Sea and Aleutian passes. The rate of radon evasion could be used to calculate exchange rates of carbon dioxide, oxygen, and other gases in these areas, if the exchange mechanism of these gases is the same as for an inert gas. The presence of ice cover during winter also offers possibilities for radon measurements. In the eastern shelf area of the Bering Sea, the sea ice would serve as a barrier to radon evasion and should offer a means by which the contributions of radon from the sea floor can be evaluated without complications introduced by evasion to the atmosphere. These measurements could be used to evaluate vertical advective velocities near the bottom and to estimate contributions from the bottom. During the spring melt, opportunities to observe evasion of radon in broken ice and ice-free conditions would exist as the ice front moved northward.

Organic chemistry

The organic chemistry of the Bering Sea is largely confined to the data obtained by Loder (1971) for dissolved and particulate organic matter in a region north of Unimak and Unalaska islands in the eastern Aleutian arc. Stations were occupied as shown in Figure 9.1. Redundancy in station numbers indicates that the station was subsequently reoccupied.

The dissolved organic carbon (DOC) values varied widely between values of 1.00 and 1.85 mg C/liter in the top 50 m. The particulate organic carbon (POC) was more uniform and appeared in high concentrations in regions of high light absorbency. There was a high correlation of POC with light absorbency for sea surface stations in the Unimak Pass area. The distribution of POC along the section shown in Figure 9.1 is given in Figure 9.2. The thermal stability of this water was high at the time of sampling, and it is probable that the patches of high POC correspond to areas of high productivity. Upwelling of water from about 200 m is common north of the Aleutian Islands, and the area is known for its high fertility.

At two stations in this area, subsurface profiles (0–1 m), sampled with a multiple-outlet fixed-distance sampling device, were found to be very uniform with respect to POC but varied greatly in DOC (Table 9.1). These results show no consistent trends in the top meter of the ocean surface.

Loder (1971) conducted a detailed study in the Chukchi Sea, for which the station pattern is given in Figure 9.3. Data for 16 oceanographic parameters, including DOC, POC, and particulate nitrogen (PN), were correlated and subsequently submitted to factorial analysis. Four sets of factors (Table 9.2) were grouped, together accounting for about 85 percent of the variance. From these data, Loder was able to identify four different water masses in this region; the values of DOC, POC, and C/N are given in Table 9.3. It was concluded (Loder 1971) that the distribution of POC, PN, DOC, and oxygen was more strongly influenced by physical or advective variables than by nutrients; that these variables were a much better indication of productivity than the nutrients; that when factor scores were plotted against each other, four or five water-mass types were distinguished; that plotting factors against each other may be more useful in defining water masses than in plotting individual variables; that four physically and chemically identifiable water masses were found moving northward in the Bering Strait; that POC and PN showed a high degree of correlation, indicating a planktonic origin for the organic matter; and that several levels of DOC concentration occur and are semi-stable for increasing lengths of time as the concentration is reduced. Thus, in a limited sense, DOC may be used as a quasi-conservative water mass property.

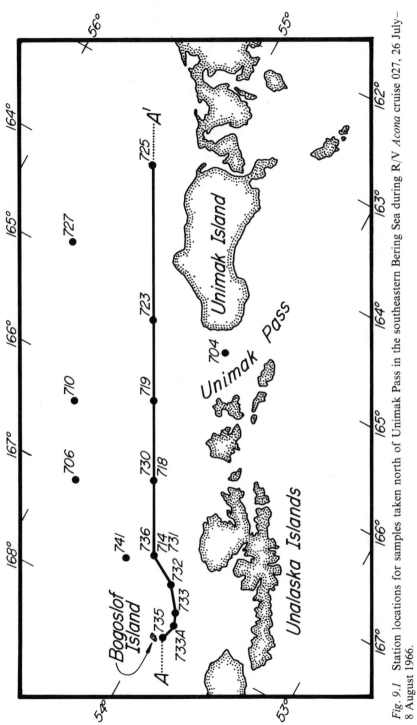

Fig. 9.1 Station locations for samples taken north of Unimak Pass in the southeastern Bering Sea during R/V *Acona* cruise 027, 26 July– 8 August 1966.

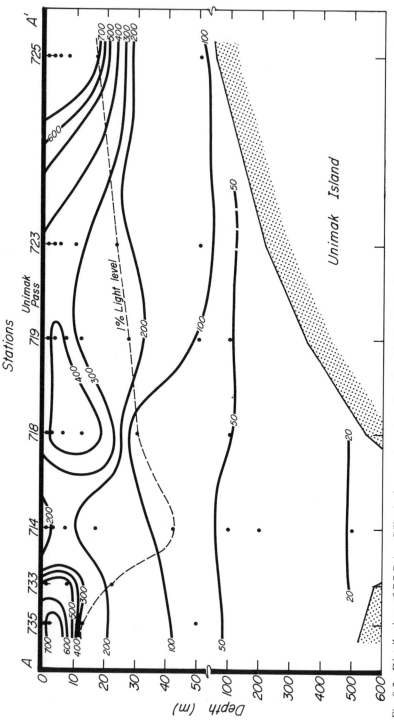

Fig. 9.2 Distribution of POC (µg C/liter) along east-west section A—A' (Fig. 9.1). The depth of the euphotic zone is shown by the dotted line representing the 1% light penetration level.

TABLE 9.1 DOC and POC data for the samples taken from the surface to 1 meter depth
at stations 730 and 741

Station 730, 8-6-66

Depth (cm)	DOC (mg C/liter)	POC (μg C/liter)
0–1	1.20	756
3	1.00	701
6	1.40	769
10	1.45	831
20	1.10	721
50	1.25	741
100	1.20	831

DOC average: 1.25 mg C/liter
Standard deviation: 0.16 (12.8%)

POC average: 764 μg C/liter
Standard deviation: 51 (6.6%)

Station 741, 8-11-66

Depth (cm)	DOC (mg C/liter)	POC (μg C/liter)
0–1	1.40	963
3	1.25	1071
6	1.40	1111
10	1.15	1021
20	1.65	1071
50	1.50	1011
100	1.40	963

DOC average: 1.40 mg C/liter
Standard deviation: 0.16 (11.6%)

POC average: 1030 μg C/liter
Standard deviation: 57 (5.5%)

Growth factors

The concentrations of vitamins or other growth factors have not been examined in the Bering Sea, although Natarajan and Dugdale (1966) have found that thiamine in the coastal waters of the North Pacific is present at the 100 ng/liter level.

Stable isotope analysis

The use of stable isotopes in environmental studies has been useful in solving problems related to sources of organic matter, carbon dioxide system kinetics, and gas-water phase phenomena in the oceans. The ratios $^{12}C/^{13}C$ have been found (Parker and Calder 1970) to be -5 to -14 for organic matter in marine plants and between -20 and -28 in land plants. In crude oil and hydrocarbon gases, the ratios range between -20 and -46. Techniques utilizing stable

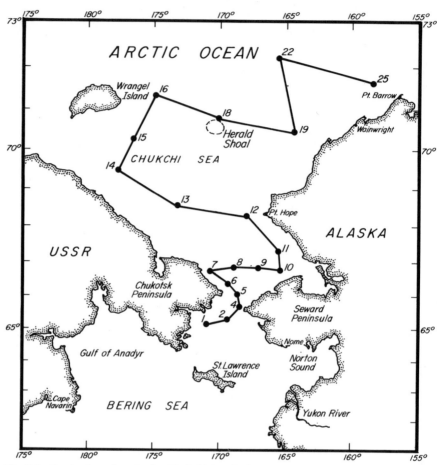

Fig. 9.3 Station locations in the Chukchi Sea during USCG Icebreaker *Northwind* cruise 751, 28 July–2 August 1968.

TABLE 9.2   Primary composition of four factors for Chukchi Sea data based on the varimax factor matrix. The percentages of the total variance "explained" by each factor are given.

Factors			
I (29.6%)	II (26.3%)	III (9.6%)	IV (19.3%)
NO₃ 0.928	POC −0.997	C/N 0.764	SIGT 0.753
SIO₃ 0.881	PN −0.976	(PM) 0.547	DEP 0.751
NO₂ 0.865	DOC −0.782		TEMP −0.721
PO₄ 0.852	OXY −0.757		SAL 0.693
(NH₃) 0.671			

TABLE 9.3   Average values for DOC, POC, and C/N data for four groups of samples from surface and depth in the Chukchi Sea

Sample Locations	Surface		Deep	
	No. of samples	Average DOC (mg C/liter)	No. of samples	Average DOC (mg C/liter)
Central Bering Strait	7*	1.62	5**	0.83
Stations 1 to 12	11	1.51	10	0.92
Stations 13 to 25	9	0.94	8	0.85
Stations 1 to 25	20	1.25	18	0.89
	No. of samples	Average POC (μg C/liter)	No. of samples	Average POC (μg C/liter)
Central Bering Strait	7*	990	5**	156
Stations 1 to 12	11	733	10	213
Stations 13 to 25	9	133	8	172
Stations 1 to 25	20	463	18	194
	No. of samples	Average C/N	No. of samples	Average C/N
Central Bering Strait	7*	5.7	5**	5.9
Stations 1 to 12	11	5.8	10	5.9
Stations 13 to 25	9	5.4	18	5.5
Stations 1 to 25	20	5.6	18	5.7

*   These seven samples were from stations 2, 5, 6, 8, 9, and 12.
**   These five samples were from stations 2, 5, 6, 9, and 12.

isotopes have not been used in Bering Sea studies, but such methods should be as powerful a tool in this area of research as they have proven in other applications.

Trace metals

The biologically mediated cycling of trace metals copper, lead, and zinc in the Bering Sea were studied (Barsdate and Nebert 1971; Barsdate et al., chapter 28, this volume) by anodic stripping techniques, coupled with radioisotopic traces and dialysis experiments. Seldom over 20 percent of the total trace metals were found associated with large molecules in upwelled regions of the Bering Sea, and this amount co-varied with primary productivity. Additional data on zinc are available for the Chukchi Sea (Burrell et al. 1968). The Bering Sea, with its generally high productivity and its dynamic physical and geological regime, should be a fertile area for the study of trace-metal chemistry underlying primary productivity of the oceans.

REFERENCES

ASSUR, A.
    1958   Composition of sea ice and its tensile strength. In *Arctic sea ice*. Nat. Acad. Sci.-Nat. Res. Counc. Publ. 598, pp. 106–138.

BARSDATE, R. J., and M. NEBERT
1971   Biologically mediated trace metal cycles. Rep. 71–2, Inst. Mar. Sci., Univ. Alaska, Fairbanks, pp. 58–101.

BENNINGTON, K. O.
1962   Some chemical composition studies on arctic sea ice. In *Ice and snow: properties, processes and application*, edited by W. D. Kingery. MIT Press, Cambridge, Mass., pp. 248–257.

BROECKER, W. S.
1964   An application of natural radon to problems in ocean circulation. In *Symposium on diffusion in oceans and fresh waters*, edited by T. Ichige. Lamont Geol. Observ., Palisades, N. Y., pp. 116–145.

BROECKER, W. S., Y. H. LI, and J. CROMWELL
1967   Radium-226 and radon-222: concentration in Atlantic and Pacific oceans. *Science* 158: 1307–1310.

BROECKER, W. S., J. CROMWELL, and Y. H. LI
1968   Rates of vertical eddy diffusion near the ocean floor based on measurements of the distribution of excess Rn-222. *Earth and Planetary Science Letters* 5: 101–105.

BROECKER, W. S., and A. KAUFMAN
1970   Near-surface and near-bottom radon results for the 1969 North Pacific GEOSECS station. *J. Geophys. Res.* 75: 7679–7681.

BROECKER, W. S., and T. H. PENG
1971   The vertical distribution of radon in the BOMEX area. *Earth and Planetary Science Letters* 11: 99–108.

BURKHALTER, A. C.
1967   The effect of freezing on the sulfate-chloride and density-chloride ratios of seawater. M. S. Thesis, Texas A&M Univ., College Station, 37 pp.

BURRELL, D. C., C. G. WOOD, and P. J. KINNEY
1968   Direct spectrophotometric determination of zinc in summer Chukchi Sea waters. *Trans. Amer. Geophys. Un.* 49: 759 (Abstract) and Rep. 69–10, Inst. Mar. Sci., Univ. Alaska, Fairbanks.

CHAVE, Y. E.
1970   Carbonate organic interactions in seawater. In *Organic matter in natural waters*, edited by D. W. Hood. Occas. Publ. No. 1, Inst. Mar. Sci., Univ. Alaska, Fairbanks, pp. 373–385.

COACHMAN, L. K., and K. AAGAARD
1966   On the water exchange through Bering Strait. *Limnol. Oceanogr.* 11: 44–59.

COACHMAN, L. K., and R. B. TRIPP
1970   Currents north of Bering Strait in winter. *Limnol. Oceanogr.* 15: 625–632.

DODIMEAD, A. J., F. FAVORITE, and T. HIRANO
    1963   Review of oceanography of the subarctic Pacific region. *Bull. Int. North Pac. Fish. Comm.* 13: 1–195.

FUKAI, R., and F. SHIOKAWA
    1955   On the main chemical components dissolved in the adjacent waters to the Aleutian Islands in the North Pacific. *Bull. Chem. Soc. Japan* 28: 636–640.

GROSS, M. G.
    1966   Distributions of radioactive sediment derived from the Columbia River. *J. Geophys. Res.* 71: 2017–2021.

GROSS, M. G., C. A. BARNES, and G. K. RID
    1965   Radioactivity of the Columbia River effluent. *Science* 149: 1088–1090.

HOOD, D. W., D. BERKSHIRE, R. ADAMS, and I. SUPERNAW
    1963   Calcium carbonate saturation level of oceans. Data Rep., Texas A&M Univ. Proj. 295, Ref. 63–3D.

KNUDSEN, M.
    1901   Hydrographical tables [transl. from Danish]. Repr. 1946, Woods Hole Oceanogr. Inst., Woods Hole, Mass.

LAKTIONOV, A. F.
    1931   The properties of sea ice [in Russian]. *Tr. Nauchno-Issled. Inst. Izuch. Sev.*, 49: 71–96. Cited in N. N. Zubov (1943), *Arctic ice* (English transl. by U. S. Navy Electronics Laboratory), U. S. Navy Oceanogr. Off. and Amer. Meteor. Soc., p. 148.

LEWIS, G. J., and T. G. THOMPSON
    1950   The effect of freezing on the sulfate chlorinity ratio of seawater. *J. Mar. Res.* 9: 211–217.

LODER, T. C.
    1971   Distribution of dissolved and particulate organic carbon in Alaskan polar, subpolar and estuarine waters. Ph.D. Thesis, Univ. Alaska, Fairbanks.

MALMGREN, F.
    1927   On the properties of sea ice. In *Norwegian North Polar Expedition with the Maud 1918–1925. Scientific Results* 1(5): 67 pp.

McRoy, C. P.
    1966   The standing stocks and ecology of eelgrass (*Zostera marina* L.) in Izembek Lagoon, Alaska. M. S. Thesis, Univ. Washington, Seattle, 138 pp.
    1970   On the biology of eelgrass in Alaska. Ph.D. Thesis, Univ. Alaska, Fairbanks, 156 pp.

McRoy, C. P., and R. J. BARSDATE
    1970   Phosphate absorption in eelgrass. *Limnol. Oceanogr.* 15: 6–13.

MCROY, C. P., R. J. BARSDATE, and M. NEBERT
   1972   Phosphorus cycling in an eelgrass (*Zostera marina L.*) ecosystem. *Limnol. Oceanogr.* 17: 58–67.

NATARAJAN, K. V., and R. C. DUGDALE
   1966   Bioassay and distribution of thiamine in the sea. *Limnol. Oceanogr.* 11: 621–629.

NELSON, K. H., and T. G. THOMPSON
   1954   Reposition of salts from seawater by frigid concentration. *J. Mar. Res.* 13: 166–182.

PARK, K.
   1966   Columbia River plume identification by specific alkalinity. *Limnol. Oceanogr.* 11: 118–120.

PARKER, P. L., and J. A. CALDER
   1970   Stable isotope ratio variations in biological systems. In *Organic matter in natural waters*, edited by D. W. Hood. Occas. Publ. No. 1, Inst. Mar. Sci., Univ. Alaska, Fairbanks, pp. 107–122.

REEBURGH, W. S.
   1969   Observations of gases in Chesapeake Bay sediments. *Limnol. Oceanogr.* 14: 368–375.

RINGER, W. E.
   1906   Changes in the composition of seawater salts during freezing. *Chem. Week Blad.* 3: 223–249.

RODEN, G. I.
   1967   On river discharge into the Northeastern Pacific Ocean and the Bering Sea. *J. Geophys. Res.* 72: 5613–5629.

SUGIURA, Y., E. R. IBERT, and D. W. HOOD
   1963   Mass transfer of CO_2 across sea surfaces. *J. Mar. Res.* 21: 11–24.

SVERDRUP, H. U.
   1929   The waters of the North-Siberian shelf. In *Norwegian North Polar Expedition with the "Maud", 1918–1925. Scientific Results* 4(2): 131 pp. Geofysisk Inst., Bergen.

WIESE, W.
   1930   Zur Kenntnis der Salze des Meereises. *Ann. Hydrogr. Mar. Meteor, Jahrg.*, pp. 282–286.

Part 3

INVENTORY OF
RENEWABLE RESOURCES

CHAPTER 10

Plankton of the Bering Sea

SIGERU MOTODA

Faculty of Marine Science and Technology, Tokai University, Shimizu, Japan

TAKASHI MINODA

Faculty of Fisheries, Hokkaido University, Hakodate, Japan

Abstract

Based largely on material obtained during cruises of the *Oshoro Maru* during 1954 to 1970, a list of phytoplankton and zooplankton species in the Bering Sea was recorded and compiled with reference to published papers and unpublished manuscripts. An inventory of more than 300 species resulted.

In early to mid-summer, boreal-oceanic diatom communities occupy a large part of the western and central Bering Sea and eastern shelf; temperate-neritic diatoms are distributed along the Aleutian Islands. Subarctic oceanic copepods are predominantly distributed in the western and central Bering Sea; an arctic copepod species *Calanus glacialis* is present on the eastern shelf, along the continental shelf edge and around Bowers Bank; and *Centropages abdominalis* is found in the neritic water around Unimak Pass. Most copepods show clear depth preference, 80 percent of the biomass in the 0–150 m water column occurring in the upper 80 m.

The density of the surface diatom standing crop ranged from 1×10^5 to 10^9 cells/m^3. The summer zooplankton biomass in the Bering Sea has varied from year to year (large in 1958, 1962, 1965 and 1968), showing certain periodical variations with intervals of two to three years. The average zooplankton biomass was 36.8 g/m^2, except in shallow parts of Bristol Bay. Fifty percent of the zooplankton biomass observed in the upper 80 m of the Bering Sea in early to mid-summer consisted of two herbivorous copepods, *Calanus cristatus* and *C. plumchrus*. There is a tendency of inverse relationship in the regional abundance reported between phytoplankton and zooplankton in the upper layer, except off the coast of the Kamchatka Peninsula.

Surface primary production in summer is extremely high (5 mg C/m^3-hr) in an area east of Bower's Bank and poor (< 1 mg C/m^3-hr) on the northern shelf and in the western Bering Sea Gyre. Surface primary production in the western and central Bering Sea is 1–2 mg C/m^3-hr. The ratio of zooplankton to phytoplankton production (based on carbon) is estimated to be 1.25, which is higher than in other regions. The turn-over rate of diatoms in summer is calculated as 0.93.

INTRODUCTION

According to the comprehensive compilation of hydrographic data of the Subarctic North Pacific Ocean, 1887–1953, published by Watanabe (1954), more than 20 cruises covering the Bering Sea for hydrographic or fisheries surveys were undertaken by the USSR, USA, and Japan before the outbreak of World War II. In 1958 the Institute of Oceanology, Academy of Sciences of the USSR, conducted large-scale fisheries and oceanographic expeditions in the Bering Sea. The United States also conducted several research cruises to the Bering Sea in recent years. The first Japanese oceanographic cruise to the Bering Sea after the war was made by R/V *Tenyo Maru* of the Tokai Regional Fisheries Research Laboratory, Japan Fisheries Agency, in the summer of 1952, followed by similar cruises of the same vessel in the subsequent two summers. The *Oshoro Maru II* (replaced by the *Oshoro Maru III* in 1963), Training Ship of the Faculty of Fisheries, Hokkaido University, sailed to the Bering Sea in the early summer of 1953 for the first time, and thereafter training cruises combined with fisheries and oceanographic surveys in the Bering Sea have been conducted annually in early to mid-summer.

Standard samplings of zooplankton with a conical NORPAC net (45 × 180 cm, 0.33 or 0.35 mm mesh openings, 0–150 m vertical haul) have been made since 1956 on the *Oshoro Maru* (Motoda 1957; Motoda et al. 1957; Morioka 1965). The schedule of the *Oshoro Maru* in the Bering Sea was confined to early and mid-summer, with different cruise tracks each year. A total of 734 stations were occupied for standard plankton hauls during the 15 years from 1956 to 1970, covering almost the whole Bering Sea and area north of 50°N and west of 170°W. Of these stations, 621 were located in the Bering Sea in the area north of Attu and the Komandorskii islands (Fig. 10.1). Data on hydrography and preliminary plankton processing have appeared in the *Data Record of Oceanographic Observations and Exploratory Fishing*, published annually since 1957 by the Faculty of Fisheries, Hokkaido University.

BERING SEA PLANKTON SPECIES

Species of major plankton groups recorded in the Bering Sea are listed in Appendix 10.1. There will be certain disagreement on the species names listed with respect to synonymy or separation of species.

A total of 74 species of pelagic diatoms (Bacillariophyceae) are listed. Among the dominant species during early through mid-summer in the offshore waters of the Bering Sea are: *Chaetoceros convolutus, Ch. concavicornis, Ch. debilis, Ch. compressus, Ch. radicans, Ch. didymus, Ch. seiracanthus, Ch. furcellatus, Ch. constrictus, Rhizosolenia hebetata* forma *hiemalis, Denticula seminae, Nitzschia seriata, N. delicatissima, N. longissima, Fragilaria* spp. and *Thalassiothrix longissima.* The list includes 28 species of diatoms adhering to

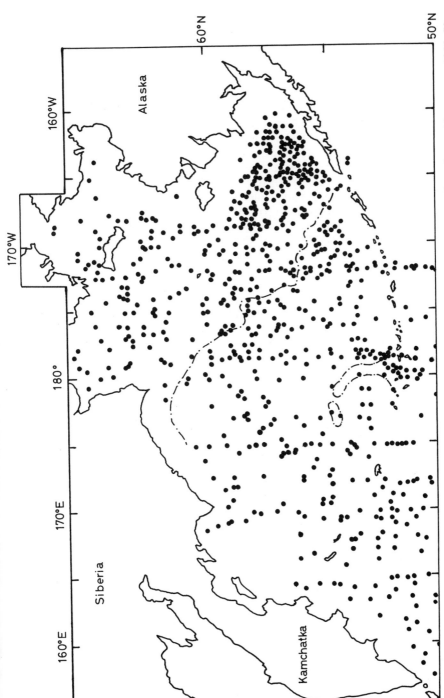

Fig. 10.1　Location of stations occupied by the *Oshoro Maru* for zooplankton standard hauls (0–150 m) in early to mid-summers, 1956–70.

the skin film of whales (Nemoto 1956). Kisselev (1937, cited from Zenkevitch 1963) recorded 2 flagellates, 55 peridinians, 104 diatoms and 2 green algae in the northern part of the Bering Sea and in the Okhotsk Sea. Dinoflagellata have not been studied satisfactorily in the Bering Sea; however, 16 important species are listed (Kisselev 1937, cited from Zenkevitch, 1963; Tsuruta 1963). Five species of Foraminifera (Yamade, ms, 1971) and 24 species of Radiolaria (Yamade, ms, 1971; Z. Nakai, personal communication) and six species of Tintinnoinea (Tsuruta 1963) are listed. Dogiel and Reschetnjak (1952) reported 89 species of Radiolaria off the east coast of Kamchatka, but probably not in the Bering Sea; 21 of these species were previously unreported. Later Reschetnjak (1955) reported Radiolaria in the Kurile-Kamchatka Trench. Z. Nakai (personal communication) made collections by divided net hauls from 1050 m at three stations in the northern part of the North Pacific, including one location in the western Bering Sea. One species, collected at 590–800 m, is closely related to *Challengeria xiphodon* but is distinguishable slightly by its notably short adoral teeth and flat shell. Another species (from 580–1050 m) shows close relationship to *Protocystis harstoni*, differing from it by having a raised test. In the upper 200 m a large number of a species related to *Protocystis vicina* and a species related to *P. ornithocephala* were collected. The former species is common in the deep layer of the Sea of Japan, where the latter species has never been found despite its presence in the subarctic North Pacific waters. Nakai mentions that there are a relatively few species of Radiolaria in the Bering Sea and Sea of Japan, in contrast with a large number of species in the subarctic North Pacific.

Ten species of Siphonophora (Stephanynthes 1967) and 10 species of Scyphomedusae (Naumov 1961) are cited. Another Hydromedusa (Kurioka, ms, 1963) and one Ctenophora species (Takeuchi, ms, 1972) are listed. A schyphomedusa, *Chrysaora helvora*, migrates upward at night and is notorius for entangling with fishing nets in Bristol Bay. No Heteropoda are found in the Bering Sea. Two species of Pteropoda are important (Yamamoto, ms, 1959; Tanaka, ms, 1969) Representing molluscan Decapoda, young specimens of *Gonatus* and *Gonatopsis* are taken with larva nets (Takeuchi, ms, 1972); such micronekton are not included in the list. One pelagic Polychaeta was recorded (Takeuchi, ms, 1972). Six species of Chaetognatha are listed (Tchindonova 1955; Alvariño 1962, 1964; Kotori 1969).

Copepoda collected in the Bering Sea during cruises of the U. S. Fisheries USS *Albatross*, in 1893 and 1895 were studied by Wilson (1950), together with numerous collections from other regions. Due to uncertain records of the locality of collection for some species in Wilson's report, however, his data have been excluded from the list. There is a valuable publication by Brodskii (1950) on Calanoida copepods in the Far-Eastern seas of the USSR. Tanaka (1960) is of the opinion that *Calanus tonsus* is a species originating in the southern hemisphere of the Pacific, and that a closely associated species, *C. plumchrus*, which is distributed in the northern hemisphere of the Pacific, can be morphologically differentiated from *C. tonsus*. *Calanus finmarchicus* recorded in the Bering Sea should be *C. glacialis*, which can be distinguished from *C. finmarchicus* s. str. in the Atlantic (Jaschnov 1955, 1963, 1970). *Calanus pacificus*, differing from *C. helgolandicus* s. str. in the Atlantic (Brodskii 1948), is widely distributed in the subarctic North Pacific water south of the Aleutian arc but does not penetrate into the Bering Sea. *Centropages mcmurrichi* was synonymized with *C. abdominalis* by Brodskii (1950). A total of 97 species of Calanoida, two species of Mormonillidae, nine species of Cyclopoida and two species of Harpacticoida can be listed in the Bering Sea. Species important by their abundance in the Bering Sea are: *Calanus cristatus*, *C. plumchrus*, *Eucalanus bungii bungii*, and *Pseudocalanus minutus* in the offshore water. *Calanus glacialis*, *Centropages abdominalis* and *Acartia longiremis* are regionally important.

Among six species of Hyperiidea amphipods, *Parathemisto pacifica* and *P. libellula* are important inhabitants of the shelf. Sixteen species of Mysidacea and seven species of Euphausiacea are listed. Predominant among the Euphausiacea are *Thysanoessa longipes*, *Thy. inermis* and *Thy. raschii* (Anadyr Bay, Bristol Bay). *Euphausia pacifica*, an important species of the North Pacific, occurs near the Aleutian Islands in the Bering Sea.

REGIONAL CHARACTERISTICS OF PLANKTON COMMUNITIES DURING EARLY TO MID-SUMMER

The coastal Alaskan Stream flows westward along the southern Aleutian arc, with northward branches flowing through island passes. When the branches disperse, the water circulates anticlockwise in the central to western Bering Sea and loses its warm neritic nature (Ohtani 1965). On the wide continental shelf of the northern and eastern Bering Sea the water which had been cooled greatly during the winter still remains in the summer. There is a cold current close to the coasts of Siberia and the Kamchatka Peninsula in the western Bering Sea. These surface current systems govern primarily the distribution of plankton communities in the upper layer of the Bering Sea. Temperate-neritic communities characterize the water near the Aleutian Islands, boreal-oceanic communities typify the wide area of the deep western and central Bering Sea and shelf waters of the eastern Bering Sea; and arctic communities characterize the water near the coasts of Siberia and the Kamchatka Peninsula. The inner shallow basin of Bristol Bay in the eastern Bering Sea is warmed in summer and contains temperate-neritic communities.

Diatoms

In the deep basin of the western Bering Sea, the surface water is predominated by boreal-oceanic diatom communities (*Thalassiothrix longissima*, *Corethron hystrix*, *Denticula seminae*; *Chaetoceros-Phaeoceros* group—including *Chaetoceros atlanticus*, *Ch. convolutus*, and *Ch. concavicornis*; and *Nitzschia seriata*). The water along the Aleutian arc as far as Attu Island in the west is characterized by temperate-neritic communities which are transported by the Alaskan Stream from the east. In this process, some members of the *Chaetoceros-Hyalochaete* group (including *Chaetoceros debilis*, *Ch. decipiens*, and *Ch. radicans*) occur as far as Attu Island, while *Chaetoceros-Hyalochaete* of warmer-water forms of this group (*Chaetoceros didymus*, *Ch.*

constrictus, etc) do not appear west of the Unimak Pass area. On the shelf of the eastern Bering Sea (except for the inner shallow portion of Bristol Bay), the *Chaetoceros-Phaeoceros* group and *Thalassiothrix longissima* are found predominantly as in the western Bering Sea, indicating the cold oceanic nature of the water in this region (Motoda and Kawarada 1955; Marumo 1956; Kawarada 1957; Kawarada and Ohwada 1957; Karohji 1958, 1959; Ohwada and Kon 1963) (Fig. 10.2). Kisselev (1937, cited from Zenkevitch 1963) mentions that Anadyr phytoplankton is mainly arctic or arctic-boreal, the boreal forms being predominant only in the surface layer in summer, and that phytoplankton of the eastern northern half of the Bering Sea is characterized by the predominance of boreal forms with a mixture of brackish-water and neritic species. Aikawa (1932, 1940) indicates that the neritic *Chaetoceros-Hyalochaete* group are abundantly distributed around Karaginskii Island, along the coast of Siberia, around St. Lawrence Island and north of Unimak Pass, and that the *Chaetoceros-Phaeoceros* group and *Thalassiothrix longissima* are main components in the area west of the Aleutian Islands. Generally speaking, summer diatom communities in the Bering Sea can be represented by the *Chaetoceros-Hyalochaete* group along the coasts and by *Thalassiothrix longissima* and the *Chaetoceros-Phaeoceros* group in the offshore regions. In the area of the Western Subarctic Domain (Dodimead et al. 1963), boreal oceanic diatoms predominate, especially in the area of Bering Sea Gyre, as characterized by the predominance of *Thalassiothrix longissima*.

Copepods

Soviet scientists (Gerschanovich et al., chapter 17 this volume) state that in summer the Nunivak Island-Cape Navarin line is the border separating the distribution area of the north Bering Sea group from that of the south Bering Sea group. The waters of the shallow eastern part of the sea are inhabited by the neritic group. Along with the neritic forms, however, some typically oceanic species such as *Pseudocalanus elongatus (P. minutus)* and *Calanus finmarchicus (C. glacialis)* often occur in the eastern part. In the continental slope area, the major part of the concentrations consists of the bathypelagic *Eucalanus bungii*, while neritic species of *Acartia* and *Centropages* predominate in the low-salinity waters of the eastern shallow region.

Among 110 species recorded in the Bering Sea, 27 species are found in the epipelagic zone (0–200 m). Subarctic oceanic species such as *Calanus plumchrus*, *C. cristatus*, *Eucalanus bungii*, and *Metridia pacifica* are predominantly distributed in the central and western part of the Bering Sea in early summer. They are widely distributed in the subarctic oceanic region in the North Pacific Ocean. On the continental shelf of the eastern Bering Sea, copepod communities composed mainly of arctic *Calanus glacialis* and arctic-boreal *Acartia longiremis* are present. Jaschnov (1963, 1970) states that *Calanus glacialis* is an arctic species whose production area in the Bering Sea is confined to the shelf region and Gulf of Anadyr and that around Unimak Pass the water is characterized by a neritic copepod *Centropages abdominalis* (syn. *C. mcmurrichi*). Along the edge of the continental shelf of the eastern Bering Sea, as well as around Bowers Bank north of the Aleutian Islands, mixed communities of subarctic oceanic

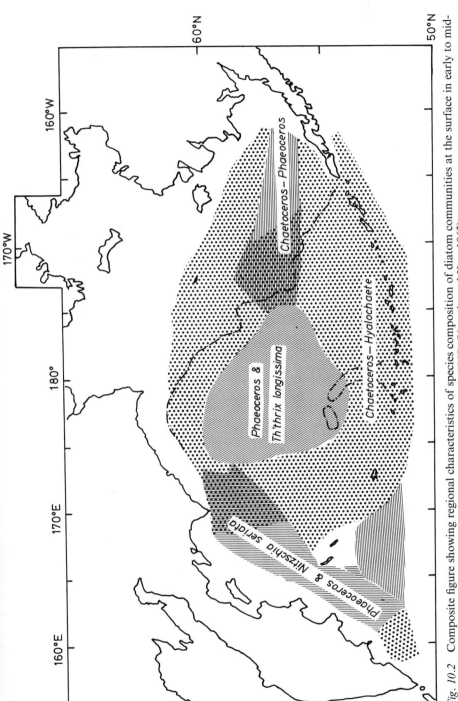

Fig. 10.2 Composite figure showing regional characteristics of species composition of diatom communities at the surface in early to mid-summer (Motoda and Kawarada 1955; Marumo 1956; Kawarada 1957; Ohwada and Kon 1963).

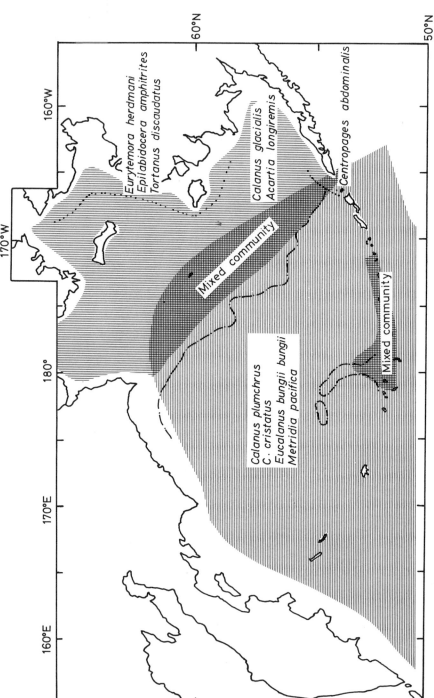

Fig. 10.3 Composite figure showing regional characteristics of species composition of copepod communities in the upper water in early to mid-summer (Johnson 1953; Anraku 1954; Vinogradov 1956; Minoda 1958; Nishio, ms, 1961; Kawamura, ms, 1962; Koseki, ms, 1962; Morioka, ms, 1963; Yamazaki, ms, 1963; Omori 1965; Matsumura, ms, 1966).

species such as those observed in the central and western Bering Sea are found, together with *Calanus glacialis* and *Acartia longiremis* in less numbers. *Pseudocalanus minutus* and *Oithona similis* are widely distributed in the Bering Sea, the former increasing in numbers around Bowers Bank (Anraku 1954; Vinogradov 1956; Ito, ms, 1957; Minoda 1958; Nishio, ms, 1961; Kawamura, ms, 1962; Koseki, ms, 1962; Morioka, ms, 1963; Yamazaki, ms, 1963; Omori 1965; Matsumura, ms, 1966) (Fig. 10.3). Johnson (1953) found *Eurytemora herdmani*, *Centropages mcmurrichi (C. abdominalis)*, *Epilabidocera amphitrites*, *Acartia longiremis*, *A. clausi*, and *Tortanus discaudatus* in the area east of Nunivak Island and Norton Sound.

Vertical distribution of copepods
A large part of the southwestern Bering Sea is at least as deep as 3500 m (Zenkevitch 1963). Vertical distribution of Copepoda in the central Bering Sea was studied on cruises in 1961, 1962, 1966 (Minoda 1971), in 1967 (Morioka, ms. 1970), in 1969 (Hamaoka, ms, 1970) (Fig. 10.4), and in 1970 (Fig. 10.9). Only one species, *Acartia longiremis*, is confined to the epipelagic zone (0–200 m) throughout the day and night. Another 26 species appearing in the epipelagic zone extend their vertical distribution to the transition zone (200–500 m) or to the bathypelagic zone (below 500 m). Twenty-two of these species are distributed down to 2000 m depth. On the other hand, 42 species do not appear in the epipelagic zone, but always inhabit the transition or bathypelagic zones; for instance, adults of *Calanus plumchrus* and *C. cristatus* appear only below 200 m. *Calanus glacialis* also is not found in the upper 200 m in this case, although it is collected occasionally in a 0–150 m haul in other cases. *Calanus glacialis* is reported to be distributed in the surface layer throughout the entire region of the Bering Sea in winter (Zenkevitch 1963). Fifteen species inhabit only the zone below 1000 m *(Bathycalanus bradyi, Spinocalanus magnus, Undinopsis pacificus, Pseudochirella spectabilis, Pareuchaeta brevirostris, Valdiviella brevicornis, Undinella oblonga, Metridia ornata, Haloptilus longicirrus, Augaptilus glacialis, Euaugaptilus graciloides, E. mixtus, Centraugaptilus horridus. Mormonilla minor*, and *Aegisthus aculeatus)* (Hamaoka, ms, 1970; Morioka, ms, 1970; Minoda 1971). Morioka (ms, 1970) mentioned that all copepods inhabiting the deep zone from 800–1700 m in the Bering Sea were also found in the deep zone of the subarctic and in subtropical regions of the Pacific Ocean, while some of bathypelagic species in the subtropical Pacific were endemic. Contrary to this, epipelagic species and transition zone species in the Bering Sea *(Scolecithricella minor, Pleuromamma scutulata, Heterorhabdus tanneri)* are boreal forms, never occurring in the subtropical water. Stage III copepodites of *Calanus plumchrus* exist at the surface and are not seen below 140 m. Stage V copepodites are distributed from the surface to 1300 m, increasing in numbers in the upper 280 m. Stage VI adults are present below 280 m but never appear at the surface. The same manner of vertical distribution is observed in copepodites and adults of *Calanus cristatus* (Morioka, ms, 1970) (Fig. 10.5). Phytoplankton productivity increases from spring to summer in the Bering Sea, resulting in an increase in the phytoplankton standing crop in the upper layer as the thermocline develops (Semina 1960; Taniguchi 1969;

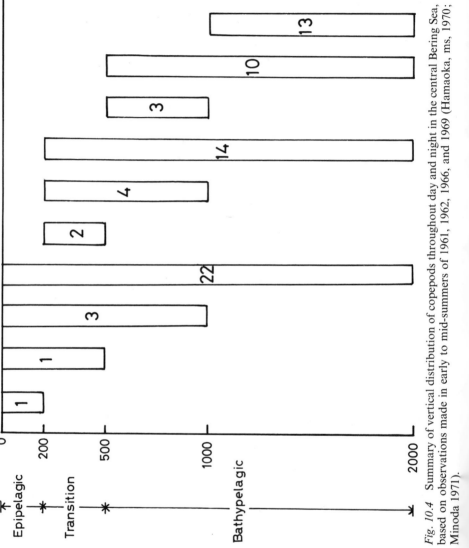

Fig. 10.4 Summary of vertical distribution of copepods throughout day and night in the central Bering Sea, based on observations made in early to mid-summers of 1961, 1962, 1966, and 1969 (Hamaoka, ms, 1970; Minoda 1971).

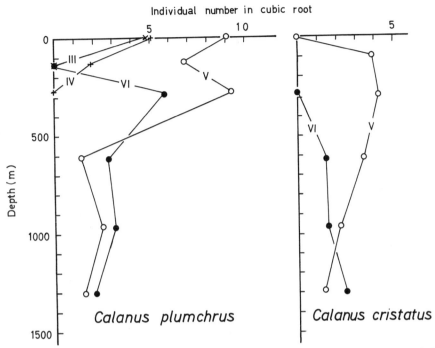

Fig. 10.5 Vertical distribution of copepodites and adults of *Calanus cristatus* and *C. plumchrus* at a station in the Bering Sea, June 1967 (Morioka, ms, 1963).

Taguchi 1972). If a rise in temperature exceeds the range of preference of some species of subarctic copepods or certain stages of a species, it forces them to migrate downwards. *Metridia pacifica*, whose remarkable diel vertical migration has been reported in various localities, does not migrate up to the water in which temperature exceeds 15 C in the Kuroshio region (Furuhashi 1965). The distribution zone of *Calanus plumchrus*, *C. cristatus*, *Eucalanus bungii*, and *Metridia pacifica* in Sagami Bay is limited to water below 12 C (Omori and Tanaka 1967). Disappearance of subarctic oceanic species from the surface layer and replacement by southern forms such as *Calanus pacificus* are observed as seasons progress from summer to autumn at station P in the northeastern Pacific (LeBrasseur 1965). However, the summer temperature at the surface in the central Bering Sea still remains below 10 C.

Other zooplankton

In Euphausiacea, *Tessabrachion oculatus* is distributed only along the Aleutian Islands, being absent in the central Bering Sea (Brinton 1962). The dominant species of the central Bering Sea is *Thysanoessa longipes*, while *Thy. raschii* is found in the shallow water of Bristol Bay and the Gulf of Anadyr. *Euphausia pacifica* hardly extends its distribution into the Bering Sea, mostly being distributed in the subarctic water south of the Aleutian Islands. *Thysanoessa*

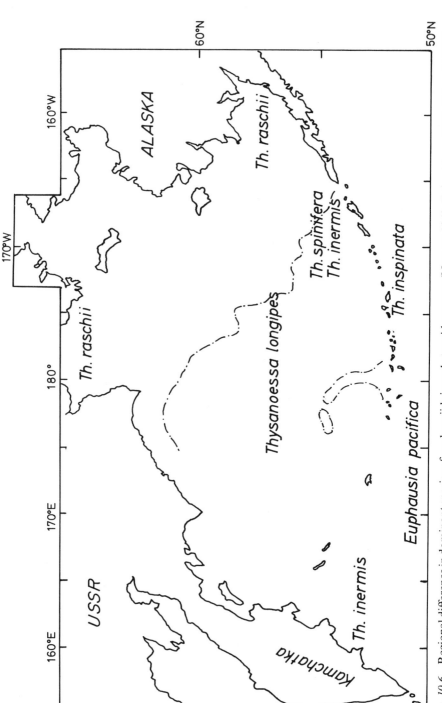

Fig. 10.6 Regional difference in dominant species of euphausiids in early to mid-summer (Nemoto 1962; Fukuchi, ms, 1970; Takeuchi 1972).

inspinata appears in the area east of Attu Island. This species inhabits the Gulf of Alaska and the Pacific coast of the Alaska Peninsula, so that the species would be transported by the Alaskan Stream to the Bering Sea area. *Thysanoessa inermis* is found on the edge of the eastern shelf and on the southern coast of Kamchatka Peninsula. Thus, *Thysanoessa longipes* prefers the considerably high salinity water than *Thy. spinifera* and *Thy. inermis*, and *Thy. raschii* prefers low salinity water (Nemoto 1962). Fukuchi (ms, 1970) observed a similar distribution of *Thy. longipes* and *Thy. raschii* on the edge of the eastern shelf and on the shelf (Fig. 10.6). Salinity of water inhabited by *Thysanoessa raschii* observed by Fukuchi (ms, 1970) is slightly higher than in observations of Nemoto (1962) (Table 10.1).

Takeuchi (ms, 1972) reports that *Thysanoessa longipes* is found in the Bering Sea in June through August in highest abundance among euphausiids; *Thy. inermis* appears in the Bering Sea in April through June in lower numbers; *Thy. spinifera* appears only in June; and *Euphausia pacifica* is found in the central Bering Sea during June through July, though only in low numbers.

Amphipoda *Parathemisto pacifica* and *P. japonica* are distributed in the western and central Bering Sea in early to mid-summer, while *P. libellula* is distributed on the eastern shelf (Fukuchi, ms, 1970; Takeuchi, ms, 1972) (Fig. 10.7). *Parathemisto libellula* is distributed from the Arctic Ocean to the Bering Sea (Bowman 1960). The distribution of *P. libellula* in the Bering Sea is not always confined to the shelf area but sometimes extends to the western deep basin in lower numbers (Yabuguchi, ms, 1957; Takeuchi, ms, 1972). Although euphausiids and amphipods are in an important ecological niche as food resources of pelagic fish (Isobe, ms, 1966; Takeuchi, ms, 1972), the NORPAC plankton net hauls do not yield large numbers of these animals; due to their net avoidance, they occupy only less than 10 percent (wet wt.) of total specimens (Yamahana, ms, 1961).

The major part of the Chaetognatha community in the Bering Sea is represented by *Sagitta elegans*. This species is widely distributed in the western and central Bering Sea, as well as on the eastern shelf and in Bristol Bay in lesser numbers (Kusajima, ms, 1959; Koyama, ms, 1967).

Pteropoda representatives in the Bering Sea are *Limacina helicina helicina* and *Clione limacina limacina*. They are distributed in the western and central Bering Sea and also on the eastern shelf, found frequently in the stomachs of salmon (Isobe, ms, 1966; Takeuchi, ms, 1972).

A pelagic Polychaeta, *Tomopteris septentrionalis*, appears in the western and central basin of the Bering Sea in June through July (Takeuchi, ms, 1972).

TABLE 10.1 Salinity of water inhabited by four euphausiid species in the Bering Sea

Species	Nemoto (1962)	Fukuchi (ms, 1970)
Thysanoessa longipes	34.1–32.6	33.4–32.5
Thysanoessa inermis	34.0–32.4	33.4–32.1
Thysanoessa spinifera	33.6–32.1	33.4–32.0
Thysanoessa raschii	32.7–30.0	33.4–31.5

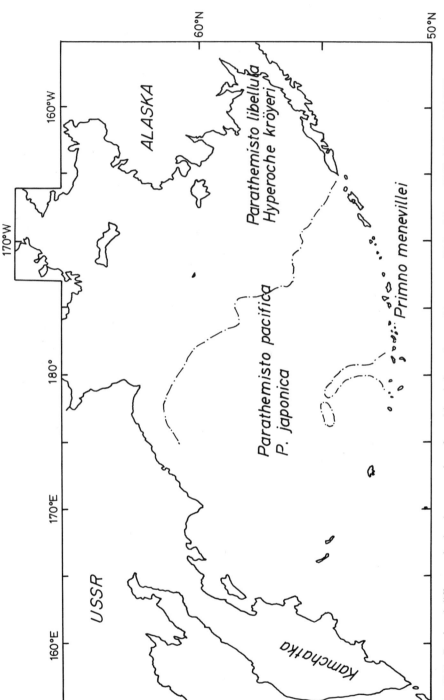

Fig. 10.7 Regional difference in dominant species of amphipods in early to mid-summer (Fukuchi, ms, 1970; Takeuchi 1972).

Appendicularia species *Oikopleura labradoriensis* and *Fritillaria borealis* are collected in the western and central Bering Sea, the former extending its distribution to the eastern shelf (Yamamoto, ms, 1959; Shiga, ms, 1971).

PLANKTON STANDING STOCKS

Diatoms

The diatom standing crop at the surface was determined by cell count on samples obtained both by dipping the surface water and by vertical haul (0–50 m) with a fine mesh net (>0.1-mm). In general, early to mid-summer standing crops were $1 \times 10^5 - 10^7$ cells/m^3 in the 0–50 m net haul (Motoda and Kawarada 1955; Karohji 1958), while it was $1 \times 10^5 - 10^9$ cells/m^3 (Kawarada 1957) or $1 \times 10^3 - 10^9$ cells/m^3 (Ohwada and Kon 1963) in surface water samples. It was indicated in 1967 and 1968 by water samples collected from several depths in the euphotic zone (0–50 m) that every depth contained $> 10^7$ cells/m^3 without marked variations throughout the depths (Faculty of Fisheries, Hokkaido University 1969, pp. 66–68 and 395–398): On the east coast of Nunivak Island peridinian groups were dominant in 1968, differing from other regions where phytoplankton communities are composed mainly of diatoms. The regional difference in diatom standing crops in 1955 is illustrated in Figure 10.8, based on observations of Kawarada (1957), supplemented by data near Nunivak Island obtained by Ohwada and Kon (1963) in 1960. The poorest area (10^5 cells/m^3) is shown to be the offshore water of Bristol Bay. In the central Bering Sea and off Kamchatka Peninsula in the western Bering Sea, diatom crops are also poor (10^6 cells/m^3). A remarkably dense standing crop (10^9 cells/m^3) is seen in the area north of the central Bering Sea.

Cupp (1937) observed seasonal distribution and occurrence of marine diatoms and dinoflagellates at Scotch Cap Light on Unimak Island in the Aleutian arc by daily samplings (dipping the surface water and filtering with fine mesh net) for seven years from August 1926 to June 1933. The largest catch, 6.3×10^8 cells/m^3, was obtained in the last week of April. She did not find any sign of more abundant diatom production at Scotch Cap Light compared with that of the California coast and Friday Harbor, Washington. The yearly cycle of diatom abundance at Scotch Cap Light closely resembles that usually found in temperate and northerly regions. The spring increase appears suddenly in April and continues until May; numbers then decline through June, July, and August. The fall maximum occurs in September, and numbers again decline through October and November with December to March being the months of poorest productivity. Karohji (1958) obtained a diatom catch at a station north of Unimak Pass on 20 July 1955 with as many as 9959 cells/m^3 by vertical haul in the upper 50 m; he states that his record is somewhat smaller than the summer collection of Cupp (1937) at Scotch Cap Light.

Zooplankton biomass

Zooplankton standard sampling has been continued in the Bering Sea every early summer or mid-summer for 15 years from 1956 to 1970 with 0–150 m

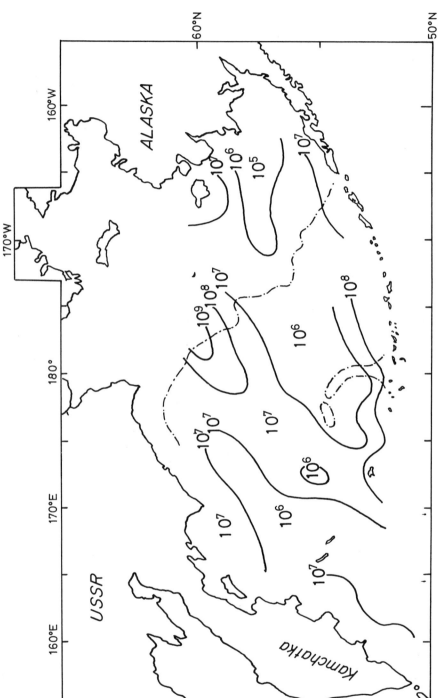

Fig. 10.8 Regional distribution of diatom standing crops at the surface in early to mid-summer (cells/m³) (Kawarada 1957; Ohwada and Kon 1963).

hauls in the deep basin and 0-m to bottom hauls on the shelf. Zooplankton biomass is given by wet weight of samples (g/1000 m³), averaging the water column sampled (Faculty of Fisheries, Hokkaido University, 1957–70). It is unreasonable to use this value, however, to compare the biomass obtained by a 0–150 m haul in the deep basin to that obtained by a shallow haul on the shelf without knowing the vertical variation of biomass in the deep basin. In 1970 samplings with multiple horizontal nets were made through 19 strata from the surface to 1000 m both during the day and at night at a station (Os 27) north of the Aleutian Islands at 172°W (Motoda 1971). Day and night variations of biomass in this range are illustrated in Figure 10.9, which indicates that the major portion of biomass is located in the upper 80 m; 61.6 percent of the total biomass (in the 0–150 m layer) are present in the upper 50 m, and 80.3 percent of the total biomass (in the upper 150 m) occur in the upper 80 m. Although the data were obtained at only one station in the Bering Sea, this value was used to estimate the biomass in the upper 150 m in the deep basin (g/1000 m³) for expressing in g/m² the biomass in the upper 80 m. On the shelf, where hauls were made from shallow bottom depths less than 80 m, the biomass was compared without any correction to the biomass in the upper 80 m in the deep basin. Values obtained in the 0–150 m haul in the deep basin were multiplied by 0.8.

The regional difference in zooplankton biomass, thus adjusted to g/m² (0–80 m) in each grid of 5 degrees during early to mid-summer on an average for 15 years, is illustrated in Figure 10.10. Values from stations located just on the degree lines are included to the left and lower area enclosed by the degree lines. It is shown that the highest value of biomass (67.1 g/m², 0–80 m) is found around the Pribilof Islands. In the central part of the Bering Sea the biomass is poor; 31.0–33.3 g/m² in the Bering Sea Gyre, and 22.8–28.4 g/m² around Bower's Bank. On the eastern shelf, biomass is poor in the northern part but rich in the southern part. In the inner part of Bristol Bay, the biomass is not rich (37.2 g/m², 0-m to bottom). In the water off Kamchatka Peninsula, the biomass is as high as 44.0 g/m² (0–80 m) but becomes poor on the coast of Siberia south of Cape Navarin (24.8 g/m², 0–80 m). The average biomass for 1956–70 in the upper 80 m in the deep basin and on the eastern shelf (except the shallow inner part of Bristol Bay) is calculated as 36.8 g/m². From a glance at this value and Figure 10.8, it is noticed that phytoplankton is rich in the central part of the Bering Sea where zooplankton is poor. Similar inverse conditions are observed in the northeastern and southern parts near the Aleutian Islands. The water off Kamchatka is rich in both phytoplankton and zooplankton. As a whole, it could be said that the level of zooplankton abundance in the Bering Sea in early to mid-summer is considerably high compared with that in the subarctic water south of the Aleutian Islands, Odate (1966) reports similar values of plankton biomass in the Bering and Okhotsk seas, about four times those in the Oyashio Current off the Japan islands.

Year-to-year variation of zooplankton biomass

Because different stations are occupied by the *Oshoro Maru* on its cruises each year, it is impossible to trace the year-to-year variation of zooplankton biomass throughout the entire area of the Bering Sea. Stations in a grid of

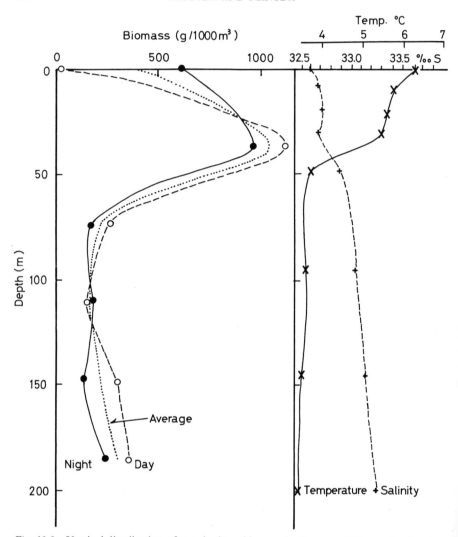

Fig. 10.9 Vertical distribution of zooplankton biomass in the upper 200 m in the daytime and at night at 54°N, 172°W, 24 June 1970 (g wet wt/1000 m³). Eighty percent of biomass in 0–150 m are located in upper 80 m.

55°–60°N, 170°E–175°W are present in certain numbers every year from 1956–68 except 1957, however, which allows comparison of biomass in different years within this particular grid (Fig. 10.11). The biomass of all four grids together varied from 27.9 g/m² (0–150 m) in 1956 to 111.6 g/m² (0–150 m) in 1965, averaging 74.1 g/m² (0–150 m) over 12 years. The biomass was large in 1958, 1962, 1965, and 1968, while it was small in 1956, 1961, 1964, and 1967. This would suggest that there is an apparent periodical variation of biomass with

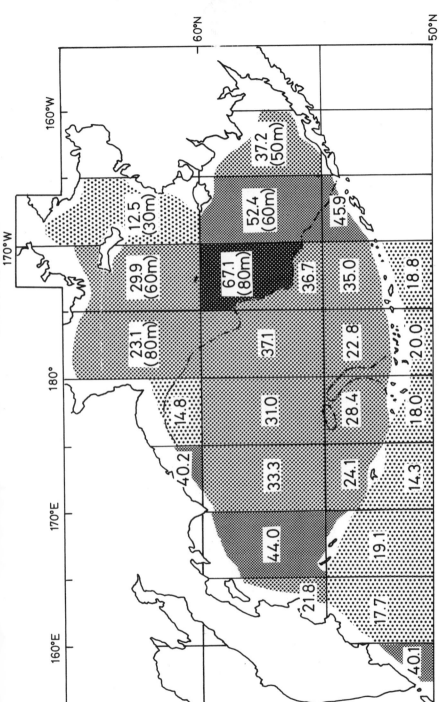

Fig. 10.10 Average summer zooplankton biomass for 15 years from 1956 to 1970 in each 5-degree grid. Values are expressed in wet wt g/m² in 80-m water column.

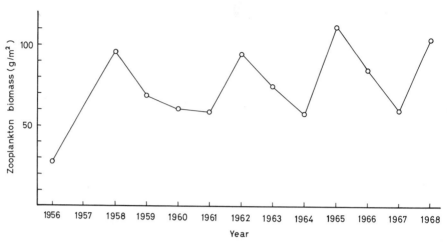

Fig. 10.11   Year-to-year variations in mean zooplankton biomass in upper 150 m of grid 55°–60°N, 170°E–175°W (excluding shelf area), 1956–68.

an interval of two to three years, which may reflect periodical variation of meteorological conditions.

Phytoplankton chlorophyll and primary production

Estimation of phytoplankton chlorophyll and primary production in the Bering Sea has been recently developed. Holmes (1958) observed surface chlorophyll *a* and primary production in the eastern Pacific Ocean and recorded values of 10 mg Chl a/m^3 and 71 mg C/m^3-day, respectively, in the area just north of the Aleutian Islands (about 165°W). Motoda and Kawamura (1963) reported light assimilation curves of surface phytoplankton in the Bering Sea on a cruise of the *Oshoro Maru*, June–July 1960. Kawamura (1963) later reported chlorophyll *a* and primary production at the surface in more detail on the same cruise, obtaining average values of 0.62 mg Chl a/m^3 and 0.79 mg C/m^3-hr in the western Bering Sea; 0.29 mg Chl a/m^3 and 1.14 mg C/m^3-hr in the central Bering Sea north of the Aleutian Islands (about 172°W); 0.22 mg Chl a/m^3 and 0.44 mg C/m^3-hr on the northern shelf; and 2.49 mg Chl a/m^3 and 2.93 mg C/m^3-hr on the southeastern shelf off Bristol Bay. McAlister et al. (1968) obtained average values of 6.6 mg Chl a/m^3 and 38 mg C/m^2-day near Amchitka in the Aleutian Islands. Taniguchi (1969) estimated chlorophyll *a*, primary production rates at the surface, and water column rates of primary production on a cruise of the *Oshoro Maru* in summer 1967. He found chlorophyll *a* and photosynthesis at the surface to be 0.23 mg Chl a/m^3 and 1.71 mg C/m^3-hr, respectively, in the central Bering Sea. *In situ* primary production measurement was also undertaken on this cruise, with results of 340 mg C/m^2-day in the Bering Sea Gyre and 630 mg C/m^2-day east of Bowers Bank. Taguchi (1972) measured primary production on an *Oshoro Maru* cruise in summer 1968, estimating values of 0.458 g C/m^2-day in the eastern shallow water and 0.33 g C/m^2-day in the central Bering Sea. McRoy et al. (1972) observed primary production in February–March and June–July in the eastern Bering Sea, reporting surface photosynthesis rates in the Aleutian area in summer of 0.15–11.0 mg C/m^3-hr (2.2–165 mg C/m^3-day) and corresponding chlorophyll values of 0.2–9.9 mg Chl a/m^3. The average integrated water column rate was 243 mg C/m^2-day. They describe the Aleutian passes as not very productive in comparison to high rates observed along the lee side of the Aleutian Islands. The most productive area (410 mg C/m^3-day at the surface) was found in the Bering Strait in summer. Larrance (1971) demonstrates a seasonal change of primary production in the Alaskan Stream south of the Aleutians; estimated values were 229 mg C/m^2-day in March, 290 mg C/m^2-day in June–July, and 71 mg C/m^2-day in January–February. On the basis of these fragmented data, it can be assumed that primary productivity at the surface in summer is about 0.8 mg C/m^3-hr in the western Bering Sea (in which the Bering Sea Gyre area is most productive with 1.6 mg C/m^3-hr); 1.1–1.7 mg C/m^3-hr in the central Bering Sea; on the continental shelf of the northeastern Bering Sea, the northern part is far less productive (0.4 mg C/m^3-hr) than the southeastern part (3.0 mg C/m^3-hr); Bristol Bay is relatively productive (1.5 mg C/m^3-hr), and the Aleutian area rate is 0.15–11.0 mg C/m^3-hr, being especially highly productive east of Bowers Bank (5.0 mg C/m^3-hr). Water column rates of productivity are 330 mg C/m^2-day in the central Bering Sea; 340 mg C/m^2-day in the area of the Bering Sea Gyre in the central Bering Sea; very productive (630 mg C/m^2-day); 243 mg C/m^2-day in the Aleutian area (38 mg C/m^2-day in Amchitka Pass); and 458 mg C/m^2-day on the eastern shelf. Regional variations in primary productivity generally coincide directly with the grades of phytoplankton standing crops (Fig. 10.8) and inversely with the abundance of zooplankton biomass (Fig. 10.10).

PHYTOPLANKTON—ZOOPLANKTON RELATIONSHIP

On a cruise in 1968, four stations were occupied on a line from 52°30′N to 59°30′N along 178°W in the Bering Sea, for estimation of phytoplankton chlorophyll and phytoplankton primary productivity in addition to the standard hauls of zooplankton sampling during the period from June 10 to 16 (Faculty of Fisheries, Hokkaido University 1969, pp. 1–135). Chlorophyll a and primary productivity were measured (Taguchi 1972; Taguchi and Ishii 1972). Phytoplankton carbon was converted from phytoplankton chlorophyll a (52.96 mg Chl a/m^2 in the euphotic water column) by multiplying by 30, since the community was composed almost entirely of diatoms (cf. Strickland 1960). Therefore, the turnover rate (Cushing et al. 1958), calculated as primary productivity (1.48 g C/m^2-day)/standing stock (1.589 g C/m^2) by the carbon method was an average of 0.93 over four stations. A factor of 100:5 was given by M. Kotori (unpublished) to convert zooplankton wet weight to zooplankton carbon on R/V *Hakuho Maru* cruise KH-69-4 in the North Pacific. The ratio of zooplankton (g C/m^2, 0–80 m) to phytoplankton (g C/m^2, 0–50 m) becomes 1.25, which is much higher than commonly recognized in the other oceanic regions.

SUMMARY

1. A list of phytoplankton and zooplankton species recorded in the Bering Sea was compiled by referring to accessible literature and manuscripts. The compilation is tentative, and the list should be supplemented and revised as possible in the future.

2. The surface water of the western and central Bering Sea in early to mid-summer is characterized by the predominance of boreal-oceanic diatom communities, while the water along the Aleutian Islands east of Attu is characterized by temperate-neritic communities. On the continental shelf of the eastern Bering Sea (except the inner shallow basin of Bristol Bay), boreal-oceanic diatom communities predominate.

3. Subarctic oceanic copepods, such as *Calanus plumchrus*, *C. cristatus*, *Eucalanus bungii bungii*, and *Metridia pacifica*, are predominantly distributed in the upper layer of the western and central parts of the Bering Sea in early to mid-summer. On the continental shelf of the eastern Bering Sea, communities composed mainly of arctic *Calanus glacialis* and arctic-boreal *Acartia longiremis* are present. Along the edge of the continental shelf of the eastern Bering Sea and around Bowers Bank north of the Aleutian Islands, mixed copepod communities of subarctic oceanic species and *Calanus glacialis* with *Acartia longiremis* are found. *Pseudocalanus minutus* and *Oithona similis* are widely distributed in the Bering Sea.

4. Only one species, *Acartia longiremis*, is confined to the upper 200 m in early to mid-summer. Another 26 species of copepods appearing in the epipelagic zone (0–200 m) extend their vertical distribution down to the transition zone (200–500 m) or bathypelagic zone (below 500 m). Twenty-two species are distributed from the epipelagic zone to 2000 m. On the other hand, 42 species always inhabit the transition zone or bathypelagic zone, never appearing in the epipelagic zone. Fifteen species occur only below 1000 m.

5. Dominant species of euphausiids in the Bering Sea in early to mid-summer are *Thysanoessa longipes*, which is distributed in the central Bering Sea, and *Thy. raschii* which occurs in shallow Bristol Bay and in the Gulf of Anadyr. Amphipods *Parathemisto pacifica* and *P. japonica* are distributed in the western and central Bering Sea, while *P. libellula* is found on the eastern shelf, sometimes in the western deep basin. *Sagitta elegans* is a representative species of chaetognaths in the Bering Sea from early to mid-summer. Two pteropods, *Limacina helicina helicina* and *Clione limacina limacina*, are distributed in the western and central Bering Sea and also on the eastern shelf. A pelagic polychaete *Tomopteris septentrionalis* appears in the same regions as the above pteropods. Two appendicularians, *Oikopleura labradoriensis* and *Fritillaria borealis*, are collected in the western and central Bering Sea. The former extends its distribution to the eastern shelf.

6. Diatom standing crops at the surface of the Bering Sea in early to mid-summer are generally within 10^5–10^9 cells/m^3, showing nearly similar density from the surface to 50 m depth. Peridinians were dominant on the east coast of Nunivak Island in 1969. There is a general tendency of decreasing diatom standing crops in the offshore water of Bristol Bay (10^5 cells/m^3), in

the central Bering Sea and off Kamchatka Peninsula (10^6 cells/m³) and an increasing tendency in the area north of the central Bering Sea (10^9 cells/m³).

7. Zooplankton biomass in the upper 80 m in early to mid-summer is considerably poor in the area of the Bering Sea Gyre in the central Bering Sea and around Bowers Bank (28.4–33.3 g wet wt./m²). The biomass off Kamchatka Peninsula is very rich (44.0 g wet wt./m²). The biomass in the upper 80 m in the Bering Sea as a whole (except for shallow inner Bristol Bay) is 36.8 g wet wt./m² on an average for 1956–70. It is noticed that phytoplankton is poor in the central Bering Sea, while zooplankton biomass is rich. A similar inverse relationship is observed in the northeastern and southern parts of the Bering Sea. Both phytoplankton and zooplankton are rich off the Kamchatka coast, however.

8. There is an apparent periodical variation with intervals of two to three years in zooplankton biomass in the Bering Sea. The biomass was large in 1958, 1962, 1965, and 1968, while it was small in 1956, 1961, 1964, and 1967.

9. Citing the observations in the Bering Sea (178°W) in early summer of 1968, it is calculated that the turnover rate of phytoplankton is 0.93 and that the ratio of zooplankton in the upper 80 m to phytoplankton in the upper 50 m is 1.25 based on carbon.

Discussion

GIRS: Your illustration showed a plankton variation from year to year. Do you concur with me that this might correlate with the annual atmospheric circulation?

MOTODA: I have not investigated that aspect, but it sounds interesting.

APPENDIX 10.1

List of plankton species recorded in the Bering Sea

BACILLARIOPHYCEAE pelagic diatoms

Porosira glacialis	Thalassiosira gravida
Melosira sulcata	T. hyalina
Coscinodiscus excentricus	T. baltica
C. concinnus	T. condensata
C. marginatus	T. japonica
C. centralis	Coscinosira polychorda
C. asteromphalus	Lauderia glacialis
C. granii	Bacterosira fragilis
C. oculus-iridis	Skeletonema costatum
Actinoptychus undulatus	Stephanopyxis nipponica
Asteromphalus heptactis	Leptocylindrus danicus
Actinocyclus ehrenbergi	L. minimus
Thalassiosira nordenskiöldii	Corethron hystrix
T. decipiens	Rhizosolenia alata

Rhizosolenia alata f. curvirostris
R. alata f. inermis
R. alata f. gracilima
R. obtusa
R. fragilissima
R. imbricata
R. imbricata f. shrubsolei
R. hebetata f. hiemalis
R. hebetata f. semispina
Chaetoceros atlanticus
C. atlanticus v. neapolitana
C. densus
C. borealis
C. concavicornis
C. convolutus
C. decipiens
C. decipiens f. singularis
C. teres
C. compressus
C. didymus
C. constrictus
C. laciniosus
C. distans
C. brevis
C. debilis
C. subsecundus
C. seiracanthus

Chaeroceros radicans
C. furcellatus
C. socialis
Biddulphia aurita
Bellerochea malleus
Ditylum brightwellii
Eucampia zoodiacus
E. groenlandica
Fragilaria inlandica
F. oceanica
F. cylindrus
F. striatula
Thalassionema nitzschioides
Thalassiothrix longissima
T. frauenfeldii
Achnanthes longipes
Navicula granii
N. vanhöffenii
Pleurosigma sp.
Amphiprora hyperborea
Amphora sp.
Denticula seminae
Nitzschia seriata
N. longissima
N. closterium
N. delicatissima

BACILLARIOPHYCEAE adhering to whale skin film

Melosira sulcata
Coscinodiscus anguste-lineatus
C. kützingi
C. radiatus
C. weilesii
Detonula confervacea
Leptocylindrus minimum
Rhizosolenia styliformis
Biddulphia aurita
Licmophora abbreviata
Raphoneis amphiceros
Synedra camtschatica
S. tabulata**
S. hennediana
S. karcheri**

Thalassiothrix longissima
Cocconeis costata
C. costata v. pacifica f. plana**
C. scutellum v. stauroneiformis
C. cetiola*
C. cetiola f. constrica*
Rhoicosphenia pullus
Stauroneis olympia*
S. aleutica*
S. omurai*
Navicula ammophila v. intermedia
N. arenaria
Gomphonema kamtschaticum**
G. kamtchaticum v. californica**
Nitzschia tubicole

* parasitic
** symbiotic

DINOFLAGELLATA

Peridinium thorianum
P. pallidum
P. depressum

Peridinium ovatum
P. pellucidum
P. granii

Peridinium excentricum
P. roseum
P. catenatum
Dinophysis acuta
Ceratium pentagonum

Ceratium longipes
C. lineatum
C. fusus
C. tripos v. *atlanticum*
C. furca

FORAMINIFERA

Globigerina bulloides
G. dubia
G. pachyderma

Globigerina quinqueloba
Globigerinella aequilateralis

RADIOLARIA

Challengeria xiphodon?
Protocystis vicina?
P. harstoni?
P. ornithocephala?
Rhizoplegma boreale
Xiphosphaera planeta
X. sp.
Xiphostylus sp.
Hexalonche brevicornis
H. sp.
Coscinomma ectosiphon
Cladococcus sp.

Cormyomma circumtextum
Heliodiscus echiniscus
Stylochlamidium venustum
Litharachnium araneosum
Cornutella sp.
Lychnocanium sigmopodium
Sethphormis sp.
Sethoconus pileus
Lophoconus sp.
Pterocorys hirundo
Aulastrum triceros
Porosphathis holostoma

TINTINNOINEA

Acanthostomella norvegica
Ptychocylis obtusa
Parafavella subrotendata

Parafavella jørgensenii
P. ventricosa
Codonellopsis frigida

SIPHONOPHORA

Marrus antarcticus pacifica
Bargmannia elongata
Ramosia vitiazi
Vogtia serrata
Rosacea plicata

Chuniphyes moserae
Lensia achilles bigelowi
L. reticulata
Muggiaea bargmannae
Dimophyes arctica

TRACHYLINA

Aglantha sp.

CTENOPHORA

Beroë cucumis

SCYPHOMEDUSAE

Atolla wyvillei
Periphylla hyacinthina
Chrysaora helvola
C. melanaster
Cyanea capillata

Phacellophora camtschatica
Aurelia aurita
A. limbata
Haliclystus auricula
H. stejnegeri

PTEROPODA

Limacina helicina helicina

Clione limacina limacina

POLYCHAETA

Tomopteris septentrionalis

CHAETOGNATHA

Sagitta elegans
S. maxima
S. macrocephala

Eukrohnia hamata
E. bathypelagica
E. fowleri

COPEPODA

Calanus glacialis (= *C. finmarchicus)*
C. cristatus
C. plumchrus (= *C. tonsus)*
Bathycalanus bradyi
Eucalanus bungii bungii
Paracalanus parvus
Pseudocalanus minutus
    (P. elongatus)
Microcalanus pygmaeus
M. pusillus
Spinocalanus magnus
S. stellatus
S. spinipes
S. abyssalis
Mimocalanus cultrifer
M. distinctocephalus
Aetideus armatus
A. pacificus
Gaidius tenuispinus
G. variabilis
G. brevispinus
Undinopsis pacificus
Aetideopsis multiserrata
Gaetanus armiger
G. simplex
Euchirella sp.
Pseudochirella polyspina
P. pacificus
P. spectabilis
P. spinifera
Pareuchaeta elongata
    (= *P. japonica)*
P. birostrata
P. brevirostris
P. rubra
P. crassa
P. californica
Valdiviella brevicornis
Xanthocalanus kurilensis
Onchocalanus magnus
Cornucalanus indicus
Scottocalanus securifrons

Scaphocalanus magnus
S. major
S. brevicornis
S. subbrevicornis
S. insignis
Scolecithricella valida
    (= *Amallothrix valida)*
S. emarginata (= *A. inornata)*
S. minor
S. ovata
S. globulosa
Racovitzanus antarcticus
R. erraticus
Undinella brevipes
U. oblonga
Eurytemora pacifica
E. herdmani
E. thompsoni
E. kieferi
E. hirundoides
Metridia pacifica (M. luccens)
M. okhotensis (M. longa)
M. asymmetrica
M. curticauda
M. brevicauda
M. gurjanovae
M. ornata
Pleuromamma robusta
P. scutulata
Centropages abdominalis
    (= *C. mcmurrichi)*
Lucicutia pacifica
L. ellipsoidalis
L. gradis
L. frigida
L. polaris
L. ovaliformis
Heterorhabdus papilliger
H. abyssalis
H. robstoides
H. tanneri
Heterostylites major

Haloptilus oxycephalus
H. pseudooxycephalus
H. longicirrus
Augaptilus glacialis
Euaugaptilus graciloides
E. mixtus
Centraugaptilus horridus
Pachyptilus pacificus
Arietellus simplex
Pseudaugaptilus sp.
Candacia columbiae
C. parafalcifera
Epilabidocera amphitrites
Acartia longiremis
A. clausi

Acartia tumida
Tortanus discudatus
Mormonilla minor
M. phasma
Oithona plumifera
O. similis
Oncaea conifera
O. borealis
O. notupus
O. ornata
Sapphoncaea moria
Lubbockia glacialis
Danodes plumata
Microsetella norvegica
Aegisthus aculeatus

AMPHIPODA—Hyperiidea

Parathemisto japonica
P. pacifica
P. libellula

Hyperia galba
H. medusarum
Hyperoche kroyeri

MYSIDACEA

Boreomysis inermis
B. californica
Archaeomysis grebnitzkii
Amblyops abbreviata
Holmesiella anomala
Inusitatomysis serrata
Stilomysis camtschatca
Neomysis awatschensis

Neomysis intermedia
N. mirabilis
N. rayii
N. czerniawskii
Acanthomysis stelleri
A. dybowskii
A. pseudomacropsis
Paracanthomysis kurilensis

EUPHAUSIACEA

Euphausia pacifica
Thysanoessa spinifera
Thy. longipes
Thy. inermis

Thysanoessa raschii
Thy. inspinata
Tessarabrachion oculatus

APPENDICULARIA

Oikopleura labradoriensis

Fritillaria borealis f. typica

DATA SOURCE (*unpublished manuscripts) FOR ABOVE COMPILATION:

Bacillariophyceae: Kisselev 1937 (cited from Zenkevitch 1963); Motoda and Kawarada 1955; Marumo 1956; Nemoto 1956; Kawarada 1957; Kawarada and Ohwada 1957; Karohji 1958, 1959; Kawamura 1963; Ohwada and Kon 1963; Tsuruta 1963.

Dinoflagellata: Kisselev 1937 (cited from Zenkevitch 1963); Tsuruta 1963; Faculty of Fisheries, Hokkaido University (identified by Taguchi).

Foraminifera: Ch. Yamade* (1971).

Radiolaria: Ch. Yamade* (1971); Z. Nakai (personal communication).

Tintinnoinea: Tsuruta 1963.

Siphonophora: Stephanynthes 1967.

Trachylina: S. Kurioka* (1963).

Ctenophora: Takeuchi 1972.

Scyphomedusae: Naumov 1961.

Pteropoda: S. Yamamoto* (1959); T. Tanaka* (1969).

Polychaeta: Takeuchi 1972.

Chaetognatha: Tchindonova 1955; Alvariño 1962, 1964; Kotori 1969, 1972.

Copepoda: Brodskii 1950; Johnson 1953; Anraku 1954; J. Ito* (1957); Minoda 1958, 1971; K. Nishio* (1961); A. Kawamura* (1962); K. Koseki* (1962); Y. Morioka* (1963, 1970); S. Yamazaki* (1963); Omori 1965; T. Matsumura* (1966); S. Hamaoka* (1970).

Amphipoda: Y. Yabuguchi* (1957); Bowman 1960; M. Fukuchi* (1970); Takeuchi 1972.

Mysidacea: Ii 1964.

Euphausiacea: Brinton 1962; Nemoto 1962, 1963; M. Fukuchi* (1970); Takeuchi 1972.

Appendicularia: S. Yamamoto* (1959); N. Shiga* (1971).

REFERENCES

AIKAWA, H.
  1932　On the summer plankton in the waters of the west Aleutian Islands in 1928 [in Japanese]. *Bull. Jap. Soc. Sci. Fish.* 1: 70–74.
  1940　On the plankton associations in the Bering Sea and the Okhotsk Sea [in Japanese]. *Kaiyo-Gyogyo* 5: 20–31.

ALVARIÑO, A.
  1962　Two new Pacific chaetognaths, their distribution and relationship to allied species. *Bull. Scripps Inst. Oceanogr.* 8: 1–50.
  1964　Bathymetric distribution of chaetognaths. *Pac. Sci.* 18: 64–82.

ANRAKU, M.
  1954　Gymnoplea Copepoda collected in Aleutian waters in 1953. *Bull. Fac. Fish., Hokkaido Univ.* 5: 123–136.

BOWMAN, T. E.
  1960　The pelagic amphipod genus *Parathemisto* (Hyperiidea-Hyperiidae) in the North Pacific and adjacent Arctic Ocean. *Proc. U. S. Nat. Mus., Smithsonian Inst., Bull.* 112: 343–392.

BRINTON, E.
  1962　The distribution of Pacific euphausiids. *Bull. Scripps Inst. Oceanogr.* 2: 51–126.

BRODSKII, K. A.
  1948　The pelagic Copepoda in the Japan Sea [in Russian]. *Bull. Pac. Sci. Inst. Fish. Oceanogr. Vladivostock* 26: 3–130.
  1950　Calanoida of the far eastern seas and polar basin of the USSR [in Russian]. *Izd. Akad. Nauk SSSR*; 440 pp. (English translation, 1967).

CUPP, E. E.

1937 Seasonal distribution and occurrence of marine diatoms and dinoflagellates at Scotch Cap, Alaska. *Bull. Scripps Inst. Oceanogr. Tech. Serv.* 4: 71–100.

CUSHING, D. H., G. F. HUMPHREY, K. BANSE, and TAIVO LAEVASTU.

1958 Report of the committee on terms and equivalents. *Rapp. P.-v. Réun. Cons. perm. int. Explor. Mer.* 144: 15–16.

DODIMEAD, A. J., F. FAVORITE, and T. HIRANO

1963 Review of oceanography of the subarctic Pacific region. *Bull. Int. North Pac. Fish. Comm.* 13, 195 pp.

DOGIEL, V. A., and V. V. RESCHETNJAK

1952 Material of Radiolaria in the North-western part of the Pacific Ocean [in Russian]. *Res. Far-Eastern Seas USSR*, 3: 5–19.

FACULTY of FISHERIES, HOKKAIDO UNIVERSITY

1957 Data record of oceanographic observations and exploratory fishing. *Bull. Fac.*
–70 *Fish. Hokkaido Univ.*, Nos. 1–14.

FUKUCHI, M.

1970 Regional distribution of Euphausiacea and Amphipoda collected with a high speed sampler at the surface in the Bering Sea and northern North Pacific in summer of 1969. Manuscript, Fac. Fish., Hokkaido Univ.

FURUHASHI, K.

1965 Occurrence of a cold water copepod, *Metridia lucens* Boeck, in the Enshunada of the Kuroshio region [in Japanese, English abstract]. *Inform. Bull. Planktol. Japan* 12: 49–50.

HAMAOKA, S.

1970 Vertical distribution of copepods between surface through abyssal zone in the Bering Sea and northern North Pacific (Oshoro Maru cruise 1969, Palumbo-Chun-Petersen type double closing net). Manuscript, Fac. Fish., Hokkaido Univ.

HOLMES, R. W.

1958 Surface chlorophyll *a*, surface primary production and zooplankton volumes in the eastern Pacific Ocean. *Rapp. P.-v. Réun. Cons. perm. int. Explor. Mer.* 144: 109–116.

II, N.

1964 Mysidae (Crustacea). In *Fauna Japonica*. Biogeogr. Soc. Japan, Nat. Sci. Mus., Tokyo, 610 pp.

ISOBE, H.

1966 Note on the stomach contents of salmonid fishes in the eastern Bering Sea. Manuscript, Fac. Fish., Hokkaido Univ.

ITO, J.

1957 Regional distribution of copepods collected with a fish larva net at the surface in the Bering Sea and the northern North Pacific in 1955. Manuscript, Fac. Fish., Hokkaido Univ.

JASCHNOV, W. A.

  1955   Morphology, distribution, and systematism of *Calanus finmarchicus* s. [in Russian]. *Zool. Zhur.* 34: 1210–1222.

  1963   Water masses and plankton. 2. *Calanus glacialis* and *Calanus pacificus* as indicators of definite water masses in the Pacific [in Russian]. *Zool. Zhur.* 42: 1005–1021.

  1970   Distribution of *Calanus* species in the seas of the northern hemisphere. *Int. Revue ges. Hydrobiol.* 55: 197–212.

JOHNSON, M. W.

  1953   Studies on plankton of the Bering and Chukchi seas and adjacent area. *Proc., Seventh Pac. Sci. Cong.* 4: 480–500.

KAROHJI, K.

  1958   Diatom standing crops and the major constituents of the populations as observed by net sampling. 4. Report from the *Oshoro Maru* on oceanographic and biological investigations in the Bering Sea and northern North Pacific in the summer of 1955. *Bull. Fac. Fish., Hokkaido Univ.* 9: 243–252.

  1959   Diatom associations as observed by underway sampling. 6. Report from the *Oshoro Maru* on oceanographic and biological investigations in the Bering Sea and northern North Pacific in the summer of 1955. *Bull. Fac. Fish., Hokkaido Univ.* 9: 259–267.

KAWAMURA, A.

  1962   Distribution of copepods in the Bering Sea and the northern North Pacific in the summer of 1961 (Oshoro Maru cruise, standard net haul). Manuscript, Fac. Fish., Hokkaido Univ.

KAWAMURA, T.

  1963   Preliminary survey of primary production in the northern North Pacific and Bering Sea, June–August 1960 [in Japanese]. *Inform. Bull. Planktol. Japan* 10: 28–35.

KAWARADA, Y.

  1957   A contribution of microplankton observations to the hydrography of the northern North Pacific and adjacent seas. 2. Plankton diatoms in the Bering Sea in the summer of 1955. *J. Oceanogr. Soc. Japan* 13: 151–155.

KAWARADA, Y., and M. OHWADA

  1957   A contribution of microplankton observations to the hydrography of the northern North Pacific and adjacent seas. 1. Observations in the western North Pacific and Aleutian waters during the period from April to July 1954. *Oceanogr. Mag.* 14: 149–158.

KISSELEV, J.

  1937   Composition and distribution of phytoplankton in the northern part of the Bering Sea and in the southern part of the Chukchi Sea [in German]. *Tr. Inst. Okeanol. Akad. Nauk SSSR*, 25 pp.

KOSEKI, K.

  1962   Relationship between the distribution of copepods and water masses in the north-western North Pacific and the Bering Sea in summer of 1961 (Oshoro Maru cruise, fish larva net). Manuscript, Fac. Fish., Hokkaido Univ.

KOTORI, M.

1969   Vertical distribution of chaetognaths in the northern North Pacific and Bering Sea [in Japanese, English abstract]. *Bull. Plankt. Soc. Japan* 16: 52–57.

1972   Vertical distribution of chaetognaths in the northern North Pacific Ocean and the Bering Sea. In *Biological oceanography of the northern North Pacific Ocean* [Motoda commemorative volume], edited by A. Y. Takenouti et al. Idemitsu-shoten, Tokyo, pp. 291–308.

KOYAMA, A.

1967   Zooplankton abundance in the central Bering Sea in the summer of 1966 (Oshoro Maru cruise, standard net haul). Manuscript. Fac. Fish., Hokkaido Univ.

KURIOKA, S.

1963   Note on zooplankton in the Bering Sea and the northern North Pacific in the summer of 1962 (Oshoro Maru cruise, fish larva net). Manuscript, Fac. Fish., Hokkaido Univ.

KUSAJIMA, M.

1959   Zooplankton collected with a fish larva net at the surface in the Bering Sea and the northern North Pacific in the summer of 1957. 2. Chaetognatha. Manuscript, Fac. Fish., Hokkaido Univ.

LARRANCE, J. D.

1971   Primary production in the mid-subarctic Pacific region, 1966–68. *Fish. Bull.* 69: 595–613.

LEBRASSEUR, R. J.

1965   Seasonal and annual variations of net zooplankton at Ocean Station P 1956–64. Ms. Rep. Serv., Fish. Res. Bd. Can. 202: 163 pp.

MARUMO, R.

1956   Diatom communities in the Bering Sea and its neighboring waters in the summer of 1954. *Oceanogr. Mag.* 8: 69–73.

MATSUMURA, T.

1966   Observations on plankton biomass and copepods in the Bering Sea and the northern North Pacific in the summer of 1965. Manuscript, Fac. Fish., Hokkaido Univ.

MCALISTER, W. B., C. MAHNKEN, R. C. CLARK, JR., W. J. INGRAHAM, J. LARRANCE, and D. DAY

1968   Oceanography and marine ecology in the vicinity of Amchitka Island. Final Rep., Battelle Mem. Inst., Contr. #AT(26–1)–353, 146 pp.

MCROY, G. P., J. J. GOERING, and W. E. SHIELS

1972   Studies of primary production in the eastern Bering Sea. In *Biological oceanography of the northern North Pacific Ocean* [Motoda commemorative volume], edited by A. Y. Takenouti et al. Idemitsu-shoten, Tokyo, pp. 199–216.

238   MOTODA AND MINODA

MINODA, T.

  1958   Observations on copepod community. 5. Report from the *Oshoro Maru* on
         oceanographic and biological investigations in the Bering Sea and northern
         North Pacific in the summer of 1955. *Bull. Fac. Fish., Hokkaido Univ.* 8:
         253–263.
  1971   Pelagic Copepoda in the Bering Sea and the northwestern North Pacific
         with special reference to their vertical distribution. *Mem. Fac. Fish., Hokkaido
         Univ.* 18: 1–74.

MORIOKA, Y.

  1963   Notes on the copepods collected from the Bering Sea and the northern
         North Pacific in the summer of 1962 (Oshoro Maru cruise, standard net
         haul). Manuscript, Fac. Fish., Hokkaido Univ.
  1965   Intercalibration of catch efficiency between bolting silk net and pylen net
         [in Japanese, English abstract]. *Inform. Bull. Planktol. Japan* 12: 54–60.
  1970   Vertical distribution of Calanoida Copepoda in the northern and south-
         western North Pacific. Dissertation. Hokkaido Univ.

MOTODA, S.

  1957   North Pacific standard net [in Japanese, English abstract]. *Inform. Bull.
         Planktol. Japan* 4: 13–15.
  1971   Devices of simple plankton apparatus. *Bull. Fac. Fish., Hokkaido Univ.*
         22: 101–106.

MOTODA, S. M. ANRAKU, and T. MINODA

  1957   Experiments on the performance of plankton samplings with net. *Bull. Fac.
         Fish., Hokkaido Univ.* 8: 1–22.

MOTODA, S., and T. KAWAMURA

  1963   Light assimilation curves of surface phytoplankton in the North Pacific
         42°N–61°N. In *Proceedings of symposium on marine microbiology*, edited by
         C. H. Oppenheimer. Springfield, Ill., pp. 251–259.

MOTODA, S., and Y. KAWARADA

  1955   Diatom communities in western Aleutian waters on the basis of net samples
         collected in May–June 1953. *Bull. Fac. Fish., Hokkaido Univ.* 6: 191–200.

NAUMOV, S. V.

  1961   Scyphomedusae of seas of USSR [in Russian]. *Izd. Akad. Nauk SSSR* 75:
         1–98.

NEMOTO, T.

  1956   On the diatoms of the skin film of whales in the northern Pacific. *Sci. Rep.
         Whales Res. Inst.* 11: 99–132.
  1962   Distribution of five main euphausiids in the Bering Sea and the northern
         part of the North Pacific [in Japanese, English abstract]. *J. Oceanogr. Soc.
         Japan* [20th Aniv. Vol.]: 615–627.
  1963   A new species of Euphausiacea. *Thysanoessa inspinata*, from the North
         Pacific. *J. Oceanogr. Soc. Japan* 19: 41–47.

NISHIO, K.

1961 Regional distribution of copepods collected with a fish larva net at the surface in the Bering Sea and the northern North Pacific in the summer of 1959. Manuscript, Fac. Fish., Hokkaido Univ.

ODATE, K.

1966 On the comparative study of volumes of zooplankton distributing in the Oyashio area and in the seas adjacent to that area [in Japanese, English abstract]. *Bull. Tohoku Reg. Fish. Res. Lab.* 26: 45–53.

OHTANI, K.

1965 On the Alaskan Stream in summer [in Japanese, English abstract]. *Bull. Fac. Fish., Hokkaido Univ.* 15: 260–273.

OHWADA, M., and H. KON

1963 A microplankton survey as a contribution to the hydrography of the North Pacific and adjacent seas. 2. Distribution of the microplankton and their relation to the character of water masses in the Bering Sea and northern North Pacific Ocean in the summer of 1960. *Oceanogr. Mag.* 14: 87–99.

OMORI, M.

1965 The distribution of zooplankton in the Bering Sea and northern North Pacific as observed by high-speed sampling of the surface waters with special reference to the copepods. *J. Oceanogr. Soc. Japan* 21: 18–27.

OMORI, M., and O. TANAKA

1967 Distribution of some cold-water species of copepods in the Pacific water off east-central Honshu, Japan. *Oceanogr. Mag.* 21: 63–73.

RESCHETNJAK, V. V.

1955 Vertical distribution of the Radiolaria of the Kurile-Kamchatka Trench., *Zool. Inst. Akad. Nauk SSSR* 21: 94–101.

SEMINA, H. J.

1960 The influence of vertical circulation on the phytoplankton in the Bering Sea. *Int. Revue ges. Hydrobiol.* 45: 1–10.

SHIGA, N.

1971 Quantitative distribution of the appendicularian *Oikopleura labradriensis* Lohman in the eastern Bering Sea in summers 1966–1970. Manuscript, Fac. Fish., Hokkaido Univ.

STEPHANYNTHES, S. D.

1967 Siphonophores in the sea of USSR and North Pacific Ocean [in Russian]. *Izd. Akad. Nauk SSSR* 96: 1–216.

STRICKLAND, J. D. H.

1960 Measuring the production of marine phytoplankton. *Bull. Fish. Res. Bd. Can.* 122: 1–172.

TAGUCHI, S.

1972 Mathematical analysis of primary production in the Bering Sea in summer. In *Biological oceanography of the northern North Pacific Ocean* [Motoda commemorative volume], edited by A. Y. Takenouti et al. Idemitsu-shoten, Tokyo, pp. 253–262.

TAGUCHI, S., and H. ISHII

1972 Shipboard experiments on respiration, excretion, and grazing of *Calanus cristatus* and *C. plumchrus* (Copepoda) in the northern North Pacific. In *Biological oceanography of the northern North Pacific Ocean* [Motoda commemorative volume], edited by A. Y. Takenouti et al. Idemitsu-shoten, Tokyo, pp. 419–431.

TAKEUCHI, I.

1972 Food animals collected from the stomachs of three salmonid fishes (*Oncorhynchus*) and their distribution in the natural environments in the northern North Pacific [in Japanese, English summary]. *Bull. Hokkaido Reg. Fish. Res. Lab., Fisheries Agency*, No. 38: 112.

TANAKA, O.

1960 Pelagic Copepoda. Biol. Res. Japanese Antarc. Res. Exped. 10. *Spec. Publ., Seto Mar. Biol. Lab.*, 98 pp.

TANAKA, T.

1969 Distribution of the pteropods and heteropods in the Pacific Ocean. Manuscript, Fac. Fish., Hokkaido Univ.

TANIGUCHI, A.

1969 Regional variations of surface primary production in the Bering Sea in summer and vertical stability of water affecting the production. *Bull. Fac. Fish., Hokkaido Univ.* 20: 169–179.

TCHINDONOVA, T. G.

1955 Chaetognatha of the Kurile-Kamchatka Trench [in Russian]. *Tr. Inst. Okeanol. Akad. Nauk* 12: 298–310.

TSURUTA, A.

1963 Distribution of plankton and its characteristics in the oceanic fishing grounds with special reference to their relation to fishery. *J. Shimonoseki Univ. Fish.* 12: 13–214.

VINOGRADOV, M. E.

1956 Distribution of zooplankton in the western region of the Bering Sea [in Russian]. *Tr. Vsesoju-nogo Gidrobiologceslogo o-Ba* 7: 173–203.

WATANABE, N.

1954 Hydrographic data in the subarctic North Pacific Ocean, 1887–1953. *Jap. Ass. Agr. Tech.*, 556 pp.

WILSON, C. B.

1950 Copepods gathered by the United States Fisheries Steamer "Albatross" from 1887 to 1909, chiefly in the Pacific Ocean. *U. S. Nat. Mus., Smithsonian Inst. Bull.* 100: 1–141.

YABUGUCHI, Y.

1957 On amphipods, mysids and euphausiids collected with a larva net at the surface in the Bering Sea and the northern North Pacific in the summer of 1955. Manuscript, Fac. Fish., Hokkaido Univ.

YAMADE, CH.

1971 Foraminifera and Radiolaria in the Bering Sea and the northern North Pacific. Manuscript, Fac. Fish., Hokkaido Univ.

YAMAHANA, K.

1961 Biomass of copepods and other zooplankton in the Bering Sea and the northern North Pacific in the summer of 1960 (Oshoro Maru cruise, standard net haul). Manuscript, Fac. Fish., Hokkaido Univ.

YAMAMOTO, S.

1959 Zooplankton collected with a fish larva net at the surface in the Bering Sea and the northern North Pacific in the summer of 1957. 5, Appendicularia. Manuscript, Fac. Fish., Hokkaido Univ.

YAMAZAKI, S.

1963 Distribution of copepods in the northern North Pacific and the Bering Sea in the summer of 1962 (Oshoro Maru cruise, high-speed sampling). Manuscript, Fac. Fish., Hokkaido Univ.

ZENKEVITCH, L.

1963 Biology of the sea of the USSR [in Russian]. (English transl. by S. Botchrskaya). George Allen & Unwin, London, 955 pp.

CHAPTER 11

Fishes of the Bering Sea: the state of existing knowledge and requirements for future effective effort

NORMAN J. WILIMOVSKY

Institute of Animal Resource Ecology, University of British Columbia, Vancouver, B. C., Canada

Abstract

The ichthyofauna of the Bering Sea comprises about 300 species distributed among 45 families; eight families make up over 70 percent of the total. The numbers and kinds of benthic and pelagic fishes reflect the natural environmental divisions of the Bering Sea, which closely parallel its geographic boundaries. Two major environmental temperature regimes can be distinguished—the polar "Anadyr cold region" lying to the northwest of St. Matthew Island and the remaining more uniform boreal environment.

The cold region ichthyofauna for the most part are benthic species, related primarily to the Central Arctic fishes; the remainder of the Bering Sea fish fauna is composed mainly of boreal forms whose origins can be most easily traced in relation to their depth distribution. There is strong evidence for a filter bridge effect in diversity of shore fishes from east to west along the Aleutian Islands. A highly endemic fish fauna has been found to inhabit the Semisopochnoi Island-Petrel Bank area.

With only a few exceptions, our biological knowledge of the fish fauna of the Bering Sea is limited to forms exploited commercially. Assembly of information on the fishes of the Bering Sea is hampered not only by the scattered nature of the literature but by the fact that it is written in at least three languages. There is an overriding need for gathering together available information into a synthesized accessible form.

INTRODUCTION

Our first recorded knowledge of the fishes of the Bering Sea stems from the work of G. W. Steller, the naturalist who accompanied Bering's ill-fated second voyage in the middle of the eighteenth century. In spite of several waves or periods of investigations, our knowledge of the fishes of the Bering Sea—their distribution, origins, and biology—is still rudimentary.

The distribution of fishes, like that of other marine life, follows natural environmental features rather than the political boundaries of man's maps. Conveniently, the biogeographical limits of the Bering Sea biota closely parallel the geographic boundaries of the Bering Sea except in the far southwest portion. For the purposes of this analysis, the Bering Sea fauna is considered to extend from the Kamchatka Peninsula on the west to the Bering Strait in the north and Alaska in the east. The southern limit is defined by the Aleutian arc, including the Komandorskii Islands. The exact biogeographic limits in the southwest area of the Bering Sea between the Komandorskii Islands, Kamchatka, and the Kuriles has yet to be determined.

Historical summary

Exploratory period

Ichthyological explorations of the Bering Sea appear to have progressed in a series of stages in the advancement of our knowledge of the area. Investigations began with Vitus Bering's crossing of the sea which bears his name in 1732–49. Georg Wilhelm Steller's notes (Stejneger 1936, published posthumously by Pallas 1811 and Tilesius 1813) formed the first substance of our marine biological knowledge of the area. Bering's voyage was followed by a series of explorations by several nations of the world. The Russians, English, French, and Spaniards all sent ships staffed with naturalists and artists into the area. The total contribution of this "exploratory period" to our scientific knowledge was slight, and only a sketchy picture of the fish resources developed.

Albatross period

The second wave of explorations took place about the turn of the century, from the late 1800s through the early 1900s, motivated in part by the fur-seal question. Because of the fur-seal controversy and expanding fishery interests, the United States Fish Commission's steamer USS *Albatross* worked extensively in the Bering Sea and adjacent parts of the North Pacific. Her collections formed the basis for more than a score of ichthyological publications, chiefly by D. S. Jordan and colleagues (e.g., Jordan and Gilbert 1899).

Additional data were published as a result of Russian and Swedish work, the most notable being Shmidt's (1904).

In spite of increasing questions concerning the salmon and cod resources, most work in the Bering Sea following the " *Albatross* period" was very limited and often confined to coastal waters.

Soviet period

A third wave of data resulted from efforts of the Soviet oceanographic explorations of the late 1930s. One of the most significant works published to date appeared as a consequence of the "Soviet period": Andriashcv's (1939)

Zoogeography and origin of the fish fauna of Bering Sea, a thesis that was printed in the Russian language in small numbers with only limited distribution. During this time period, Japan also had numbers of ships operating in the Bering Sea. Unfortunately, I have been unable to trace any published reports of Japanese exploratory operations during this period.

Post World War II period

Following World War II, increased interest in commercial harvest resulted in a number of commercial exploratory ventures to the Bering Sea by the United States (Ellson et al. 1949) followed by a series of American cruises of broad oceanographic objectives. Then followed several extensive fishery expeditions by the Soviet Union and Japan, which resulted in subsequent distant-water fishery operations by both nations. Much valuable literature resulted from these efforts of the recent period (Romanov 1959; Moiseev 1963).

THE FISH FAUNA AND ITS ORIGINS

The ichthyofauna of the Bering Sea comprises about 300 species distributed among 45 families. These gross taxonomic values mean little until examined in light of their natural environmental groupings as well as their probable origins.

Habitat characterization

Within the multi-layered water masses of the Bering Sea, two major environmental temperature regimes can be distinguished—the polar "Anadyr cold region" lying to the northwest of St. Matthew Island and the remaining more uniform boreal environment. Superimposed on this gross temperature pattern are widely varying salinity values expressed primarily in the surface layers that range from the relatively high salinity of the western Bering Sea to the river-influenced low salinity estuarine environment of the eastern Bering Sea shelf. Three major groupings are suggested by bottom morphology: the immediate inshore regions throughout the periphery of the sea, the eastern shelf area, and the deep basins separated by the Olyutorskii Ridge. There is strong evidence that a fourth area should be distinguished—the submarine plateau forming the arcuate bank north of Semisopochnoi Island, including Petrel Bank. Finally, there is the specialized environmental region in close proximity to the drifting pack ice that supports a distinct biota. The geographical domain of this ice cover varies with season (Dunbar 1967).

Origin of distribution patterns

The numbers and kinds of benthic and pelagic fishes reflect these broad environmental divisions of the Bering Sea. Their limits bear out Ekman's law in that the zones of strongest faunal change coincide with the aforementioned geographical categories.

The negative temperatures of the "Anadyr cold region" support a very depauperate fauna of about 15 species. These are mainly benthic species whose origins lie primarily with the Central Arctic fishes. The remainder of the Bering

Sea fish fauna is composed primarily of boreal forms. The origins of the boreal Bering Sea fishes are most easily traced in relation to their depth distribution. The inshore fauna of the Aleutian Islands and Alaskan coasts are clearly derivable from eastern Pacific counterparts. The low-salinity water of the eastern Bering Sea shelf supports a euryhaline fauna of a very limited number of species. As elsewhere reported (Wilimovsky 1964) there is strong evidence for a filter-bridge effect along the Aleutian Islands as one examines the diversity of shore fishes from east to west. The inshore fauna of the western Bering Sea appears to be an extension of the lower Kamchatkan-Kurile fauna, although there is only a scant amount of material available from the Kamchatkan and Koriak shores. The shore fauna of the Anadyr Gulf is a non-distinctive mixture of arctic stragglers and anadromous forms.

The shore fishes as a group are dominated by the cottoid and stichaeid groups. Of the 30 families recorded within the Aleutian Islands, half are represented by a single species and the remainder by only a few. The Cottidae are dominant with 45 representatives; no other fauna in the world contains such a high proportion of cottid forms. The stichaeids and liparids are very evident and have reached an evolutionary "flowering" in the area.

The benthic fishes form two broad groups: one consisting of wide-ranging species exemplified by many pleuronectids and lycodids whose species migrate vertically up and down the banks seasonally, and the other is comprised of very deep-water benthic forms represented by several families of one or two species each. A number of the latter forms appear to be endemic to the basins lying on either side of the Olyutorskii Ridge. In contrast to the shore fauna, it would appear on the basis of present evidence that the benthic fauna from modest depths was derived from the north Japanese mother fauna, with one major exception—the Gadidae. It is likely, however, that the deeper benthic fishes have evolved in the Bering area and that their occurrence elsewhere is the result of population expansion since the late Pleistocene period. The forms inhabiting the extreme depths can be further subdivided according to the Soviet scheme of primary or secondary deepwater fish. Although these major divisions are useful in an evolutionary evaluation, they do not affect our environmental allocations.

The Semisopochnoi Island-Petrel Bank area is inhabited by a highly endemic fish fauna. It has many of the characteristics of an evolving island fauna and deserves careful study. Not only are the cottids, stichaeids, liparids, and other fish distinctive, but the molluscan fauna appears equally unique.

The pelagic fauna of the Bering Sea taken as a whole is highly diverse, comprising many relatively small-sized fishes, including some species suitable for commercial exploitation. The dominant fish food of the fur seal are inhabitants of the bathypelagic realm. Though biological data are lacking, many of the pelagic species appear to carry out extensive seasonal movements both horizontally and vertically. In the uppermost layers, the salmonid groups predominate; *Osmerus*, *Mallotus*, and *Oncorhynchus* are dominant. Somewhat deeper are the scorpaenids, gadids and their relatives and, finally, the macrouroids are found in the bathypelagic or engybenthic zones. To this list must be added a host of macro-planktonic forms such as the larva of hexagrammids, cottids, pleuronectids, etc.

Although only one or two species of fishes are associated with floating ice, the structure of the community and cryobiotic cycle in relation to the age of ice warrants more detailed study (Wilimovsky 1963). (Fig. 11.1)

With the exception of some of the deep-sea argentinids, none of the pelagic forms appear to be endemic to the Bering Sea, and the adult components of the pelagic fauna can only be categorized broadly as North Pacific in character. As with most pelagic forms, the species are wide-ranging and occur throughout the water masses characterized as "salmon water" by the oceanographer. The macroplankton is, of course, reflective of the larval stages of the inshore and benthic fauna.

Studies on the interface biota, collectively termed *neuston*, have just begun and no data on the Bering Sea have been published since the work of Zaitsev 1970.

The significance of Pleistocene and possibly later sea-level change to the distribution pattern and sources of evolving populations of the Bering Sea is illustrated in Figures 11.2 (a) and (b), which show the extent of enclosure of the region at maximum sea-level depression.

Numerical distribution

Analysis of data (Bean 1880; Turner 1886; Nelson 1887; Jordan and Gilbert 1899; Evermann and Goldsborough 1907; Taranetz 1937; Andriashev 1937, 1939, 1954; Shmidt 1948, 1950; Wilimovsky 1954; Okada and Kobayashi 1969), when adjusted for differing systematic ranking, misidentifications and synonyms, suggests that the Bering Sea supports a fish fauna of approximately 300 species.[1]

Of the approximately 300 species comprising the Bering Sea fish fauna, eight families make up nearly three-fourths of the total as shown in Table 11.1 below:

TABLE 11.1   Proportion of eight predominant families to total species composition (about 300) of Bering Sea fish fauna

Family	Percentage of total fish species
Cottidae	22
Liparididae	15
Stichaeidae	8
Pleuronectidae	8
Zoarcidae	6
Agonidae	5
Scorpaenidae	5
Salmonidae	4
Total of eight dominant families	73

The Bering Sea fish fauna is unique in having a higher proportion, by a considerable extent, of cottids and liparids than any other sea.

The vertical distribution of Bering Sea fishes in terms of numbers is as

[1] A recent Alaskan list (Quast and Hall 1972), which appeared after this analysis, apparently will have no significant effect on the numbers presented herein.

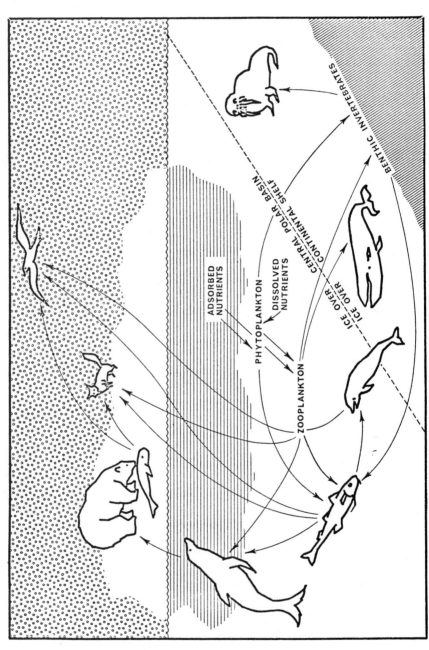

Fig. 11.1 Cryobiotic cycle (stylized) associated with floating ice (Wilimovsky 1963).

Fig. 11.2(a) Contemporary bathymetric chart of the Bering Sea (depths in meters).

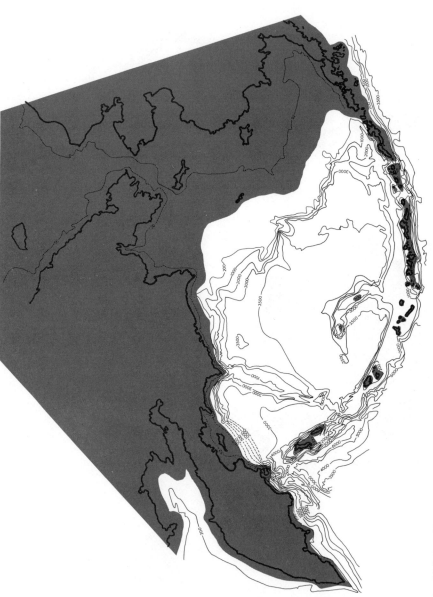

Fig. 11.2(b) Potential enclosure of Bering Sea at maximum sea level depression (depth contours in meters). The extent of sea level depression varied with time and place; this schematic represents the maximum depression and is

follows: the pelagic and bathypelagic fauna comprises about 40 species including a substantial proportion of anadromus and euryhaline forms. The inshore elements consist of about 80 species. The greatest proportion (about 150) of species are benthic fish which occur in subtidal or intermediate depths. Only about 35 species make up the deep benthic fauna.

The horizontal distribution of the various groupings can be clearly comprehended from the distributional diagrams of Shmidt (1950). These data indicate that 194 species occur in the western Bering Sea, 235 are found in the eastern Bering Sea, and the fish fauna drops to 66 species in the northern part and even fewer in the "Anadyr cold region". (As several species are wide-ranging, these totals are greater than 300.)

Short of a detailed assessment of the type to be proposed below and significant new data, any more specific analysis of the Bering Sea ichthyofauna at this time would be a simple duplication of effort of the excellent treatment of Andriashev (1939) and Shmidt (1950).

Commercial utilization of the Bering Sea fish fauna

With only a few exceptions, our biological knowledge of the fish fauna of the Bering Sea is limited to forms exploited commercially (e.g., Maeda 1963; Maeda et al. 1967). Even here many conclusions are extrapolated from data extralimital to the Bering Sea. Catch statistics have been assembled for many species, but only since the mid-1950s have such data been to any extent uniformly collected and reported. Indeed not all nations fishing in the Bering Sea report full catch data. The catch of principal ground species had risen from about 4×10^4 metric tons in 1954 to about 4×10^6 metric tons in 1969, a hundredfold increase. Population estimates of the available resources have been estimated (Fukuda 1970; Gulland 1970) on a gross basis and for some individual stocks (Alverson et al. 1964). These data suggest that the limits of harvest for several species have been attained or even surpassed. The strategies of harvest vary drastically from species to species. Attesting to the high rate of exploitation is the generally unrecognized fact that after the Peruvian anchovieta, the North Pacific hake is the next most heavily utilized species in the world catch.[2] The growth rate, migratory pattern of fish and fishermen, as well as exploitation of the region by at least five nations make the problem of base data acquisition for resource understanding and management an urgent matter. This fact prevails, despite any question of whether multi-national management is politically feasible.

Informational needs

Consideration of the state of information on the Bering Sea, as well as some of the kinds of problems concerning it, leads to the following conclusions. There are no concise summaries of the distribution of the fish fauna—but only a series of regional checklists and faunal works. There is no concise, publicly available, summary of catches by depth, season or locality—but only a series of cruise and expedition reports. There are no concise summaries of biological

[2] With the recent decline of the Peruvian stocks, the hake *is the* number one contribution to the world catch.

information, only a series of contributions scattered in the fishery and biological journals. There is an overriding need for the gathering together of
available information into synthesized form.

In order to add significant information, however, major efforts possibly
supportable only by large foundations or governments, would be required.
The day of pioneer contributions is rapidly drawing to a close. The areas of
obvious weakness in our academic knowledge of the Bering Sea fish fauna
are several. Distant-shore areas and isolated islands will require large-ship
capability for transport and helicopter support to provide adequate assessment
of such inshore faunae. Data on the deep-sea environment are particularly
weak, and the ship requirements here are well established. A third significant
and special area for study are the winter conditions and faunal shifts under the
Bering Sea ice cover. Work on commercial stocks requires valid stock assessment and much better information on the catch and effort being applied to
Bering Sea waters.

Requirements for effective effort

Clearly, definition of national and multi-national resource objectives and
priorities involves consideration of the needs, feasibility and possible support
for individual programs. In order to delimit such goals and objectives of both
long-range and short-term efforts of any Bering Sea Study, there would be a
requirement to assess existing data. This would necessarily include ongoing
and unpublished efforts. Significant amounts of proprietary fishery statistics
are reportedly maintained by major companies operating in the Bering Sea,
and conceivably these data could be made available in a selective fashion.

Assembling information on the fishes of the Bering Sea is hampered not
only by the scattered nature of the literature but by the fact that it is written
in at least three languages. Not only is the fish and fishery literature comprised
of at least three or four times the page volume of the oceanographic literature,
but it is published in perhaps 50 more journals than its sister contributions.
The last general bibliography on the region was prepared by Grier in 1941,
and the AINA *Arctic bibliography* is falling behind in its annual coverage. There
has been a growing trend for more and more of the Japanese material to be published in that language with only brief English summaries, and an increasing
amount of the Soviet literature is without any English summaries. Although
it is true that most papers published in English do not carry Japanese or Russian
summaries, there is no escaping the reality that the body of the world's scientific
literature is in English.

The problem of data assembly entails the general question of information
handling, storage and retrieval. While the oceanographers have made considerable progress in this area (and the biologists virtually none), there are
still serious problems concerning the problem of data resolution. How does
one handle, in a compatible information system, one set of data that are applicable to a 5-mile^2 area as opposed to another set of data applicable to 500-
miles^2? Possibly one solution is the use of integer identifications (Fig. 11.3)
to indicate degree of geographic coverage or resolution.

While it will be considered heretic by some, and admitting the personal

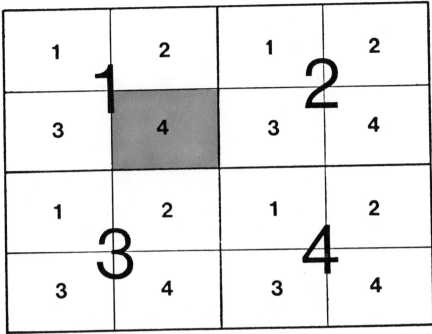

Fig. 11.3 Integer code technique for area resolution. For example, the upper left main quadrant would be given a value of 1. The lower right section of the upper left quadrant (shaded) would have a value of 14, and so on, to any desired degree of division. Retrieval resolution is then controlled by the number of digits used in scanning the data base.

pleasures of ichthyological exploration, I would urge against any *new* fish or fishery oriented activity in the Bering Sea until an analysis of existing data is carried out. The research problems are so vast, the implications so complex, the time so short, the public confidence in resource managers so low and competition for funds so high, that anything less than a well-planned effort is totally indefensible. My comment applies specifically to using this analytical approach on fishery problems, but I suspect it could be profitably implemented by most disciplines before any further pleas are made for ship time or field budgets.

The usual reaction to such a suggestion is the objection that such cooperation is impossible to obtain and that there are sometimes overriding pressures to withhold certain types of information. This approach fails to recognize that as long as the *planning* for what is to be assessed and the *format* for reporting can be agreed upon, the efforts can be carried out on a national basis with whatever mechanism best suits that country. Likewise, as information can be made public, it is acceptable to the common data base because of format compatibility.

Perhaps a useful first step would be the drawing together of the information suggested above so that one can separate fish and fishery programs clearly attainable under the terms of the International Decade of Ocean Exploration, as opposed to either those of a longer range basis or those that might better be done in the various laboratories of the supporting countries.

Discussion

KASAHARA: General research workers have little or no access to a large body of detailed data which have been collected over the years by various national agencies. In considering a consolidated system of data collection, analysis and retrieval, we need to distinguish the question of data *existence* from that of their *availability* at national and international levels.

WILIMOVSKY: Agreed. Also we must recognize the additional problem of screening information of a proprietary nature, such as that furnished by oil companies. It is very possible to code this type of material by a password file system to permit use only by those authorized. Such a consideration is essential to any information system we should develop.

REFERENCES

ALVERSON, D. L., A. T. PRUTER, and L. L. RONHOLT
    1964   A study of demersal fishes and fisheries of the Northeastern Pacific Ocean. In *H. R. McMillan lectures in fisheries*. Univ. British Columbia, 190 pp.

ANDRIASHEV, A. P.
    1937   K. poznaniyu ikhtiofauny Beringova i Chukotskogo morey [English summary]. [The ichthyofauna of the Bering and Chukchi seas]. *Explor. Mers USSR* 25: 292–355.
    1939   Ocherk zoogeografii i proiskhozhdeniya fauny ryb Beringova morya i sopredeinykh vod. Izd. Leningr. Gos. Univ., 187 pp.
    1954   *Fishes of the northern seas of the U.S.S.R.* Israel Prog. for Scientific Transl., 1964, 617 pp.

BEAN, T. H.
    1880   A preliminary catalogue of the fishes of Alaskan and adjacent waters. *Proc. U. S. Nat. Mus.* 4: 239–272.

DUNBAR, M.
    1967   The monthly and extreme limits of ice in the Bering Sea. In *Physics of snow and ice*, edited by H. Oura, Hokkaido Univ., Japan, pp. 687–703.

ELLSON, J. G., B. KNAKE, and J. DASSOW
    1949   Report of Alaska exploratory fishing expedition, fall of 1948, to northern Bering Sea. *U. S. Fish Wildl. Serv., Fish. Leafl.* 342: 1–25.

EVERMANN, B. W., and E. L. GOLDSBOROUGH
    1907   The fishes of Alaska. *Bull. Bur. Fish.* 26: 219–360.

FUKUDA, Y.
    1970   Northwest Pacific. In *The fish resources of the ocean*, edited by J. A. Gulland. FAO Fish. Tech. Paper No. 97: 56–82.

GRIER, M. C.
1941   Oceanography of the North Pacific Ocean, Bering Sea, and Bering Strait. *Univ. Washington Library Series* 2: 1–290.

GULLAND, J. A.
1970   *The fish resources of the ocean.* FAO Fish. Tech. Paper No. 97, 425 pp.

JORDAN, D. S., and C. H. GILBERT
1899   The fishes of Bering Sea. In *The fur-seals and fur-seal islands of the North Pacific Ocean*, Part 3, by D. S. Jordan. Rep. Fur-Seal Invest. 1896–1897, pp. 433–492.

MAEDA, H.
1963   Distribution pattern of ground-fishes hooked along a row of setline in the shallower part of the continental slope in the Bering Sea. 2. Bathymetric difference in relative abundance in relation to distribution pattern. *Res. Popul. Ecol.* 5: 74–86.

MAEDA, H., H. FUGII, and K. MASUDA
1967   Studies on the trawl fishing grounds of the E. Bering Sea. 1. On the oceanographical condition and the distribution of the fish shoals in 1963. *Bull. Jap. Soc. Scient. Fish.* 33: 713–720.

MOISEEV, P. A.
1963   *Soviet fisheries investigations in the northeast Pacific*, Part 1. Israel Prog. for Scientific Transl., 1968, 333 pp.

NELSON, E. W.
1887   Fishes. In *Report upon natural history collections made in Alaska between the years 1877 and 1881*, edited by E. W. Nelson, pp. 297–322.

OKADA, S., and K. KOBAYASHI
1969   *Colored illustrations of pelagic and bottom fishes in the Bering Sea* [in Japanese]. Nat. Resources Sec., Japan, 179 pp.

PALLAS, P. S.
1811   *Zoographia Rosso-Asiatia*, Vol. 3. St. Petersburg, 428 pp.

QUAST, J. C., and E. L. HALL
1972   List of fishes of Alaska and adjacent waters with a guide to some of their literature. U. S. NOAA Tech. Rep. NMFS SSRF-658, pp. 1–47.

ROMANOV, N. S.
1959   *Annotated bibliography on Far Eastern aquatic fauna, flora and fisheries.* Israel Prog. for Scientific Transl., 1966, 391 pp.

SHMIDT, P.
1904   *Pisces Marium Orientalium*, St. Petersburg, 466 pp.
1948   Ryby Tikhogo Okeana, Ocherk sovremennykh teorii i vozzreniy na rasprostranenie i rasvitie fauny ryb Tikhogo Okeana, Pishchepromizdat. 124 pp.
1950   Ryby Okhotskogo Morya. Izd. Akad. Nauk SSSR, pp. 1–370. (Israel Prog. for Scientific Transl., 1965, "Fishes of the Sea of Okhotsk," 392 pp.

STEJNEGER, L.

1936   *Georg Wilhelm Steller. The pioneer of Alaskan natural history.* Harvard Univ. Press, Cambridge, Mass., 623 pp.

TARANETZ, A. YA

1937   Handbook for identification of fishes of Soviet Far East and adjacent waters [in Russian, English summary]. *Bull. Pac. Scient. Inst. Fish. Oceanogr.* 11: 1–200.

TILESIUS, W. G.

1813   Iconum et descriptionum piscium Kamkchaticorum. . . Mem. Acad. Sci. St. Petersburg, 4: 406–478.

TURNER, L. M.

1886   Contributions to the natural history of Alaska, (Fishes). In *Report upon natural history collections made in Alaska between the years 1877 and 1881,* edited by E. W. Nelson, pp. 87–113.

WILIMOVSKY, N. J.

1954   List of the fishes of Alaska. *Stanford Ichthyol. Bull.* 4: 279–294.

1963   Discussion of ecology of ice substrates. In *Proc. Proceedings of Arctic Basin Symposium,* pp. 249–252.

1964   Inshore fish fauna of the Aleutian Archipelago. In *Proc. Proceedings of 14th Alaskan Science Conference,* pp. 172–190.

ZAITSEV, YU. P.

1970   *Morskaya neustonologiya.* Akad. Nauk Ukrain. SSR. 264 pp.

CHAPTER 12

Bering Sea benthos as a food resource for demersal fish populations

MILES S. ALTON

NOAA, National Marine Fisheries Service, Northwest Fisheries Center, Seattle, Washington

Abstract

Certain benthic animals such as the Tanner and king crabs, pandalid shrimp and mollusks are of direct importance to man; other organisms associated with the bottom of the Bering Sea compete with or prey upon animals of use to man. Many members of the macrobenthos, however, provide a nutritional base for fish and crustaceans of commercial importance.

Density of the benthos is highest in the western and northern parts of the shelf region, reaching a maximum average figure of 905 g/m² in the Chirikov Basin. The lowest value is 55 g/m² for the broad shelf region of the southeastern Bering Sea, where major fisheries take place. In total biomass by region, the Chirikov Basin alone has an estimated 40.5 million metric tons, or almost twice that of the western Bering Sea. The northern Bering Sea, in fact, accounts for 80 percent of the total benthos biomass of the shelf region. In the southeastern sector, where over one million metric tons of bottom fish have been removed annually in recent years, the amount of benthos is less than 10 percent of the estimated total for the Bering Sea. The distribution of the food benthos parallels somewhat that of the total biomass, but the proportion of food benthos to total benthos is highest (over 50 percent) in the Gulf of Anadyr and the southeastern shelf region. The western Bering Sea lacks a developed food benthos but is exceedingly rich in epibenthic animals such as sand dollars, barnacles, sea anemones, and sponges.

Of the total estimate of food benthos in the Bering Sea (64 million metric tons), only 17 percent (or 11 million metric tons) are accessible to commercial concentrations of demersal fish because of the cold temperatures that prevail in many parts of the sea. It appears that these temperature barriers limit the bottom-feeding fish from invading the richly concentrated benthos of the northern regions.

Although the benthos plays some part in supporting fish populations, it must be viewed within the total framework of the nutritional dependence of the demersal fishes on the nekton and plankton as well.

INTRODUCTION

Our knowledge of the Bering Sea bottom fauna is mainly of a variety of familiar and conspicuous forms of macrobenthos such as crabs, clams, starfish, urchins, sponges, and polychaete worms. The smaller elements of the benthos—microbenthos and meiobenthos— are poorly known. The microbenthos consist of bacteria and protozoans which are essential for the mineralization of organic debris that has settled to the sea bottom. The meiobenthos are comprised of small animals such as minute crustaceans, nematode worms, and young stages of benthic invertebrates, which fall within the size range of 0.1–1.0 mm.

Animals of the Bering Sea macrobenthos that are presently of direct use to man are the Tanner crab (*Chionoecetes*), king crab (*Paralithodes*), and pandalid shrimp (*Pandalus*). Others, such as starfish and snails, compete with or prey upon animals of value to man. Many members of the macrobenthos provide a nutritional base for fish and crustaceans of commercial importance, and in this paper our present knowledge of this important food resource for demersal fish populations of the Bering Sea is examined. Since our information on Bering Sea food benthos has been obtained mostly through the efforts of Soviet investigators, these contents are essentially a summary and discussion of their findings.

Benthos investigations in the Bering Sea

Only within the past 40 years have there been any serious studies of the quantitative distribution of benthic invertebrates in the Bering Sea. These studies were part of Soviet investigations of the commercial resources and hydrology of various sectors of the Bering Sea. The earliest investigations took place in 1932 and 1933 during the *Krasnoarmeets* expedition to the northern Bering Sea (Fig. 12.1). A total of 37 benthic stations were occupied, and a report of the findings was given by Makarov (1937). About 20 years later, Bering Sea benthos studies were resumed in connection with the 1950–52 *Vityaz* scientific explorations of the northwestern Pacific and connecting sea. One hundred seventy-seven benthic stations were sampled in the Bering Sea during what was then the most ambitious of Soviet efforts in that part of the world. Benthos collections were obtained throughout the western and northwestern sectors and from shallow to abyssal depths (less than 50 m to 5000 m) (Belyayev 1960). The main effort was in the region of Olyutorskii Bay (44 stations) and on the broad shelf area of the northwestern Bering Sea, which included 79 stations in the Gulf of Anadyr. Seventy-four percent of the 177 *Vityaz* stations were located on the shelf (200 m and less). In addition to the *Vityaz* collections, 49 quantitative benthic samples were obtained from the research vessel *Akademik Shuleikin* operating in the Karaginskii region of the western Bering Sea in 1956 (Lus and Kuznetsov 1961).

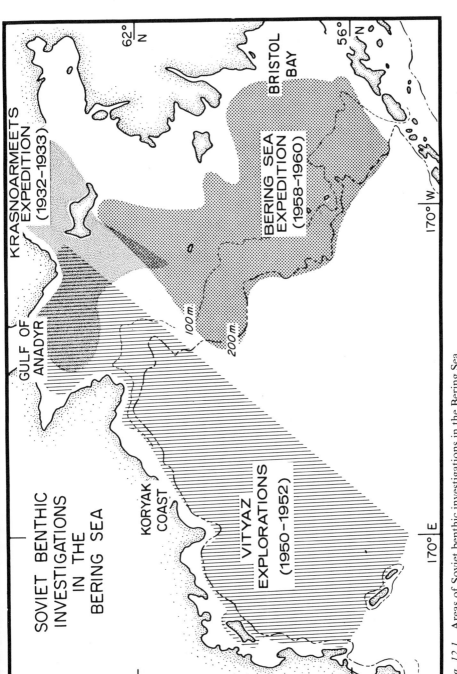

Fig. 12.1 Areas of Soviet benthic investigations in the Bering Sea.

In 1957 the Bering Sea Expedition was organized for the purpose of determining the extent of the demersal fishery resources of the eastern Bering Sea and the environmental and behavioral factors associated with the distribution and abundance of these resources. Major emphasis was directed towards the benthos. During the Expedition (1958–60) a total of 280 bottom grab samples were obtained from the broad shelf area and upper slopes (20–500 m) extending from the entrance to the Gulf of Anadyr southeastward towards Bristol Bay (Neiman 1960, 1963). Benthos data obtained from Bristol Bay and along the northern coast of the Alaska Peninsula were then analyzed by Semenov (1964).

In this review of the Soviet findings, only the quantitative sampling data will be considered; that is, results obtained by the use of a bottom grab that samples a given area of the sea bottom. During the *Krasnoarmeets* Expedition, a 0.2 m² grab was used. Three types were used in the *Vityaz* investigations: 0.4 m² Petersen grab; 0.25 m² prism grab, and 0.25 m² Okean-50 grab (Belyayev 1960). The latter type grab was employed in the Bering Sea Expedition and in the studies by Semenov (1964). Animals were separated from the sediment by washing the samples through screens having a minimum mesh size of approximately 1 mm. The collected animals were then fixed in 80 percent alcohol in the case of the *Vityaz* samples or in 4 percent formalin for the Bering Sea collections and later weighed and enumerated (Neiman 1960; Filatova and Barsanova 1964). No attempt is made in this discussion to adjust the weights of animals fixed in one preservative to weights of animals fixed in the other preservative.

The portion of the benthos biomass considered as potential food for commercially valuable fish included all worms, thin-shelled mollusks (under 4–5 cm), small crustaceans, brittlestars, sea cucumbers, young sea urchins, and early stages of echinoids (Neiman 1963; Belyayev 1960). Their selection was based on the studies of the diet of cod and halibut in the Bering Sea (Gordeeva 1952, 1954) and of flatfish from off Sakhalin and the Southern Kurile Islands (Mikulich 1954).

Zoogeography of Bering Sea benthos

The eastern Bering Sea and particularly the waters overlying the shelf in the southeastern area are warmer than other parts of the Bering Sea because of the influence of the relatively warm Alaska current that enters through the eastern gateways of the Aleutian Islands from the Pacific. During the summer months (July—August), bottom-water temperatures above 0°C occur over a wide area of the southeastern Bering Sea shelf, whereas sub-zero temperatures prevail in many of the remaining shelf regions (Gulf of Anadyr, Olyutorskii Bay, and some sectors north of St. Matthew Island). Benthos distribution and composition are associated with these temperature regimes.

Neiman (1960) noted that a complex of numerous low-arctic animals predominated by the clam *Macoma calcarea* was widely distributed in the colder regions of the shelf (northern Bering Sea southeastwards to the latitude of St. Matthew Island) (Fig. 12.2). South of St. Matthew Island the depth distribution of the *Macoma calcarea* complex narrow to 50–70 m and remain confined to regions of cold bottom waters in the southeastern shelf region.

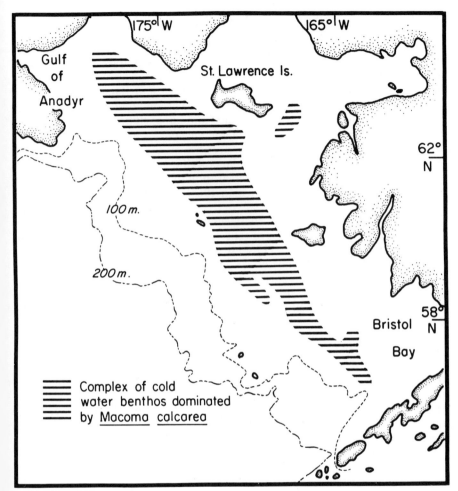

Fig. 12.2 Distribution of low-Arctic benthos complex in which the clam *Macoma calcarea* predominates (Neiman 1963).

Associated with the narrowing depth range of this cold water animal complex was a decrease in the biomass of low-arctic forms from the northwest toward the southeast. In the southeastern sector a temperate-boreal complex of animals was prevalent principally at depths of 50 m and less where water temperatures reach 4 C and above during the summer.

As a result of her findings, Neiman (1960) considered the southern boundary of the low-arctic in the eastern Bering Sea as being south of St. Matthew Island at 59°N. This falls within the transitional zone of mixed molluscan fauna which lies between the arctic province (low-arctic) north of 62°N and a temperate-boreal (Aleutian) province south of 58°N as established by Schenck and Keen (1936).

Another aspect of the zoogeography of the Bering Sea benthos is the presence of subarctic-boreal fauna throughout the outer shelf and upper slope (Zenkevitch 1963). Heart urchins of the genera *Brisaster, Brissopsis,* and *Spatangus* are important elements in benthos communities of the edge and upper slopes of the continental shelves in north boreal and subarctic waters. The distribution of communities dominated by *Brisaster* extends to upper bathyal depths in both the western side (Filatova and Barsanova 1964) and eastern side (Neiman 1960) of the Bering Sea. The subarctic-boreal complex is associated with that part of the outer shelf and upper slope (150–500 m) which is in contact with the permanent *warm intermediate layer* (Neiman 1963) of the Bering Sea. Temperatures of this layer are above 0°C (about 2 to 4°C) and lack a seasonal aspect. The upper boundary of the subarctic-boreal complex is governed apparently by the depth penetration of a mixed cooled surface layer that persists during the winter (Neiman 1963).

Benthos biomass distribution on continental shelf and upper slope (20–500 m)
From her own studies and those of others (Makarov 1937; Belyayev 1960; Lus and Kuznetsov 1961) Neiman (1963) discussed the distribution of the benthic biomass in the Bering Sea. She found that the density of the benthos was highest in the western and northern parts of the shelf (Fig. 12.3), reaching a maximum average figure of 905 g/m² in the Chirikov Basin. The lowest value of 55 g/m² was found in the broad shelf region of the southeastern Bering Sea, site of major trawl fisheries.

In terms of total biomass, the Chirikov Basin has an estimated 4.05 × 10⁷ metric tons (MT) of benthos—almost twice that of the western Bering Sea, which includes Olyutorskii Bay, the Korf-Karaginskii area, and the shelf off the Koryak coast (Fig. 12.4). The northern Bering Sea, in fact, accounts for 80 percent of the total benthos biomass of the shelf region. The remainder lies in the only areas of sizable demersal fisheries. In the southeastern sector, where over 1 × 10⁶ MT of bottom fish are removed annually, the amount of benthos is less than 10 percent of the estimated total for the Bering Sea.

There are regional differences also in the bathymetric distribution of benthos density (Fig. 12.5). In the eastern Bering Sea, the highest biomass occurs at intermediate depths of 50–150 m, whereas at shallow depths the quantity of benthos is small and decreases at depths greater than 150 m. In the western and northwestern Bering Sea, the highest density benthos is found at depths less than 50 m. Belyayev (1960) suggests that the development of such a rich benthos at shallow depths of these parts of the Bering Sea may be related to a high supply of nutrients from continental runoff and the decomposition of macrophytes. For example, sand dollars (discoidal urchins) occur in large concentrations in the Gulf of Anadyr and Olyutorskii Bay as one of the main contributors to the biomass at shallow depths.

Benthos as a food source for economically important fish species
Although the benthos biomass in the main region of the trawl fisheries may be relatively low, its proportion of possible use as food for bottom fish is much higher than in any of the other shelf regions excepting the Gulf of Anadyr (Fig. 12.4). Food-benthos in the southeastern region exceeds 50 percent of

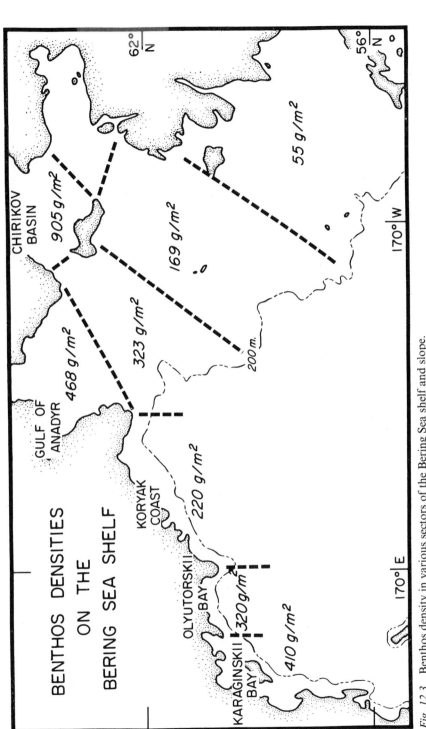

BENTHOS DENSITIES
ON THE
BERING SEA SHELF

62°
N

56°
N

CHIRIKOV
BASIN

905 g/m²

55 g/m²

GULF OF
ANADYR

468 g/m²

169 g/m²

323 g/m²

200 m.

170° W

KORYAK
COAST

220 g/m²

OLYUTORSKII
BAY

320 g/m²

KARAGINSKII
BAY

410 g/m²

170° E

Fig. 12.3   Benthos density in various sectors of the Bering Sea shelf and slope.

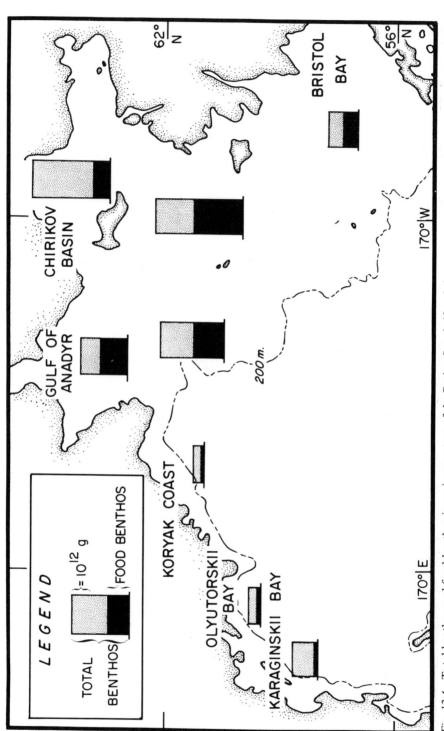

Fig. 12.4   Total benthos and food-benthos in various sectors of the Bering Sea shelf and slope.

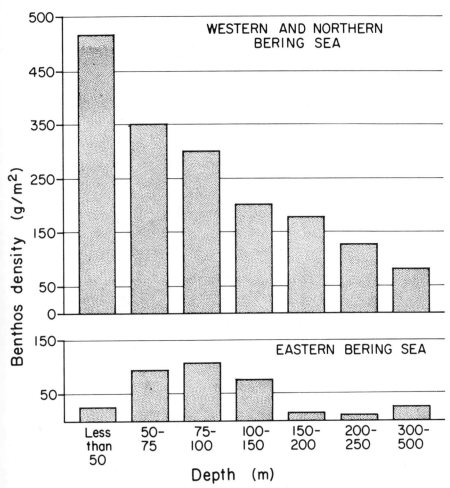

Fig. 12.5   Bathymetric trends in benthos density in the Bering Sea.

total benthos and consists predominantly of small clams, polychaete worms, and brittlestars. The western Bering Sea lacks a developed food benthos but is exceedingly rich in epibenthic animals such as sand dollars, barnacles, sea anemones, and sponges.

Of the total estimated food benthos on the Bering Sea shelf (6.4×10^7 MT), only 17 percent, or 1.1×10^7 MT, are accessible to bottom fish because of the cold temperatures prevailing in many parts of the Bering Sea (Moiseev 1964). About 7.5×10^6 MT of this potential food are located in the southeastern shelf region, but not all of this may be available at times because of the temporary flooding of portions of the shelf with waters too cold for the existence of sizable concentrations of flatfish. Associated with these regions of cold water are benthos characterized by low-arctic animals and relatively high food benthos.

The presence not only of older bivalves but the increase in their biomass from spring to fall suggests that fish are unable to fully utilize this food source in these cold regions (Neiman 1964); in years when the shelf waters are extensively heated, the entire food benthos becomes exposed to the flatfish at the end of summer (Neiman 1963). In cool years, the areas with abundant food resources may be inaccessible to fish populations.

Neiman (1963) tentatively considers the possible yield of flatfish in the Bering Sea based on the estimated biomass of benthos food organisms in regions accessible to fish. She assumes that the relation of benthos production to its biomass is 1 to 4 and that the efficiency at which benthic fish convert their food to new tissue is in the ratio of 15 to 1 or, in terms of a growth coefficient,[1] about 7 percent. Therefore, with these assumptions and a standing crop of food organisms of 1.11×10^7 MT, she estimates the annual production of flatfish at about 2×10^5 MT.

DISCUSSION

Neiman (1963) estimates about 1.78×10^8 MT of benthos biomass on the Bering Sea shelf and upper slope; yet, she points out that much of this lies in regions that are inaccessible—presumably because of the extreme cold—to commercial size populations of fish. The Gulf of Anadyr has a large food-benthos biomass but relatively low population sizes of commercial fish such as flatfish, cod, and pollock. The Chukchi Sea, which in many respects has a similar climatic regime as parts of the northern Bering Sea, is another case in point. There exists a high benthic biomass (Sparks and Pereyra 1966), but the flatfish populations (*Limanda aspera* and *Hippoglossoides robustus*) are small and probably not self-sustaining because of the extreme temperature conditions. Recruitment to the populations may come from Bering Sea stocks through transport of eggs, larvae, and young fish (Pruter and Alverson 1962). Moiseev (1964) has speculated on the possibility of stocking these cold regions of apparent high food biomass with commercially useful benthophage fish having demersal eggs, since many of these regions have current systems which would prevent the confinement of pelagic eggs and larvae to the favorable feeding grounds.

The presence alone, therefore, of an abundant supply of benthic organisms does not ensure corresponding large populations of demersal fish. The southeastern Bering Sea has the largest accumulations of demersal fish but the lowest density of benthos. One possible explanation for this low density is that the grazing of such large numbers of fish may be keeping the benthic populations at low numbers relative to other areas. Hayne and Ball (1956), working in a freshwater environment, found experimentally that not only was the biomass of benthos lower during the presence of fish, but that production of the benthos

---

[1] Growth coefficient $= \dfrac{\text{predator's annual increment of weight}}{\text{quantity of food consumed}}$

(Ricker 1969)

was higher than when the fish were absent. A small standing stock may possibly be composed of fast-growing but short-lived species which, though present at a low density as a result of predation, are highly productive.

Rather than a positive relation between total benthos biomass and demersal fish yield, we may find a negative one or perhaps none at all. For example, when the benthos density of the southeastern Bering Sea is compared with that of other major demersal fishing areas of subarctic and boreal regions, the largest yields of demersal fish are generally associated with low benthos densities (Table 12.1). The association is maintained if only benthophage fish (flatfish) are considered. Of course, such comparisons have serious deficiencies; these major fishing areas are ecologically dissimilar, and the yield may not represent the actual production of fish. The benthos densities presented in Table 12.1 are only averages, with no account given of temporal and spatial variations nor of the completeness of the areas surveyed. Information is insufficient also on the benthos that is actually accessible and of use to fish populations. Yet even with these shortcomings, it is tempting to speculate that perhaps the benthos of highly productive fishing areas may have a lower benthos biomass but higher production than areas of low yield of benthophage fish. There is also the possibility that disturbance of the sea bottom from trawling may actually increase benthos production (Marshall 1970).

Neiman (1963) has evaluated some of the problems of relating benthos biomass to fish yield and has tentatively estimated an annual flatfish production for the Bering Sea at 2×10^5 MT, with a possible catch limit of 2.5×10^5 MT, which does compare with the actual annual yields. In the early years 1960–61) of the fisheries, approximately $5–6 \times 10^5$ MT of flatfish, mostly yellowfin sole (*Limanda aspera*), were removed annually. The yield in recent years has fluctuated between $1–2 \times 10^5$ MT. Gulland (1970) considers the potential annual yield of the principal flatfish species (yellowfin sole and rock sole, *Lepidopsetta bilineata*) from the Bering Sea to be about 2.9×10^5 MT. Neiman's estimate therefore seems reasonable, although it is based on certain questionable data and assumptions. Major problems are a lack of quantitative data on trophic interactions and accurate estimates of food-benthos biomass from bottom grab samples.

No replicate samples were obtained at each station in the Soviet studies, which precluded any measure of variance, and such nektobenthic animals as mysids, euphausiids, and shrimp, which were found by Skalkin (1963) to be important in the diet of yellowfin sole, were either entirely missed or inadequately sampled by the grab. Neiman's assumption that all the food-benthos production is utilized by flatfish is obviously inaccurate, since there are no doubt other users of the food-benthos. Petersen (1918) estimated the food-benthos in the Kattegat at about 1×10^6 MT; yet this had to be shared among an estimated 5×10^3 MT of flatfish and 7.5×10^4 MT of invertebrates. Our ignorance of these trophic interactions complicates any attempt to assess the food available to an animal population. Even the production of flatfish must be shared between man and other animals.

Neiman assumed also that the annual production of the Bering Sea benthos was one-fourth the biomass as given by Zenkevitch (1947) for the Barents Sea.

TABLE 12.1 Yield of demersal fish and density of benthos in the southeastern Bering Sea compared with other major trawl fishing areas.

Area	Demersal fish yield MT/km²		Density of benthos (g/m²) by depth intervals		
	All species	Flatfish			
S.E. Bering Sea[1]	3.7	0.58	100 (50–100 m), 45 (100–200 m)	Nesis	1965
North Sea[2]	2.5	0.34	73[3] (11–70 m)	McIntyre	1958
			125[3] (15–70 m)	Holme	1953
Newfoundland[2]	2.0	0.27	312 (50–100 m), 76 (100–200 m)	Nesis	1965
Georges Bank[2]	1.7	0.29	157 (50–200 m)	Wigley	1961
Barents Sea[2]	1.1	0.01	210[4] (75–124 m), 150[4] (125–174 m)	Zatsepin	1970
World			200 (0–200 m)	Bogorov	1969

[1] 1969 catch statistics
[2] 1968 catch statistics
[3] Adjusted to wet weight from weight in alcohol
[4] Weight of animals preserved in alcohol

Neiman used this 1 : 4 ratio because of the similarity in temperature between the Barents and Bering Seas. Estimates of benthos production elsewhere in the world have often been based on the growth, mortality, and recruitment of a few selected populations of the communities under study. Jensen (1920), for example, obtained annual production values in Danish waters that were one to three times the benthos biomass. Sanders (1956), working in Long Island Sound, considered the production of both long-lived and short-lived species in the community and concluded overall production to be 2.4 times the biomass. Of course, Jensen and Sanders were examining communities in climates warmer than the Bering Sea, and therefore a lower ratio might be expected for colder water communities (Thorson 1957). There is no basis on which to even estimate the ratio of production to biomass in the Bering Sea.

Neiman (1963) assumed that the Bering Sea flatfish convert their food to flesh at an efficiency of about 7 percent. Dawes (1930, 1931), in his experimental studies of adult European plaice (*Pleuronectes platessa*), found growth efficiencies from 6 to 15 percent with an average of 9 percent. For the same species Petersen (1918) estimated a 6 percent efficiency. The growth efficiency of *Limanda yokohamae*, of the same genus as yellowfin sole, ranged from 16 percent in zero age fish to 6 and 7 percent in five-year-olds (Hatanaka et al. 1956). Regardless of the reasonableness of Neiman's figures, however, the growth efficiency of the Bering Sea flatfish is not known and, moreover, cannot be expected to remain constant year after year. Growth efficiency will depend largely on the age composition of the population and will increase as the average age decreases, a common trend in exploited populations. Young fish are able to convert a greater proportion of their food to new tissue than older fish, as demonstrated by Hatanaka and his coworkers.

In the discussion of Neiman's findings, several problems have been emphasized in predicting fish production from the supply of food organisms. Although the focus has been on the benthos and its production of food, demersal fish populations are not completely dependent on the benthos for food. Some investigators (Vinogradov 1966) estimate that only 10 percent of the net primary production goes directly to the benthos, another 10 percent becomes stored in the sediment, and the remaining 80 percent is used by herbivorous zooplankton.

The seemingly minor role of the benthos in the economy of the sea appears to be reflected in the relative contribution of benthic organisms to the sustenance of important fish populations. Aside from the clupeoid species which are strictly feeders of pelagic organisms, the abundant species in the world fisheries (mostly gadoids: cod, whiting, hake, pollock, Norway pout) either obtain all or much of their food directly from non-benthic organisms. Even flatfish, which are considered benthophagic, obtain some part of their nutrition from off-bottom animals. In the Bering Sea it appears that perhaps only 3 percent of the yield of demersal and semi-demersal fish species are of strictly benthophagic forms (rock and flat-head sole) (Table 12.2). Pollock, the main component of the Bering Sea trawl catch, feeds on small nektonic crustaceans (amphipods, euphausiids, copepods), small fish, and young squid. This is based on studies

TABLE 12.2   Major species and their relative contribution to the total demersal fish yield from the southeastern Bering Sea and their preference for certain food groupings.

Species	Contribution to total demersal fish yield[1] (%)	Principal food organisms
Pollock (Theregra chalcogramma)	73	Small nektonic and nektobenthic organisms
Pacific Cod (Gadus macrocephalus), turbot (Atheresthes stomias), and sablefish (Anoplopoma Eimbria)	6	Small to large nektonic and nektobenthic organisms
Yellowfin sole (Limanda aspera)	13	Benthic and nektobenthic organisms
Other flatfish: mainly rock sole (Lepidopsetta bilineata) and flathead sole (Hippoglossoides)	3	Benthic organisms
Pacific ocean perch (Sebastes alutus)	2	Small nekton

[1] 1969 statistics; total yield equals 1.2×10^6 MT.

of pollock in the Sea of Japan and the Okhotsk Sea (Iizuka et al. 1954), and it is assumed that the Bering Sea population has a similar diet.

Although the Bering Sea benthos appears to contribute relatively little to the production of useful fish populations, nothing definite can be said until we know more about the pathways by which energy and material flow to useful animal populations in the Bering Sea. Nektobenthic organisms such as amphipods, mysids, and various shrimps are important in the diet of many demersal fish populations and may obtain a substantial part of their nutrition from the small organisms (meiobenthos), a group often overlooked in benthic studies. We also have no information on the role of the benthos in supplying food for the young stages of useful animal species of the Bering Sea. Although net primary production may contribute only 10 percent directly to the benthos, the amount of organic detritus reaching the benthos may be large.

An important but little understood energy pathway to useful fish populations may be through the smaller elements of the benthos; that is, the microbenthos and meiobenthos (Fig. 12.6).

The role of the benthos is only one aspect of the general problem of linking primary production at higher trophic levels. The production of useful animals in any region of the sea requires a knowledge of food-web structure and energy and material transfer rates, information which is difficult to pursue. A first step in this understanding would be to elucidate the diet of the important fish populations of the southeastern Bering Sea, and then, based on these findings, work up the food-chain to determine the relative importance of the various pathways of energy flow from primary production to production of useful animals.

Studies of the food of fish populations should be designed to disclose not only the changes in diet related to age, season, and area, but also to allow us to characterize the feeding pattern of the species under study (Keast 1970). Many abundant demersal fish species appear to be opportunistic because of the variety of their prey, but their preferences are limited. They may select only prey of a certain size range, density, motility, and behavior pattern, and these prey may represent various trophic levels. Characterizing predators according to their food preferences permits us flexibility in our interpretation of potential food supplies to the predator. To say that an animal feeds predominantly on a herbivore may be true for a specific time and place, but when that prey becomes scarce, a similar size carnivore may become the animal's chief prey. This consideration also eliminates the error of proportioning out a predator's energy requirements to various trophic levels based on short-term studies.

The influence of man's activities must be included in the study of the food-web structure in the Bering Sea. He is making changes which may be deleterious or beneficial. What effect does his annual removal of some 1×10^6 MT of biomass have on the ecosystem? Does the disturbance of the benthos through trawling actually change the structure of the benthos so that it becomes more productive? Is more food made available to younger fish because of the cropping of older individuals, or is the food surplus taken up by other components in the community? One cannot help but be impressed by the somewhat constant yield of demersal fishes from the North Sea (Steele 1965) and wonder whether

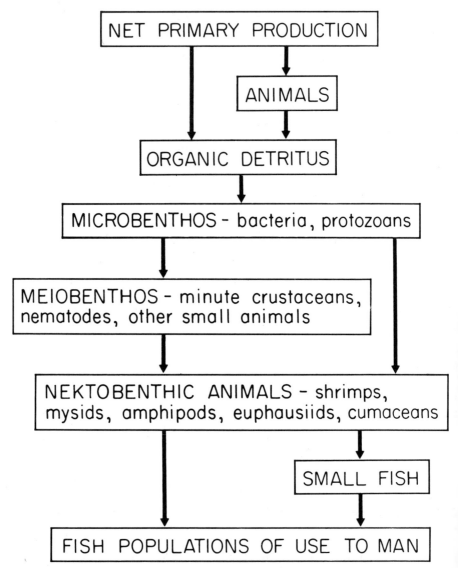

Fig. 12.6 A possibly important pathway by which energy and materials reach demersal fish populations.

this is some dynamic equilibrium established and maintained by man's activities. Man may be preventing the overexpansion and subsequent collapse of fish populations that may have been a natural occurrence before man's intervention. Exploitation rather than food supply may be the major factor keeping population sizes in check. Hopefully, through rational fishing, a predictive element may be introduced into man's use of the Bering Sea biological resources. We certainly hope that we are not laying the basis for the replacement of useful animal populations with ones of little or no value through fishing pressure that decreases predation on certain species or brings about a shift in competitive advantage from one species to another. Only through an understanding of the food-web structure can we anticipate some of these changes.

The Bering Sea benthos plays a certain part in the production of fish populations of use to man, but it must be viewed within the framework of the system. The Soviet studies have contributed significantly to the knowledge of the Bering Sea benthos—its biomass, biocoenosis, and trophic structures—but these findings must be supplemented by studies to better define the nutritional needs of demersal fish populations in relation not only to the benthos but also to the nekton and plankton.

Discussion

NAZAROVA: I would like to make one minor comment: Dr. Neiman, to whom reference is made in Mr. Alton's paper, is a *she* (not a *he*).

KASAHARA: That is certainly a very important clarification to what can sometimes be a dangerous assumption!

McROY: The food-web structure which has been presented here is not the only system operating in the Bering Sea. Along the Alaskan coast there are many lagoons and two large rivers that contribute substantial quantities of particulate organic material. The lagoons contain dense stands of *Zostera*, which are known to contribute particulate organic matter to the continental shelf. This particulate detritus may be directly nourishing the benthos.

KASAHARA: This is quite possible. One must qualify any food-web theory at this stage of our knowledge as sheer speculation. Except for some very simple cases, our knowledge of the actual food-web structure is still extremely poor.

UDA: In contrast to the notable decline of flatfish catches, particularly yellowfin sole, this paper cites the rising harvest of Alaskan pollock which has occurred in recent years. Has a corresponding natural fluctuation in benthic populations been observed in the Bering Sea, enhanced perhaps by the great climatic changes and mass mortality of fishes noted in the winter of 1963?

KASAHARA: I recall that yellowfin sole remained predominant in Japanese fishing at the time of the last Bering Sea Expedition, and the shift of emphasis to pollock took place at a later time. To my knowledge there has been no assessment made of either the impact of this succession on fish stocks or the effect of climatic changes on benthic populations.

FADEEV: Alaskan pollock do not feed heavily on benthos; 90 percent of their food consists of plankton.

REFERENCES

BELYAYEV, G. M.

1960   Quantitative distribution of bottom fauna in the northwestern part of the Bering Sea [in Russian]. *Tr. Inst. Okeanol. Akad. Nauk SSSR* 34: 85–103 (Transl. 337, 1968, U. S. Naval Oceanogr. Office).

BOGOROV, V. G.

1969   Productivity of the ocean: primary production and its utilization in food chains. In *Proceedings of Second International Oceanographic Congress*, UNESCO, pp. 117–124.

DAWES, B.

1930   Growth and maintenance in the plaice (*Pleuronectes platessa* L.), Parts 1–3.
–31   *J. Mar. Biol. Ass. UK* 17: 103–174, 877–947, 949–975.

FILATOVA, Z. A., and N. G. BARSANOVA

1964   Communities of benthic fauna in the western Bering Sea [in Russian]. *Tr. Inst. Okeanol. Akad. Nauk SSSR* 69: 6–97. (Transl. 459, 1969, U. S. Naval Oceanogr. Office).

GORDEEVA, K. T.

1952   The feeding of cod in the northern part of the Bering Sea [in Russian]. *Izd. Tikhook, In-Ta Ryb. Khoz. Inst. Okeanogr.*, Vol. 37.

1954   The feeding of halibuts in the Bering Sea [in Russian]. *Izd. Tikhook, In-Ta Ryb. Khoz. Inst. Okeanogr.*, Vol. 34.

GULLAND, J. A.

1970   The fish resources of the ocean. *FAO Fisheries Tech. Rep.* 97: 1–425.

HATANAKA, M., M. KOSAKA, and Y. SATO

1956   Growth and food consumption in plaice, Part 1. *Limanda yokohamae* (Gunther). *Tohoku J. Agr. Res.* 7(2): 151–162.

HAYNE, D. W., and R. C. BALL

1956   Benthic productivity as influenced by fish predation. *Limnol. Oceanogr.* 1(3): 162–175.

HOLME, N. A.

1953   The biomass of the bottom fauna in the English Channel off Plymouth. *J. Mar. Biol. Ass. UK* 32(1): 1–49.

IIZUKA, A., T. KUROHAGI, K. IKUTA, and S. IMAI

1954   Composition of the food of Alaska pollack (*Theregra chalcogramma*) in Hokkaido with special reference to its local differences. *Bull. Hokkaido Reg. Fish. Res. Lab.* 11: 7–20.

JENSEN, P. B.
  1920 Valuation of the Limfjord 1. Studies on the fish-food in the Limfjord 1909–1917, its quantity, variation, and annual production. *Rep. Dan. Biol. Sta.* 26: 1–44.

KEAST, A.
  1970 Food specialization and bioenergetic interrelations in the fish faunas of some small Ontario waterways. In *Marine food chains*, edited by J. H. Steele. Univ. Calif. Press, Berkeley, pp. 377–411.

LUS, V. YA., and A. P. KUZNETSOV
  1961 Data on the quantitative calculation of bottom fauna in the Korfo-Karaginskii region of the Bering Sea [in Russian]. *Tr. Inst. Okeanol. Akad. Nauk SSSR,* Vol. 46.

MAKAROV, V. V.
  1937 Data on the quantitative calculation of benthic fauna of the northern Bering Sea and southern Chukotski Sea [in Russian]. *Issled. Morey SSSR,* Vol. 25.

MARSHALL, N.
  1970 Food transfer through the lower trophic levels of the benthic environment. In *Marine food chains*, edited by J. H. Steele. Univ. Calif. Press, Berkeley, pp. 52–66.

McINTYRE, A. D.
  1958 The ecology of Scottish inshore fishing grounds. 1. The bottom fauna of East Coast grounds. *Scot. Mar. Res.* 1: 1–24.

MIKULICH, L. V.
  1954 Diet of flatfishes off the shores of southern Sakhalin and southern Kuriles [in Russian]. *Izd. TINRO,* Vol. 39.

MOISEEV, P. A.
  1964 Some results of the work of the Bering Sea expedition. In *Soviet fisheries investigations in the northeast Pacific*, Part 3, pp. 1–21. (Israel Program for Scientific Translations, 1968).

NEIMAN, A. A.
  1960 Quantitative distribution of benthos in the eastern Bering Sea [in Russian]. *Zoologicheskiy Zhurn.* 39(9): 1281–1292. (Transl. 402, U. S. Naval Oceanogr. Office, 1968).
  1963 Quantitative distribution of benthos on the shelf and upper continental slope in the eastern part of the Bering Sea. In *Soviet fisheries investigations in the northeast Pacific*, Part 1, pp. 143–217. (Israel Program for Scientific Translations, 1968).
  1964 Age of bivalve mollusks and the utilization of benthos by flatfishes in the southeastern Bering Sea. In *Soviet fisheries investigations in the northeast Pacific*, Part 3, pp. 191–196. (Israel Program for Scientific Translations, 1968).

NESIS, K. N.
   1965   Biocoenoses and biomass of benthos of the Newfoundland-Labrador region [in Russian]. *Tr. VNIRO* 57: 453–489. (Transl. 1375, Fish. Res. Bd. Can., 1970).

PETERSEN, C. G. JOH.
   1918   The sea bottom and its production of fish food. *Rep. Dan. Biol. Sta.* 25: 1–62.

PRUTER, A. T., and D. L. ALVERSON
   1962   Abundance, distribution, and growth of flounders in the southeastern Chukchi Sea. *J. Conseil perm. intern. Explor. Mer* 27(1): 81–99.

RICKER, W. E.
   1969   Food from the sea. In *Resources and man*. Nat. Acad. Sci.-Nat. Res. Council. W. H. Freeman & Co., pp. 87–108.

SANDERS, H. L.
   1956   Oceanography of Long Island Sound, 1952–1954. 10. Marine bottom communities. *Bull. Bingham Oceanogr. Coll.* 15: 345–414.

SCHENCK, H. G., and M. KEEN
   1936   Marine molluscan provinces of western North America. *Proc. Amer. Phil. Soc.* 76(6): 921–938.

SEMENOV, V. N.
   1964   Quantitative distribution of benthos on the shelf of the southeastern Bering Sea. In *Soviet fisheries investigations in the northeast Pacific*, Part 3, pp. 167–175. (Israel Program for Scientific Translations, 1968).

SKALKIN, V. A.
   1963   Diet of flatfishes in the southeastern Bering Sea. In *Soviet fisheries investigations in the northeast Pacific*, Part 1, pp. 235–250. (Israel Program for Scientific Translations, 1968).

SPARKS, A. K., and W. T. PEREYRA
   1966   Benthic invertebrates of the Chukchi Sea, (Chapter 29). In *Environment of the Cape Thompson region, Alaska*, edited by N. J. Wilimovsky. Div. Tech. Inf., U. S. Atomic Energy Comm., Oak Ridge, Tenn., pp. 817–838.

STEELE, J. H.
   1965   Some problems in the study of marine resources. *Spec. Publ., Int. Comm. N.W. Atlantic Fish.* 6: 463–467.

THORSON, G.
   1957   Bottom communities (sublittoral or shallow shelf). In *Treatise on marine ecology and paleoecology*, edited by J. W. Hedgpeth. Mem. Geol. Soc. Amer. 67(1): 461–534.

VINOGRADOV, L. G.
   1966   The trophodynamics of marine associations. 1968 FAO, ICES, and ICNAS Symp. *Oceanol. Okeanologiia* 6(6): 894–900.

WIGLEY, R. L.

1961   Benthic fauna of Georges Bank. In *Proceedings of the Twenty-sixth North American Wildlife Conference*, pp. 310–317.

ZATSEPIN, V. I.

1970   On the significance of various ecological groups of animals in the bottom communities of the Greenland, Norwegian, and the Barents Sea. In *Marine food chains*, edited by J. H. Steele. Univ. Calif. Press, Berkeley, pp. 207–221.

ZENKEVITCH, L. A.

1947   *Marine fauna and biological productivity*, Vol. 2 [in Russian]. Morya SSSR.

1963   *Biology of the seas of the USSR*. Interscience Publishers, New York, 955 pp.

CHAPTER 13

Status of marine mammals in the Bering Sea

MASAHARU NISHIWAKI

Ocean Research Institute, University of Tokyo, Tokyo, Japan

Abstract

The following species of marine mammals have been reported in the Bering Sea:

CETACEA

Bowhead (Greenland right whale): *Balaena mysticetus*
Right whale: *Eubalaena glacialis*
Gray whale: *Eschrichtius gibbosus*
Fin whale: *Balaenoptera physalus*
Sei whale: *Balaenoptera borealis*
Minke whale: *Balaenoptera acutorostrata*
Humpback whale: *Megaptera novaeangliae*
Sperm whale: *Physeter catodon*
Cuvier's beaked whale: *Ziphius cavirostris*
Baird's beaked whale: *Berardius bairdi*
Stejneger's beaked whale: *Mesoplodon stejnegeri*
White whale (white dolphin): *Delphinapterus leucus*
Common porpoise (harbor porpoise): *Phocoena phocoena*
Dall's porpoise: *Phocoenoides dalli*

PINNIPEDIA

Walrus: *Odobenus rosmarus*
Steller's sea lion: *Eumetopias jubata*
Northern fur seal: *Callorinus ursinus*
Common seal (harbor seal): *Phoca vitulina*
Zenigata seal: *Phoca insularis*
Ringed seal: *Pusa hispida*
Ribbon seal: *Histriophoca fasciata*
Bearded seal: *Erignatus barbatus*

CARNIVORA

Sea otter: *Enhydra lutris*

(SIRENIA)

(Steller's sea cow: *Rhytina stelleri*)

Detailed investigations of larger species of whales have been carried out both from whaling vessels and land-based stations. Based on the results of these studies, the International Whaling Commission (IWC) has recommended certain measures of conservation. The IWC regulatory policies have been based more upon mathematical predictions, however, than upon accumulated practical evidence.

A problem still open to basic investigation are the smaller species of toothed whales, the populations of which have not been ascertained. An understanding is needed of the feeding habits of these smaller species and whether their impact on prey populations is in any way reflected in the commercial fisheries harvest.

About 1000 Dall's porpoises (*Phocoenoides dalli*) are accidentally entangled in Japanese fishing nets each year. If such mishaps are inevitable, methods should be developed for making optimum use of the carcasses.

Among Bering Sea pinnipeds, only the fur seal has been the subject of any detailed study; as a consequence, this species has been protected by international treaty. Some aspects of the biology of the fur seal still remain to be studied adequately, however, for elucidation of the species' distribution, migration, and feeding habits. Application of satellite telemetry to the tracking of animals at sea would offer a substantial advancement to our knowledge of their movements.

Steller's sea lion (*Eumetopias jubata*) is at present the victim of conflicting national attitudes. Whereas the State of Alaska has executed a protection policy in its behalf, this alert animal is unfortunately regarded by the Japanese government as a competitor of the Hokkaido fishing industry. The Hokkaido Prefectural Office therefore views it as a fisheries pest and is attempting to decrease its numbers. Steller's sea lion probably breeds not in the adjacent waters of northern Japan, but in the Bering Sea. Its population size, distribution, and migratory routes remain little known.

Verification of Steller's sea cow (*Rhytina stelleri*) in the Bering Sea would mean the recovery of the species once close to complete extermination in this area.

A new species of *Phoca*, quite distinct from *P. vitulina*, has been identified as *P. insularis* (Zenigata seal). Not much is yet known about the distribution of this species, but it is possible that it may occur along the Bering Sea coast. There some groups of *Phoca* that are clearly different from *P. vitulina richardi* have the same fur pattern and coloration as *P. insularis* found of Japan may represent the most easterly representatives of this newly defined species.

The biology of the sea otter has been studied most intensively by K. W. Kenyon in the Aleutian Islands, but its status in the Bering Sea is still open to speculation.

Even though the study of marine mammals in the Bering Sea is of unquestionable importance, many major aspects of biological study have been left completely untouched. This significant gap in knowledge may be explained not only by the harsh climate and sparcity of human inhabitants and boats in the area, but also by administrative restrictions between countries that do not permit scientists to cross political boundaries as freely as the marine animals they pursue. A plan for long-term conservation of valuable species of marine mammals is dependent on more data which can only be acquired through cooperative efforts.

Discussion

WILIMOVSKY: A few years ago a Soviet publication reported that Steller's sea cow exists in limited populations on the coast of the Kamchatka Peninsula. Has this been confirmed?

NISHIWAKI: USSR scientists A. A. Berzin, E. A. Tikhomirov, and V. I. Troinin reported that about six large animals of strange appearance were sighted by men from a whale-catcher in the region of Cape Navarin in July 1962. These animals were presumed to be Steller's sea cow (*Rhytina stelleri*), which was formerly abundant in the Bering Sea until apparently exterminated by man shortly after it was discovered about 200 years ago. If this recent report is true, it is very welcome news.

KASAHARA: What is the present status of the Greenland whale?

NISHIWAKI: The Greenland right whale (*Balaena mysticetus*) is now hunted mainly by the Eskimos at the rate of perhaps 10 individuals per year typically, but annual catches sometimes number as high as 38 or 40 kills. One very young animal of unknown origin was found to have migrated as far as Osaka Bay. It is possible that the Greenland whale population may be very slowly in the process of recovery, although unrestricted slaughter by the Eskimos may be a significant deterrent.

KASAHARA: Has there been any indication of recovery of the gray whale on the Asian side?

NISHIWAKI: A juvenile gray whale (about two years old) was caught by accident two years ago near Shingu, on the coast of southern Japan. This might be considered as a sign of very slow recovery. The Asian gray whale is possibly a subspecies of the California gray whale (*Eschrichtius gibbosus*) but is clearly distinguishable by the shape of its nasal bone.

FAVORITE: Your slide showing the distribution of the California gray whale indicated that these mammals move only through the Commander-Near Strait in the course of their north-south migration. Is this the only pass in the Aleutian Islands through which they move?

NISHIWAKI: The gray whale is reported to move predominantly through this particular pass.

MOTODA: You showed in your illustration that one species of whale, *Balaenoptera acutorostrata*, migrates up to the shelf water in the eastern Bering Sea. Does this species exhibit particular habits, such as temperature preference or selective feeding?

NISHIWAKI: This species migrates only in the shallow waters around the Aleutian Islands, not far into the eastern Bering Sea or in the shallower waters of Bristol Bay.

Part 4

DYNAMICS OF
RENEWABLE RESOURCES

Ecology and behavior of juvenile sockeye salmon (*Oncorhynchus nerka*) in Bristol Bay and the eastern Bering Sea

RICHARD R. STRATY

NOAA, National Marine Fisheries Service, Auke Bay Fisheries Laboratory, Auke Bay, Alaska

Abstract

Sockeye salmon originating in western Alaska spend at least half of their first year of marine life in the Bering Sea and adjacent waters. As a result, the oceanic climate of this region during the period of seaward migration may be expected to have a significant effect on the abundance of sockeye that reach maturity and return to spawn.

The seaward migration route of the sockeye, extending from the river mouth to the southeast side of Bristol Bay as far seaward as Port Moller, is largely coastal; movement during continued seaward migration beyond Port Moller is offshore.

Major sockeye salmon stocks are well separated at the time of their entry into Bristol Bay, and this separation is maintained as far seaward as Port Moller. Migration through inner Bristol Bay is rapid but proceeds at a more leisurely rate farther seaward in the outer bay. This more leisurely migration rate is cited as the cause for the progressive mixing of sockeye stocks that had been well separated at the time of outmigration from major river systems.

Juvenile sockeye occur in small scattered schools during their seaward migration. They are most abundant in the upper 1 m of water at night and at a depth of 2 m during the day.

Predation on juvenile sockeye by fish did not appear to be a significant cause of mortality. Juvenile sockeye were found in the stomachs of several species of diving birds that are abundant along the seaward migration route; the importance of the bird predation requires further research.

Sockeye undergo little or no growth for a period of four to six weeks after they enter Bristol Bay. Rapid marine growth does not begin until the sockeye have reached the outer bay.

Zooplankton, the food of the juvenile sockeye salmon, was in low abundance in inner Bristol Bay and the inshore waters of the outer bay. Low food abundance during the sockeye's early residence in Bristol Bay is suggested as the principal cause for the lack of growth during this period.

The seaward migration route in inner Bristol Bay for all sockeye stocks studied is through the region having the steepest salinity gradients. This supports previous hypotheses based on laboratory experiments, which state that juvenile salmon use estuarial salinity gradients as a directive cue in seaward migration.

Sea temperature, which shows great annual variation in Bristol Bay and the eastern Bering Sea for the same monthly periods, determines the sockeye's period of residence in Bristol Bay and the Bering Sea through its influence on swimming speed, and thus the rate of migration. Sea temperatures prevailing during seaward migration will also determine the rate and the amount of growth completed before the onset of the sockeye's first winter at sea. In addition, sea temperature is responsible for keeping the seaward migration route largely coastal through outer Bristol Bay.

INTRODUCTION

The life cycle of the sockeye salmon (*Oncorhynchus nerka*) may be conveniently divided into four phases: freshwater life, seaward migration,[1] ocean life, and spawning migration. Much research on the life of sockeye salmon in freshwater has been carried out in the United States, Canada, and the Soviet Union. In addition, considerable knowledge has been gained on this species' life in the ocean from research conducted by the United States, Canada, and Japan under the auspices of the International North Pacific Fisheries Commission. The seaward migratory phase of the sockeye salmon's life cycle, however, has received little attention.

The major portions of both the seaward and spawning migrations of the sockeye salmon stocks of western Alaska pass through the central and eastern Bering Sea and the adjacent waters—Bristol and Kuskokwim bays.

The duration of the spawning migration of these stocks through the Bering Sea from their North Pacific Ocean feeding grounds is relatively short (40–50 days) when compared with the period of their freshwater and ocean life (one to four years). The distribution of the stocks in the Bering Sea during the spawning migration, as well as some aspects of their behavior and ecology, is fairly well known, but little information is available on the life of these stocks during seaward migration. We do know, however, that the seaward migration probably lasts longer than six months, and we can therefore expect that early marine growth and survival will be influenced greatly by environmental conditions prevailing in the Bering Sea during this period.

---

[1] Seaward migration as considered here is that period of the sockeye salmon's life from the time it leaves fresh water and arrives in the North Pacific Ocean, i.e., its juvenile life in the Bering Sea and Bristol Bay.

Many fishery scientists agree that the estuarine and early marine phases of a salmon's life may be a critical period in terms of the magnitude of the mortality incurred. For example, Parker (1965) estimated that a pink salmon stock in British Columbia, Canada, had a much greater natural mortality rate during the initial period of life (40 days) in the coastal marine waters than during the remaining period of life (410 days).

For a complete understanding of the dynamics of the sockeye salmon stocks of western Alaska, it is necessary to determine the importance of this species' early life in the Bering Sea in terms of the mortality incurred during this period with respect to its total marine life. Obtaining this information is one of the ultimate goals of the salmon research carried out in the Bering Sea since 1966 by the National Marine Fisheries Service. We must understand the relationship between critical environmental factors and the well-being of the salmon during this period so that environmental influence upon these stocks in a given year may be ultimately predicted.

This paper summarizes the results of our research to date on the ecology and behavior during seaward migration of the major stocks of sockeye salmon produced in the watersheds draining into Bristol Bay.

Origin and migratory pattern of sockeye salmon stocks in Bristol Bay

The smolts leaving the Kvichak, Naknek, Wood, Egegik, and Ugashik river systems in Bristol Bay (Fig. 14.1) constitute the major portion of the annual population of juvenile sockeye that migrate seaward through the eastern Bering Sea.

The seasonal migrations of sockeye smolts from these systems begin May 15 and are complete by July 15. The exact date depends on whether the systems are composed of one lake, several lakes, or lakes with several separate basins. The time that smolts leave appears to be strongly related to the distance they must travel to reach the trunk river outlet (Hartman et al. 1967). The migration is rapid from single-lake systems and extends over a longer period of time in multilake and multibasin systems. Hartman et al. (1967) discuss in detail the migratory pattern of sockeye salmon smolts from major North American river systems and list the average dates by which 50 percent of the smolts have left the systems (Table 14.1).

Several days after they leave the lake outlets, smolts enter Bristol Bay and begin their seaward migration through the bay and Bering Sea enroute to the North Pacific Ocean. Before our studies in 1966, almost no information was available on the life of these fish after they left the mouths of the major river systems.

Our research has been carried out primarily in the region between the mouths of the major river systems and Unimak Pass (Fig. 14.2). For reference purposes, this region has been divided into two areas—inner and outer Bristol Bay. Inner Bristol Bay includes that portion of the bay northeast of a line between Port Heiden and Hagemeister Island. This area receives the freshwater discharge from most of the major sockeye salmon-producing river systems entering the bay and can be characterized as estuarine. Outer Bristol Bay includes the area seaward of the line that delineates inner Bristol Bay to a line between Cape Sarichef on Unimak Island and the Kuskokwim River.

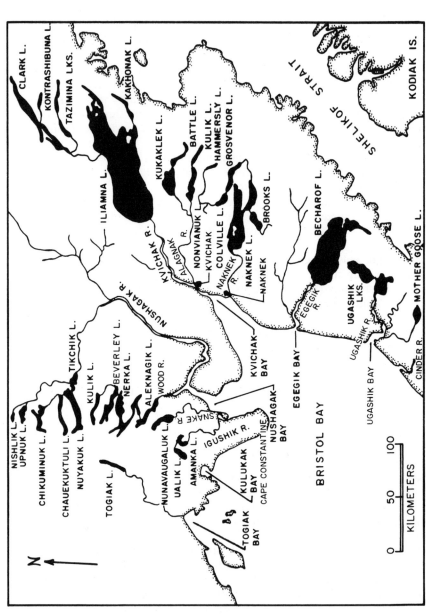

Fig. 14.1 Bristol Bay, showing principal sockeye salmon-producing river systems.

TABLE 14.1   Average dates by which 50 percent of sockeye smolts leave major Bristol Bay river systems

River system	Date	Number of years of observation
Ugashik	May 30	6
Egegik	June 1	3
Kvichak	June 2	9
Naknek	June 16	7
Wood	June 22	7

SOURCE: Hartman et al. (1967).

SEAWARD MIGRATION IN BRISTOL BAY

A few scattered observations before 1965 give some indication of the route followed by sockeye salmon during their seaward migration. In 1962, Hartt et al. (1967) caught "fingerling" sockeye salmon in a purse seine 121 km north of the north side of the Alaska Peninsula. Aspinwall and Tetsell (1966) observed seaward-migrating sockeye in Bristol Bay and obtained samples of some which had died as a result of seismic operations in the area in August 1965. These investigators stated that "the red salmon were found no closer than 19 miles [31 km] from Cape Seniavin nor farther than 35 miles [56 km]." Most of the fish were observed 35 to 38 km offshore from Cape Seniavin (see map, Fig. 14.2), which suggests that sockeye salmon may follow a rather discrete migration route seaward through Bristol Bay and the Bering Sea.

We attempted to define this route by exploratory fishing in the inner and outer areas of Bristol Bay in 1966, 1967, and 1969. Several types of fishing gear were used: circular tow nets 2.1 m in diameter, a 100-fathom small mesh lampara seine, a 60-fathom small mesh round haul seine, variable mesh monofilament nylon gill nets (55 mm, 42 mm, and 30 mm stretch measure), and a 200-fathom small mesh purse seine. The purse seine proved to be the most effective gear for capturing seaward-migrating sockeye and has therefore been used in the studies we have conducted since 1969.

The variable mesh monofilament nylon gill nets were fished experimentally in Bristol Bay in 1969 and 1970 in a joint study by Japan and the United States on the distribution and migratory habits of seaward-migrating sockeye salmon. These nets, which were of Japanese manufacture, were fished from the *Oshoro Maru* of the Faculty of Fisheries, Hokkaido University.

In addition to defining the seaward migration route of sockeye, we also studied some characteristics of the migration: period of residence in inner and outer Bristol Bay, order of seaward migration or extent of stock segregation during migration, vertical distribution, and schooling behavior.

We marked large numbers of sockeye salmon smolts on various major river systems and recaptured them later during exploratory fishing in Bristol Bay. The date of capture and river system of origin of each marked fish were recorded. A total of 1.4 million sockeye smolts were marked on the Kvichak,

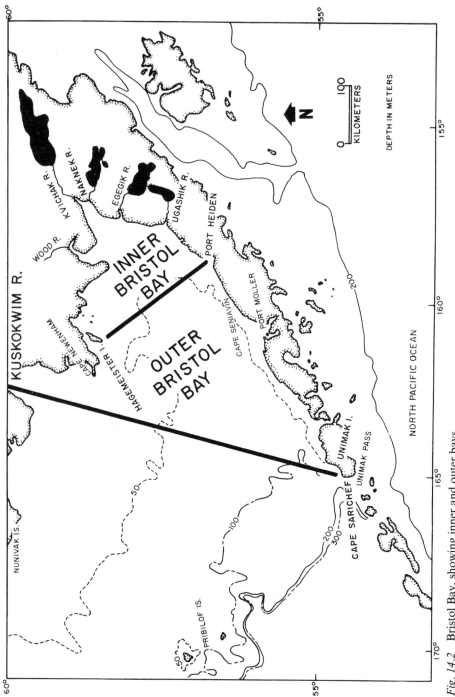

Fig. 14.2 Bristol Bay, showing inner and outer bays.

Naknek, Wood, and Ugashik rivers in 1967 and 1969, and an additional 379,000 smolts were marked on the Ugashik and Wood rivers in 1970. The fish were marked by spraying them with a colored fluorescent grit; this enabled large numbers of them to be marked in a short time. The grit was sprayed from paint canisters under an air pressure of 100 lb/inch^2 and impregnated the skin and scales of the fish. A different colored grit was used on smolts from each river system. The marked fish were viewed under ultraviolet light.

Seaward migration routes in Bristol Bay

The locations of the exploratory fishing with the various types of gear and the locations of capture of seaward-migrating sockeye for 1966, 1967, and 1969 are illustrated in Figures 14.3–14.6. The distribution of these catches in inner and outer Bristol Bay shows the seaward migration route through the area. In the figures, the catches are given for the entire period of fishing for each year. As a result, the distribution of seaward-migrating sockeye salmon for 1966 may be somewhat misleading (Fig. 14.3); the apparent absence of seaward migrants across inner Bristol Bay may be due to the fact that fishing was carried out in the inner bay primarily in June and early July with the result that the catches were probably composed mostly of stocks that migrated early, such as those from the Kvichak River system (Table 14.1). Later outmigrating stocks (Wood River) would not have had time to move into the fishing area by June or early July.

The fishing effort in 1969 (Fig. 14.5) was more complete and systematic than in 1966 and 1967. The combined catches for 1969, which show that juvenile sockeye were present across most of the width of the inner bay (Fig. 14.5), are the results of fishing at most stations across the inner bay in both early and late summer and therefore depict more accurately the true distribution of sockeye during their seaward migration. The results of exploratory fishing in 1966 and 1967 (Figs. 14.3 and 14.4) are presented only because they show a pattern of distribution that is similar to that in 1969.

Juvenile sockeye salmon were most abundant during their seaward migration along the southeast side of inner and outer Bristol Bay and least abundant on the north side of the outer bay (Figs. 14.3 and 14.5). The 1967 lampara seining (Fig. 14.4) and the 1969 fishing with the Japanese gill nets (Fig. 14.6) also indicate that juvenile sockeye were more abundant along the southeast side of the bay. Abundance decreased offshore toward the north side of outer Bristol Bay (Fig. 14.5). This decrease was also shown by results of purse seining off Port Moller in the outer bay in 1970, when seaward-migrating sockeye were found most abundant between the coast and 40 km north offshore. The size of the catches decreased farther north offshore, and no sockeye were captured beyond 56 km offshore. Data on a few fish captured during purse seining in outer Bristol Bay in 1971 also show that juvenile sockeye were most abundant nearer the southeast coast.

The distribution of juvenile sockeye described thus far implies that the seaward migration route of all major stocks of Bristol Bay sockeye is from the river mouths of the producing river systems and largely along the coast of the southeast side of the outer bay as far as Port Moller and perhaps Unimak

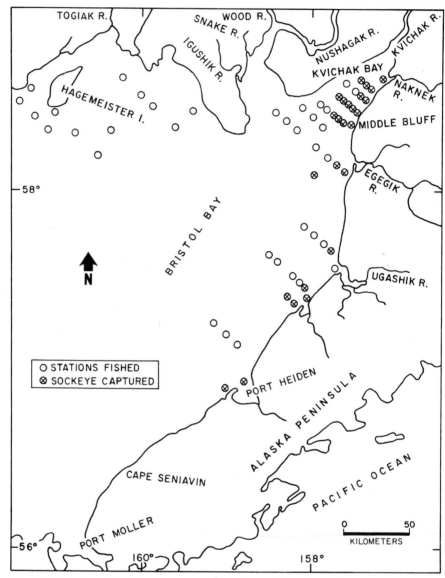

Fig. 14.3 Stations fished by tow nets and seining and locations of capture of seaward-migrating sockeye salmon in inner Bristol Bay, June–September 1966.

Pass. The recapture during seining of juvenile sockeye marked as smolts on the Kvichak, Naknek, Wood, and Ugashik river systems in 1967, 1969, and 1970 (Figs. 14.7, 14.8, and 14.9) supports the conclusion that all major sockeye stocks follow this migratory route. One important feature shown by these figures is that Wood River sockeye which entered Bristol Bay on the north side appeared to move across the bay and then seaward on the southeast side.

From this distribution of Wood River fish and the absence of fish during seining on the north side of the outer bay, we conclude that all sockeye stocks originating in rivers entering inner Bristol Bay on the north side must follow a route similar to that of the Wood River stock. Some possible factors influencing this route of migration will be discussed in a later section of this paper.

The results discussed thus far indicate that the seaward migration route

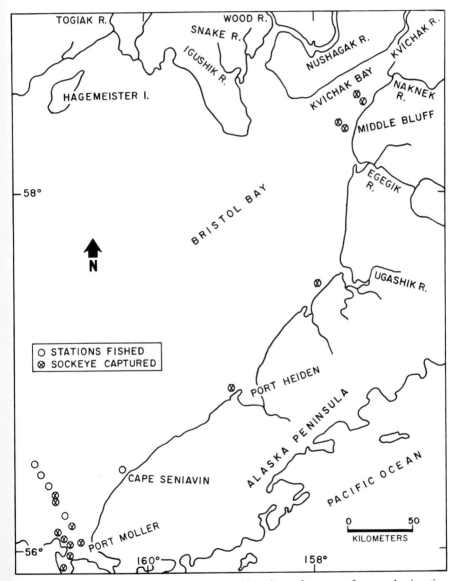

Fig. 14.4 Stations fished by lampara seine and locations of capture of seaward-migrating sockeye salmon in inner and outer Bristol Bay, June–August 1967.

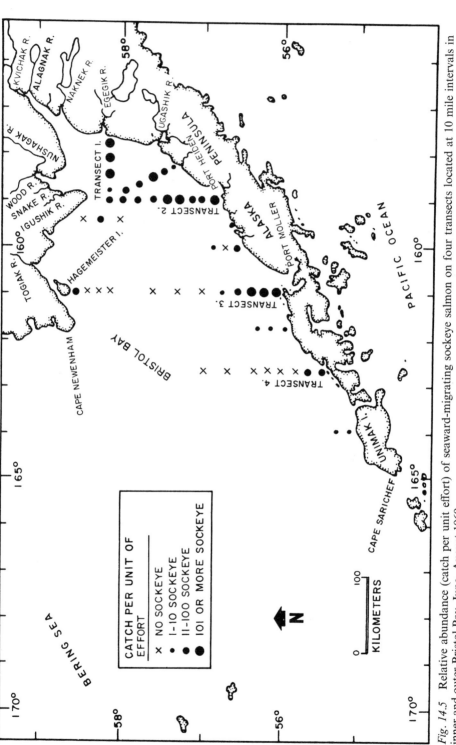

Fig. 14.5 Relative abundance (catch per unit effort) of seaward-migrating sockeye salmon on four transects located at 10 mile intervals in inner and outer Bristol Bay, June–August 1969.

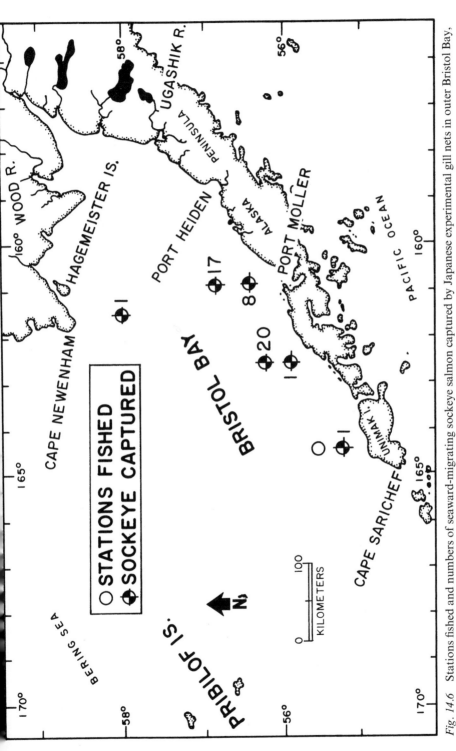

Fig. 14.6 Stations fished and numbers of seaward-migrating sockeye salmon captured by Japanese experimental gill nets in outer Bristol Bay, 21–30 July 1969.

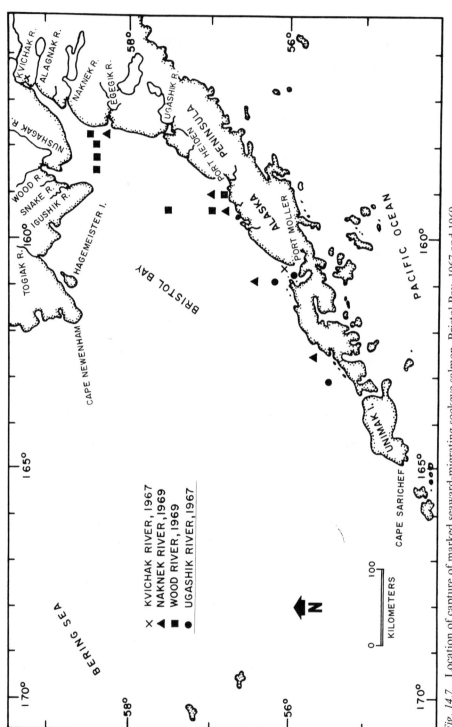

Fig. 14.7 Location of capture of marked seaward-migrating sockeye salmon, Bristol Bay, 1967 and 1969.

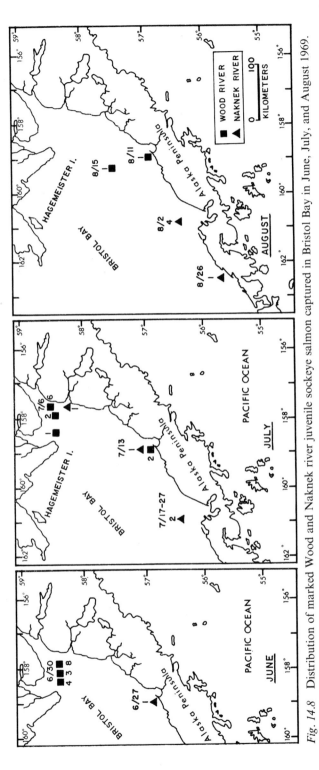

Fig. 14.8 Distribution of marked Wood and Naknek river juvenile sockeye salmon captured in Bristol Bay in June, July, and August 1969.

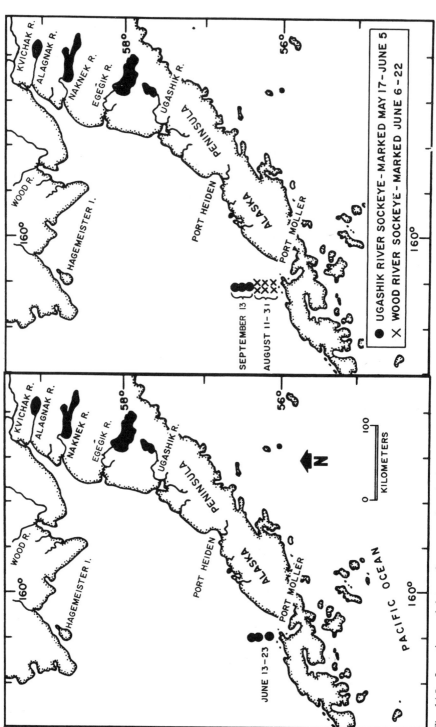

Fig. 14.9 Location and dates of capture of marked seaward-migrating sockeye salmon, Bristol Bay, 1970.

of sockeye salmon through outer Bristol Bay is essentially coastal. Hartt et al. (1966), Hartt, Smith, Dell (1967) and Hart, Smith and Kilambi (1967) reported similar coastal movement for juvenile sockeye in the Gulf of Alaska. Evidence from exploratory fishing in Bristol Bay and from experimental gill netting during the winter in the central Bering Sea suggests that juvenile sockeye migrate to waters farther offshore seaward of Port Moller after late July.

In 1967 the size of the seine catches of juvenile sockeye at Port Moller declined after mid-July. On the basis of purse seine catches in outer Bristol Bay between 17 August and 15 September 1967, Hartt et al. (1968) concluded that juvenile sockeye were present east of 164° W at this time (see Fig. 14.2). This conclusion and the decline in near-shore abundance suggest an offshore movement.

In 1969 the size of the purse seine catches of juvenile sockeye in the near-shore waters off Port Moller declined only after mid-August. This indicates a continued seaward movement and a decline in the total population of sockeye toward the head of Bristol Bay. Although juvenile sockeye were caught at the few near-shore stations fished seaward of Port Moller after mid-August, the catches were not large enough to indicate the presence of a sizable portion of the seaward migrant population near shore, as was found off Port Moller.

Further evidence supporting the offshore movement of seaward migrating sockeye salmon beyond Port Moller is the distribution of age 0.1[2] sockeye salmon (age 0.0 prior to January 1) captured in gill nets in winter (February and March) in the central Bering Sea in 1963 and 1965 (Bakkala 1969). These fish would have left their river systems of origin (most probably in Bristol Bay) the previous May–July. Age 0.1 sockeye were captured in the Bering Sea as far as 241 km north of the Aleutian Islands. If these age 0.1 fish were of Bristol Bay origin, then a gradual and continued offshore movement occurred during seaward migration until the southward movement into the North Pacific Ocean, which is most probably through passes east of 175° E, primarily in the central Aleutian Islands area.

From the evidence discussed, the seaward migration route of sockeye salmon from the river mouths in Bristol Bay to the North Pacific Ocean appears to be on the extreme southeast and south side of Bristol Bay and the eastern and central Bering Sea respectively. Verification of the implied migration route seaward of Port Moller, i.e., movement from coastal to more offshore waters of the Bering Sea before southerly migration into the North Pacific Ocean, will require additional exploratory fishing beyond Port Moller in the fall and winter seasons.

Characteristics of seaward migration

The abundance of juvenile sockeye salmon in the seine catches in Bristol Bay in 1967, 1969, 1970, and 1971 and the occurrence of marked fish of several major stocks in these catches indicate the relative length of time fish spent in the inner and outer bays. In addition, these data show how well the order of

---

[2] The use of this method (European method) for designating the age of adult Pacific salmon in reference to marine life only was proposed by Koo (1962). An age 0.1 sockeye is thus considered to have spent one winter of marine life.

the outmigration of smolts of various stocks is maintained during their seaward migration.

Another factor contributing to stock separation is the travel distance from the lake system of origin via major trunk rivers to any selected location in Bristol Bay. For example, sockeye from the Ugashik and Egegik River systems are well separated from those of the Naknek and Wood River systems not only by time of outmigration (Table 14.1) but also by distance of travel, say to Port Moller (see Fig. 14.2). Kvichak River system sockeye are separated from Ugashik and Egegik stocks primarily by travel distance and from Naknek and Wood river stocks by time of outmigration.

A third factor which may contribute to stock separation and perhaps even to separation between members of the same stock is the age (and therefore the size) of the smolts when they migrate. Usually the older and larger smolts in each group are the first to migrate seaward in spring (Hartman et al. 1967). In some years smolts migrating from one river system may be largely age I, while smolts of another stock may be mainly age II. A factor contributing to the separation of stocks because of age and size differences will be the faster swimming speed of the larger fish.

The distribution of juvenile sockeye salmon, as shown by fish captured in the variable mesh gill nets, in Bristol Bay in 1969 and 1970 combined with the echograms produced by our fish finders indicates the size of schools during seaward migration and their depth of travel.

Order of seaward migration

Several factors theoretically should act to keep stocks of sockeye smolts from the various river systems separated during their seaward migration. One factor, which has already been mentioned, is the difference in the time of outmigration (Table 14.1). If seaward migration continues with little delay after the smolts of each stock enter Bristol Bay, then the stocks should remain somewhat separated from one another.

The distribution of marked sockeye captured in Bristol Bay in the early and late summer of 1969 (Fig. 14.8) and 1970 (Fig. 14.9) illustrates the progression of seaward migration for several major stocks.

In 1967 fishing for smolts in the Port Moller area began on July 2, and almost immediately (July 4 and 9) marked fish from the Kvichak and Ugashik rivers were captured. This indicated that these stocks may have been present before fishing began. The average time of outmigration of sockeye smolts from both river systems is about the same (Table 14.1). These stocks are separated, however, by more than 160 km travel distance from Port Moller. Because only a few marked fish were captured in 1967, little can be said about the relative abundance of juvenile sockeye from the Kvichak and Ugashik rivers in the Port Moller area at this time. The capture of a marked Kvichak River sockeye about the same time that a Ugashik sockeye was captured, however, implies that Ugashik sockeye may either migrate seaward at a more leisurely rate than Kvichak sockeye or perhaps tend to linger or slow their rate of migration when farther seaward in outer Bristol Bay. These factors would allow the Kvichak stocks to eventually overtake and become mixed with the Ugashik stock, thus destroying the separation which existed at the time of outmigration. The mixing of other major sockeye stocks well separated at the time of outmigration is also indicated by the distribution of marked fish in the 1969 and 1970 seine catches (Figs. 14.8 and 14.9).

The absence of marked Wood and Naknek river smolts in the large seine catches in the Port Moller area in late June and early July 1969 (Fig. 14.8) indicates that stocks from these rivers were not abundant in the area at this time. The first marked Naknek River sockeye smolt was captured in the Port Moller area on July 17, whereas marked Wood River sockeye were recovered no farther seaward than Port Heiden by mid-August. Naknek and Wood river sockeye both have about the same travel distance from the river location where they were marked to the fishing area off Port Moller. The average time of outmigration for Naknek sockeye, however, is about one week earlier than for Wood River sockeye (Table 14.1). The distribution of marked members of each stock captured in Bristol Bay (Fig. 14.8) indicates that a degree of separation exists between these stocks during their seaward migration through the inner and into the outer bay. Marked Naknek River sockeye were more abundant farther seaward in June, July, and August than were marked Wood River sockeye. This apparent separation may be due to the difference in the time of outmigration of each stock or to the average swimming speed. Among smolts entering Bristol Bay, the Wood River system consistently produces the smallest fish and the Naknek River system the largest. As a result, the larger and therefore faster swimming Naknek fish might be expected to out-distance the smaller Wood River fish, thereby increasing the separation.

Figure 14.9 shows the distribution of marked Wood and Ugashik river sockeye captured in the Port Moller area in June, August, and September 1970. No fishing was done in July. Ugashik sockeye, which are among the earliest to enter Bristol Bay, are separated from Wood River sockeye by more than three weeks in their average time of outmigration (Table 14.1) and by a difference of 92 km travel distance to Port Moller. As would be expected, no marked Wood River sockeye were captured in seining off Port Moller in June. Marked members of both the Ugashik and Wood river stocks, however, were captured off Port Moller in August and September—an indication that Ugashik sockeye stocks, and probably other stocks as well, may move seaward at a more leisurely pace after entering the outer region of Bristol Bay.

Period of residence in inner and outer bay

The distribution in inner and outer Bristol Bay of marked members of several sockeye stocks captured during seining suggests that juvenile sockeye move rapidly through the inner and into the outer bay. This rapid movement is also apparent in the purse seine catch data for 1969.

The average catches of juvenile sockeye salmon per seine haul at all pro-ductive fishing stations (catch of one or more sockeye) located along four transects across Bristol Bay (Fig. 14.5) are listed in Table 14.2 for the early period (late June and July) and the late period (August). Up to the end of July, sockeye were abundant toward the head of Bristol Bay from Port Moller; they declined during August in the inner bay (transects 1 and 2—Table 14.2). Based on the average time of outmigration for various sockeye stocks (Table 14.1) and the 1969 purse seine catch data, seaward-migrating sockeye will be most

TABLE 14.2　Average purse seine catch per set of sockeye salmon by transect, June–August 1969

Transect number (see Fig. 14.5 for location)	Early period (late June–July)		Late period (August)	
	Number of seine sets	Average catch per set	Number of seine sets	Average catch per set
1	10	231	4	3
2	12	248	9	69
3	19	319	12	179
4	10	1	5	25

abundant in inner Bristol Bay from late May through late July, or for about two months. The purse seine catch data indicate that after early August 1969 the major population of juvenile sockeye was farther seaward. The period of residence of a particular stock in inner Bristol Bay will depend not only on the travel distance from river of origin to the outer bay but also on the size of individual fish and (most probably) water temperature, both of which will affect swimming speed.

As shown by the recapture of marked sockeye, the first stocks to arrive at Port Moller in outer Bristol Bay are those that enter the inner bay the earliest in spring and have the shortest travel distance to the outer bay, i.e., stocks from the Ugashik and probably the Egegik river systems.

The mark and recapture dates for Ugashik and Wood river sockeye in 1970 (Fig. 14.9) give a rough approximation of the time taken by these stocks to move through the inner and into the outer bay. Ugashik River smolts are among the earliest to enter Bristol Bay (Table 14.1) and have the shortest distance (145 km) to travel to the outer bay; Wood River smolts are the latest to enter Bristol Bay and have a much greater distance (273 km) to travel to reach the outer bay.

In 1970, smolts marked on the Ugashik River between May 17 and June 5 were first recaptured between June 13 and 23 off Port Moller (Fig. 14.9). The period between the midpoint of marking (May 27) and June recaptures (June 18) off Port Moller was 23 days. Travel distance from the site of marking at the outlet of Ugashik Lake to Port Moller (Fig. 14.2) is 306 km. About half of this distance (145 km) is in the Ugashik River and through what we have defined as the inner bay. If the seaward migration proceeds without delay through the inner bay, as our data indicate, the Ugashik seaward migrants would take roughly an average of two weeks to cover this distance.

If these same calculations are applied to the marked Wood River sockeye captured off Port Moller in 1970 (Fig. 14.9), six weeks is the average period of time members of this stock would take to move through the inner bay. Kvichak and Naknek river sockeye stocks, with only a slightly greater distance to travel than the Wood River stock, would be expected to take about the same time to move through the inner bay.

If the latest outmigrating stock, i.e., that from the Wood River system, spent six weeks in the trunk stream and the inner bay, the population of seaward migrants in the inner bay should decline sharply after early August. Such a decline was indicated by the data from our purse seine catches in 1969.

The above results show that the inner bay will contain a major portion of the total population of seaward-migrating sockeye salmon for 2 to 2½ months. In 1967, 1969, and 1970, this would have been between late May and early August.

The results of the seining in Bristol Bay in 1971 suggest that the timing and the rate of seaward migration through the inner bay may vary considerably in some years. The size of the seine catches of sockeye made off Port Moller before 1971, particularly those in 1970, indicated that fish were present in early June. In 1971, however, seaward-migrating sockeye were virtually absent from the Port Moller area in early July; they were found no farther seaward than Port Heiden (Fig. 14.2).

The apparent absence of significant numbers of seaward-migrating sockeye in outer Bristol Bay off Port Moller in late June and early July of 1971 can be attributed to two factors: (1) a later than usual outmigration of smolts and (2) reduced swimming speed of these fish as a result of unusually cold sea water temperatures in Bristol Bay and the Bering Sea. The overall effect of the cold temperatures would be to prolong the period of residence of seaward-migrating sockeye in inner and outer Bristol Bay and the Bering Sea, unless compensated for later by other factors.

The length of time seaward-migrating sockeye remain in outer Bristol Bay is not known. Our data indicate that in 1967, 1969, and 1970, the major population of seaward migrants was in the outer bay after mid-August. Purse seining has not been carried out in Bristol Bay beyond mid-September, and our catch data in 1970 indicate that the major population of seaward migrants was within the defined limits of the outer bay at this time. To obtain information on the migratory pattern of juvenile sockeye seaward of Port Moller, seining will have to be carried out during the fall and early winter.

Depth of travel

In 1966 and 1967, we used a ranging echo sounder to locate concentrations of seaward-migrating sockeye salmon in inner Bristol Bay. The sounder was operated while circular tow nets and lampara and round haul seines were being used. The echograms from the sounder and the corresponding tow net catches showed that juvenile sockeye were abundant in the surface 2.1 m of water. Echograms obtained during seine fishing indicated that juvenile sockeye were most abundant during daylight hours at depths of 1 to 3 m; a few fish schools believed to be juvenile sockeye were recorded as deep as 6 m.

The variable mesh nylon experimental gill nets fished from the Japanese training ship *Oshoro Maru* in outer Bristol Bay in 1969 and 1970 support our findings of a vertical distribution for seaward-migrating sockeye (Table 14.3); most of the sockeye were distributed in the upper 2 m of net. Sockeye captured in nighttime sets were most abundant in the top 1 m of water, while those caught in daytime sets were most abundant at a depth of 2 m (Table 14.3). This apparent vertical change in depth of travel during light and dark hours is consistent with what was reported for young sockeye movements in open-lake waters by Hartman et al. (1967). We caught a few sockeye in gill nets at depths of 5 m during daylight hours.

TABLE 14.3   Numbers of seaward-migrating sockeye salmon caught in experimental gill nets by period and depth of capture, Bristol Bay, during 1969 and 1970

	Number of fish captured			
	1969		1970	
Depth of net (meters)	Night	Day	Night	Day
1	19	0	8	5
2	0	21	1	28
3	1	1	2	11
4	0	2	0	0
5	0	4	0	1
6	0	0	0	0

In only one instance in the six years this investigation was underway were juvenile sockeye salmon seen at the surface—in 1966 near the mouth of the Naknek River and several miles seaward. Large numbers of small fish were seen breaking the water surface, and in addition, tow net and seine catches in the area yielded large numbers of juvenile sockeye salmon.

Schooling behavior

Sockeye smolts are known to be schooled during their migration through lakes to the lake outlets and in rivers (Hartman et al. 1967). McInerney (1964) states that "laboratory measurements show schooling intensity to be at the peak of its seasonal development during the time young salmon are moving through the estuary." Both the echograms produced by our echo sounder and the distribution of juvenile sockeye in experimental gill nets showed that smolts remained schooled during their seaward migration through Bristol Bay.

Most of the times when sockeye were captured in inner and outer Bristol Bay the echograms indicated that fish appeared to be in small scattered schools. The largest catch of juvenile sockeye in the inner bay, where the schools appeared to be more concentrated, was 1900. This catch was made in 1966 with a 60-fathom-long round haul seine. In outer Bristol Bay, where the schools appeared to be less concentrated, the largest catch was about 2400 sockeye. This catch was made in 1970 with a 200-fathom-long purse seine held open for 30 min.

The clustered appearance of the juvenile sockeye caught in the variable mesh gill nets in outer Bristol Bay verified further that juvenile sockeye remain in small scattered schools during seaward migration.

GROWTH OF SOCKEYE SALMON

Samples of juvenile sockeye were obtained from each major river system during the period of outmigration and from each seine catch made in Bristol Bay. These fish were measured for fork length (tip of snout to fork of tail), and a scale was removed to determine their freshwater age. In addition, scales of juvenile sockeye taken at sea were examined to determine the amount of marine growth completed up to the time of capture. Early marine growth on the scales of juvenile sockeye is defined here as that area beyond the last annulus formed in fresh water and is characterized by wide spacing of the individual scale circuli (Fig. 14.10).

We do not have enough data to calculate early marine growth rates of sockeye salmon while they are in Bristol Bay and the Bering Sea. The differences in the times of outmigration and the sizes of the members of various stocks (and therefore the period of residency in various regions of Bristol Bay and the Bering Sea) dictate that we work with individual stocks. This requires the collection and positive identification of adequate samples of the sockeye stocks of interest during the period of seaward migration through Bristol Bay and the Bering Sea. As yet we have found no positive method of identifying the various stocks of juvenile sockeye at sea, other than through large-scale mark-and-recapture experiments.

Scales from some of the marked members of individual stocks and un-marked members of presumably mixed stocks show two interesting charac-teristics about the nature of early marine growth in Bristol Bay: (1) little or

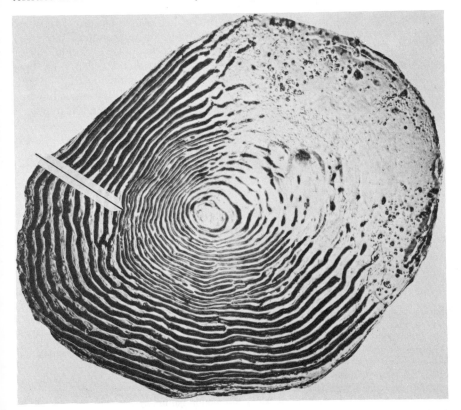

Fig. 14.10 Scale of a marked Ugashik juvenile sockeye salmon captured in outer Bristol Bay on 13 September 1970, showing 10 marine circuli (fork length, 205 mm).

no growth in fish from the inner bay and (2) rapid growth in those from the outer bay and farther seaward. The scales show that true marine growth (wide spacing of scale circuli) does not begin until after mid-July, when most of the fish have left the inner bay. Table 14.4 lists the first dates when the catches of juvenile sockeye contained some fish that had wide-spaced circuli on their scales. Although some marine growth was underway in mid-July, the proportion of fish in each catch showing marine growth at this time was less than 2 percent; and only a few marine circuli had been formed, which indicates that the growth had been underway for a very short period.

Table 14.5 gives the average number of marine circuli by freshwater age group for marked sockeye of various stocks captured in Bristol Bay in 1967, 1969, and 1970. Although few marked fish were recaptured, the time when marine type circuli first appeared on the scales indicates that all stocks, regardless of their time of outmigration, do not undergo true marine growth until they have been at sea at least four to six weeks. Among the marked fish recaptured in Bristol Bay, none had marine circuli on their scales before July 27 (Table 14.5). The number of circuli also indicate that marine growth did not begin to any extent until juvenile sockeye had reached outer Bristol Bay. In 1966 several sockeye captured off Port Heiden (transect 2), the inner boundary of outer Bristol Bay, showed marine growth (Table 14.4); and in 1969 two Wood River sockeye captured in the same area showed marine growth (Table 14.5).

From the appearance of the scales of juvenile sockeye caught during seaward migration and a comparison of the lengths of these fish with those of individual stocks at the time of outmigration, we conclude that juvenile sockeye grow very little (if at all) while migrating through the inner bay and into the outer bay. This period usually lasts four weeks, although it may vary annually and for certain stocks.

Once the juvenile sockeye enter outer Bristol Bay, marine growth occurs at a rapid rate. The number of circuli on the scales of marked Wood and Ugashik river sockeye captured off Port Moller in August and September 1970 (Table 14.5) and the size of these fish, compared with their size at the time of

TABLE 14.4   Date and location of capture of first seaward-migrating sockeye salmon with marine circuli[1] on their scales in 1966, 1967, 1969, and 1970

	Average number of marine circuli		
Date of capture	Age I	Age II	Location of capture
July 12, 1966	—	4	Port Heiden
July 17, 1967	0	1	Port Moller
July 17, 1969	2	2	Port Moller
August 9, 1970[2]	5	4	Port Moller

[1] A marine circulus is defined as that area of the salmon's scale beyond the last freshwater annulus consisting of wide spaced circuli (see Fig. 14.10).

[2] August 9 was the first day of fishing this month. No fishing was done between June 26 and August 8.

TABLE 14.5 Average number of marine circuli[1] on scales of marked juvenile sockeye from Kvichak, Naknek, Wood, and Ugashik River systems recaptured in inner and outer Bristol Bay, June to August, 1967, 1969, and 1970

Year, river system, and marking dates	Date of recapture	Location of recapture	Number of fish captured	Average number of marine circuli	
				Age I[2]	Age II[2]
1967					
Kvichak					
May 26–June 11	July 4	Outer Bay (Port Moller)	1	—	0
Ugashik					
May 27–30	July 9	Outer Bay (Port Moller)	1	0	
May 27–30	July 19	Outer Bay (Port Moller)	1	—	0
May 27–30	September 7	Outer Bay (55°36′N, 162°57′W)	1	([3])	—
1969					
Naknek					
May 30–June 14	June 27	Inner Bay (Transect 2)[4]	1	0	—
May 30–June 14	July 6	Inner Bay (Transect 1)	1	0	—
May 30–June 14	July 13	Inner Bay (Transect 2)	1	0	—
May 30–June 14	July 17	Outer Bay (Transect 3)	1	0	—
May 30–June 14	July 27	Outer Bay (Transect 3)	1	2	—
May 30–June 14	August 2	Outer Bay (Transect 3)	1	—	2
May 30–June 14	August 2	Outer Bay (Transect 3)	3	3	—
May 30–June 14	August 26	Outer Bay (Transect 4)	1	5	—

[1] A marine circulus is defined as that area of the salmon's scale beyond the last freshwater annulus consisting of wide spaced circuli (see Fig. 14.10).

[2] Refers to the numbers of years spent in the freshwater environment.

[3] No scale was obtained from this fish. Its fork length was 170 mm, or about twice the length of age 1 and 67% greater than the length of an age II smolt at the time of outmigration.

[4] See Figure 14.5 for location of transects and Figure 14.9 for precise location of recapture of marked fish.

TABLE 14.5 (continued)

Year, river system, and marking dates	Date of recapture	Location of recapture	Number of fish captured	Average number of marine circuli	
				Age I[2]	Age II[2]
Wood					
June 14–July 1	June 30	Inner Bay (Transect 1)	15	0	—
June 14–July 1	July 6	Inner Bay (Transect 1)	8	0	—
June 14–July 1	July 9	Inner Bay (Transect 1)	2	0	—
June 14–July 1	July 13	Inner Bay (Transect 2)	2	0	—
June 14–July 1	August 11	Inner Bay (Transect 2)	1	2	—
June 14–July 1	August 15	Inner Bay (Transect 2)	1	3	—
1970					
Ugashik					
May 17–June 5	June 13	Outer Bay (Port Moller)	1	—	0
May 17–June 5	June 15	Outer Bay (Port Moller)	1	—	0
May 17–June 5	June 23	Outer Bay (Port Moller)	1	—	0
May 17–June 5	September 13	Outer Bay (Port Moller)	1	—	10
May 17–June 5	September 13	Outer Bay (Port Moller)	2	10	—
Wood					
June 6–June 22	August 11	Outer Bay (Port Moller)	1	6	—
June 6–June 22	August 16	Outer Bay (Port Moller)	4	7	—
June 6–June 22	August 31	Outer Bay (Port Moller)	1	9	—

[1] A marine circulus is defined as that area of the salmon's scale beyond the last freshwater annulus consisting of wide spaced circuli (see Fig. 14.10).

[2] Refers to the numbers of years spent in the freshwater environment.

[3] No scale was obtained from this fish. Its fork length was 170 mm, or about twice the length of age I and 67% greater than the length of an age II smolt at the time of outmigration.

[4] See Figure 14.5 for location of transects and Figure 14.9 for precise location of recapture of marked fish.

marking, substantiate this initial rapid growth. If we can assume that marine growth for both stocks began in mid-July of 1970, Ugashik fish would have more than doubled their size (fork length) between the time marine growth began and the time they were captured—eight weeks. Wood River sockeye would have increased their length by more than 50 percent during the intervening four weeks.

FOOD HABITS OF SOCKEYE SALMON

In this paper, only those features of feeding behavior which appear to illustrate the influence of environment on the migratory pattern and growth of the sockeye during seaward migration are discussed.

Stomachs from more than 1200 juvenile sockeye salmon from the purse seine catches in Bristol Bay in 1969 and 1970 were analyzed to determine their contents and degree of fullness. For comparison, the samples are grouped by month (June to September) and by the place where the fish were captured, i.e., Kvichak Bay or off Port Heiden in the inner bay and in the inshore (<20 fathoms) or offshore (>20 fathoms) waters off Port Moller in the outer bay. The number of fish sampled from each seine haul was based on the ratio of the total number of fish caught in the seine haul to the number of fish taken in all seine hauls included in a particular group. The fork length of each fish was recorded; a visual estimate was made of the degree of stomach fullness; the contents were measured volumetrically; and organisms were identified to broad taxonomic groups. Volumetric measurement was carried out by a modification of the method of Yentsch and Hebard (1957).

The major food items in order of their relative importance were as follows:
Pacific sand lance, *Ammodytes hexapterus*—larval and young stages
Euphausiids—mainly larval stages
Copepods—all stages
Cladocera, *Podon* sp.
Pteropods
Decapod larvae
Other fish, cods and cottids—larval stages
Eggs, mainly invertebrates
Insects
The dominance of particular organisms varied with the location and time of capture. Limited evidence suggests that the diet of sockeye at any particular location or time reflects the relative abundance of food items available.

A comparison of the degree of fullness of the stomachs of sockeye salmon by location and time of capture for 1969 and 1970 is shown in Figure 14.11. In 1969, in June a larger proportion of the sockeye captured in the inner bay and the inshore waters of the outer bay had empty stomachs than did those captured in the offshore waters of the outer bay. In July the proportion of fish with empty stomachs decreased in all areas. Almost all fish from the offshore catches of the outer bay had full stomachs. This apparent increase in the degree of fullness of the stomachs of the sockeye between June and July implies

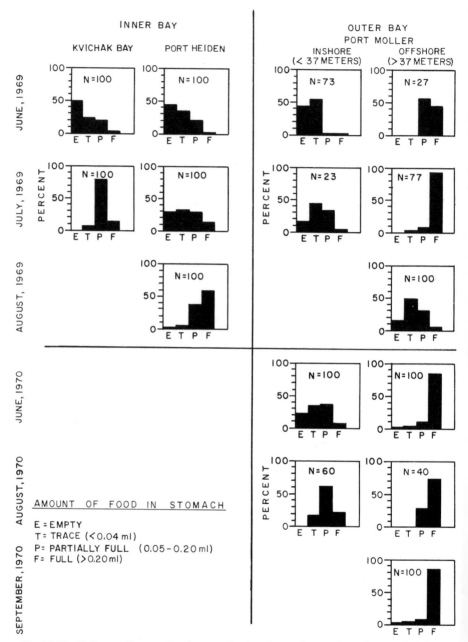

Fig. 14.11 Fullness of stomachs of seaward-migrating sockeye salmon by month in inner and outer Bristol Bay, 1969 and 1970.

that (1) sockeye for some reason do not actively feed for a period after entering, and during early residence in, the inner bay, or (2) food may be in lower abundance in June in the inner bay and inshore waters of the outer bay. In August, the proportion of sockeye with empty stomachs captured off Port Heiden at the outer boundary of inner bay had declined to almost zero. In view of the trend in the degree of stomach fullness in the inner bay from June through August and in the outer bay in June and July, the decrease in the proportion of fish with full stomachs which were sampled from catches off Port Moller in the outer bay in August (Fig. 14.11) was unexpected. This decline may be real and may reflect a decrease in food abundance, or it may simply be due to the vagaries of sampling.

The results in 1970 of a similar comparison of fullness of stomach of juvenile sockeye captured off Port Moller in the outer bay support the analysis of the 1969 samples. A greater proportion of sockeye captured in early summer had empty stomachs than did those captured in late summer (Fig. 14.11). Most of the sockeye captured in the offshore waters of the outer bay throughout the summer had full stomachs compared with the fish captured in the inner bay.

The similarities in the trends in fullness of stomachs in 1969 and 1970 imply that changes are occurring in feeding behavior which may be physiologically or environmentally induced. The distribution and abundance of food items (zooplankton) (to be discussed in a later section of this paper) tend to support the latter view.

PREDATION ON SOCKEYE SALMON

Predation by fish and birds on sockeye salmon smolts migrating downstream to Bristol Bay is well documented, but little information is available on predation that occurs once the sockeye have entered the bay and the Bering Sea.

The beluga, or white whale (*Delphinapterus leucas*), which frequents inner Bristol Bay during the spring and summer months, is one reported source of predation. The stomach contents of 101 beluga whales collected in 1954 and 1955 by the Alaska Department of Fish and Game near the mouth of the Kvichak River contained 20,000 sockeye salmon smolts (Anon. 1956).

We have examined the stomach contents of several species of fish taken in the same seine hauls as seaward-migrating sockeye salmon but found juvenile sockeye in only one species, the sandfish (*Trichodon trichodon*). The proportion of sandfish containing juvenile sockeye was estimated at less than 1 percent. The predation probably occurred while the seine was being hauled aboard ship and the sockeye and sandfish were forced into close association with one another. The juvenile sockeye were fresh and showed little evidence of digestion.

Diving birds, which are extremely abundant in Bristol Bay and the eastern Bering Sea, are also a potential source of predation to seaward-migrating sockeye. The stomach contents of samples of diving birds captured in the experimental gill nets fished in Bristol Bay in 1969 and 1970 are being examined by scientists of the Faculty of Fisheries of Hokkaido University in Hakodate, Japan. Thus far, remains of juvenile sockeye have been reported in the stomach contents of the common murre (*Uria oalge*) and a guillemot (*Cepphus* sp.).

Other possible predators on juvenile sockeye salmon in Bristol Bay and the Bering Sea include seals, whales, porpoises, and adult chinook and coho salmon.

ENVIRONMENTAL FACTORS ASSOCIATED WITH SEAWARD MIGRATION

To describe the environment common to the sockeye salmon during seaward migration, oceanographic surveys of the inner and outer bays were conducted monthly from spring through fall in most years of this investigation. Preliminary results of these surveys indicate that a number of environmental factors influence the migratory pattern, growth, and behavior of the sockeye during their seaward migration through Bristol Bay.

Salinity

In general, from spring through fall salinity is highest in the middle of Bristol Bay and decreases steadily toward the north side and the southeast side of the bay. The circulation pattern during most of this period is counterclockwise, i.e., northeastward along the Alaska Peninsula toward the inner bay and northwestward from the inner bay past Cape Newenham. The northwesterly current carries with it most of the freshwater arising from the major river systems discharging into inner Bristol Bay, which accounts for the lower salinities on the northwest side of the bay.

The inner bay and the lower portions of most major river systems are well mixed and extremely turbid. The pronounced horizontal salinity gradients in this area (Fig. 14.12) may serve as directive cues to juvenile and perhaps adult sockeye salmon in their migrations through the turbid water.

On the basis of laboratory experiments with five species of Pacific salmon, including sockeye, McInerney (1964) proposed that juvenile salmon are able to use estuarial salinity gradients as one of the directive cues in their seaward migration. He showed that juvenile salmon displayed a temporal progression of salinity preference changes which paralleled closely the salinity gradients typical of river outflows through which young salmon pass on their way to the ocean. The terminal preference shown by young salmon was for water of open-ocean concentration.

The seaward migration route of sockeye from the major Bristol Bay river systems, which is from the river mouth to the southeast side of the bay, supports McInerney's experimental conclusions. This route, which is through the region of the most pronounced salinity gradients for all stocks (Fig. 14.12), was particularly apparent in the distribution of marked Wood River sockeye that were captured across the inner bay in 1969 (Fig. 14.9). Seaward migration by this route is therefore the most direct route to the clearer waters of higher salinity in the outer bay. Seaward of the turbid waters of the inner bay, where salinity gradients are less pronounced, other environmental cues must be used to direct further migration, which remains essentially coastal as far seaward as Port Moller.

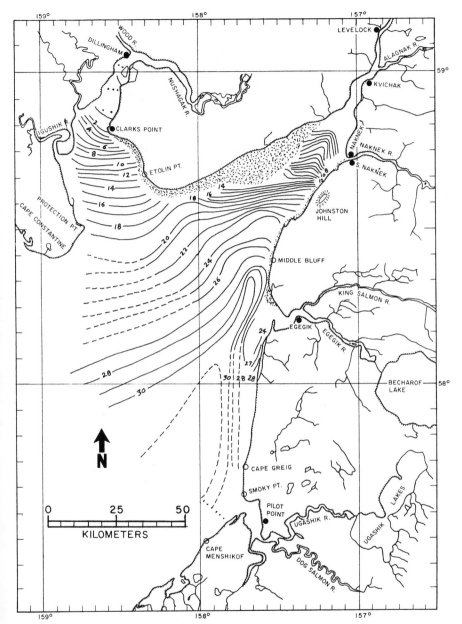

Fig. 14.12   Surface salinity (‰) at low tide, Bristol Bay, during July and August 1966.

Sea temperature

Our monthly oceanographic surveys of Bristol Bay since 1969 show that from spring through early fall, hydrographic conditions do not remain stationary for periods longer than a few weeks. In general, sea temperatures from May through September in the surface waters occupied by juvenile sockeye tended to be warmer along the southeast side, colder in the middle, and somewhat warmer again along the north side of the inner and outer bays. Temperature data collected toward the head of Bristol Bay from Port Moller in 1967 were not adequate for plotting the temperature distribution in this region.

Figure 14.13 illustrates that sea temperatures in Bristol Bay and the eastern Bering Sea vary greatly in about the same monthly period each year. Sea temperatures from mid-June to early July in 1969 and 1970 were much warmer than for the same period in 1971. Even warmer temperatures prevailed during this period in 1967 in the outer bay and the eastern Bering Sea. Surface temperatures 16 and 24 km off Port Moller on 2 July 1967, were 11.6 and 10.3 C respectively, indicating the usual pattern of warmer water near the coast.

Sea temperature data available for the same periods in outer Bristol Bay and the eastern Bering Sea before 1967 indicate that temperatures as cold as those which prevailed in mid-June to early July 1971 are not unusual but are less common than those that prevailed in 1967, 1969, and 1970. The available data indicate also that cold temperatures similar to those in 1971 occurred during approximately the same period in 1956 and 1959.

If annual differences in sea temperature of the magnitude shown in Figure 14.13 persisted throughout the sockeye's residence in Bristol Bay in a given year, they would probably result in annual differences in the temperature-dependent activities of the young fish. Temperature, through its influence on the rate of metabolism, governs the relative energy demanded for maintenance and for performing essential activities such as swimming and feeding. For example, Brett et al. (1958) found that temperature had a profound effect on the swimming speed of sockeye yearlings and underyearlings. Donaldson and Foster (1941) discovered that growth in young sockeye tends to be poor to nonexistent at temperatures below 4 C.

Sea temperature then may be expected to be a dominant environmental factor determining the length of the sockeye's residence in Bristol Bay through its influence on swimming speed and therefore the rate of seaward migration. Temperature will also govern the rate of growth and thus the amount of marine growth completed before the sockeye's first ocean winter. Finally, sea temperature may be an important factor controlling, for a time, the route of seaward migration.

Sockeye salmon have undoubtedly evolved behavior characteristics with respect to temperature which serve to keep them in a thermal range most conducive to their survival during seaward migration. Extremely low sea temperatures which inhibit activities such as feeding and therefore growth may be avoided if warmer waters are present. The existence of cold water offshore and warmer waters nearer the coast on the southeast side of outer Bristol Bay affords juvenile sockeye a range of temperatures which may include those best suited to its thermal requirements at the time. Avoidance of the colder offshore

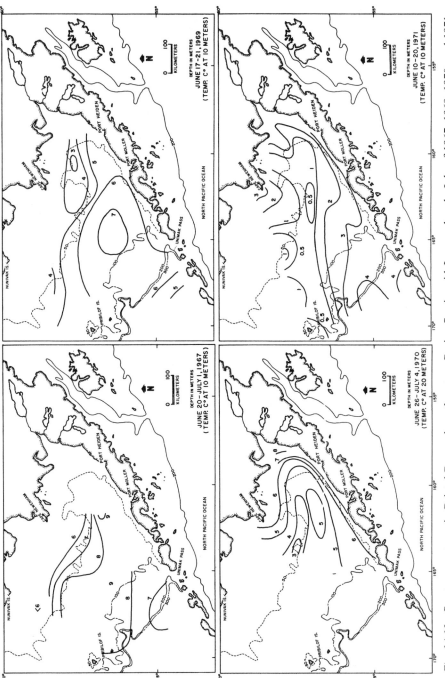

Fig. 14.13 Sea temperatures in Bristol Bay and southeastern Bering Sea in mid-June and early July of 1967, 1969, 1970, and 1971.

water is offered here as one possible explanation for the more coastal migration route through the inner portion of outer Bristol Bay as far seaward as Port Moller.

Variations in what we have come to regard as the more common seaward migration pattern were apparent in the results of our 1971 purse seining in Bristol Bay. As mentioned above, sea temperatures in mid-June to early July of 1967, 1969, and 1970 were much warmer than in 1971. Much of the discussion on migratory pattern and characteristics and growth has been based on the results obtained in these three years. The cold sea temperature in Bristol Bay during mid-June of 1971, which persisted through September, may have inhibited or reduced the rate of seaward migration and undoubtedly accounts for the lack of juvenile sockeye in the seine catches made off Port Moller in late June and early July. As discussed above, juvenile sockeye were found in abundance in this area in early June in previous years. At this time in 1971, seaward-migrating sockeye were found no farther seaward than Port Heiden. In addition, juvenile sockeye were captured only at fishing stations nearest the coast and in the shallowest water in which we were able to seine. Unfortunately, we were not able to determine the effect the colder sea temperatures prevailing in 1971 had upon the commencement and extent of marine growth, because we could not obtain samples of juvenile sockeye throughout the summer period. A study of the scales of adults from the stocks which migrated seaward in 1971, however, will provide insight into the influence of temperature on marine growth when they return to spawn one to four years hence.

With our present state of knowledge, we can only speculate on the assets or liabilities imposed on the juvenile sockeye by cold or warm sea temperatures during seaward migration. Warm sea temperatures will, of course, be conducive to faster growth, more rapid seaward migration, and perhaps to earlier entrance into the North Pacific Ocean. Cold sea temperatures would have the reverse effect. We do not know if early entrance into the North Pacific and rapid growth (and therefore a larger size) before the onset of winter are conducive to a higher survival rate than would occur if the situation were reversed. By reducing the metabolic demand and thus inhibiting feeding activity, cold temperatures could be an asset to survival if food abundance were low during some period of seaward migration.

Zooplankton abundance

Samples of zooplankton, the principal food of juvenile sockeye, were obtained during the monthly oceanographic surveys of Bristol Bay in 1969–71. To obtain the samples an 8-inch diameter bongo-net plankton sampler was towed in a direct oblique path from bottom to surface as the vessel traveled at a speed of 3 to 4 knots. The sampler had a mesh of #3 Nitex with a 0.33 mm average aperture dimension. Plankton abundance was determined volumetrically by the mercury immersion method described by Yentsch and Hebard (1957) and expressed as milliliters of plankton per cubic meter of water strained.

To determine if zooplankton abundance and distribution may be an important factor in the early marine growth and distribution of sockeye salmon during seaward migration through Bristol Bay, a comparison was made of the relative abundance of plankton by month in the inner and outer bays. Only zooplankton samples collected in the area encompassing the seaward migration route of the sockeye through the inner and outer bays are included in this comparison. Table 14.6 lists the average volumes of zooplankton determined from samples collected in the inner and outer bays for 1969–71.

Zooplankton were considerably more abundant in outer Bristol Bay than in the inner bay. Although the sample sizes were too small for an adequate comparison of abundance between the inshore and offshore areas, they do show that in the outer bay outside the primary seaward migration route of the sockeye, zooplankton were considerably more abundant than in the more coastal inshore waters (< 20 fathoms deep). The inner bay and coastal waters of the outer bay are both more turbid than the offshore waters. Therefore, in these turbid waters primary production and thus secondary production might be expected to be less than in the clearer offshore waters of the bay.

In the discussion on food habits of the sockeye, it was shown that in general, more of the fish captured in the offshore waters of the outer bay had full stomachs than did those captured in the inshore waters and inner bay (Fig. 14.11).

Zooplankton abundance in the inner bay may be at levels that would be measurably reduced by the influx of a large population of juvenile sockeye, and this, in turn, would be reflected in the fullness of stomachs of these fish. The volumes of zooplankton listed in Table 14.6 for the inner and outer bays are probably indicative of the relative secondary productivity of these regions, as well as the extent of grazing. Grazing may account for the absence of an expected increase in zooplankton abundance with time, i.e. from June to September in the inshore waters and in the inner bay (Table 14.6). The population of juvenile sockeye steadily increases in the inner bay from June through early July and then declines. Figure 14.11 showed that the proportion of fish with full stomachs also increased during the same period. An increase in zooplankton was not apparent in the inner bay during this period except in August of 1969 after the sockeye population in this area had declined.

If food abundance is at such low levels in the inner bay at the time of the sockeye's migration through this area that it is reflected in the fullness of their stomachs, this should have an influence on their growth during this period. As discussed above, rapid marine growth was almost nonexistent during the sockeye's residence in the inner bay. None of the sockeye stocks studied showed rapid marine-type growth on their scales before mid-July to late July, when they were in the outer bay. The combination of a large population of sockeye in this area of apparent low food abundance could account for the lack of growth during this period.

The evidence presented, although rather limited, strongly suggests that food abundance at the time the sockeye enter the bay and throughout their residence in the inner bay and the inshore areas of the outer bay may be a critical factor in their early growth and perhaps survival. The influence of food abundance on survival remains speculative at the present time, but it will, of

TABLE 14.6 Zooplankton volumes (ml m³) by month in inner and outer Bristol Bay, 1969–71

	Inner Bristol Bay				Outer Bristol Bay			
Year	June	July	August	September	June	July	August	September
1969	0.15	0.18	0.43	—	0.74	0.61	0.59	—
1970	0.23	—	0.28	0.16	0.36	—	0.29	0.33
1971	0.22	—	0.14	0.24	0.48	—	0.53	0.57

course, depend in a given year on such things as the size and age structure of the juvenile sockeye population and the dynamics of the zooplankton population.

A final subject of discussion is the influence of zooplankton abundance on the rate and route of the seaward migration. In the discussion thus far, the various sockeye stocks studied have been shown to migrate rather rapidly through the inner and into outer Bristol Bay, generally maintaining their order of outmigration as far as Port Moller. Thereafter the various early- and late-outmigrating stocks become mixed and move into the more offshore waters.

The apparent low abundance of food in the inner bay may be a factor that stimulates the sockeye to move rapidly through the inner bay and into the outer bay, where food is in greater abundance. Once they arrive in an area where food is abundant, the juvenile sockeye appear to grow rapidly and proceed seaward at more leisurely rates. This would allow late-outmigrating stocks to overtake and become mixed with those stocks which entered the bay much earlier.

The question may be asked as to why juvenile sockeye do not move immediately into areas of heavy zooplankton abundance offshore as soon as they enter the outer bay. Temperature has been suggested as a possible factor that may keep the seaward migration route largely coastal through the outer bay as far seaward as Port Moller.

Presumably, the movements of juvenile sockeye which have evolved and have been preserved are those that are most conducive to survival while the young fish are in Bristol Bay. Coastal migration through the outer bay as far seaward as Port Moller and beyond, theoretically might benefit the juvenile sockeye in two ways. Much of the juvenile sockeye seaward migration through this area occurs at a time when the adults of five species of Pacific salmon are on their spawning migrations and are abundant in the adjacent offshore waters. If the young salmon remained close to shore through this area at this time, they would avoid potential competition for similar food items with the migrating adults. In addition, the juvenile sockeye would not be preyed upon by such animals as seals, sea lions, and sharks, which are prevalent among adult salmon. The movement of juvenile sockeye into the offshore waters seaward of Port Moller was suggested to occur after mid-August. By this time most adult salmon have entered their spawning streams or the inner bay and they are no longer abundant in the offshore waters.

PROBLEMS AND FUTURE OPERATIONS

In our present studies we have been handicapped by the lack of sufficient environmental data on the Bristol Bay region. Much of the oceanographic data used to describe conditions and occurrences in this region are fragmental and span too long a period of time to be of maximum benefit to the biologist. This is particularly true for the near coastal and the inner regions of the bay. As a result, we have been forced to make rather broad general statements about the dynamics of the oceanic climate of this region.

The results of our monthly surveys of Bristol Bay since 1969 have shown that within the period of spring through fall, oceanographic conditions change rapidly in a short period of time. The biologist interested in predicting the distribution, abundance, and growth of organisms residing in this region must understand the causes of environmental variation and must be able to describe and measure their effects upon the organisms. This will require not only a more accurate description of the dynamics of the oceanic climate of Bristol Bay but also an understanding of the interrelationship of atmosphere and ocean-ographic conditions existing in the entire Bering Sea. The latter can best be accomplished by coordinating the data-collecting activities and by data ex-change between those nations having interests in the living resources of this productive area.

Our future plans are tentative. We hope to continue our monthly ocean-ographic surveys of Bristol Bay from spring through fall, increasing the number of oceanographic stations in some areas and if possible the frequency of coverage of all stations. In conjunction with the collection of environmental data, we tentatively plan to collect samples of juvenile sockeye by means of variable fine mesh gill nets. These samples will provide material for the con-tinued study of migratory characteristics and growth of the juvenile sockeye during a variety of environmental conditions. Only after acquiring a con-siderable time series of data will we be able to predict the influence variations in the oceanic climate of Bristol Bay and the Bering Sea will have upon the early growth and survival of the sockeye salmon.

Acknowledgments

I wish to acknowledge the contributions of Herbert W. Jaenicke and Malin B. Murray. Jaenicke supplied the zooplankton data and Mrs. Murray the food habit data reported on in this paper.

REFERENCES

ANONYMOUS

    1956   Beluga investigations. Alaska Dept. Fish and Game, Annual Rep. 1955, pp. 98–106.

CHAPTER 15

Energy requirement of Bristol Bay sockeye salmon in the central Bering Sea and Bristol Bay

Tsuneo Nishiyama

Faculty of Fisheries, Hokkaido University, Hakodate, Japan

Abstract

Mature sockeye salmon bound for western Alaska prevail as one of the most densely distributed fish stocks in the Bering Sea in summer. The fish migrate from the northern North Pacific to the central Bering Sea, passing through several channels of the Aleutian Islands, and move easterly to the slope area and finally the continental shelf of Bristol Bay. During their month in the Bering Sea, sockeye salmon feed actively on fish larvae, squid, euphausiids, and other zooplankton to accomplish the growth, gonad development, and tissue accumulation of energy-generative materials necessary for their migratory functions.

The daily food intake of a 2.2 year-old sockeye salmon equivalent to the energy required for swimming, growth and gonad development during its Bering Sea migration has been estimated to be 34 to 115 kcal in studies conducted in 1965 and 1967 to 1971. This estimate is based on the assumption (Hartt 1966) that the fish move at a speed of 30 n. miles/day, that their daily growth rate is 0.1 to 0.7 percent of their body weight, and that their daily gonad development is 0.6 to 2.0 percent of their gonad weight.

INTRODUCTION

The Bering Sea is an ecologically significant habitat both for numerous kinds of marine fish and for a number of anadromous species. Most of the Pacific salmon (*Oncorhynchus* sp.) originating in either Siberian or Alaskan streams enter much of the Bering Sea to feed during late spring and summer. Although demersal fish such as Alaskan pollock, yellowfin sole, and ocean perch spend their entire life in the Bering Sea, the several species of *Oncorhynchus* are only seasonal or transient inhabitants, resembling in this respect the sea mammals such as whales and fur seals that visit the Bering Sea only periodically to feed or reproduce.

Bristol Bay sockeye salmon (*Oncorhynchus nerka* Walbaum), which are considered here to be those salmon originating in northwestern Alaskan waters, occupy the most conspicuous place both in population size and commercial importance among the species of fish found in the Bering Sea. From 1964 to 1971 the total population (based on catch and escapement numbers) ranged from $6-60 \times 10^6$ (average 22×10^6) fish with wide annual fluctuations.

Important findings in the last 10 years have advanced our knowledge of the ocean distribution and migration of Bristol Bay sockeye salmon (Bakkala 1971; French and McAlister 1970; Royce et al. 1968) and enable us to form certain hypotheses about their migratory pattern in the Pacific Ocean and Bering Sea.

Bristol Bay sockeye salmon do not live throughout the year in the Bering Sea but migrate seasonally between the northern North Pacific, the Gulf of Alaska, and the Bering Sea. In their final year of ocean life, maturing sockeye salmon from the southern or southeastern North Pacific and Gulf of Alaska waters move westward or northwestward to the south side of the Aleutian Islands, entering the Bering Sea through the Aleutian passes from early June to early July. The distribution of the schools of fish extends over about two-thirds of the Bering Sea to 60°N and 165°E, depending strongly on oceanographic conditions. The fish feed actively for at least a month in the Bering Sea before converging in a clockwise pattern toward the mouth of Bristol Bay.

In addition to population size, Bristol Bay sockeye salmon are characterized by their consistent arrival in Bristol Bay during a very brief period of time each year; invariably most of the population arrives at the river mouth between the last week of June and the first week of July, and about 50–80 percent of the migration is concentrated in only one week (Royce et al. 1968). Another feature, as shown by the recovery of tags, is a faster speed of migration in the Bering Sea than in other areas; the average rate of migration of Bristol Bay sockeye salmon is estimated to be about 30 n. miles per day for long distances (Hartt 1966). Also, in common with other salmonid species, sockeye salmon are found usually near the sea surface.

In connection with the above behavioral peculiarities of mature Bristol Bay sockeye salmon during their month in the Bering Sea prior to entering the rivers to spawn, the relationship between their energy demand and supply was investigated (Nishiyama 1970). Many characteristics associated with feeding behavior and food requirements are closely related to food supply and competition with other marine organisms for food within a rather limited period of time, season, and space.

Materials and methods

The data and samples for this study were obtained mainly from the *Oshoro Maru*, a fisheries training ship of Hokkaido University. Each summer since

1964 research cruises have been made aboard the vessel to investigate sockeye salmon in the Bering Sea and Bristol Bay (Faculty of Fisheries, Hokkaido University 1965–71.

Drifting gill nets of various mesh sizes were used in the exploratory salmon fishing in selected locations (Fig. 15.1). The distribution pattern of sockeye salmon was clearly shown for Bristol Bay, where the fishing effort was concentrated, but the concentration values remained relatively obscure for the central and northwestern Bering Sea.

The relative density of Bristol Bay sockeye salmon was expressed in terms of *catch per unit effort* (CPUE) per *tan* (standard gill net area). Since one tan is about 5 m high by 50 m long and is fished for about 12 hours (overnight), the CPUE is actually a measure of the relative density of fish caught in an area of gill net about 250 m² during a half-day period; the *average* CPUE is defined as the average number of fish per tan in a one-degree quadrangle.

Specimens of Bristol Bay sockeye salmon were sampled at random from the gill net catch. The caudal fork length and body-gonad weights were noted, and the age was determined from the scales. A linear scale was used to estimate the body and gonad growth rates of fish taken in the Bering Sea between early June and early July.

The stomach contents of 8–60 specimens of Bristol Bay sockeye salmon taken in 1964–66 (Fig. 15.2), and in 1969 were examined to determine the average seasonal and geographical differences in food composition. Food organisms were identified and placed into one of nine groups (copepods, euphausiids, amphipods, pteropods, zoea, squids, fish larvae, appendicularia and others), and wet weight was determined for each category. Prey organisms (which corresponded to the dominant species found in the stomach contents) were collected concurrently at each fishing location with a Norpac plankton net or fish larvae net. The prey organisms obtained were immediately divided into two parts, one of which was preserved in 10% formaldehyde solution for taxonomic study and the other deep-frozen for caloric analysis. Similarly, samples of muscles and gonads were prepared for caloric determination by a combustion bomb method.

The environmental temperature for sockeye salmon was based on mean seawater temperature data collected by Japanese salmon factory ships and research vessels operating in the Bering Sea salmon fishing grounds. Temperature data were available from 1965 to 1969 but were incomplete for more recent years.

RESULTS

Relative abundance and age composition of sockeye salmon

The relative abundance of mature Bristol Bay sockeye salmon is expressed in Figure 15.3 as catch per unit effort (CPUE) during the summer season (1964–71); the values vary from 0.01 to over 100.00 during this seven year period. High values were found usually in the area south of the Pribilof Islands, on the edge of the continental shelf, and along the Alaska Peninsula. In contrast, relatively low CPUE values were recorded in the northeastern Bering Sea and north of Adak. The concentrations of fish were similar for each year.

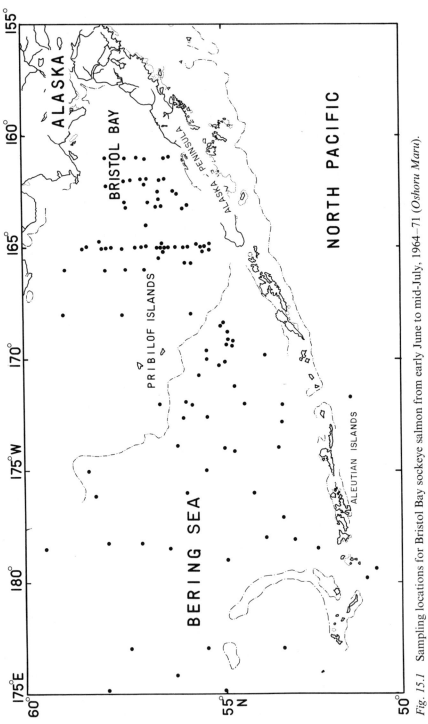

Fig. 15.1　Sampling locations for Bristol Bay sockeye salmon from early June to mid-July, 1964–71 (*Oshoru Maru*).

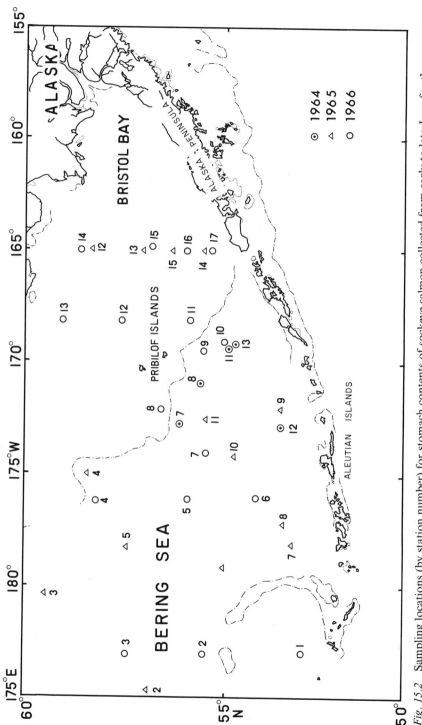

Fig. 15.2 Sampling locations (by station number) for stomach contents of sockeye salmon collected from early to late June for three years (1964–66) by the *Oshoro Maru*.

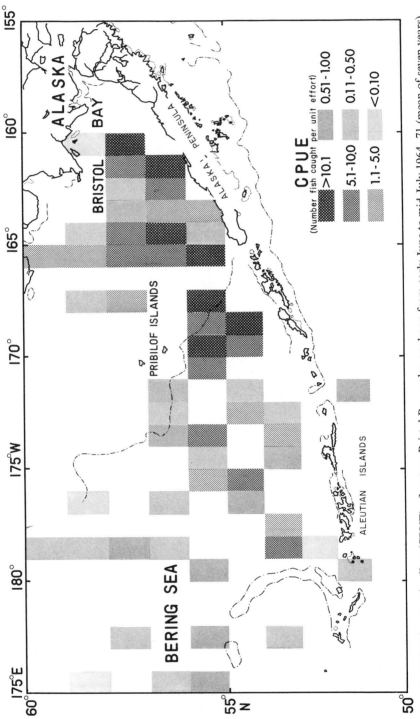

Fig. 15.3 Catch per unit effort (CPUE) of mature Bristol Bay sockeye salmon from early June to mid-July 1964–71 (mean of seven years).

CPUE values remained generally low in early to mid-June and became higher in late June to early July. Superficially, this evidence may suggest that sockeye salmon are dispersed in the Bering Sea in the early part of the season, gradually gather at the mouth of Bristol Bay, and pass along the southern part of the Bay as the season progresses.

According to population numbers and food concentration, the distribution of sockeye salmon has been arbitrarily divided into three geographic regions: the central Bering Sea, the slope area, and Bristol Bay.

The age composition of Bristol Bay sockeye salmon specimens taken on board ship is shown in Table 15.1. Although the composition fluctuated annually, most of the fish could be classified into age groups of 1.3, 2.2, 2.3, and 1.2 years, respectively. The age composition of samples from the *Oshoro Maru* and the catch in the Bristol Bay were nearly identical except for certain differences in the younger age groups in 1968 and 1969.

It was observed also that the older and larger Bering Sea fish tended to migrate earlier and to range further west and north than the younger, smaller fish.

Environmental temperature of sockeye salmon

Since sockeye salmon are essentially surface swimmers inhabiting water less than 10–20 m deep, the surface water temperature is used in this study as an indicator of the fishes' environmental temperature. Although there are not records for all areas of the Bering Sea, the available data offer considerable information on the temperature of the sea where sockeye salmon are found.

The summer temperatures for the seven year period studied ranged from 4 to 12 C, and the mean temperature changed with each 10-day period

TABLE 15.1 Age composition of Bristol Bay sockeye salmon sampled from the *Oshoro Maru* in the Bering Sea and Bristol Bay, mid-June to early July 1965–71

| Year | Age 2.3 (yrs) | Number of specimens by age groups (% of annual total for age group) | | | Other | Annual total |
		2.2	1.2	1.3		
1965	34	336	6	17	6	399
	(8.5)	(84.2)	(1.5)	(4.3)	(2.5)	(100.0)
1966	737	197	42	126	16	1117
	(66.0)	(17.6)	(3.8)	(11.2)	(1.4)	(100.0)
1967	507	1010	22	36	68	1643
	(30.9)	(61.5)	(1.3)	(2.2)	(4.1)	(100.0)
1968	488	452	275	236	59	1513
	(32.3)	(29.9)	(18.2)	(15.6)	(3.9)	(100.0)
1969	329	996	795	313	44	2477
	(13.3)	(40.2)	(32.1)	(12.6)	(1.8)	(100.0)
1970	119	1149	103	110	44	1525
	(7.8)	(75.3)	(6.8)	(7.2)	(2.9)	(100.0)
1971	305	651	116	742	3	1817
	(16.8)	(35.8)	(6.4)	(40.8)	(0.1)	(100.0)

(Table 15.2). For example, in 1969 fish were found in early June in water temperatures of about 5 C near the Aleutian Islands, 6 to 7 C in the central continental slope area in mid-June, 7 to 8 C in late June, and finally 8 to 10 C on the continental shelf in early July. Monthly differences in temperature ranged from 1.0 to 2.8 C. This is evidence that fish migrating from the North Pacific to the Bering Sea and Bristol Bay are found in different temperature zones both in time and space. Further, it is apparent that seawater temperatures differ considerably from year to year.

TABLE 15.2   Surface seawater temperatures (°C) in areas of Bristol Bay sockeye salmon occurrence in the Bering Sea and Bristol Bay

Year	Early June	Middle June	Late June	Early June
1965	4.9 ± 0.48 (18)	5.5 ± 0.83 (19)	5.9 ± 1.31 (5)	
1966	5.2 ± 0.47 (24)	5.3 ± 0.85 (37)	6.4 ± 1.10 (9)	
1967	5.7 ± 0.78 (33)	6.2 ± 0.69 (34)	8.5 ± 1.02 (19)	
1968	5.1 ± 0.49 (8)	5.5 ± 0.68 (15)	6.5 ± 0.83 (17)	
1969	5.3 ± 0.38 (20)	5.9 ± 0.56 (33)	7.2 ± 1.15 (11)	8.7 ± 1.68 (8)
1970		5.6 ± 0.81 (13)	6.3 ± 1.05 (18)	
1971			4.6 ± 0.74 (12)	4.5 ± 1.76 (10)

Food organisms

The occurrence and composition of food organisms varied with area and season. In 1964 squid and euphausiids were dominant in the stomachs of fish from the central Bering Sea, while squid prevailed in the slope area (Table 15.3). In 1965 the occurrence pattern of prey animals was quite similar to that of the previous year; in the central Bering Sea, squid and euphausiids were dominant, followed by amphipods (Table 15.4). On the continental shelf of Bristol Bay, the dominant prey animals were almost exclusively euphausiids, with some fish larvae. In 1966 the stomach contents consisted largely of squid in the central Bering Sea and euphausiids in Bristol Bay, with the slope area being intermediate between these two (Table 15.5). A similar pattern was also observed

TABLE 15.3   Number of prey animals found in stomach contents of Bristol Bay sockeye salmon (20–29 June 1964)

Organism	Bering Sea			Slope area	
	Station 11	12	13	7	8
Copepods	4	1	4	—	—
Euphausiids	10	12	6	4	2
Amphipods	1	5	2	2	—
Pteropods	3	—	2	—	—
Squid	10	1	11	7	20
Fish larvae	2	3	8	7	2
No. of samples	28	22	21	19	26
Empty stomachs	4	4	3	2	4

TABLE 15.4 Prey animals (frequency of occurrence) found in stomachs of Bristol Bay sockeye salmon (5–27 June 1965)

Organisms	Bering Sea								Slope area		Bristol Bay (shelf area)			
	Station 2	3	5	6	7	8	9	10	4	11	12	13	14	15
Copepods	—	—	—	—	4	—	1	17	—	2	8	—	—	—
Euphausiids	5	12	10	2	4	18	12	17	4	9	1	30	29	—
Amphipods	14	14	22	14	16	15	14	3	8	3	—	—	4	8
Pteropods	2	—	—	19	1	6	11	10	—	2	—	—	7	1
Polychaeta	—	—	—	1	—	1	3	4	—	1	—	—	2	—
Squid	3	3	3	18	4	6	8	16	—	16	—	—	—	—
Fish larvae	9	11	8	14	9	10	12	11	8	12	3	—	14	12
Crab zoea	—	—	—	—	—	—	—	—	—	11	—	—	4	18
Appendicularia	—	—	—	—	18	2	—	—	—	—	—	—	1	—
No. of samples	18	30	30	30	28	31	30	30	13	30	8	30	29	30
Empty stomachs	2	4	3	5	4	4	0	0	3	3	0	0	0	7

TABLE 15.5   Mean weight composition (g) of stomach contents per fish for Bristol Bay
sockeye salmon (14–29 June 1966)

Organisms	Bering Sea Stations 1–3, 5–7	Slope area 4, 8–11	Bristol Bay 12–17
Euphausiids	1.2 g	0 g	11.3 g
Amphipods	0.5	1.9	3.0
Pteropods	0	0	1.1
Squid	57.8	0	0
Fish larvae	1.7	1.2	0.3
Others	0.1	1.1	0.1
No. of samples	95	22	94

in 1969 when the collection of samples from Bristol Bay was increased:
euphausiids comprised about 70 percent of the total weight of the stomach
contents, followed by 20 percent fish larvae and other planktonic animals.

Thus, throughout the four-year period, it was generally found that in the
central Bering Sea the main food organisms consisted of squid, fish larvae,
euphausiids, amphipods, and other planktonic animals, while euphausiids and
fish larvae were dominant in Bristol Bay. In the slope area, the occurrence of
food organisms in the stomachs was intermediate between the other two areas.

Ignoring annual fluctuations, the weight of stomach contents of fish from
the central Bering Sea was relatively low compared with those from the western
or northern Bering Sea. The weight of stomach contents was low also for fish
from the southern Bering Sea and from the middle of Bristol Bay. Stomach
contents varied from scanty to full in areas of high CPUE value. Conversely,
fish from areas of low CPUE value consistently had relatively large amounts
of food in their stomachs.

Fish larvae, as well as euphausiids and squid, form a significant part of
the sockeye salmon diet. To examine the distribution of fish larvae, extensive
samples were collected from the Bering Sea and Bristol Bay. Although sampling
was inadequate for the north side of the Pribilof Islands and north of the
Aleutian Islands, there were sufficient samples taken to show the distribution
of fish larvae in the Bering Sea. More than 20 species of fish were found in these
samples, about half of which could be identified (Table 15.6). The dominant
species of fish larvae consisted of Alaskan pollock, Pacific sandlance, and many
kinds of sculpins. In general, most kinds of larval fish found in the samples
were found also in the stomachs of sockeye salmon.

Caloric values of food organisms and sockeye salmon
The potential energy of each category of food organisms for sockeye salmon
is shown in Table 15.7. In this analysis, copepods included *Calanus cristatus*
and *Eucalanus bungii bungii*; euphausiids considered were *Euphausia pacifica*,
Thysanoessa longipes, *Thy. raschii*, and *Thy. inermis*; amphipods were represented by *Parathemisto pacifica*, *P. japonica*, *Euthemisto libellula*, and *Hyperoche*
kroyeri. Pteropods consisted of *Clione limacina* and *Limacina helicina*, and

TABLE 15.6   Fish larvae (number of individuals) collected by surface tow from early June to early July

Species of fish	Bering Sea				Slope area				Bristol Bay			
	1966	1967	1968	1969	1966	1967	1968	1969	1966	1967	1968	1969
Theragra chalcogramma	2	1	10	4	6	24	328	21	1	1	176	667
Ammodytes hexapterus	40	—	27	—	30	1	—	—	1412	—	2	2890
Ptilichthys goodei	—	—	—	—	1	—	—	—	—	—	2	2
Bathymaster sp.	—	16	2	8	1	—	—	27	29	—	1	27
Stichaeus sp.	64	1	15	—	99	2	1	—	99	—	—	—
Lumpenes sp.	5	—	—	—	—	—	—	3	—	—	—	6
Sebastes sp.	—	1	—	—	—	—	—	—	—	—	—	—
Pleurogrammus monopterygius	11	5	2	—	95	—	1	26	1	—	6	20
Hexagrammos sp.	20	5	11	—	160	—	11	48	3862	—	—	31
Hemilepidotus sp.	50	287	55	11	3	—	1	6	—	—	—	—
Myoxocephalus sp.	1	—	—	—	—	—	17	—	38	—	111	1802
Cyclopteridae gen. sp.	1	—	—	—	—	—	—	—	—	—	—	—
Liparis sp.	—	—	—	1	—	1	—	—	13	—	6	677
Podothecus sp.	—	—	—	—	—	—	—	—	—	—	3	37
Atheresthes sp.	—	2	1	—	20	19	—	—	—	—	—	—
Reinhardtius hippoglossoides	—	—	—	—	8	1	5	—	2	—	4	—
Hippoglossoides elassodons	—	—	—	—	2	2	—	—	16	—	—	3
Aptocyclus ventricosus	—	—	—	—	—	—	—	—	—	—	—	—
Pleuronectidae	—	—	6	—	—	—	—	—	—	—	17	2
Anoplopoma fimbria	—	—	—	18	—	—	—	—	—	—	45	2
Myctophum californiensis	—	—	—	7	—	—	—	—	—	—	—	—
Chauliodus macouni	—	—	—	—	—	—	—	—	—	—	—	—
Number of samplings	8	7	5	2	4	2	4	1	6	1	7	13

TABLE 15.7   Caloric values of food organisms collected in the Bering Sea and Bristol Bay from early June to early July in 1966, 1967, and 1968

Organisms	No. of samples	Moisture (%)	Calories in dry weight per gram	Calories in wet weight per gram
Copepods	7	83.0	5114 ± 509	872
Euphausiids	11	76.3	5494 ± 520	1303
Amphipods	5	81.9	4517 ± 89	320
Pteropods	5	85.7	5502 ± 549	768
Polychaeta	3	85.1	5032 ± 313	752
Crab zoea	3	80.8	4186 ± 312	804
Squid	5	84.6	5578 ± 737	857
Fish larvae	9	73.8	5246 ± 289	1374
Mean value		81.4	5084 ± 499	946

there were larval forms of Alaskan pollock, Pacific sandlance, and sculpin. The caloric values varied from 320 to 1400 cal/g wet weight, differing for each group of food organisms. It is noted that euphausiids, fish larvae, squid, and copepods have a higher energy potential than amphipods and pteropods. Accordingly, the total potential energy may be quite different depending upon the *kinds* of food organisms eaten.

As the fish grows, the muscle tissue in the body wall increases in proportion to that in other body parts such as the head, skin, skeleton, and viscera. Muscle tissue finally comprises 65 to 80 percent of the total body weight, and in this sense the growth of sockeye salmon can be regarded simply as an increase in muscle tissue (T. Nishiyama, unpublished). The mean caloric value for dorsal and ventral muscle tissue combined in mature Bristol Bay sockeye salmon was found to be about 2000 cal/g wet weight (Table 15.8). These values were extremely high due to the high content of fat which accumulates in the mature sockeye salmon before it enters freshwater to spawn. From the caloric values for ovaries (25 to 250 g wet weight) previously presented (Nishiyama 1970), a linear relationship has been shown to exist between the female gonads and either the size of the eggs, maturity, or the gonad index. This relationship can be applied to calculate the caloric values for ovaries of any weight. From the

TABLE 15.8   Caloric values of muscle tissue from Bristol Bay sockeye salmon collected in the Bering Sea and Bristol Bay from mid-June to early July in 1970 and 1971

	Dorsal	Ventral	Combined
Calories			
Whole	5749 ± 147 (12)	6188 ± 249 (12)	5971 ± 294 (24)
Ash-free	6033 ± 168 (10)	6453 ± 221 (10)	6245 ± 281 (20)
Moisture (%)	67.5 ± 4.52 (10)	67.0 ± 2.08 (10)	67.3 ± 3.44 (20)
Ash (%)	4.3 ± 0.26 (10)	3.6 ± 2.08 (10)	4.0 ± 0.45 (20)
Lipids (%)	13.2 ± 2.82 (12)	21.9 ± 4.82 (12)	17.3 ± 5.73 (24)

present analysis, the caloric value for testes (20 to 70 g wet weight) was found to be 4948 $\pm$ 122 per gram dry weight in nine replicate samples.

As a general approximation, the caloric values per gram wet weight is 1000 for food organisms, 2000 for muscle tissue, 3000 for ovaries and 1100 for testes.

Method of estimating energy requirement

The energy which is released from food organisms in the sea provides for growth of the fish (production of muscle tissue, viscera, and sex products) and for metabolism.

Based upon laboratory and field studies, Winberg (1956) was able to develop the following general formula to express the relationship between the nutritional demand and supply of fish:

$$0.8R = T + P$$

where R is food intake, T expenditure for metabolism, and P growth in weight. The author has used this equation to estimate the energy requirements of Bristol Bay sockeye salmon in their final month of body growth and gonad development. The relationship between sample date and body weight or gonad index (gonad weight $\times$ 100/ body weight) was determined for several age groups by year (1967– 71) and by sex. The body and gonad weight, rate of growth, and growth increment per day are given in Tables 15.9 and 15.10.

The energy expenditure for active metabolism of Bristol Bay sockeye salmon is shown in Tables 15.11 and 15.12. These calculations are based on two important experimental results made by Brett (1965), who measured the maximum sustained swimming speed of sockeye salmon and determined empirically the following relationship between speed and body length:

$$\text{Log } Y = 1.29 + 0.50 \log X$$

where Y is speed in cm/sec and X body length in cm. Next he developed the following equation which permits computation of the maximum active metabolic rate at 15 C:

$$\text{Log } Y = -0.064 + 0.963 \log X$$

where Y is mg oxygen/hour and X body weight in grams. By further use of these formulas, one can also estimate the maximum swimming speed of the fish.

Since the velocity of 30 n. miles per day corresponds to 42–46 percent of the maximum swimming speed, the amount of oxygen consumed at a speed of 30 n. miles/day will be proportional also to the amount of oxygen consumed at the maximum sustained swimming activity. Since the environmental temperature for salmon differs each year, a temperature coefficient is used to adjust the metabolism to the mean temperature for late June or early July of the respective years. The total amount of oxygen consumption per fish per day was then calculated (last column, Tables 15.11 and 15.12). Estimates of the total

TABLE 15.9 Growth and gonad development of 2.2 year old Bristol Bay sockeye salmon

Year	Sex	Body weight (g)					Gonad weight (g)				
		Initial	End	Mean	Rate of growth per day (%)	Increment of growth	Initial	End	Mean	Rate of development (%)	Increment of gonad
1965 June 12–July 15	Female	1694	1744	1677	0.089	1.49	71.5	95.9	83.7	0.813	0.893
	Male	1775	1885	1830	0.218	4.00	26.8	77.4	52.1	2.830	2.973
1967 June 18–July 1	Female	2106	2412	2259	0.753	17.00	85.0	109.5	97.3	0.648	0.631
	Male	2426	2750	2588	0.696	18.00	37.1	64.6	50.9	2.315	3.002
1968 June 20–July 4	Female	2139	2223	2181	0.275	6.00	85.5	96.9	91.2	0.622	0.567
	Male	2593	2747	2670	0.412	11.00	44.9	57.2	51.1	1.311	0.700
1969 June 29–July 9	Female	2108	2274	2193	0.775	17.00	114.1	150.1	132.1	1.996	2.637
June 29–July 4	Male	2385	2455	2420	0.579	14.01	65.1	73.6	69.3	1.886	1.307
1970 June 13–July 3	Female	1929	2069	1999	0.350	7.00	47.0	55.1	51.1	0.316	0.035
	Male	2189	2409	2299	0.478	10.98					
1971 June 25–July 2	Female	2086	2156	2121	0.471	9.98	82.3	87.1	84.7	0.316	0.268
	Male	2343	2413	2378	0.438	10.41	39.9	50.2	45.1	2.854	1.287

TABLE 15.10 Growth and gonad development of Bristol Bay sockeye salmon of age groups 2.3, 1.2, and 1.3 years

Year	Sex	Body weight (g)					Gonad weight (g)				
		Initial	End	Mean	Rate of growth per day (%)	Increment of growth	Initial	End	Mean	Rate of development (%)	Increment of gonad
Age 2.3 years											
1966											
June 19–28	Female	2907	2980	2944	0.276	8.20	125.5	152.0	138.8	1.848	2.565
1967											
June 13–July 1	Female	2516	3110	2813	1.173	34.00	99.3	142.8	121.1	0.843	1.020
	Male	2776	3586	3181	1.415	45.00	41.8	74.6	58.2	1.786	1.040
1968											
June 20–July 4	Female	3056	3098	3077	0.097	2.98	181.0	198.2	189.6	0.552	1.050
1971											
June 25–July 2	Female	2693	2763	2727	0.367	10.00	141.0	158.4	149.7	1.297	1.940
Age 1.2 years											
1968											
June 21–July 4	Female	2030	2160	2095	0.477	9.99	91.6	112.2	102.0	1.071	1.092
1969											
June 23–July 4	Female	1973	2050	2012	0.348	7.00	106.8	125.2	116.0	1.094	1.269
Age 1.3 years											
1969											
June 20–July 4	Female	2362	2922	2642	1.514	40.00	125.7	194.3	160.0	1.587	2.540
1971											
June 25–July 5	Female	2703	2823	2763	0.434	12.00	144.9	192.2	168.5	2.383	4.020
	Male	3215	3595	3405	1.116	38.00	60.3	74.6	67.4	1.013	0.683

TABLE 15.11 Energy expenditures for active metabolism in 2.2 age Bristol Bay sockeye salmon

Year	Sex	Size of fish (g)	Max. swimming speed cm/sec.	Rate for 30 miles/day	Max. active metabolic rate at 15 C (mg O_2/kg-hr)	Active metabolic rate at 30 miles/day	Temp. zone (C)	Temp. coefficient	Active metabolic rate for body weight/day (g)	(liters)
1965	Female	1677	141.6	45.4	1099.4	499.1	5.9	2.929	6.858	4.800
	Male	1830	143.7	44.7	1196.1	534.7	5.9	2.929	8.018	5.612
1967	Female	2259	151.0	42.5	1465.0	622.6	8.5	2.076	16.260	11.382
	Male	2588	153.4	41.9	1669.9	687.8	8.5	2.076	17.355	12.148
1968	Female	2181	150.2	42.8	1416.2	606.1	6.5	2.713	11.694	8.185
	Male	2670	153.9	41.8	1698.3	709.5	6.5	2.713	16.758	11.731
1969	Female	2193	150.3	42.8	1423.7	609.1	8.7	2.019	15.878	11.115
	Male	2420	151.4	42.5	1565.4	664.8	8.7	2.019	19.124	13.387
1970	Female	1999	147.4	43.6	1302.0	567.9	6.3	2.791	9.765	6.836
	Male	2299	150.0	42.9	1489.9	638.6	6.3	2.791	12.629	8.840
1971	Female	2121	149.1	43.1	1378.9	594.5	4.5	3.310	9.143	6.400
	Male	2378	151.0	42.6	1539.2	655.4	4.5	3.310	11.300	7.910

TABLE 15.12 Energy expenditures for active metabolism in Bristol Bay sockeye salmon of age groups 2.3, 1.2, and 1.3 years

Year	Sex	Size of fish (g)	Max. swimming speed cm/sec	Rate for 30 miles/day	Max. active metabolic rate at 15 C (mg O_2/kg-hr)	Active metabolic rate at 30 miles/day	Temp. zone (°C)	Temp. coefficient	Active metabolic rate for body weight/day (g)	(liters)
Age 2.3 years										
1966	Female	2944	158.8	40.5	1890.5	765.5	5.9	2.713	19.936	13.956
1967	Female	2813	157.2	40.9	1809.5	740.1	8.5	2.076	24.068	16.848
1969	Female	2741	156.2	41.2	1764.8	726.4	8.7	2.019	23.668	16.567
	Male	3040	158.4	40.6	1949.9	791.5	8.7	2.019	28.602	20.021
Age 1.2 years										
1968	Female	2095	150.4	42.8	1362.4	582.4	6.5	2.713	10.794	7.555
1969	Female	2012	148.6	43.3	1310.4	567.0	8.7	2.019	13.568	9.498
	Male	2312	149.7	43.0	1498.2	643.5	8.7	2.019	17.685	12.380
Age 1.3 years										
1969	Female	2642	155.1	41.5	1703.4	706.1	8.7	2.019	22.175	15.523
	Male	2954	157.5	40.8	1896.8	774.3	8.7	2.019	27.189	19.032

amount of oxygen consumed by 2.2 year-old fish, for instance, ranged from 13 liters.

Body and gonad growth increments and active metabolic rates were converted into caloric values by using the results of the caloric analysis (Tables 15.13 and 15.14). The energy requirement value derived from these procedures was 27–29 kcal for 2.2 year-old fish, which means that 34–115 kcal in terms of food intake would be required to satisfy both growth and metabolism.

CONCLUDING REMARKS

The distinguishing features of Bristol Bay sockeye salmon are that they occur in large numbers in a rather limited surface layer of water, and they move rapidly toward freshwater in a short and fixed period of time in the late spring or early summer.

The mature sockeye salmon considered in the present study were caught in the area south of 60°N and east of 175°E from mid-June to early July. In separate studies of the distribution of sockeye salmon (Fukuhara et al. 1962; Margolis et al. 1966; Royce et al. 1968), it was established that sockeye salmon in the above areas were mostly of western Alaskan or Bristol Bay origin. Accordingly, the sockeye salmon used in this study are regarded as fish from Bristol Bay, even though there still remains some possibility of stocks of mixed origins, particularly in samples from the northwestern Bering Sea.

From the CPUE values it was found that relatively large, dense schools of fish occur along the edge of the continental shelf south of the Pribilof Islands and in the southern part of Bristol Bay close to the Alaska Peninsula. Seasonally, the sockeye salmon were abundant in the central Bering Sea in early and mid-June, in the slope area from mid-June to early July, and on the continental shelf in early and mid-July. It is possible that the distribution of sockeye salmon may be related to such environmental factors as water temperature and the availability of food for the fish.

Seawater temperatures in areas where salmon are found varied considerably with season and even within each 10-day period of the month. Although higher temperatures generally facilitated growth of the fish, the same high temperatures acted also adversely to bring about an excessive expenditure of energy. The reverse was true for low temperatures. Temperature, therefore, showed an important effect on growth by controlling the balance between these two factors.

Since food provides potential energy for use in fish growth and metabolism, the quantity and quality of food governed the growth rate. Examination of the stomach contents revealed certain aspects of the feeding behavior of salmon. Bristol Bay sockeye salmon appeared to feed essentially on plankton and fish—mainly squid, euphausiids, and fish larvae. The composition of food in the stomachs of sockeye salmon was nearly the same as that reported by Favorite (1970) for the Bering Sea and northern Pacific Ocean and by LeBrasseur (1966) for the northeastern Pacific Ocean. Differences in food organisms depend upon the area or season, and it is particularly important to know how the different food items influence the energy supply of the fish.

TABLE 15.13   Daily energy and food requirements for 2.2 year old Bristol Bay sockeye salmon

Year	Sex	Size of fish		Increment of body and gonad weight per day (g)	Active metabolic rate per day (O₂ liters)	Conversion into caloric value			Ration required in caloric value (kcal)
		Weight (g)	Gonad (g)			Growth (kcal)	Metabolism (kcal)	Total (kcal)	
1965	Female	1677	83.7	1.49	4.801	3.89	23.48	27.37	34.21
	Male	1830	52.1	4.00	5.612	5.39	27.45	32.84	41.05
1967	Female	2259	97.3	17.00	11.381	33.97	55.66	89.63	112.04
	Male	2588	50.9	18.00	12.148	32.79	59.41	92.20	115.25
1968	Female	2181	91.2	6.00	8.186	12.36	40.03	52.39	65.49
	Male	2670	51.1	11.00	11.731	20.97	57.36	78.33	97.91
1969	Female	2139	132.1	17.00	11.115	36.03	54.35	90.38	112.98
	Male	2420	69.3	14.01	13.387	26.37	65.46	91.83	114.79
1970	Male	2299	51.1	10.98	8.840	21.48	37.00	58.48	73.10
1971	Female	2121	84.7	9.98	6.400	19.84	26.79	46.63	58.29
	Male	2378	45.1	10.41	7.910	19.33	33.11	52.44	65.55

TABLE 15.14 Daily energy and food requirements for Bristol Bay sockeye salmon of age groups 2.3, 1.2, and 1.3 years

Year	Sex	Size of fish		Increment of body and gonad weight per day (g)	Active metabolic rate per day (O_2 liters)	Conversion into caloric value			Ration required in caloric value (kcal)
		Weight (g)	Gonad (g)			Growth (kcal)	Metabolism (kcal)	Total (kcal)	
Age 2.3 years									
1966	Female	2944	138.8	8.20	11.036	18.71	68.24	86.95	108.69
1967	Female	2813	121.1	34.00	16.848	64.59	82.39	146.98	183.73
Age 1.2 years									
1968	Female	2095	102.0	9.99	7.556	21.40	36.95	58.35	72.94
1969	Female	2012	116.0	7.00	9.498	15.05	46.44	61.49	76.86
	Male	2312	71.0	4.99	9.318	9.32	60.54	69.86	87.33
Age 1.3 years									
	Female	2642	160.0	40.00	15.523	84.93	75.91	160.84	201.05

The caloric values of main groups of food of sockeye salmon are not given as net energy, but as *potential* energy: the *availability* of food seemed to be different for each food organism. Provided that the availability of food is the same, however, then the difference in caloric values would imply a different *net* energy for the various groups of food organisms. Based upon such an assumption, it is believed that euphausiids, squid, and fish larvae provide higher potential energy than amphipods and pteropods. The differences in caloric values were due mainly to the variations in ash and fatty substances in the food organisms.

When the fish feed on different food organisms, they obtain a different potential energy for growth and metabolism. In this study it was shown that potential energy was highest in Bristol Bay, then the slope area, followed by the central Bering Sea. This, in turn, implies a different potential energy from area to area and seems to be a significant factor in the growth of sockeye salmon.

In applying Winberg's equation to the present study, body and gonad growth were treated separately, since muscle tissue and gonads exhibit a different increment of body weight and have different caloric values. Because of the difficulty in working aboard ship, an estimate of active metabolism was derived from Brett's (1965) empirical equations. Thus, the procedures used in this study were based upon many assumptions and are open to some criticism and argument. With these reservations in mind, the energy requirement of mature Bristol Bay sockeye salmon was estimated. The values were found to vary widely from year to year but inadequately to make a valid comparison of annual fluctuation.

The distribution and production of the main food organisms varied independently of salmon population size. Growth and distribution of food organisms is probably governed by primary production and other oceanographic conditions such as currents, temperature, gyres, and turbulent mixing. There is a lack of information on the distribution, abundance, and dynamics of the respective food organisms for salmon throughout the year.

The estimate of energy requirement was confined to mature Bristol Bay sockeye salmon, which spend only a month in the Bering Sea. Similar estimates should be made either on other populations of sockeye salmon, on other stages of their marine life, or on other *Oncorhynchus* species. Both interspecific and intraspecific aspects of food and energy requirements are likely major factors in biological production. Further, the food and energy requirements of important marine fish, birds, and marine mammals native to the Bering Sea should be studied in relation to competition for food and to the total potential energy supply of the Bering Sea.

Discussion

SATOH: Does the direct line drawn between release and recapture points on your data slide represent a constant rate of migration for the mature sockeye salmon?

NISHIYAMA: Although the fish move quite continuously, it is obvious that a constant swimming speed would not be natural. I merely presented the results of data obtained by A. C. Hartt as an indication of active metabolism of the Bristol Bay sockeye salmon in the ocean.

SATOH: Have you observed any change in swimming speed according to maturity in the spawning fish?

NISHIYAMA: I have seen no evidence of acceleration in migration rate as the fish matures.

STRATY: According to Hartt, 30 miles per day is the average rate of migration of the adult sockeye salmon. There is evidence, however, that there may be an increase in the migration rate with maturity which occurs toward the latter half of the spawning migration.

KASAHARA: You mentioned two dominant forms of larval fish found in the stomachs of sockeye salmon. Was one of these *Ammodytes*?

NISHIYAMA: Yes. In 1969, when our group examined specimens taken in Bristol Bay, 70 percent of the stomach contents were euphausiids and 20 percent fish larvae; most of the latter were Pacific sandlance.

Acknowledgment

The author wishes to express gratitude to Professor Tokimi Tsujita, Director of the Research Institute of North Pacific Fisheries, Hokkaido University, for his invaluable advice and encouragement in the course of his work.

REFERENCES

BAKKALA, R.
    1971   Distribution and migration of immature sockeye salmon taken by U.S.
            research vessels with gillnets in offshore waters, 1956–1967. *Bull. Int. North
            Pac. Fish. Comm.* 27: 1–69.

BRETT, J. R.
    1965   The relation of size to rate of oxygen consumption and sustained swimming
            speed of sockeye salmon (*Oncorhynchus nerka*). *J. Fish. Res. Bd. Can.* 22(6):
            1491–1501.

FACULTY OF FISHERIES, HOKKAIDO UNIVERSITY
    1965   The *Oshoro Maru* cruise 9 to the northern North Pacific, Bering Sea, and
            Chukchi Sea in June–August 1964. *Data Rec. Oceanogr. Ob. Expl. Fish.* 9:
            220–315.
    1966   The *Oshoro Maru* cruise 14 to the northern North Pacific and Bering Sea in
            May–August 1965. *Data Rec. Oceanogr. Ob. Expl. Fish.* 10: 250–354.
    1967   The *Oshoro Maru* cruise 19 to the northern North Pacific and Bering Sea in
            June–August 1966. *Data Rec. Oceanogr. Ob. Expl. Fish.* 11: 166–261.

1968   The *Oshoro Maru* cruise 24 to the northern North Pacific and Bering Sea in June–August 1967. *Data Rec. Oceanogr. Ob. Expl. Fish.* 12: 292–400.

1969   The *Oshoro Maru* cruise 28 to the northern North Pacific, Bering Sea, and the Gulf of Alaska in June–August 1968. *Data Rec. Oceanogr. Ob. Expl. Fish.* 13: 2–137.

1970   The *Oshoro Maru* cruise 32 to the northern North Pacific, Bering Sea, and Bristol Bay in June–August 1969. *Data Rec. Oceanogr. Ob. Expl. Fish.* 14: 2–125.

1971   The *Oshoro Maru* cruise 37 to the northern North Pacific, Bering Sea, and the Gulf of Alaska in June–August 1970. *Data Rec. Oceanogr. Ob. Expl. Fish.* 15: 2–97.

FAVORITE, F.

1970   Fishery oceanography 6. Ocean food of sockeye salmon. *Comm. Fish. Rev. Rep.* 861: 45–50.

FRENCH, R. R., and W. B. McALISTER

1970   Winter distribution of salmon in relation to currents and water masses in the northeastern Pacific Ocean and migrations of sockeye salmon. *Trans. Amer. Fish. Soc.* 99(4): 649–663.

FUKUHARA, F. H., S. MURAI, J. J. LALANNE, and A. SRIBHIBHADH

1962   Continental origin of red salmon as determined from morphological characters. *Bull. Int. North Pac. Fish. Comm.* 8: 15–109.

HARTT, A. C.

1966   Migrations of salmon in the North Pacific Ocean and Bering Sea as determined by seining and tagging, 1959–1960. *Bull. Int. North Pac. Fish. Comm.* 19: 1–141.

LeBRASSEUR, R. J.

1966   Stomach contents of salmon and steelhead trout in the northeastern Pacific Ocean. *J. Fish. Res. Bd. Can.* 23(1): 85–100.

MARGOLIS, L., F. C. CLEAVER, Y. TUKUDA, and H. GODFREY

1966   Salmon of the North Pacific Ocean—Part VI. Sockeye salmon in offshore waters. *Bull. Int. North Pac. Fish. Comm.* 20: 1–70.

NISHIYAMA, T.

1970   Caloric values of ovaries of sockeye salmon at last stage of marine life. *Bull. Jap. Soc. Sci. Fish.* 36(11): 1095–1100.

ROYCE, W., L. S. SMITH, and A. C. HARTT

1968   Models of oceanic migrations of Pacific salmon and comments on guidance mechanisms. *Fish. Bull.* 66(3): 441–462.

WINBERG, G. G.

1956   Rate of metabolism and food requirement of fishes. Fish. Res. Bd. Can., Trans. Ser. No. 194, 251 pp.

CHAPTER 16

Movement of baleen whales in relation to hydrographic conditions in the northern part of the North Pacific Ocean and the Bering Sea

KEIJI NASU

Far Seas Fisheries Research Laboratory, Shimizu, Japan

Abstract

The movement of whales in the feeding area appears to be generally influenced by the oceanographic environment. For the purpose of analyzing the movement among the whaling grounds, examples of comparatively long distances between positions of marking and recapture were obtained. According to these charts, the fin whales which moved towards the west along the Aleutian Islands came to the northern area of Unalaska Island, and their movement may be related to the eastward current along the northern area of the Aleutian Islands. The results of marking indicate that some fin whales which came to the east area off Kamchatka Peninsula moved toward the eastern Bering Sea, and the routes are considered to be related to the eastward oceanic current.

The fin whales which came to the northern area of Unalaska Island are divided into two groups—one which closely follows the oceanic front between the water masses covering the continental shelf of Alaska and associated shelf-edge waters as far northwest as Cape Navarin, and another group which stays in the northern sea region of Unalaska Island.

The group which stayed behind was characterized by a high percentage of young, sexually immature individuals (41.5% of males and 54.6% of females) and lactating whales (12.5%), while the migrating group was comparatively high in sexually mature individuals (immaturity rates of 20.6% and 23%, male and female, respectively) with few lactating whales (3.8%).

The northbound fin whales which traveled up to the Bering Sea in summer were seen further north in the Chukchi Sea, where a few whales were observed as early as July and some remained until October. It appears that the number of fin whales in the Chukchi Sea decreases from mid-August, while the number of gray whales increases.

It is clear from the distribution of physical (temperature and salinity) and chemical (dissolved oxygen and hydrogen ion concentration) properties that a dense area in the distribution of sighted gray whales coincides with a conspicuous oceanic front characterized by high marine productivity.

INTRODUCTION

Whaling in the northern part of the North Pacific Ocean and the Bering Sea has been carried out by several nations for a relatively long period.

The catch compositions originally consisted of blue, fin, humpback, and sei whales; since 1966, however, the hunting of blue and humpback whales has been prohibited, and only fin and sei whales continue to be harvested. In this paper the relationship between hydrographic conditions and the movement of baleen whales (*fin whales*, specifically) is discussed on the basis of whale marking experiments, whale sightings, and commercial catches.

Distribution of baleen whales

In the North Pacific Ocean, blue whales occur only in the waters south of the Aleutian and Komandorskii islands. No blue whales were taken by Japanese expeditions in the Bering Sea, and sei whales have been collected only rarely. It may therefore be considered that the main herds of these two species do not penetrate the Bering Sea.

Fin whales are very widely distributed in the northern part of the North Pacific Ocean and disperse far into the Arctic Ocean through the Bering Strait; the author sighted a fin whale also in the Chukchi Sea in 1958. Although the fin whale was the most important species in years past, since 1966 the catch of sei whales has rapidly increased to become now the more prominent. From catch data prior to 1966, the distribution of humpback whales is believed to be similar to that of fin whales.

From the above, it is clear that fin whales are found in the whaling ground of other species. Baleen whaling grounds have been determined from data obtained by Japanese expeditions operating in the northern part of the North Pacific Ocean from 1952 to 1970 (Fig. 16.1).

Whale marking studies

Whale marking in the northern part of the North Pacific and the Bering Sea has been carried out since 1953, and the movement of larger whales is reported by Nemoto (1959) and Nishiwaki (1961). Examples of the comparatively long distances between the positions of marked and recaptured whales are shown in Figure 16.2. In some instance fin whales have moved from off Vancouver Island in June towards the Gulf of Alaska. One fin whale migrated from Baja California waters in January to the Queen Charlotte Islands in June. It is clear that fin whales in the North Pacific Ocean migrate between low and high

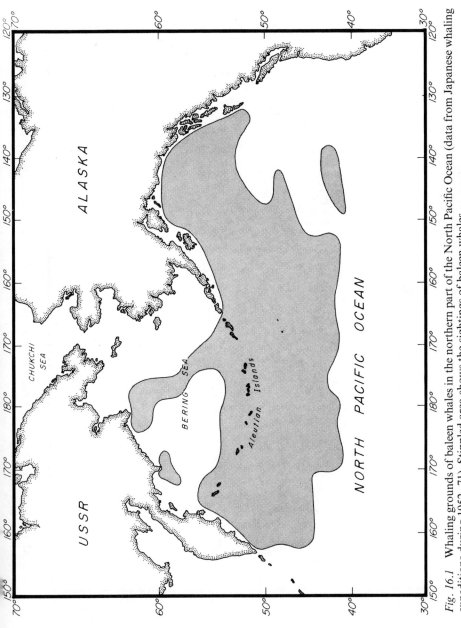

Fig. 16.1 Whaling grounds of baleen whales in the northern part of the North Pacific Ocean (data from Japanese whaling expeditions during 1952–71). Stippled area shows the sightings of baleen whales.

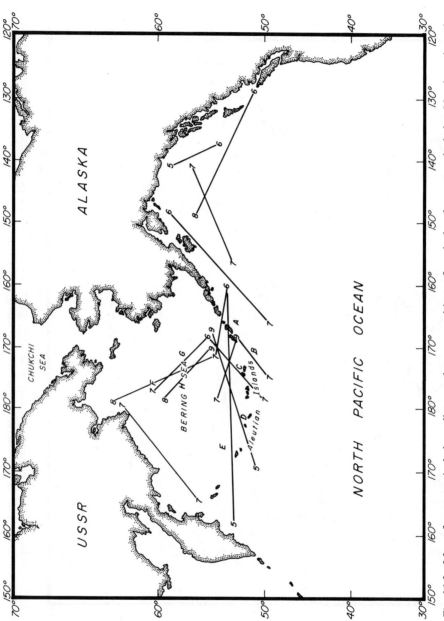

Fig. 16.2 Map of comparatively long distances between positions of marked and recaptured whales. Numbers show the month of marking and recapture.

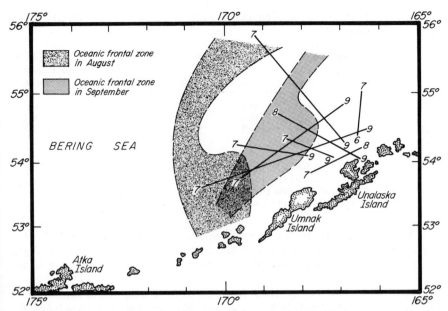

Fig. 16.3 Results of marking survey of fin whales north of Unalaska Island. Numbers show the month of marking and recapture.

latitudes. Also, Kellogg (1929) stated that fin whales may move from Baja California to the Bering Sea.

The fin whales which move westward along the Aleutian Islands probably go north or to Unalaska Island (A, B, C, D, and E in Fig. 16.2).

The marking results indicate that some fin whales which came to the east area off Kamchatka Peninsula moved towards the eastern Bering Sea (E, Fig. 16.2). The fin whales which came north of Unalaska Island earlier in the season were differentiated into two groups. One group (area II) moved toward the waters adjacent to Cape Navarin; the second group (area III, Fig. 16.3) approached Unalaska Island later in the season. Biological characteristics of fin whales in the subarctic Pacific Ocean are given in Table 16.1 by area in 1958. Immature male specimens were defined on the basis of testes weight, and the age of females was estimated from the number of ovulation scars. The group which remained north of Unalaska Island (III) was characterized by a high percentage of young, lactating whales, whereas the area II (Cape Navarin) group which moved further north consisted of comparatively older whales. The reasons for the distribution of biological features characterizing the whales north of Unalaska Island are obscure.

The distribution of fin whaling grounds by months (Fig. 16.4) has been described for the Chukchi Sea, based on observations during 1940 and 1941 by Japanese expeditions, and data for the Vancouver Island area were obtained from Canadian whaling reports.

TABLE 16.1   Physical analysis of fin whales caught in four areas of the subarctic Pacific during 1958.

Areas of catches*	Average length (meters)		Specimens immature (%)		Average number of ovulations	Percentage of mature females lactating
	male	female	male	female		
I	18.39	19.18	18.8	27.6	8.1	5.6
II	18.42	19.21	20.6	23.0	8.0	3.8
III	18.06	18.48	41.5	54.6	5.6	12.5
IV	18.48	19.09	25.0	32.4	6.3	3.2

   * I Area east of Kamchatka Peninsula
     II Off Capes Navarin and Olyutorskii
   III Northern side of Unalaska Island
   IV Southern side of Aleutian Islands

The distribution of fin and sei whale sightings is shown in Figures 16.5–16.7. From maps of the whaling grounds and of the distribution of sighted whales (Figs. 16.5 and 16.6) it is clear that fin whales move northward as the season progresses.

Northbound fin whales which travel up to the Bering Sea in summer have been observed in the Chukchi Sea. Nikulin (1964) reported that a few fin whales were observed in the Chukchi Sea in July and that some remained until October. According to the results of an oceanographic survey made by a Japanese vessel in 1937 and from whaling operations conducted by Japanese expeditions in 1950, the number of fin whales in the Chukchi Sea decreased after mid-August, although gray whales increased in number. Such a tendency was indicated also by the 1958 survey carried out from 16–20 August. Also, most of the gray whales seen during this survey were sighted in the whaling grounds operated by the 1940 Japanese expeditions.

The distribution of fin whales in relation to hydrographic conditions
The water mass in the vicinity of Cape Navarin can be divided into zones of melted-ice water, oceanic water, and mixed water (Fig. 16.7), typified as follows:
  (1) Melted-ice water is relatively cold (less than 2.2 C) and of low salinity (less than 26.0‰).
  (2) Oceanic water is relatively warm (over 6.0 C) and of high salinity (more than 30.0‰).
  (3) Mixed water is located between the melted-ice water and oceanic water.
The main whaling grounds near Cape Navarin were located near the intermediate zone of mixed water.

From surface data obtained during the whale-marking survey of August 1958 (Fig. 16.8), isotherms are located parallel to Cape Navarin, and the cold water mass extends to the south. The distribution of salinity indicates that the cold water mass which extended to the south was caused by melted-ice water in the vicinity of the Siberian continent; that is, the distribution of cold water

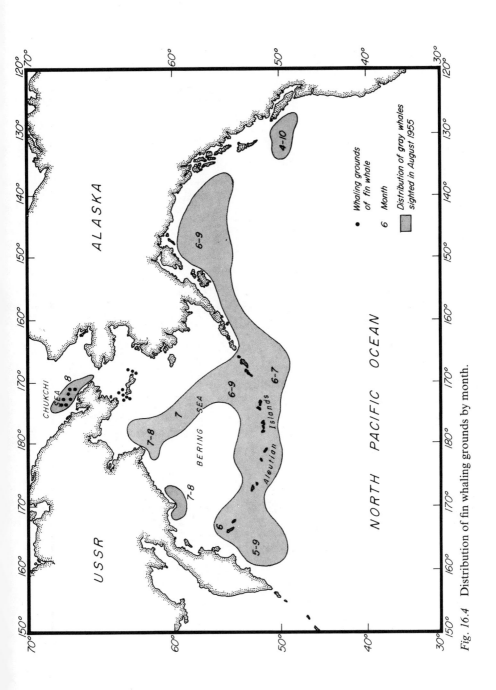

Fig. 16.4 Distribution of fin whaling grounds by month.

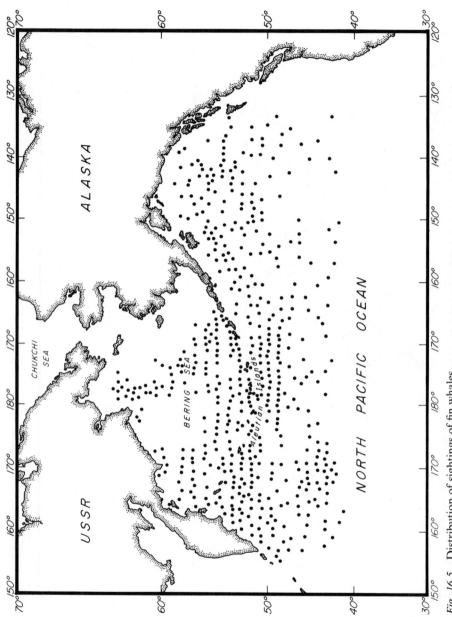

Fig. 16.5   Distribution of sightings of fin whales.

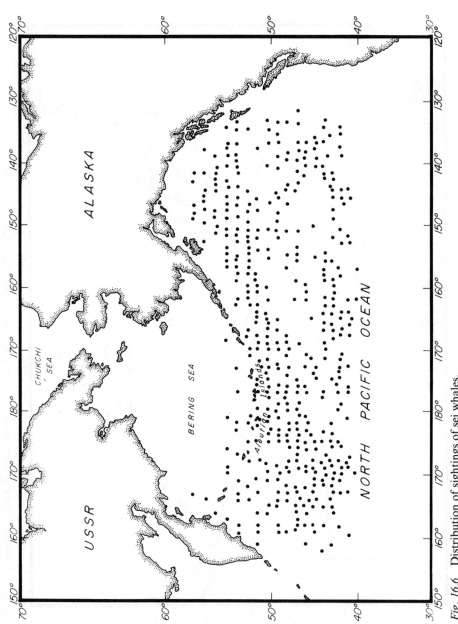

Fig. 16.6 Distribution of sightings of sei whales.

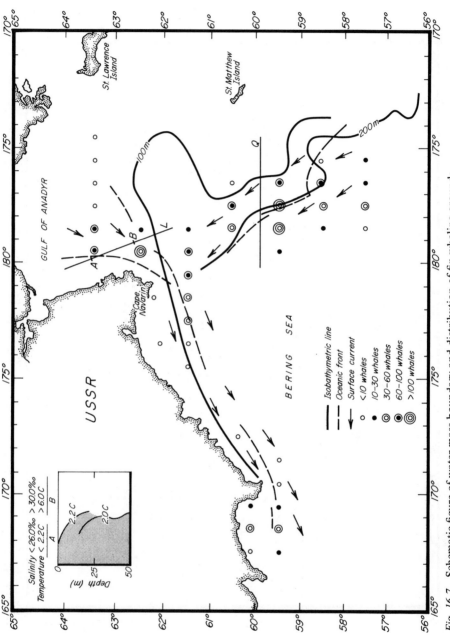

Fig. 16.7 Schematic figure of water mass boundary and distribution of fin whaling grounds.

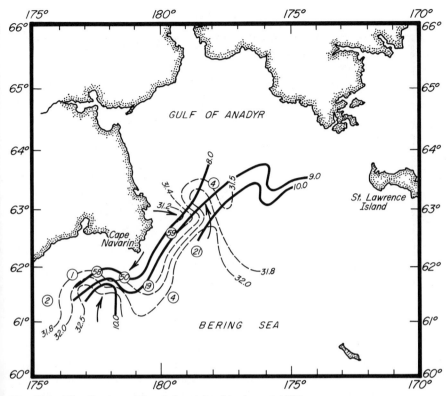

Fig. 16.8   Distribution of fin whales sighted in August 1958.

mass of temperature less than 8.0 C at the surface corresponds to water with
salinity less than 31.40‰.

Many fin whales were sighted near the mixing zone between the water
along the Siberian continent and the oceanic water mass, a phenomenon similar
to conditions observed near the whaling grounds (Fig. 16.7). A vertical profile
of sigma-t along section Q (Fig. 16.7) is shown in Figure 16.9. This data was
obtained by the *Oshoro Maru* in the summer of 1969. The sea region of salinity
less than 32.4‰ in the surface layer was designated the *coastal domain*
(Dodimead et al. 1963) and corresponds to the area east of the boundary of
the water mass shown in Figure 16.9.

In the better whaling years, 163 fin whales were caught in 1957 and 399
in 1959 (Fig. 16.10). The fin whaling grounds in both peak years were located
roughly in the same region, and the area of heavy catch follows the edge of
the continental shelf. The whaling grounds were located at the water mass
boundary between the coastal domain, covering the continental and the oceanic
shelf of Alaska, and the oceanic water mass or western subarctic domain
(Dodimead et al. 1963). The water mass boundary ran mostly along the 200
isobathymetric line. Since the whaling grounds in this region were located

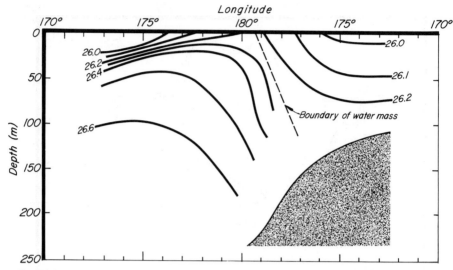

Fig. 16.9   Vertical profile of sigma-*t* along section Q.

close to the same area every year, it is possible that the whaling grounds at the oceanic front were influenced by topographic conditions.

In a gray whale observation area in the Chukchi Sea, an August 1958 Japanese whaling survey obtained surface temperature values ranging from 3.7 to 11.2 C (Fig. 16.11). In general, surface water temperature in the Chukchi Sea decreased and salinity increased from east to west. Salinity decreased toward the Siberian coastal region, where the water mass was less than 5.0 C with 30.0‰ salinity, due perhaps to dilution by ice thaw and inflow of coastal water. A maximum salinity region found in the central Chukchi Sea was attributed to the possible influence of a northward flowing current from the Bering Sea. An area of high dissolved oxygen concentration was noted along the observed current rips (Fig. 16.11) and is thought to represent the convergence between Siberian coastal waters and the central Chukchi Sea. The pH concentration corresponded to those of high dissolved oxygen concentration in the sea area along the observed current rips.

Sightings in the Chukchi Sea of fin, humpback, bowhead, right, and gray whales were reported by the Japanese survey vessel *Yuki Maru* in 1937 and by Nikulin in 1947. During the 1958 survey, fin, right, and gray whales were sighted, but many of the latter were seen along the water mass boundary and appeared to be feeding along the current rips.

DISCUSSION

In Figure 16.12 the above features are summarized (Nemoto 1959), showing the presumed migration route of baleen whales and prevailing hydrographic conditions. The migration route of gray whales is taken from Gilmore (1955).

From the distribution of main whaling grounds and from whale sightings in the region between Kodiak Island and the south Aleutian Islands, it appears that baleen whales move along the marginal zone of the Alaskan Stream. Ohtani (1966) noted an abundant fishing ground of red salmon along the marginal zone of the inner Alaskan Stream.

Whale marking experiments showed the movement of fin whales from the west to the east in the vicinity of the Aleutian Islands, and the movement of these whales may be related in general to the east current along the north side of the Aleutian Islands.

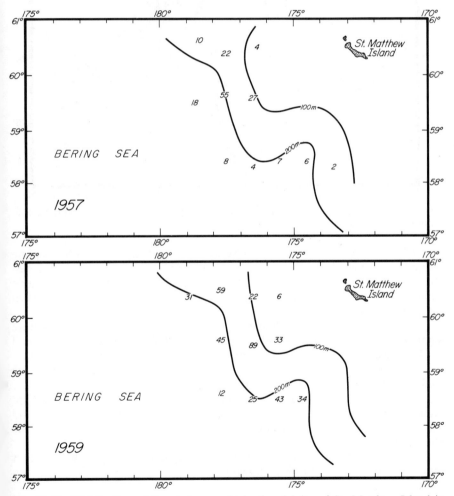

Fig. 16.10 Distribution of fin whales caught in the vicinity of St. Matthew Island in favorable whaling years of 1957 (top) and 1959 (bottom).

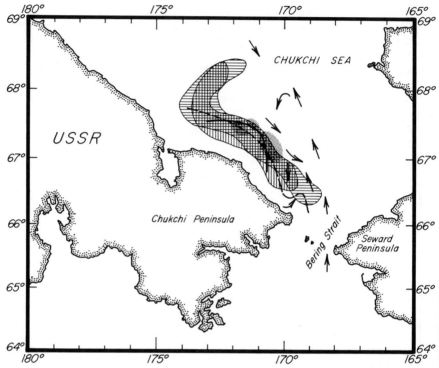

Fig. 16.11 Distribution of gray whale sightings and a summary of hydrographic conditions in August 1958.

There is evidence from whale markings and the distribution of fin whaling grounds that fin whales move along the boundary between the Bering Sea coastal water and the oceanic water. Moreover, it is obvious that the favorable whaling grounds off Cape Navarin were located in the mixing zone between the coastal water and the oceanic water.

In the Chukchi Sea both the region in which many gray whales were sighted and the area where good catches of fin whales were made in 1940 and 1941 coincided with the oceanic front. It is apparent also (Gilmore 1955) that the California gray whales migrate from the breeding area in Baja California to the feeding area in the Bering and Chukchi Seas.

In analyzing data obtained by Japanese whale expeditions in the antarctic Atlantic sector, the author found that the regions of heavy catches of fin and sei whales were located along the southerly tongue-shaped projection of the Brazil current.

To the south of the Antarctic Convergence, many fin whales were caught along the tongue of southward-flowing water characterized by 2 to 3 C isotherms at the surface.

Uda (1954) also noted relationships between the whaling grounds and the water mass boundary; namely, the whaling grounds off northeast Japan

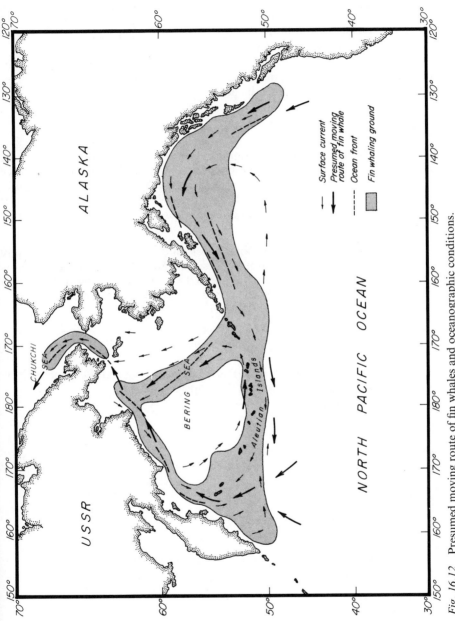

Fig. 16.12 Presumed moving route of fin whales and oceanographic conditions.

(Sanriku Sea region) corresponded to the water mass boundary between the Oyashio cold current and the northern branch of the Kuroshio warm current; and the whaling grounds off Hokkaido corresponded to the polar front between the Oyashio cold current flowing to the southwest and the northward warm branch of the Kuroshio current. Uda (1954) explained that the water mass boundary corresponded to the most abundant area of whale food.

The author has no available data on the relation between the abundance of whale food and the oceanic front; however, it appears that baleen whales have a tendency to move along the oceanic front or current.

Discussion

HUGHES: What are your hydrographic data and other evidence for the circulation you describe in the vicinity of Cape Navarin? The circulation shown by you differs from that reported by Natarov (1963), who observed no southward flow past Cape Navarin in the summer over a two-year period, and it does not coincide with the results of direct measurements in this area as reported in my Symposium paper here.

NASU: I have no hydrographic observation data for the current in the vicinity of Cape Navarin. My speculation on a southward current is based on information received from a whaler who reported that a whale-chaser drifted south in the sea. The information for the current specifically in the vicinity of Cape Navarin is based on your data, however.

Acknowledgment

Thanks are expressed to Mr. P. Robert, Fisheries Research Division in New Zealand, who assisted in reviewing this paper.

REFERENCES

DODIMEAD, A. J., F. FAVORITE, and T. HIRANO
  1963   Salmon of the North Pacific Ocean. Part 2. Review of oceanography of the subarctic Pacific region. *Bull. Int. North Pac. Fish. Comm.* 13, 195 pp.

GILMORE, R. M.
  1955   The return of the gray whale. *Sci. Amer.*, p. 192.

KELLOGG, R.
  1929   What is known of the migration of the whalebone whales. Annual Rep. Smithsonian Inst., 1928.

NATAROV, V. N.
  1963   Water masses and currents of the Bering Sea [in Russian]. *Trudy VNIRO* 48: 111–133. (Transl., 1968, in *Soviet fisheries investigations in the northeastern Pacific*, Part 1, pp. 110–130, avail. Nat. Tech. Inf. Serv., Springfield, Va., TT 67-51203).

NEMOTO, T.
 1959 Foods of baleen whales with reference to whale movements. *Sci. Rep. Whales Res. Inst.*, No. 14.

NIKULIN, V. G.
 1946 On the distribution of whales in the adjacent seas to Chukchi Peninsula. Cited from translation by H. Sakiura in *Geiken Tsushin* No. 77.

NISHIWAKI, M.
 1967 Distribution and migration of marine mammals in the North Pacific Area. *Bull. Ocean. Res. Inst., Univ. Tokyo*, No. 1.

OHTANI, K.
 1966 The Alaskan Stream and the sockeye salmon fishing grounds. *Bull. Fac. Fish., Hokkaido Univ.* 16(4).

UDA, M.
 1954 Studies of the relation between the whaling grounds and the hydrological conditions. 1. *Sci. Rep. Whales Res. Inst.*, No. 9.

CHAPTER 17

Principal results of Soviet oceanographic investigations in the Bering Sea

DAVID E. GERSHANOVICH, NIKOLAI C. FADEEV, TATIYANA G. LIUBIMOVA, PETER A. MOISEEV, and VALERY V. NATAROV

All-Union Research Institute of Marine Fisheries and Oceanography (VNIRO), Moscow, USSR

Abstract

The main characteristics of marine fauna in relation to environmental conditions in the Bering Sea show a response primarily to the effect of warm Pacific water rich in vital nutrients. An important role in the development of biological processes is played by the winter vertical circulation observed in the continental slope area.

The geological history of the Bering Sea, the bottom relief, the pattern of water dynamics, the temperature regime, and the distribution of nutrients are all factors which determine the composition and quantitative distribution of marine organisms as well as the specific characteristics which promote concentrations of bottom, bathypelagic and pelagic species.

Since the biological resources of the Bering Sea have been exploited by fishermen of several countries for a number of years, it is essential that principles of rational exploitation be formulated and that research and commercial efforts of all countries concerned be closely coordinated.

INTRODUCTION

History of Soviet research

Soviet oceanographic investigations in the Bering Sea have been conducted since 1931, focusing until 1955 mainly on the living resources and ecology of the western region. From the earlier studies were determined the characteristic biology, distribution, and migration habits of flounder, cod, and certain other species of commercially important fish in relation to their environmental conditions. This information and hydrographic knowledge of the area indicated that dense populations of fish might be found in the eastern Bering Sea, and a recommendation was made for further investigations to assess the commercial importance of this body of water and the possibilities of its rational exploitation. On the basis a Bering Sea Research Expedition was established in 1958 by the All-Union and Pacific Research Institutes of Marine Fisheries and Oceanography (VNIRO and TINRO). During the period of its activities, the Expedition employed many research and exploratory vessels and a team of more than 60 scientists to undertake extensive fisheries investigations in the Bering Sea and adjacent waters. Significant theoretical and practical conclusions resulted from these studies.

The scope of investigations conducted by the Expedition included study of the sea bottom topography and geologic character; hydrographic regime; composition and productivity of benthos and plankton; biology, distribution, and abundance of ichthyofauna; biochemical composition and processing of the species collected; and the technique and opganization of fisheries management. A uniquely important feature of the Expedition was the complexity of research conducted on a year-round basis. The geomorphological hydrographic, hydrochemical, and biological studies resulted not only in the refinement of existing data but also in the acquisition of new information applicable to several important problems. The Bering Sea Research Expedition worked the waters of the northern and northeastern Pacific Ocean, including the Bering Sea proper, the Gulf of Alaska, and the waters off the Aleutian Islands.

BERING SEA HYDROGEOMORPHOLOGY

The largest part of the area studied is under permanent and intensive influence of the Pacific Ocean water masses, to which it is connected by deep and wide passes between the Aleutian Islands, and of the Arctic Ocean waters that enter only by the narrow and shallow Bering Strait. The Bering Sea's geologic history of physical connection with the Pacific Ocean since the end of Tertiary time explains its predominant inhabitation by boreal Pacific fauna, particularly benthic and near-bottom organisms, which are characterized by higher abundance than those species originating from Arctic Ocean waters.

Geologic character

Of decided ecological significance to especially those marine organisms living in the lower water layers are the bottom relief and nature of the seabed in combination with the hydrographic regime and feeding conditions. The Bering Sea is characterized particularly in its eastern part by an extensive expanse of continental shelf. This, along with other factors, sustains conditions favorable to concentrations of bottom fish.

A main factor in the higher bioproductivity of shelf waters is the high gradient of the continental slope, which in some outer parts is steeper than 25 degrees. In these areas the vital source of nutrient phosphorus is the Pacific water that enters through the Aleutian passes and moves up the steep continental slope, enriching both the subsurface and intermediate water layers. This process creates condition contributing to a high bioproductivity manifested primarily in the great abundance of plankton and forage benthos. Unlike the case more typical in other seas, the supply of nutrients from the land adjacent to the Bering Sea is limited and only of minor importance. The continental slope of the Bering Sea is indented by well-defined and complex-shaped submarine canyons, in which the upwelling of deeper waters establishes intensive vertical water exchange and favors the concentration of many fish species. The sediment over most of the shelf and continental slope area consists mainly of fine fractions ranging from sand to clayed mud, with parent material outcropping only in the canyon areas.

Hydrologic structure

To understand the conditions supporting concentrated marine organism populations, it is essential to consider the hydrological properties of the Bering Sea:

The Bering Sea is characterized by the classic subarctic water structure. Vertically, the water masses can be divided into four types in summer and three types in winter. In summer the observations show a surface layer (6 to 12 C at the surface and 4 to 6 C at a depth of 50 m); a cold intermediate layer (minimum temperature in the mid-stream 0.1 to 1 C in the west and 2.5 to 3 C in the east); a warm intermediate layer (maximum 3.5 to 4 C); and a deep layer (1.8 to 2.2 C). In winter the water of the surface layer is considerably colder, and about half of the sea surface is covered with ice. Under the effect of cooling and related winter vertical circulation, the characteristics of the surface waters and the cold intermediate layer tend to merge in some regions, and the four-layer water structure now gives way to a three-layer pattern. The subsequent seasonal warming in summer and effect of the warm Pacific water are inadequate to completely destroy the cold intermediate layer in most areas of the sea; the subarctic water structure is therefore most pronounced in the western Bering Sea, where the effect of warm Pacific waters is much lower.

Hydrological properties in the continental slope zone are characterized by an almost entirely homogeneous distribution in the near-bottom layer. This important feature appears to be associated with strong turbulent mixing processes.

Circulation dynamics

The current system in the Bering Sea is complicated: permanent extensive circulation patterns are observed in the Olyutorskii and Anadyr gulfs as well as in a number of shelf areas. Specifically, there are extensive circulations southwest of St. Lawrence Island and east of the Pribilof Islands.

Studies of the winter regime showed persistence of the general cyclonic circulation. In winter most of the warm Pacific water enters through the eastern passes in the Aleutian Islands; some of this water flows onto the eastern Bering

Sea shelf, where it causes melting of the edge ice. As in the summer time, local circulations were observed near the Rat Islands, Andreanof Islands, and certain other islands.

Summer studies of the deep Bering Sea confirmed the existence of cyclonic water motion. Pacific waters enter the Bering Sea through the eastern channel between Medny and Attu islands and certain other major Aleutian passes. Thence the water mass moves across the central deep area to either flow in part onto the Bering shelf or proceed on to the Asian coast in the northwestern Bering Sea, from where some of these waters turn southward. A branch of these waters turns to the west in the northern part of the sea, thereby strengthening the main westward circulation. It has been concluded that in the summer there is no strong stream of cold water arising from the Anadyr Gulf such as is observed in the winter along the Asiatic coast.

Tidal currents play an important role in some areas of the Bering Sea. Current velocities around the Aleutian passes may be as high as 8–10 knots. Appreciable tidal currents are observed also on the shelf near the Pribilof Islands and on the continental slope. Related to tidal currents is the phenomenon of internal waves, which account for vertical fluctuations in certain hydrological characteristics most pronounced in the western and central Bering Sea.

The circulation of surface waters in the Bering Sea is subject to significant variation in time caused by fluctuations in the atmospheric circulation and inflow of Pacific water. Circulations in the deep part of the sea are particularly variable.

In the shallow areas of the Bering Sea, local distribution of water masses occurs in summer. So-called cold patches appear as a result of vortex water motion and are characterized by persistent cold water in the near-bottom layer throughout the year. In winter the waters of the shallow areas are isothermal. As the ice edge is approached, the temperature drops to below zero.

BIOMASS DISTRIBUTION

Phytoplankton

The high abundance of nutrients, the thawing of the ice cover, and certain other factors contribute to intensive development of phytoplankton, represented mainly by species of *Diatomea* and *Peridinea*. In most cases the distribution of the zones of phytoplankton bloom is patchy and associated with the distribution pattern of the most highly concentrated nutrients. The densest phytoplankton biomass is found in the coastal waters of the southeastern Bering Sea.

Zooplankton

The distribution of water masses of varying origin is responsible for the fact that the zooplankton complexes are not uniformly distributed throughout the Bering Sea. In summer, the Nunivak Island-Cape Navarin line forms the geographic boundary between northern and southern distribution groups of oceanic zooplankton. The waters of the shallow eastern sea are inhabited by the neritic group, although certain typically oceanic species such as *Pseudocalanus elongatus* and *Calanus finmarchicus* often occur and in some cases even predominate in the eastern Bering Sea, attributed to the intensive effect of Pacific water on this area.

In the continental slope area the major concentrations consist of bathypelagic *Eucalanus bungii*, while neritic species of *Acartia* and *Centropages* predominate in the freshened eastern part of the shallow area. Appreciable seasonal and year-to-year fluctuations of zooplankton biomass corresponding to variations in the hydrological regime are observed within the shallow waters.

The pattern of the horizontal and vertical distribution of high zooplankton biomass is most stable over the continental slope, which ensures an abundant and stable food supply for plankton feeders such as cephalopods, ctenophores, lantern anchovy, and certain large near-bottom and bathypelagic fish that feed on plankton and small fish. Pelagic plankton feeders are greatly dependent on zooplankton concentrations in the central and the northwestern Bering Sea, where the distribution of zooplankton is restricted to surface layers due to the well developed cold intermediate layer.

Benthos

The distribution of benthos provides an even more striking confirmation of the zoogeographical division of the Bering Sea. The biomass of fodder benthos is relatively high over practically the entire Bering Sea shelf but often not accessible to the bottom fish. Although many shelf areas are characterized by low (down to sub-zero) temperatures throughout the year, they are far removed from the limited wintering areas of sole and flounder.

The southeastern Bering Sea is the area most nutritionally favorable for bottom marine organisms such as sole, flounder, crab, and shrimp. In spite of higher food availability, most other areas cannot be fully utilized by bottom fish in the years characterized by higher temperatures.

Ichthyofauna

Bering Sea ichthyofauna can be classified by environmental conditions into several ecological groups, each group occupying that niche which most optimally sustains its life functions—particularly those of feeding and perpetuation of species.

Typical bottom fishes inhabit the shallow areas of the Bering Sea. The continental slope is inhabited by bottom, near-bottom and bathypelagic fishes; typical pelagic species are found in mid-water layers. Flounder and halibut are typical of bottom species found on the shelf. Within the vast shallow eastern Bering Sea are yellowfin sole (*Limanda aspera*), rock sole (*Lepidopsetta bilineata*), Alaska plaice (*Platessa quadriturberculata*), snout sole (*Limanda punctatissima*), starry flounder (*Pleuronectes stellatus*), and flathead sole (*Hippoglossoides elassodon*).

In line with their ecological characteristics and zoogeographical distribution, the two most abundant species of flounder in the Bering Sea are the Pacific-boreal yellowfin sole and the subarctic-boreal rock sole.

In winter the flounder concentrate near the outer edge of the shelf and the upper part of the continental slope (90–200 m), influenced by the warm Pacific waters. In spring they start their northward, northwestward, and northeastward migration to the shallow area. As the near-bottom water temperature elevates, the flounder then gradually fill the feeding grounds in the shallow area, avoiding the cold patches of negative temperatures. Peak spawning occurs in summer at depths of 15–75 m and is associated with the area of circular current south of Nunivak Island. The stable circulation in the eastern Bering Sea, in combination with high biomasses of plankton on which the larvae feed, insures a high reproduction rate among flounder inhabiting this area. In autumn the flounder perform their return (wintering) migration to the regions of greater depths, concentrating along the outer edge of the shelf. For many years the eastern Bering Sea flounder stocks yielded high catches but at present are depressed as a result of many years of fishing.

Among the bottom fishes inhabiting the shelf are species of the family *Gadidae*, which are subarctic-boreal according to their ecological/zoogeographical characteristics and are most prominently represented by Alaska pollock (*Theragra chalcogramma*) and cold (*Gadus morhua macrocephalus*). Although these species are of arctic origin, particularly the pollock are as numerous as the Pacific ichthyofauna and even more widely distributed. The northern limit of the distribution range of Alaska pollock runs from Cape Chukotski to St. Lawrence Island. In the eastern part of the Sea they are distributed over the area from Cape Navarin to Unimak Island at depths from 50–500 m.

Concentrations of Alaska pollock are found year-round throughout their range of distribution in the Bering Sea, concentrating most densely at depths of 120–300 m. Depending on the season and biological condition, the concentrations move actively within the distribution range. During the warm season Alaska pollock concentrate in the shelf areas at 80–150 m depths. During the spring spawn (April to May), concentrations are distributed mainly within the eastern Bering Sea shallow area. As the water gets warmer, the area over which the concentrations are distributed extends to the productive shelf regions near the Pribilof and St. Matthew islands. Until late autumn or early winter, Alaska pollock are widely distributed over the shelf, where they feed heavily and from large concentrations at depths down to 150 m.

The winter cooling of shelf waters and formation of an ice cover over most of the shallow area drive the Alaska pollock to the continental slope. The wintering concentrations in December to March are found mainly in the southeastern sea at 200–300 m depths.

Being a cold-oriented species, the Alaska pollock live and form concentrations under conditions of low (5 to 0°C) and even sub-zero temperatures; thus severe conditions are not likely to restrict the distribution range and abundance of Alaska pollock in the Bering Sea. Biological and ecological data indicate, however, that the availability of food to adult fish may be the decisive factor controlling the dynamics of the Alaska pollock population in the Bering Sea. For example, a reduction in Bering Sea herring stock, which is the principal food competitor of Alaska pollock, resulted in an appreciable extension of the concentrating area of Alaska pollock and an increase in its numbers in the Bering Sea.

Pacific cod are found mostly in the western and central Bering Sea, but they do not form concentrations as dense and as stable throughout the year as those of Alaska pollock. In some years near-bottom cod concentrations are found in summer at depths down to 300 m in the Olyutorskii and Anadyr gulfs. These are feeding concentrations on which the fisheries industry is greatly dependent. Since the formation of dense near-bottom concentrations is governed by year-to-year variations in oceanographic conditions and food supply, however, the commercial potential of cod in the Bering Sea is low.

The waters of the continental slope of the Bering Sea within the 250–1000 m depth range are inhabited by about 150 species of fish, only 10 to 15 of which occur in higher numbers. Families represented are *Scorpaenidae*, *Anaplopomatidae*, *Pleuronectidae*, and *Macrouridae*. Among the species characterized by fairly high abundance are Pacific-boreal species of ocean perch and rockfish (*Sebastodes alutus*, *S. introniger*, *Sebastolobus alascanus*, *Sebastolobus macrochiz*), black cod (*Anaplopoma fimbria*), arrow-toothed halibut (*Atherestes evermanni*, *A. stomias*), true halibut (*Hyppoglossus hyppoglossus stenolepis*), and rattails (*Nematoronurus pectoralis*, *Coryphaenoides cinereus*). Most of these species are bathypelagic, and only the halibut live on the bottom. The bathypelagic rattails are typical deepwater fish occupying depths of 1000 m and lower. A characteristic feature of all these species is their wide geographic and vertical distribution in the Bering Sea. The distribution range of rockfish, black cod, halibut, and rattails extends over the entire Bering Sea slope. Distribution areas are particularly extensive for Pacific and black halibut, black cod, *Sebastodes alutus*, *S. introniger*, and *Nematonurus pectoralis*.

All of the slope fishes tend to favor the warm intermediate water, where the physico-chemical characteristics are relatively stable on both a seasonal and year-to-year basis. Also, nutrient-rich waters upwell in the continental slope area; and in winter the boundary between water masses of different origin runs along the slope, and a peculiar polar front sets in. As a result, the continental slope waters are characterized throughout the year by a high abundance of fodder zooplankton such as *Euphausiidae* and *Calanidae*, large pelagic molluscs (squid), and small plankton-feeding fishes such as capelin, herring, smelt, and lantern anchovy. All these are organisms utilized by bathypelagic fishes living in the slope waters. In some areas eddies develop, protectively restricting the drift of eggs and larvae.

Many fish species of the Bering Sea are eurybathic. Particularly wide depth ranges are frequented by the true halibut (20–700 m), arrow-toothed and black halibut (40–100 m), black cod and Alaska rockfish (100–1000 m). Although these fishes are permanent inhabitants of the continental slope waters, they are also successful transients in the shallower shelf area. Their fish-eating habits characterize them as predators, but they feed on pelagic and bottom invertebrates as well. Euryphagous halibut, black cod, and some rockfish are able to effectively utilize the food resources of both the continental slope and shelf areas. During the warm season their feeding areas are extended to include the highly productive shallow shelf waters, where they feed mainly on benthos and nectobenthos.

Despite their wide vertical range of distribution, this group of fish concentrates in higher numbers only over the limits of the continental slope and within a comparatively narrow depth range characteristic of each species. In winter the concentrations are found at a lower depth, and the vertical distribution is restricted to a narrower layer; summer populations occur at a higher level and are distributed over a wider depth range. Dense concentrations of bottom and bathypelagic slope dwellers are found at 250–750 m depths during the winter-spring period.

Although the biological and ecological features of slope fish determine a typically high level of abundance in the Bering Sea, certain species such as halibut, rockfish, and rattails are characterized by a comparatively low reproduction rate, not reaching sexual maturity until the age of 6 to 14 years.

The commercially most important fish of the Bering Sea pelagic species is the herring (*Clupea harengus pallasii*), which is widely distributed over the entire sea but occurs in higher numbers only in the central and western parts. The formation of dense herring concentrations in the Bering Sea is distinctly seasonal. During the warm period, in connection with spawning and feeding, the herring are widely distributed in mid-water throughout the Bering Sea and do not form large or dense concentrations. Winter cooling and the decreasing temperature of the mid-water layers cause them to concentrate in the near-bottom layers (1.5 to 3.5 C). The major wintering concentrations occur in the 110–140 m depth range northwest of the Pribilof Islands. As a result of the present sharp decrease in herring stock, the Alaska pollock continues its displacement of the herring in its ecological niche as a plankophagous species.

Discussion

COACHMAN: Have there been any direct current measurements made by the Bering Sea Expeditions?

LIUBIMOVA: Only dynamic measurements were made during these investigations.

REFERENCES

Soviet fisheries investigations in the northeastern Pacific Ocean [in Russian], 5 parts:

Part 1. 1963 | *Trudy VNIRO* Vol. 48; *Izvestiya TINRO* Vol. 50
Part 2. 1964 | *Trudy VNIRO* Vol. 49; *Izvestiya TINRO* Vol. 51
Part 3. 1964 | *Trudy VNIRO* Vol. 53; *Izvestiya TINRO* Vol. 52
Part 4. 1965 | *Trudy VNIRO* Vol. 58; *Izvestiya TINRO* Vol. 53
Part 5. 1970 | *Trudy VNIRO* Vol. 70; *Izvestiya TINRO* Vol. 72

(Transl. avail. Nat. Tech. Inform. Serv., Springfield, Va. 22151, TT-67-51204, U. S. Dep. Commerce.

Part 5

ICE AND ITS EFFECT

CHAPTER 18

Movement and deformation of drift ice as observed with sea ice radar

TADASHI TABATA

Institute of Low Temperature Science, Hokkaido University, Sapporo, Japan

Abstract

About half of the Bering Sea is covered with ice, the extreme limit of which closely coincides with the edge of the continental shelf. Only a small amount of polar ice enters the Bering Sea in the fall through northwestern Bering Strait, and even this slight contribution does not occur every year. Virtually all of the ice in the Bering Sea is of local formation, one-year ice which melts mainly within the Bering Sea. Although the average pattern of ice limits for each month in the Bering Sea has been established, little is known of the movement and deformation of the ice field itself.

A radar network was constructed during 1967–1969 to observe in detail the distribution of ice off the Okhotsk Sea coast of Hokkaido and to carry out basic research on its movements. Consisting of three radar installations constructed on mountaintops along the coast, the total radar network covers an area 70 km wide by 250 km long.

By taking photographs or sketching the ice-water boundary of the radar screen, one can easily and clearly discern the pattern of ice distribution at any given moment, even under snowstorm conditions or at night.

Several distinctive features of the ice edge and certain polynyi that are clearly identifiable in successive photographs can be used as landmarks or target points for the study of ice field drift and deformation.

It was found that the change of the drift velocity of the ice field was sensitive to that of the wind velocity. When the drift speed was 0.1–1.4 knots, the wind factor of the ice drift was 0.01–0.08. The target points did not move in a parallel manner, however; the shapes and areas of regions surrounding specified reference points within the ice field may change during the course of drift.

By tracing the deformation of imaginary tetragon and triangle lines connecting several target points, corresponding strain ellipses of the ice field deformation can be obtained. It is shown that the ice field undergoes internal deformation during its drift and that the characteristics of deformation change with time and space.

Radar network for sea ice observation

In order to observe in detail the distribution of drift ice off the coast of the Okhotsk Sea and to carry out basic research on its movement, the author constructed a radar network during 1967–1969 (Tabata et al. 1969; Tabata 1971, 1972) (Fig. 18.1). Consisting of three radar units installed along coastal mountaintops, the total system covers an area about 70 km wide and 250 km long along the northern coast of Hokkaido, the northernmost island of Japan. The radar are remote-controlled from Mombetsu by radio in the Sea Ice Research Laboratory (SIRL) of the Institute of Low Temperature Science, to which information is transmitted back.

Observation is usually made through a radar screen at SIRL, where a 35-mm still camera or 16-mm motion picture camera is used to photograph the screen imagery. The cameras are operated automatically to take successive photos of the screen at specified time intervals. Since 1972 a small electronic computer has been used to analyze the radar information directly, so that one can now obtain certain data about the surface topography of the ice as well as its distribution.

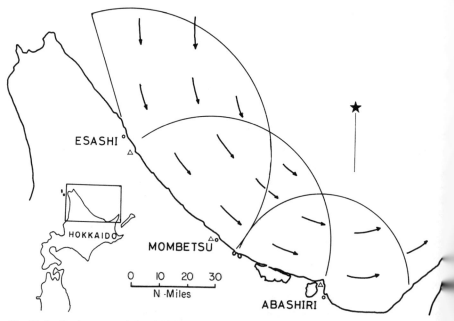

Fig. 18.1 Radar network for drift ice observation in Hokkaido, Japan. Solid arrow shows the average drift direction of sea ice.

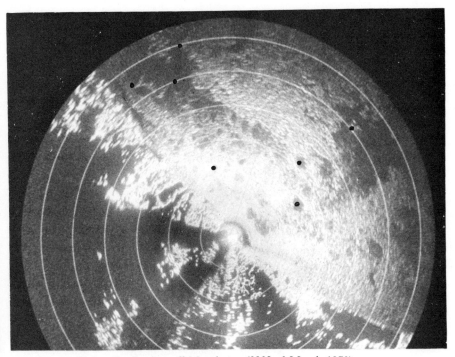

Fig. 18.2   Drift ice distribution off Mombetsu (0902, 6 March 1970).

The radar has the following specifications:

Antenna (cage type)	3.6 m
Beam width—horizontal	1°
—vertical	3°
Frequency	5540 MHz
Rated output	40 kw
Pulse width	1.0 μs
Repeating frequency	1 KHz

Although this particular radar network has the shortcoming of affording only limited coverage, it has the compensating advantage of being able to continuously monitor ice distribution. Such a system is a dynamic tool for sea ice study in the Bering Sea, especially in the Bering Strait area.

Examples of ice field radar scope

Figure 18.2 is a photograph of the Mombetsu radar screen at 0902 on 6 March 1970. The coastline runs from the upper left corner to the lower right, and the upper right portion is the sea. A white pattern in the sea is a drift ice field. Large black areas in the sea represent open water, and black spots seen within the ice field are polynyi. The radius of the outermost circle is 30 n. miles.

Figure 18.3 is a complete chart of ice distribution off the Okhotsk coast of Hokkaido obtained by a superposition of pictures from each radar screen

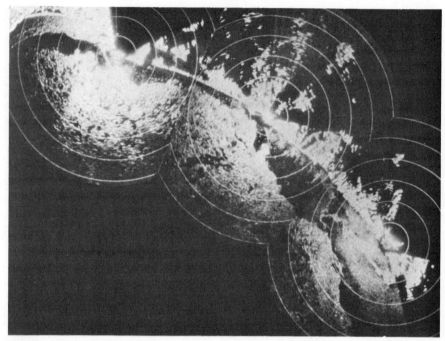

Fig. 18.3 Distribution of drift ice off the Okhotsk coast of Hokkaido (0900, 14 March 1969).

taken successively at very short intervals. Figure 18.4 is an example of the ice chart published daily.

Movement of ice field
Several distinctive features are usually apparent in a photograph of ice distribution. These special features such as polynyi and ice-water boundaries conserve their characteristics for a considerably long time interval, sometimes over several days, so that one can easily identify them in successive photographs as reference landmarks for the study of ice drift. Seven points shown in Figure 18.2 were selected as landmarks.

The solid line in Figure 18.5 is the track of the movement of seven points. To be more accurate, the tracks show the movements of polynyi, although it can be assumed that they represent the movement of an ice field as well. The dots on the line and the corresponding number refer to the times of observation as listed in the lower left of the figure. The movement of every point appears almost parallel. The average movement of the ice field as a whole, obtained by the mean of all the observed points in the figure, is shown in Figure 18.6. The wind velocity observed at the Mombetsu Meteorological Observatory is also shown.

From Figure 18.6 it is immediately recognized that the change of the drift velocity of the ice field is sensitive to that of the wind velocity.

Deformation of the ice field

The tracks of the movement of each polynya shown in Figure 18.5 are not strictly parallel, a fact indicating that the ice field underwent internal deformation during its movement. It has been ascertained that a strain of an ice field can be considered homogeneous over an area 20 km in diameter (Fig. 18.3).

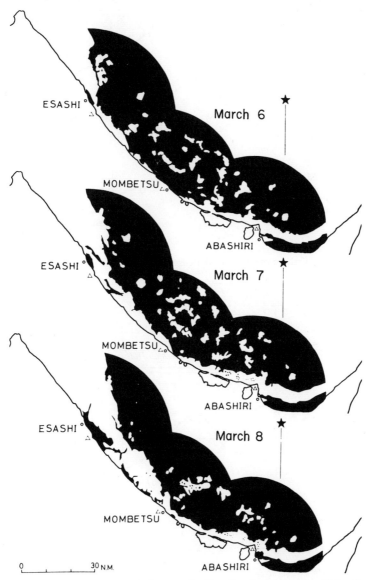

Fig. 18.4 An example of daily record of ice distribution off the Okhotsk coast of Hokkaido.

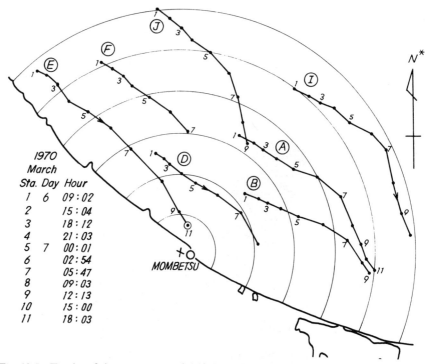

Fig. 18.5 Tracks of the movement of drift ice off Mombetsu (6 March 1968).

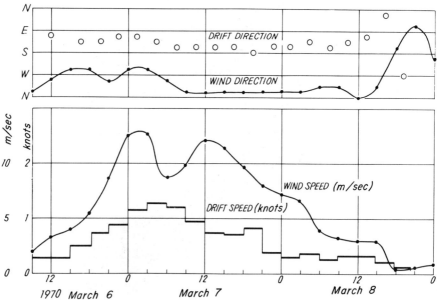

Fig. 18.6 Movement of drift ice off Mombetsu and wind velocity at Mombetsu (6–7 March 1968).

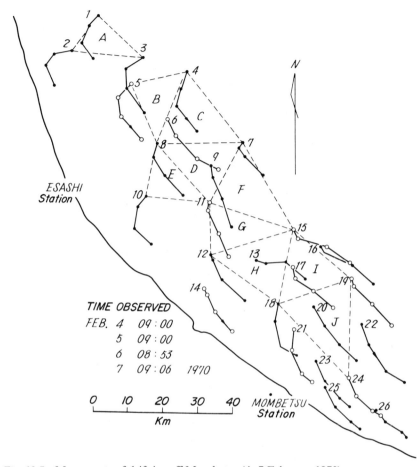

Fig. 18.7 Movements of drift ice off Mombetsu (4–7 February 1970).

Thus, the strain can be calculated from the movement of a set of three points.

Figure 18.7 shows the movement tracks of 26 polynyi during 4–7 February 1970. Each dot and open circle shows the positions of polynyi at the time listed in the lower left part of the figure. Here, 10 sets of 3 targets have been drawn; i.e., 10 triangles (A to J), and the strain of deformation was calculated on the basis of 24 successive hours.

The results are shown in Figure 18.8, where A–J corresponds to the triangles seen in Figure 18.7. The bold arrows show the direction and magnitude of the principal strain ε_1. The scale of magnitude is seen in the lower left corner of the figure. Parenthetic numbers show another principal strain ε_2 in percentage, and numbers without parentheses show the expansion in percentage.

As seen from the figure, the strain of deformation changes remarkably with time and space. The inhomogeneity of deformation over a large area is largely

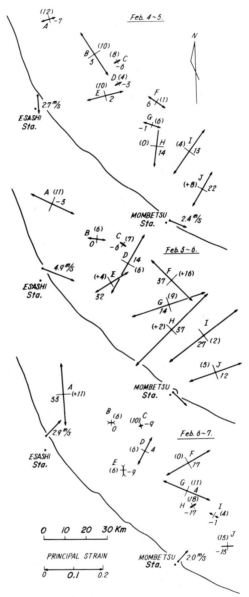

Fig. 18.8 Distribution of strains of successive daily deformation of the ice field off Mombetsu (4–7 February 1970).

dependent upon the local differences of surface and bottom features of the ice field.

Time-lapse motion picture of the radar screen

The best way to demonstrate the complicated movement and deformation of an ice field is to take a time-lapse motion picture of the radar image. Three such films were made in the spring of 1969, 1970, and 1971, each covering about 10 days' movement. The radar screen was photographed every 5 min to reveal the complicated motion of an ice field in a very dramatic manner.

Radar station near the Bering Strait

It is well known that the movement of ice in the Bering Strait is very complex (Dunbar 1966). The study of the ice regime in this area is important to an understanding of the ice problem in the Bering Sea. A sea ice radar system such as the one in Hokkaido might be very useful for ice study in the Bering Strait. By examining the map, the author has proposed two favorable locations for a radar station: Little Diomede Island or Cape Mountain near Cape Prince of Wales.

If a radar were installed on top of a 360-m mountain on Little Diomede Island, the coverage would be about 80 km in diameter and would take in the whole Bering Strait. Because the neighboring Big Diomede Island has a mountain of nearly equal height, however, one could not observe the sea west of Big Diomede. If a radar station were situated on Big Diomede Island, a problem shadow area due to Little Diomede would arise, although it would be much smaller than that due to Big Diomede.

Cape Mountain is about 670 m high and would accommodate radar coverage of about 100 km diameter. If the height of the radar antenna were about 480 m (1600 ft), the effective scanning distance would be about 90 km and adequate to just cover the entire Bering Strait. Even in this case, one could not avoid a small obstruction due to the situation of Big Diomede Island.

The author recommends the construction of a sea ice radar station in the Bering Strait area.

Discussion

NESHYBA: Has there been any attempt to produce a time-series of ice motion at a fixed point on the radar screen? Such a series, when subjected to routine Fourier analysis, would reveal the extent of inertial notion of the rafted ice. It would be most interesting to compare such an energy spectrum against those spectra obtained from fixed current meter data.

TABATA: Our present program is confined to analyzing the surface topography of icefields, since complicated ice movement might well depend upon surface features.

KUSUNOKI: Such astonishing movement of ice might be greatly dependent on the wind parameters. Has an automated network for wind system observation in the Bering Strait area been considered in conjunction with radar monitoring of the ice?

TABATA: A weather station is badly needed on the ice, but even my proposal for a radar station is only preliminary.

BENSON: Radar stations are needed to observe detailed movement of sea ice through the Bering Strait as proposed by Dr. Tabata. This has proved to be an extremely powerful tool in Japan and when combined with satellite observations, it will provide information on sea ice movement never before available and impossible to obtain by any other means. One station at Cape Prince of Wales would permit complete observation of sea motion through the Bering Strait. Other sea ice radar stations are needed along the coast near Nome and on St. Lawrence Island. Four stations would cover the entire Norton Sound and the Bering Sea between St. Lawrence Island and the Bering Strait.

REFERENCES

DUNBAR, M.
  1966   The monthly and extreme limits of ice in the Bering Sea. In *Physics of snow and ice*, edited by H. Oura, Hokkaido University, Sapporo, pp. 687–704.

TABATA, T.
  1971   Measurement of strain of ice field off the Okhotsk Sea coast of Hokkaido. *Low Temp. Sci.* Ser. A. 29: 199–211.
  1972   Radar network for drift ice observation in Hokkaido, *Sea Ice*, edited by T. Karlsson. Proc. int. conf. held in Reykjavik, Iceland, 10–13 May 1971. Nat. Res. Coun. Iceland.

TABATA, T., M. AOTA, M. ŌI, and M. ISHIKAWA.
  1969   Observation of drift ice movement with the sea ice radar [in Japanese]. *Low Temp. Sci.*, Ser. A. 27: 295–315. (Transl. by E. R. Hope, 1970, Defense Sci. Inf. Serv., DRB Canada, T-103-J).

CHAPTER 19

The role of ice in the ecology of marine mammals of the Bering Sea

FRANCIS H. FAY

Institute of Arctic Biology, University of Alaska, Fairbanks, Alaska

Abstract

The ice pack of the Bering Sea is a major component of the habitat of about one million marine mammals of 17 species, including the bowhead whale (*Balaena mysticetus*), belukha (*Delphinapterus leucas*), walrus (*Odobenus rosmarus*), bearded seal (*Erignathus barbatus*), ribbon seal (*Histriphoca fasciata*), ringed seal (*Pusa hispida*), largha harbor seal (*Phoca vitulina largha*), and the polar bear (*Ursus maritimus*).

It is widely recognized that the ice of this and other sub-polar and polar seas is important to such mammals in at least two ways: first, it serves as a substrate on which pinnipeds haul out to sleep and bear their young, and second, it forms a rigid barrier through which pinnipeds and cetaceans alike must find or make holes in order to have access to the air that they breath and the sea that holds their food. For some species of polar marine mammals, the quality and quantity of ice may be as important in habitat selection as are terrain and vegetation to terrestrial mammals. For others, the mere presence of ice may be disadvantageous and may require them to undergo extensive migrations in order to avoid it.

The study of ice as a major factor in the ecology of polar marine mammals is still in a rudimentary stage. In recent years, however, it has become increasingly apparent that ice plays many roles in the ecology of marine mammals and that its full importance to them has been greatly underestimated in the past.

Ice conditions important to marine mammals

The ice of polar seas is not stable, smooth, and uniform, like that of a freshwater lake, but is an incomplete cover, widely variable in form and structure, and typically unstable and dynamic under the influence of surface air and water currents (Armstrong and Roberts 1956).

Sea ice is classifiable from many aspects, one of which is its age or relative permanence. The Arctic Ocean, for the most part, is covered by thick, relatively permanent ice, most of which is several years old, interspersed with younger ice that has formed in the leads and polynyi. Few species of northern marine mammals penetrate far into the permanent ice pack of the Arctic Ocean, and these few do so only in summer; no species makes it its center of abundance the year round. The adjacent seas to the south have only temporary, seasonal ice that is mostly less than one year old. This ice is present in winter and spring but does not persist through the entire summer and autumn. Most of the ice-inhabiting marine mammals are found only amongst this seasonal ice. Since the Bering Sea is an area of transition between seasonal ice and ice-free open sea, its marine mammalian fauna comprises both ice-inhabiting and non-ice-inhabiting species in about equal proportions. These are mainly transient; few remain in the Bering Sea the year round.

The seasonal ice pack occurs over the intercontinental shelf of the northern and eastern portions of the Bering Sea; ice is scarce over the deep, southwestern portion. The pack persists for some six to eight months of each year and, at its maximum, usually in March or April, may extend southward to or slightly beyond the 500-m isobath at the edge of the intercontinental shelf. The dynamics of this ice are, as yet, little understood but are currently under study through analysis of satellite data (ERTS-1, Project 8. F. F. Wright, G. D. Sharma, J. J. Burns, and J. Lentfer). It seems very clear from my own experiences that the principal driving force is the prevailing northerly wind out of the cold, polar high-pressure system. The direction of ice movement, as a consequence of that wind, is mainly from northeast to southwest, throughout the winter. The ice tends to pack tightly against the north side of the major islands and peninsulas and tends to be moved away from their south side, perpetually maintaining large polynyi of open water and thin ice. The major lead systems, of great importance to both resident and transient marine mammals, are oriented in a northeasterly to southwesterly direction.

At its southern edge, the pack is intermittently affected throughout the winter also by strong, southerly winds associated with the continuous progression of North Pacific storms. These winds seem to have two major effects: temporary compaction of ice north of the edge, due to opposition to the southwesterly drift, and destruction of the edge itself, due to the heavy swells produced in the open sea. Strictly speaking, there is no actual "edge" for a 15- to 65-km wide zone at the southern periphery of the pack is alternately dispersed and compressed. American biologists working in this zone in recent years (e.g., Burns et al. 1972) have referred to it as the "front," a term borrowed from the North Atlantic sealers who recognized long ago the importance of this kind of formation to marine mammals. The southern part of the front is usually made up of broad bands and rafts of brash, small floes, and pancake ice, interspersed with areas of open water.

From the aspect of marine mammals, the seasonal ice is clearly divisible also into *fast ice* (which is attached to the shore and therefore most stable and stationary) and *moving ice* (which is perpetually shifted from place to place by winds and surface currents). The percentage of the surface covered, thickness, degree of pressure ridging, and the depth of snow on the surface seem to be important in both types (McLaren 1958; Burns 1970). In the Bering Sea, fast ice is scant or absent along the open coasts of the continents and larger islands, but it forms broad expanses in bays and other locations where it is sheltered from the powerful force of the moving ice. Only those marine mammals that are most capable of making and maintaining breathing holes are able to inhabit the fast ice. The others are obliged to reside in the moving ice, where natural openings are most numerous.

MARINE MAMMALS OF THE BERING SEA

At least 25 species of marine mammals are known to occur in the Bering Sea (Murie 1936; Tomilin 1957; Scheffer 1958, 1967; Geist et al. 1960). A list of these and their status in relation to ice is given in Table 19.1. Eight species (northern fur seal, three species of beaked whales, sperm whale, Dall porpoise, and blue and sei whales) have virtually no contact with ice, for they occur only in the southern, ice-free part of the sea or enter the northern part only in summer when it also is ice-free. Eight other species (sea otter, Steller sea lion, killer whale, harbor porpoise, gray whale, humpback whale, and fin and minke whales) have some contact with ice when it occasionally impinges on their range in the southern Bering Sea or when they advance into the northern area while some ice is still present. Some of the latter group may live in the ice front for a short time each year, but they are not usually regarded as "ice-inhabiting marine mammals," since they do not reside there for long and reproduce in open seas to the south. The remaining nine species (polar bear; walrus; harbor, ringed, ribbon, and bearded seals; and the narwhal, belukha, and bowhead whale) are the true ice-inhabitants, spending all or most of their life in the seasonal ice pack.

Polar bear and sea otter

The polar or ice bear is the only non-aquatic marine mammal. It does, however, spend the greater part of its life at sea, on the ice, where it preys on other marine mammals. The polar bear occurs mainly along the flaw between the permanent ice pack of the Arctic Ocean and the seasonal ice pack of the adjacent seas (USSR Delegation, 1966). It also wanders far southward over the seasonal ice pack and is common in the Bering Sea at least as far south as St. Lawrence Island (Rausch 1953). Ice is clearly most important to this mammal as an extension of land on which it can travel with relative ease in search of its prey; although it is not unable or unwilling to swim when circumstances require it to do so, it is fundamentally a terrestrial mammal with no special adaptations for aquatic life. Findings from recent studies in northern Alaska suggest that ice may also serve as a denning area for the polar bear when suitable sites on land are either inaccessible or unavailable (J. Lentfer, personal communication).

TABLE 19.1 The marine mammals of the Bering Sea and the degree of their contact with the ice

Species	Contact with ice		
	None	Some	Regular
Order CARNIVORA			
Suborder FISSIPEDIA			
Polar or ice bear, *Ursus maritimus* Phipps	—	—	×
Sea otter, *Enhydra lutris* Linnaeus	—	×	—
Suborder PINNIPEDIA			
Steller sea lion, *Eumetopias jubata* (Schreber)	—	×	—
Northern fur seal, *Callorhinus ursinus* (Linnaeus)	×	—	—
Walrus, *Odobenus rosmarus* (Linnaeus)	—	—	×
Harbor seal, *Phoca vitulina* Linnaeus	×	×	×*
Ringed seal, *Phoca hispida* Schreber	—	—	×
Ribbon seal, *Phoca fasciata* Zimmerman	—	—	×
Bearded seal, *Erignathus barbatus* (Erxleben)	—	—	×
Order CETACEA			
Suborder ODONTOCETI			
Bottle-nosed whale, *Berardius bairdii* Stejneger	×	—	—
Stejneger's beaked whale, *Mesoplodon stejnegeri* True	×	—	—
Goose-beaked whale, *Ziphius cavirostris* G. Cuvier	×	—	—
Sperm whale, *Physeter catodon* Linnaeus	×	—	—
Killer whale, *Orcinus orca* (Linnaeus)	—	×	—
Dall porpoise, *Phocoenoides dalli* (True)	×	—	—
Harbor porpoise, *Phocoena phocoena* (Linnaeus)	—	×	—
Narwhal, *Monodon monocerus* Linnaeus	—	—	×
Belukha, *Delphinapterus leucas* (Pallas)	—	—	×
Suborder MYSTICETI			
Bowhead whale, *Balaena mysticetus*	—	—	×
Gray whale, *Eschrichtius gibbosus* (Erxleben)	—	×	—
Humpback whale, *Megaptera novaeangliae* (Borowski)	—	×	—
Blue whale, *Balaenoptera musculus* (Linnaeus)	×	—	—
Sei whale, *Balaenoptera borealis* Lesson	×	—	—
Fin whale, *Balaenoptera physalus* (Linnaeus)	—	×	—
Minke whale, *Balaenoptera acutorostrata* Lacépède	—	×	—

*Harbor seals of the Bering Sea comprise two discrete populations, one of which (*Phoca vitulina largha*) resides in the seasonal ice pack, while the other (*P. v. richardii*) resides partly where ice occurs irregularly in winter and partly where there is no ice at all.

Ice is not a normal part of the environment of the sea otter, which resides in the southern Bering Sea and areas farther south. This animal is apparently unable to make or maintain holes in other than very thin or loose ice (Barabash-Nikiforov 1947; Reshetkin and Shidlovskaya 1947). At the northeastern edge of its range in Bristol Bay, Alaska, the sea otter is occasionally affected by extreme ice conditions from which it must either retreat or perish. Such was the case in the winters of 1970–72, when some died, apparently from inability to penetrate the ice, and several were found to have traveled far overland, apparently in an attempt to reach open water (R. D. Jones, J. Vania, personal communications).

Seals

The seals, sea lions, and walruses of the Bering Sea comprise seven species, only one of which is not ordinarily associated with ice at all—the northern fur seal usually migrates in autumn into the North Pacific Ocean well before the ice pack is formed, and it does not return again until spring, after most of the ice has disintegrated (Kenyon and Wilke 1953). A close relative and co-inhabitant of the Pribilof and Komandorskii islands, the Steller sea lion remains in the Bering Sea throughout the winter and frequently hauls out on floes in the southern part of the ice front (Tikhomirov and Kosygin 1966). During February to April, sea lions are abundant along the front, at least from Bristol Bay to the International Date Line, but they do not penetrate far into the pack where there is little open water.

The gregarious walrus is for the most part a perpetual inhabitant of the ice, spending the winter in the temporary ice pack of the Bering Sea and the summer along the edge of the permanent ice pack of the Chukchi Sea. In general, these mammals inhabit areas of moving ice, where leads and polynyi are numerous and where there is some ice thick enough to support their weight when they haul out to rest. Thus they do not ordinarily occupy the outer part of the front, where the floes are thin and small, but are found from 50 to 500 km north of it in the heavier ice. The walrus is capable of breaking holes with its head in ice up to at least 22 cm thick and of maintaining them indefinitely with the aid of its tusks (J. J. Burns and F. H. Fay, unpublished). The young walruses are born on the moving ice, mainly in May during the northward migration to the Chukchi Sea (Brooks 1954; Burns 1965). This migration is to some extent passive, in that the animals are transported much of the way by the northward-drifting ice. Occasionally in summer, more than half of the population may haul out on land when the ice recedes far from their feeding grounds (Fedoseev 1966), but they take up residence on the ice again, apparently by preference, when it returns to their area (Nikulin 1947; Gol'tsev 1968).

The four species of hair seals (Phocidae) found in the Bering Sea are closely associated with the seasonal ice pack. Of these four, the ringed seal is most dependent on ice and best adapted to it. In winter and spring, adult ringed seals are mainly solitary inhabitants of the heavy fast ice along the coasts, where they make and maintain breathing holes, apparently with the aid of the heavy claws on their foreflippers (Vibe 1950). These adults probably are highly territorial, for there is a clear segregation of age classes, with the oldest and presumably most dominant animals occupying the most stable fast ice (McLaren 1958). In a snow cave excavated by the female above one of her breathing holes, the pup is born with a white, woolly coat in late March or April, and is suckled for four to six weeks or until the fast ice breaks up. The growth and survival of the pup are positively correlated with the stability of the ice and the depth of snow upon it (McLaren 1958; G. A. Fedoseev, personal communication). The adult and young seals haul out on the fast ice in June, while completing their molt, and later may penetrate far into the permanent ice pack of the Arctic Ocean (Chapskii 1946). Ringed seals of the Bering and Chukchi seas seem to be mostly migratory, moving northward into the Arctic in summer and dispersing southward in autumn with the expanding ice pack (Johnson et al. 1966; Burns 1970). Some of the pups are pelagic in the open Bering Sea during the summer.

The bearded seal is associated with ice the year round but seems least specific of the phocids in its selection of ice quality, for its range broadly overlaps those of all other ice-inhabiting pinnipeds. Bearded seals are widely dispersed in the moving ice in winter and spring, from the Chukchi Sea to the southern part of the front (Burns 1970). They are capable of breaking holes with their head in at least 10 cm of ice and, on circumstantial evidence, are believed to be capable of maintaining breathing holes in heavier ice, perhaps with their stout claws. The young are born on the moving ice in April, just prior to the northward migration, and the bulk of the population spends the summer along the permanent ice pack of the Chukchi Sea (Burns 1967). Some of the younger seals remain in the Bering Sea throughout the summer, but unlike bearded seals of the Okhotsk Sea (Tikhomirov 1961), they do not haul out on land but are entirely pelagic. Most of the population migrates southward in autumn through Bering Strait, keeping to the coasts until the fast ice is formed. The adults are last to arrive with the heavy ice (Burns 1970).

The ribbon seal is associated with the ice front in the Bering Sea during winter and spring, ranging up to 150 km north of its southern periphery. It seems to be restricted to areas with much open water or very thin ice, however, suggesting that it is either incapable of making holes in ice more than a few centimeters thick or that it is otherwise linked to the ice front, perhaps by its feeding habits. Ribbon seals seem to be mainly solitary, coming together only during the breeding season, though they are often found in widely dispersed aggregations during the spring molt. The pups are born with a white coat in April, usually on heavy floes less than 20 m in diameter and with a substantial depth of snow (Shustov 1965; Burns 1970). In May and June, the adult seals spend much of their time on the northward-drifting ice, while they complete their annual molt. As a consequence of this drift, they undergo a passive migration, often as far as Bering Strait and the southern part of the Chukchi Sea. They are believed to be pelagic in the open Bering Sea during the Summer (Burns 1970).

The harbor seals of the Bering Sea comprise two large groups: those inhabiting the ice front and bearing white-coated pups on ice early in the spring, and those inhabiting the southern coastal areas and bearing dark-coated pups on land in late spring and summer (McLaren 1966; Burns 1970; Burns and Fay 1973). From the aspect of association with ice, the southern group is further divisible into those of the Bristol Bay area (where they have extensive contact with ice throughout the winter) and those of the Aleutian and Komandorskii islands (where sea ice is scarce or absent). The ice front harbor seal, regarded by some taxonomists as a separate species *Phoca largha* (Chapskii 1969), is found in winter and spring along the front from northern Bristol Bay to the Koryak and Kamchatka coasts, always in areas of low ice pressure with small and scattered floes. Apparently, it is unable to make or maintain holes in any but the thinnest ice, though, like the ribbon seal, its restriction to the ice front may be more related to nutrition than to ability to penetrate ice. The pups are born in late March and April, usually on heavy floes 10 to 20 m in diameter with a thick layer of snow and often with some remnant of a pressure ridge. The young seals utilize the ice blocks and caverns of the pressure ridge as shelter from the weather and as refuge from predators and scavengers. The breeding aggregations consist of isolated "family groups" of three, comprising an adult male and female with a pup, and these groups are rarely closer together than 0.2 km (Burns et al. 1972). Because of their wide spacing the adults are presumed to be territorial and to defend the area about the natal floe. These seals are highly migratory, dispersing northward into the Chukchi Sea in summer at least as far as the northern coast of Alaska, and they return again to the Bering Sea in late autumn to take up residence along the ice front.

The southern, coastal harbor seals have little contact with the ice pack, other than in Bristol Bay, and their behavior in its presence is unknown. They are believed to be non-migratory and reproductively isolated from the *largha* seals of the ice front, though there may be some possibility of their intergradation with the latter in the Bristol Bay-Kuskokwim Delta area. This is currently being investigated (Burns and Fay 1973). In general, the breeding aggregations consist of large herds, rather than pairs or family groups, and the young are born with an adult-like pelage, having shed their woolly embryonal coat in *utero*.

Whales

The Cetacea of the Bering Sea are mostly inhabitants of ice-free waters. Six species (killer whale, harbor porpoise, and fin, minke, humpback, and gray whales) migrate into the northern part of the sea in spring, before the last ice has disintegrated, and some go on into the Chukchi Sea in summer. They return southward in autumn, usually before the first ice is formed. Although some individuals may inhabit the loose ice of the receding front for three to four months each year, the majority frequents only the open sea. Three other species (bowhead, belukha, and narwhal) are the only ice-inhabiting cetaceans of the Bering Sea; the narwhal, in fact, occurs there so infrequently as to be of no importance in this review (Geist et al. 1960).

The bowhead or Greenland whale frequents the ice front of central and southwestern Bering Sea in winter and, from March to May, penetrates about 3000 km into the ice in its annual migration to the Arctic Ocean by way of Bering Strait (Tomilin 1957). The calves are apparently born along the way (Scammon 1874). During this migration, the bowheads follow the southwest-to-northwest lead system to Bering Strait and the flaw between the fast and the moving ice thereafter. This whale is not entirely dependent on natural openings for its breathing holes, for it can also break holes in ice up to perhaps 25 cm or more in thickness (Tomilin 1957). It spends the summer along the edge of the permanent ice pack in the Laptev, East Siberian, and Beaufort seas, returning to the Bering Sea in autumn.

The belukha or white whale of the Bering Sea also winters in and along the front of the seasonal ice pack and perhaps in a few more northern areas where there are persistent natural openings (Kleinenberg et al. 1964). In spring, these little whales disperse inshore and northward, some of them following in the path of the bowheads. Some small groups remain in the ice-free Bering Sea in summer, inhabiting such areas as the Gulf of Anadyr, Bristol Bay, and the Norton Sound-Yukon Delta region; others move northward through Bering Strait to summer in the Beaufort, Chukchi, and East Siberian seas. The young are born mainly from May to July in the summering areas. The belukhas return to the Bering Sea in autumn, somewhat earlier than the bowheads, before the winter ice is formed. Apparently this is necessitated by their inability to make holes in any but the thinnest ice.

DISCUSSION

Three of the eight species of marine mammals that regularly inhabit the seasonal ice pack of the Bering Sea are restricted mainly to the front zone, where thin ice and open water provide easy access between sea and air. The harbor seals, ribbon seals, and the belukha are apparently incapable of making and maintaining breathing holes in heavier ice. In summer, each resides mainly in ice-free waters, often near but seldom in the permanent polar ice pack. The remaining five species regularly reside in or travel through the heavy ice of the seasonal pack, where natural openings are scarce and quickly refrozen and the creation and maintenance of holes is a continual necessity. In summer, these five species (polar bear, walrus, ringed and bearded seals, bowhead whale) inhabit mainly the southern edge of the permanent ice pack. Excluding the bear, which is non-aquatic and not confronted with the problem of access between air and water, the prime adaptations of these mammals to ice are their ability to relocate annually the areas with types of ice most conducive to their survival and perpetuation, and their ability to make holes in unbroken ice and keep them open if necessary. These are primarily behavioral adaptations that were presumably developed quickly in an evolutionary sense, perhaps even without genetic change.

Physiologically, the five pinnipeds have been obliged to adjust their reproductive cycles and molts to coincide with the most favorable local ice conditions (Burns 1970), and presumably this too has been rapid, since the selective pressure for such change must be very strong. Morphologically, however, there seems to have been little adaptive response to the ice itself as a physical factor in their environment. Indeed, only a few possibilities come to mind:

The *massive skulls* of the walrus and bearded seal may have adaptive value for breaking through ice, but the skulls of some ancestral forms and of other living pinnipeds from ice-free waters are at least as massive, suggesting that ice alone has not been the primary agent of selection in this direction. In contrast, the ringed seal, the species most closely associated with ice and most dependent on its presence, has one of the most fragile skulls of any living pinniped.

The *claws* of all ice-inhabiting phocids of the Northern Hemisphere are appreciably larger than those of their relatives in ice-free waters to the south. That the claws of the forelimbs are used by the ringed seal for maintenance and enlargement of breathing holes is well known by the Eskimos and has apparently been confirmed by both Vibe (1950) and McLaren (1958). It is unknown, however, whether the heavy claws of the other ice-inhabiting pinnipeds are used in the same way. The contrasting structure of claws, relative to environment and ecology, is particularly notable among the three species of the subgenus *Pusa*: the ringed seal (*P. hispida*), the Baikal seal (*P. sibirica*), and the Caspian seal (*P. caspica*) (Kondakov 1960). The claws of the Baikal seal, which resides only in Lake Baikal of the eastern USSR, are the stoutest, as one might predict from its association with hard, freshwater ice. This seal winters for six to eight months in the northern part of the lake, where, like the ringed seals of the Bering Sea and Arctic Ocean, it maintains breathing holes in heavy ice and brings forth its young in a snow cave (Ognev 1935; Kozhov 1963). The claws of the Bering Sea ringed seal, which lives a similar life in the softer ice of the sea, are smaller than those of the Baikal seal but larger than those of their ancestral relict, the Caspian seal. The latter species winters (about four months) in the northeastern part of the Caspian Sea and brings forth its young on the ice without benefit of a snow cave (Ognev 1935). A comparable though smaller difference in size and structure of the claws has been found by J. J. Burns (personal communication) between the ice-inhabiting harbor seals and those of ice-free waters in the Bering Sea.

The *white natal coat* of seals of the tribe Phocini, which, in the Bering Sea, includes the ringed, ribbon, and harbor seals, is generally believed to be an adaptive specialization that has value mainly as protective coloration in an environment dominated by ice and snow. This characteristic (and others unique to this group) indicates origin from a common ancestor, in which the white coat probably had already evolved prior to diversification (McLaren 1960). Fossil seals, believed to be ancestral to this group, are known from late Miocene and Pliocene deposits (Pontian and Sarmatian seas, 10 to 12 million years B. P.) in central and southern Europe (McLaren 1960) and from Pliocene deposits (four to eight million years) in southern Alaska (C. A. Repenning, written communication, 1972). McLaren (1960), assuming that the white coat was an ice-related adaptation and that the Pontian and Sarmatian seals already possessed it, concluded that ice must have been present in the Pontian Sea "and probably in at least the innermost reaches of the Sarmatian Sea during late Miocene," which is contrary to current paleoclimatic information (see below). While it seems probable that the ancestor of the Phocini did possess a white embryonal coat, there seems to be no compelling reason for the assumption that this was an ice-related adaptation. Two other species of northern phocids and the walrus produce dark-coated young on ice with good success, while some very successful populations of gray seals, *Halichoerus grypus* (Phocini) bear white-coated pups on land (Scheffer 1958; King 1964). At least two pinnipeds, the northern fur seal and the walrus, also have a white embryonal coat that is, however, shed *in utero* several months prior to birth (Belkin 1964; F. H. Fay, unpublished) and seems to have no relationship to ice or to ice-induced adaptation. Since the ancestral Phocini were apparently circumpolar in distribution long before extensive contact with ice (see below), it is conceivable that they too possessed a whitish embryonal coat that was shed *in utero* and that its appearance in a natal and post-natal context was a more recent development linked with the physiological (reproductive) adjustment to ice, i.e., with the pupping season being shifted from summer to winter or early spring. For example, both the ice-inhabiting (*Phoca vitulina largha*) and coastal (*P. v. richardii*) harbor seals of the eastern Bering Sea complete their molt from the white embryonal pelage to an adult-type coat in May, but the ice seals are born in March and April, prior to that molt, while the coastal seals are mostly born in June and July, after the molt (Burns and Fay 1973). Thus, pups of the former possess the white coat at birth and for some weeks thereafter, whereas pups of the latter shed the white coat *in utero* and are born with the second (adult-like) pelage.

The bowhead or Greenland whale is unique among the Cetacea in having the most remarkably large and massive head with the most highly arched rostrum, both of which are relatively recent specializations (Howell, 1930: 55–56) and would seem to be advantageous in breaking through ice. This unique head is topped by a still higher promontory, in the apex of which are situated the blowholes (nasal apertures). This, too, may be an adaptation to the ice-covered habitat, for it may permit the animals to surface for air in the many small openings in the ice, i.e. openings that are large enough to accommodate the nasal promontory but not the entire head. Finally, the skin on this promontory and on the rest of the dorsal surface of the head and neck is, apparently, much thicker than it is elsewhere on the body (Scammon 1869; Tomilin 1957, p. 36), which would seem to be a protective adaptation also for its ice-breaking function.

The whiteness of the belukha has been interpreted as protective or cryptic coloration, though there seems to be little basis for that judgement (see review by Kleinenberg et al. 1964). Indeed, if it is camouflage, it is most unusual among mammals, for it is absent in the most vulnerable segment of the population, the youngest animals, which are instead dark gray or gray-brown. Conversely, the dark colored young blend well with the dark waters in which they reside, suggesting that abundance rather than scarcity of pigment has greater protective value in this case and that the whiteness of the adults has some other primary function.

The remarkably small amount of evolutionary change that is evident in the ice-inhabiting mammals as a whole, and that can be clearly related to ice, seems to indicate that, either the time available for adaptation has been relatively short (Dunbar 1968), or that ice imposes little selective pressure on its mammalian inhabitants. It is instructive to examine further the first possibility; the second cannot be tested much beyond what has already been said.

Some floating ice was present in the Arctic Ocean at least 3.5 million years ago, as indicated by the presence of ice-rafted debris in bottom sediments, but there was apparently little or no permanent ice there until the past 700,000 years of the Quarternary Period (Herman et al. 1971; Mullen et al. 1972). The implied paleoclimatic history of the Beringian area (Wolfe and Leopold 1967; Hopkins 1967) also suggests that ice probably was not present in the Bering

Sea until late Tertiary or early Quarternary times and could have been absent during at least the warmest interglacial (Sangamon) and postglacial (Hypsithermal) intervals. Thus, the marine mammals of the present arctic and subarctic regions probably had little contact with ice before about three or four million years ago, at the end of the Pliocene or beginning of the Pleistocene Epoch, and all or most of their adaptations to its presence must have been developed since that time.

The ringed seal is the most advanced in its adaptation to ice-covered seas, indicating that it has had the longest history of contact with ice (McLaren 1960). Indeed, it seems to be dependent on a particular kind of ice for maximal reproductive success (McLaren 1958). For the remainder of the ice-inhabiting marine mammals, ice is clearly an important factor in their environment, but it is not yet clear how dependent they are on its presence or whether they actually have an affinity for ice itself. Interpretations of the latter in the past have probably not been sufficiently critical, and the use of such terms as *pagophilic* (having an affinity for ice) and *pagophobic* (having an aversion to ice) for convenient categorization (McLaren 1966; Burns 1970) has been misleading, for the distinction is by no means as clear as these terms seem to indicate. For example, some Alaskan populations of the coastal harbor seals, *P. v. richardii*, usually regarded as pagophobic, regularly utilize icebergs from coastal glaciers as a place to rest and bear their young, in apparent preference to available islets and sandbars (Bishop 1967), while segments of the populations of several species that are usually regarded as pagophilic (e.g., walrus, ribbon seal, and belukha) regularly remain in summer in the ice-free waters of the Bering Sea, where food supplies or some other factors are apparently more attractive than an association with ice.

For the pinnipeds, ice serves many functions, not the least of which is its role as a solid substrate, on which they haul out to rest, molt, mate, bear their young, and suckle them. Although most of the species associated with ice probably could perform these activities on land (as some do in parts of their present range), ice is more than a substitute for land, for it offers several special advantages, such as:

Isolation. In their selection of sites on which to haul out, most pinnipeds choose isolated islets and offshore rocks, rather than the continental coasts or larger islands (Scheffer 1958, p. 7). The basis for this selection seems obvious: these isolated sites provide the best refuge from predators and other disturbing terrestrial mammals. In this respect, ice provides equal or better isolation than can be found in most terrestrial sites.

Space. The ice pack provides an enormous number of isolated islets that vastly increase the space available for pinnipeds to haul out. Many more animals are accommodated than would be feasible on the few terrestrial islets extant in the Bering Sea.

Variety. The variety of "terrain" provided by fast ice, moving ice, large and small floes, thick ice, thin ice, brash, and open water favors diversity of occupants and preferential selection of habitats. More species may be accommodated on ice than could be accommodated on land.

Food supply. For the benthic feeders in particular, the presence of ice over the entire Beringian intercontinental shelf provides easy access to a much greater and more varied food supply than would be available to them from the shores and islands. This also provides for significantly larger populations than could be supported in the absence of ice.

Transportation. Those species that inhabit the moving ice are continuously transported to new feeding areas in a passive way, which helps to distribute more evenly their influence on the sub-ice communities. This is particularly true of the benthic feeders. Much energy is conserved as a consequence of this transportation, for the animals are seldom obliged to swim to new feeding areas, and those in migration are often carried much of the way by the ice.

Sanitation. Many of the known diseases and parasites of pinnipeds are favored by crowding and continuity of site occupation. Since it is seldom necessary for pinnipeds to crowd together on the ice or to occupy the same floe more than once, the conditions for transmission of such diseases and parasites are unfavorable.

Shelter. The hummocks of pressure-ridged ice not only interfere with the predators' view but also effectively reduce wind velocity and the general severity of the weather at the seals' level. The young of the *largha* seal and harp seal take refuge from intruders and the weather among the ice blocks of a pressure ridge, while the young ringed seal is protected by a snow cave on the ice.

For the cetaceans, ice seems to provide no direct benefits other than, perhaps, protection from predation by the killer whale which, in the Bering Sea, seems not to advance far into the ice pack. Otherwise it has mainly negative value, in that it restricts their movements mostly to open waters. Only the bowhead is really effective in making breathing holes where natural openings are scarce or absent.

Finally, the seasonal ice pack probably is of great positive value to many of the marine mammals of the Bering Sea, indirectly, as a substrate for algal growth. The micro-algae that develop in extraordinary abundance at the ice-water interface comprise a major part of the total primary productivity of the area (McRoy and Goering, Chapter 21 of this volume), and a significant contribution to the support of the abundant fauna. Probably, a larger proportion of this algal production is made available to the fauna in the vicinity of the ice front than in equivalent areas elsewhere on the shelf, for the prevailing southward set of the ice should tend to carry much of the production to that area from localities farther north. The rich trophic system in and near the front, as indicated by the abundance of benthos and the important winter fisheries (Ivanov 1964a,b; Dudnik and Usol'tsev 1964), may be in large part dependent on this kind of a transport system.

Further studies of the relationships of marine mammals to ice, both in the Bering Sea and in other subpolar and polar seas, can be expected to yield much information of both academic and utilitarian value, for the mammals and the ice are parts of a particularly rich marine ecosystem, all aspects of which must be thoroughly understood before their importance to the system as a whole can be evaluated. The very qualities, quantities, and movements of ice that seem to be important to these mammals are certainly of great importance also to the primary and secondary producers and in sediment transport, and have far-reaching effects on the climate, the fisheries, and shipping, to name a few. From a navigational aspect, it may be instructive to examine the methods developed by these mammals for finding their way beneath the ice and, for example, relocating a particular breathing hole or ice floe. The turbidity of the sub-ice environment of the Bering Sea is such that the use of visual clues is practically excluded, except at very short range, and the acoustical peculiarities of ice and water also may preclude the use of sonar-like guidance systems in most cases. Since the ice is rarely stationary, some form of inertial navigation would seem to be necessary, perhaps in combination with visual and acoustical clues.

Since much of our present knowledge of the role of ice in the ecology of marine mammals is based on logic, rather than on long-term objective observation and experimentation, there is clearly a large amount of painstaking work yet to be done in this field. It is clear, also, that we need to develop a new perspective for this work, perhaps through intensive study of the sensory capabilities of these mammals and greater personal exposure to and appreciation for that vast sub-ice sector of their environment. We have scarcely begun to develop that new perspective and to question some of the points raised here.

Acknowledgments

Some of the information and much of the interpretation in this paper were derived from field studies of marine mammals of the Bering Sea supported in part by the Arctic Institute of North America, with the approval and financial support of the Office of Naval Research under Contract Nonr 1138(01), Subcontracts ONR-77 and 91. Assistance is acknowledged of the University of Alaska Sea Grant 1-36109, and of the Arctic Health Research Center, U. S. Department of Health, Education, and Welfare. Valuable logistic support was provided by the Bureau of Sport Fisheries and Wildlife, by the Bureau of Commercial Fisheries, U. S. Fish and Wildlife Service, and by the U. S. Coast Guard. The author is particularly indebted to his colleagues J. J. Burns, Alaska Department of Fish and Game, and G. C. Ray, Johns Hopkins University, with whom much of this material was discussed prior to this writing, and to R. L. Rausch, C. A. Repenning and P. D. Shaughnessy for manuscript review.

REFERENCES

ARMSTRONG, T., and B. ROBERTS
    1956   Illustrated ice glossary. *Polar Record* 8: 4–12.

BARABASH-NIKIFOROV, I. I.
    1947   The sea otter (*Enhydra lutris* L.)—Biology and economic problems of breeding. In *Kalan*. Sov. Ministrov RSFSR, Moscow, pp. 3–202. (Israel Prog. Sci. Transl., 1962).

BELKIN, A. N.

1964 Data on the foetal development of the skin and hair cover of eared seals (Otariidae). In *Fur seals of the Far East*, edited by V. A. Arseniev. Izv. TINRO 54: 97–130. (Transl. by S. Stutz and N. Wilson, Dept. Zool., Univ. British Columbia).

BISHOP, R. H.

1967 Reproduction, age determination, and behavior of the harbor seal, *Phoca vitulina* L., in the Gulf of Alaska. M. S. Thesis, Univ. Alaska, 121 pp.

BROOKS, J. W.

1954 A contribution to the life history and ecology of the Pacific walrus. Alaska Coop. Wildl. Res. Unit, Univ. Alaska, Spec. Rep. No. 1, 103 pp.

BURNS, J. J.

1965 The walrus in Alaska. Alaska Dept. Fish & Game, Juneau, 48 pp.

1967 The Pacific bearded seal. Alaska Dept. Fish & Game, Juneau, 66 pp.

1970 Remarks on the distribution and natural history of pagophilic pinnipeds in the Bering and Chukchi seas. *J. Mammal.* 51: 445–454.

BURNS, J. J., and F. H. FAY

1973 Comparative biology of Bering Sea harbor seal populations [Abstract]. Proc. 23rd Alaskan Sci. Conf. 1972, p. 28.

BURNS, J. J., G. C. RAY, F. H. FAY, and P. D. SHAUGHNESSY

1972 Adoption of a strange pup by the ice-inhabiting harbor seal, *Phoca vitulina largha. J. Mammal.* 53: 594–598.

CHAPSKII, K. K.

1946 Mammals of high latitudes of the Arctic Ocean. *Tr. Dreif. Exped. Glavsev-morput*, "*G. Sedov*" *1937–1940*, 3: 14–18.

1969 Taxonomy of seals of the genus *Phoca* sensu stricto in the light of contemporary craniological data. In *Marine Mammals*, edited by V. A. Arseniev, B. A. Zenkovich, and K. K. Chapskii. Akad. Nauk SSSR, Ikhtiol. Comm., Moscow, pp. 294–304. (Transl. Canadian Wildl. Serv., 1971).

DUDNIK, IU. I, and E. A. USOL'TSEV

1964 On the herring of eastern Bering Sea. *Tr. VNIRO* 49: 225–229.

DUNBAR, M. J.

1968 *Ecological development in polar regions: A study in evolution*. Prentice Hall, Englewood Cliffs, N. J., 119 pp.

FEDOSEEV, G. A.

1966 Aerial observations of marine mammals in the Bering and Chukchi seas. *Izv. TINRO* 58: 173–177.

GEIST, O. W., J. L. BUCKLEY, and R. H. MANVILLE

1960 Alaskan records of the narwhal. *J. Mammal.* 41: 250–253.

GOL'TSEV, V. N.
1968 Dynamics of coastal walrus rookeries in connection with distribution and numbers of walruses. *Izv. TINRO* 62: 205–215.

HERMAN, Y., C. V. GRAZZINI, and C. HOOPER
1971 Arctic palaeotemperatures in late Cenozoic time. *Nature* (London) 232: 466–469.

HOPKINS, D. M.
1967 The Cenozoic history of Beringia—a synthesis. In *The Bering land bridge*, edited by D. M. Hopkins. Stanford Univ. Press, Stanford, Calif., pp. 451–484.

HOWELL, A. B.
1930 *Aquatic mammals: Their adaptations to life in the water.* Chas C. Thomas, Springfield, Ill., 338 pp.

IVANOV, B. G.
1964a Some results of investigations on the biology and distribution of shrimp in the Pribilof area of Bering Sea. *Tr. VNIRO* 49: 113–122.
1964b Quantitative distribution of echinoderms on the shelf of the eastern Bering Sea. *Tr. VNIRO* 49: 123–140.

JOHNSON, M. L., C. H. FISCUS, B. T. OSTENSON, and M. L. BARBOUR
1966 Marine mammals. In *Environment of the Cape Thompson region, Alaska*, edited by N. J. Wilimovsky and J. N. Wolfe. U. S. Atomic Energy Comm., Oak Ridge, Tenn., pp. 877–924.

KENYON, K. W., and F. WILKE
1953 Migration of the northern fur seal, *Callorhinus ursinus. J. Mammal.* 34: 86–98.

KING, J. E.
1964 *Seals of the world.* Brit. Mus. (Nat. Hist.), London, 154 pp.

KLEINENBERG, S. E., A. V. YABLOKOK, V. M. BEL'KOVICH, and M. N. TARASEVICH
1964 *Belukha (Delphinapterus leucas): A monographic investigation of the species.* Akad. Nauk SSSR, Inst. Morfol. Zhivotnikh, Moscow, 456 pp. (Israel Prog. Sci. Transl., 1969).

KONDAKOV, N. N.
1960 On the problem of the systematic status of the Baikal seal. *Biull. Mosk. o-va icpyt prirody*, Otd. biol. 65: 120–121. (not seen; cited in Kozhov, 1963).

KOZHOV, M.
1963 *Lake Baikal and its life.* Dr. W. Junk, The Hague, 344 pp.

McLAREN, I. A.
1958 The biology of the ringed seal *(Phoca hispida Schreber)* in the eastern Canadian Arctic. Fish. Research Bd. Can., Ottawa, Bull. 118: 97 pp.

1960 On the origin of the Caspian and Baikal seas and the paleoclimatological implication. *Amer. J. Sci.* 258: 47–65.

1966 Taxonomy of harbor seals of the western North Pacific and evolution of certain other hair seals. *J. Mammal.* 47: 466–473.

MULLEN, R. E., D. A. DARBY, and D. L. CLARK

1972 Significance of atmospheric dust and ice rafting for Arctic Ocean sediment. *Bull. Geol. Soc. Amer.* 83: 205–211.

MURIE, O. J.

1936 Notes on the mammals of St. Lawrence Island, Alaska. In *Archaeological excavations at Kukulik, St. Lawrence Island, Alaska,* by O. W. Geist and F. G. Rainey. Misc. Publ. Univ. Alaska 2: 337–346.

NIKULIN, P. G.

1947 Biological characteristics of the shore aggregations of the walrus in the Chukotka Peninsula. *Izv. TINRO* 25: 226–228. (Transl. Fish. Res. Bd. Can., No. 115).

OGNEV, S. I.

1935 *Animals of the USSR and Adjacent Countries.* Vol. 3. *Carnivora* (Fissipedia *and* Pinnipedia). Gosudarst. Izdat. Biol. Med. Lit., Moscow, 743 pp. (Israel Prog. Sci. Transl., 1962).

RAUSCH, R.

1953 On the land mammals of St. Lawrence Island, Alaska. *Murrelet* 34: 18–26.

RESHETKIN, V. V., and N. K. SHIDLOVSKAYA

1947 Acclimatization of sea otters. In *Kalan.* Sov. Ministrov RSFSR, Moscow, pp. 203–262. (Israel Prog. Sci. Transl., 1962).

SCAMMON, C. M.

1874 *The marine mammals of the northwestern coast of North America.* John H. Carmany, San Francisco, 319 pp.

SCHEFFER, V. B.

1958 *Seals, sea lions, and walruses: A review of the* Pinnipedia. Stanford Univ. Press, Stanford, Calif., 179 pp.

1967 Marine mammals and the history of Bering Strait. In *The Bering land bridge,* edited by D. M. Hopkins. Stanford Univ. Press, Stanford, Calif., pp. 350–363.

SHUSTOV, A. P.

1965 Distribution of the ribbon seal (*Histriophoca fasciata*) in the Bering Sea. In *Marine mammals,* edited by E. N. Pavlovskii, B. A. Zenkovich, S. E. Kleinenberg and K. K. Chapskii. Akad. Nauk SSSR, Ikhtiol. Comm., Moscow, pp. 118–121.

TIKHOMIROV, E. A.

1961 Distribution and migration of seals in waters of the Far East. In *Transactions of the Conference on Ecology and Hunting of Marine Mammals,* edited by

E. H. Pavlovskii and S. E. Kleinenberg. Akad. Nauk SSSR, Ikhtiol. Comm., Moscow, pp. 199–210.

TIKHOMIROV, E. A., and G. M. KOSYGIN

1966 Prospects for commercial sealing in the Bering Sea. *Ryb. Khozyaistvo* 8: 25–28.

TOMILIN, A. G.

1957 Whales. In (Vol. 9) *Animals of the USSR and adjacent countries*, edited by V. G. Heptner. Akad. Nauk SSSR, Moscow, 756 pp.

USSR, DELEGATION OF

1966 The polar bear: distribution and status of stocks; problems of conservation and research. In *Proceedings of the First International Scientific Meeting on the Polar Bear*. USDI, Bur. Sport Fish. Wildl., Washington, D. C., Resource Publ. No. 16: 39–43.

VIBE, C.

1950 The marine mammals and the marine fauna in the Thule District (Northwest Greenland) with observations on ice conditions in 1939–41. *Medd. om Grønl.* 150(6): 1–117.

WOLFE, J. A., and E. B. LEOPOLD

1967 Neogene and early Quarternary vegetation of northwestern North America and northeastern Asia. In *The Bering land bridge*, edited by D. M. Hopkins. Stanford Univ. Press, Stanford, Calif., pp. 193–206.

CHAPTER 20

Underwater ice formation in rivers as a vehicle for sediment transport

CARL S. BENSON

Geophysical Institute, University of Alaska, Fairbanks, Alaska

THOMAS E. OSTERKAMP

Department of Physics, University of Alaska, Fairbanks, Alaska

Abstract

Underwater ice is a general term which includes *frazil ice* and *anchor ice*.

Frazil ice

Frazil ice consists of small disc-shaped crystals, about 0.5 mm in diameter, which form in streams that are slightly supercooled (about -0.03 C). These small crystals are suspended throughout the water because viscous drag forces exceed bouyancy forces; this is analogous to the way small silt and mica particles are suspended in flowing water. Anchor ice is fastened to the bottom and consists largely of fine frazil crystals which adhere to objects on the bottom and to each other. Supercooling is essential to the initiation and growth of the crystals because it provides a temperature gradient from the ice-water interface into the water. This allows latent heat to be removed from the growing ice crystals and to be transported by convection to the surface of the stream where it is lost to the atmosphere.

Anchor ice

Anchor ice is sometimes disturbed by the current; it can roll over while still adhering to bottom sediments. This causes it to lift sand and pebbles off the bottom, and a significant amount of sediment is incorporated into the anchor ice by this mechanism. An 8-mm motion picture was presented to show the sediment in bottom ice and the transportation of this sediment when the ice was broken from the bottom by a probe rod. The possibility of a net sediment transport into the sea by sediment-laden anchor ice was discussed. This ice forms

401

in September in the Arctic and in October in the interior part of Alaska. It enters the sea before sea ice forms and may continue all winter. Two aspects that need special investigation are the flux of sediment into the Bering Sea and Arctic Ocean by this mechanism in a wide spectrum of stream types, and the interaction between river ice with seawater and with sea ice. How does river ice affect sea ice formation? How much sediment-laden anchor ice attaches to the underside of a shore-fast and free-floating sea ice?

The structure of brackish ice, as is found in Norton and Kotzebue sounds, is less well-known than that of other ice types. Structural studies should include crystal fabric studies from top to bottom; light transmission through all types of ice, especially important to biological processes; and the sediment load in the ice and origin of these sediments.

Discussion

TAKENOUTI: What are the basic conditions and mechanisms necessary for underwater ice formation?

BENSON: Underwater ice forms in turbulently flowing water when it becomes supercooled. The turbulent flow causes the temperature to be nearly uniform from top to bottom. Supercooling is essential so that latent heat can be transported away from the ice crystals as they grow. Immediately at the crystal-liquid interface, the temperature is at the equilibrium point of $0°$ C; when the adjacent water is supercooled, there is a temperature gradient from the crystal to the water. The turbulent flow transports the latent heat to the stream surface, where it dissipates to the atmosphere. When the water is supercooled to initially about -0.05 C, small (2–3 mm) crystalline discs rapidly form to an estimated concentration of 1×10^5 crystals per m^3. They remain suspended throughout the bulk of the water, because the viscous drag exceeds the bouyancy forces on these small crystals. Since the crystals adhere to solid surfaces when encountered, they tend to concentrate on the bottom. The term *frazil ice* is frequently used for the freely suspended crystal form; the component which adheres to the bottom is commonly called *anchor ice*. In both cases, the mechanisms of underwater ice formation is the same and was first described by Altberg (1936).

REFERENCES

ALTBERG, W. J.
    1936   Twenty years of work in the domain of underwater ice formation (1915–1935). *Int. Ass. Scient. Hydrol. (IASH) Bull.* 23: 373–407.

CHAPTER 21

The influence of ice on the primary productivity of the Bering Sea

C. PETER MCROY *and* JOHN J. GOERING

Institute of Marine Science, University of Alaska, Fairbanks, Alaska

Abstract

Support is presented for the hypothesis that the ice cover of the Bering Sea increases the total annual primary productivity of the sea. For most of November through May the waters over the continental shelf, about 45 percent of the total area, are covered by sea ice. On the undersurface of this ice is a community of algae and other micro-organisms that becomes well developed in late winter and early spring. For the period March through May these algae form a clearly visible layer on the undersurface of the sea ice and constitute a major standing stock of primary producers in the Bering Sea.

Productivity studies of the ice community were conducted in 1968, 1970, and 1971. The productivity of this ice community ranged from 2.2 to 4.8 mg C/m^2-day for standing stocks of 0.34 to 2.97 mg chlorophyll a/m^2. Production in the water under the ice varied from 15 to 21 mg C/m^2-day, and it was 89 mg C/m^2 in the water column at the ice front. From studies of nitrogen uptake, it appears that the nitrogen requirement for winter productivity is mainly derived from nitrate.

The annual cycle of primary production in the Bering Sea is a complex system that begins with the development of the algal community on the undersurface of the ice followed by a bloom at the ice front, and finally by the conventional spring bloom in the open water. As a result of this system the annual increase in primary production in the Bering Sea begins in the central and northern regions rather than in the southern waters.

INTRODUCTION

Primary production in a winter ocean covered to a large extent by ice would be expected to be low and uncomplicated. Our studies reveal, however, a complex productivity system occurring in the water column and ice that makes a measurable contribution to the total annual production of the Bering Sea.

For most of the period from November through May, the waters over the continental shelf are covered by sea ice (Fig. 21.1). During March and April, ice is at its maximum and occupies nearly 75 percent of the surface area of the Sea (Lisitsyn 1969). This is seasonal ice that is usually 1 to 2 m thick in unstressed floes and much thicker (possibly 10 m or more) in pressure ridges. In varying width around the coast is a belt of shore-fast ice that remains relatively stationary during the winter. Beyond this the open sea is covered by large and small floes, separated by leads and polynyi, that are driven, primarily by winds, in unceasing motion. An estimated 97 percent of this ice is formed locally, so the contribution from rivers and adjacent seas is small (Leonov 1960). Ice is heaviest and thickest in the western sea and the northern waters east of St. Lawrence Island. Breakup begins in March or April and lasts through June. According to Soviet data, 63 percent of the ice melts *in situ*; the remainder is transported through Bering or Kamchatka straits and the Aleutian passes (Lisitsyn 1969).

Phytoplankton standing stock and production in the water under the ice are low, in fact once reported to be zero (Semina 1955). As a result of our own recent studies we know now that these quantities are not zero but are indeed low (McRoy et al. 1972). These low values in the winter ocean appear to be largely the result of low light intensities caused by a combination of the winter sun, ice cover, and strong vertical mixing.

The mechanism for the annual spring increase of arctic phytoplankton has been discussed in theoretical terms by Marshall (1957). Marshall found that the spring bloom in the western Barents Sea occurred first in the shallow water over the continental shelf in the wake of the receding ice; a later bloom occurred farther south in the deeper North Atlantic water. Both blooms were accounted for by the interaction of the critical depth and the upper homogeneous layer, based on Sverdrup's theory (1953). Our studies presented here suggest that a similar condition exists in the Bering Sea. In addition, numerous reports indicate that a high standing stock of secondary and higher producers— birds, seals—congregate at the ice front of the Bering Sea (Irving et al. 1970; Burns 1970; Fay, Chapter 19 of this volume). These populations suggest the existence of a dynamic food chain closely related to the ice front.

The low standing stock of phytoplankton in the water column in winter prior to the spring increase is in marked contrast to the intense development of micro-algae on the undersurface of the sea ice. In a previous paper (McRoy et al. 1972) we documented the development of these ice algae in the Bering Sea. Micro-algae living in sea ice have been studied in detail in the Antarctic Ocean (Burkholder and Mandelli 1965; Bunt and Lee 1970) and in the Arctic Ocean (Apollonio 1961; Meguro et al. 1966, 1967). In the Antarctic, two ice habitats have been described: the "sherbet"-like layer between the year-old ice and snow cover, and the undersurface of the sea ice. Both are aquatic habitats, but the ecological conditions for algal growth in each are evidently quite different. In the Arctic Ocean only the "bottom-type" community has been reported, and this is the only one we have observed in the Bering Sea.

Measurements of micro-algae in arctic sea ice indicate growth of a dense community with a distinct seasonal cycle. Apollonio (1961) reported an average standing stock of 89.6 μg chlorophyll *a*/liter in June, and through experiment he found this concentration to be extremely light dependent. In more recent arctic work (Meguro et al. 1966, 1967) the micro-algae of the sea ice were found to be growing in a relatively constant-temperature (−3 to 0°C) environment with good conditions of nutrient concentration and supply and little or no zooplankton grazing.

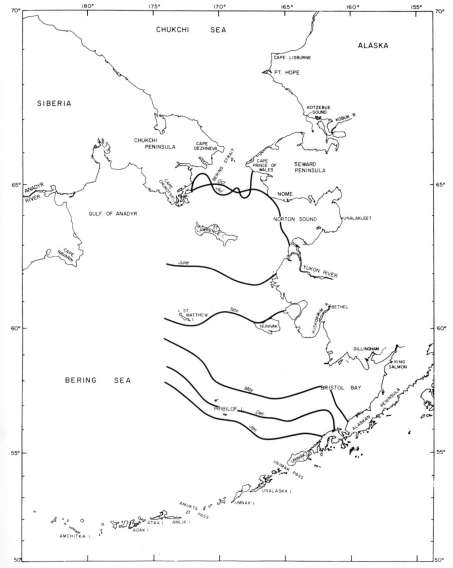

Fig. 21.1 Ice cover of the Bering Sea.

The question remains of the contribution of the sea ice community to the annual primary production. In the Antarctic, Burkholder and Mandelli (1965) report fixation of 0.19 g C/m²-day and spectulate that this could be an important contribution to the total. In the Arctic, Meguro et al. (1966, 1967) make a good case for the significance of ice algae production.

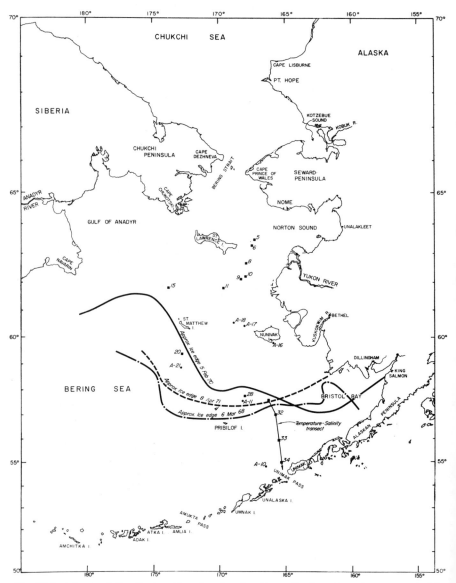

Fig. 21.2 Chart of the Bering Sea showing location of selected stations occupied in winter.

In this paper we report some of the results of three winter icebreaker cruises to the Bering Sea (Fig. 21.2). Over the years these cruises have given us the opportunity to view conditions in February, March, and April and thus put together a view of the development of primary productivity and associated nutrient cycling as the winter ocean transformed to spring seas. The results reported here, with little exception, are the only available data on the influence of ice and winter conditions on the primary production of the Bering Sea.

Methods

Primary productivity was measured using radioactive carbon, ^{14}C—HCO_3^-, in seawater samples collected with a large volume plexiglas sampler (Strickland and Parsons 1968). Water samples enclosed in light and dark bottles were incubated for six to twelve hours in an insulated box held at 0 to -1 C under natural light and neutral density filters to simulate light conditions of the environment. Hourly productivities were converted to daily values by multiplying by hours of daylight (from one hour after sunrise to one hour before sunset). Phytoplankton standing stocks are expressed as units of chlorophyll a determined by standard methods (Strickland and Parson 1968). Water samples were filtered immediately after collection through glass filters (Whitman GFC), frozen, and stored for several months before analysis.

In addition to carbon uptake, simultaneous measurements were made of nitrogen uptake using ^{15}N-labeled compounds (Neese et al. 1962; Dugdale and Goering 1967). Water samples were treated in the same manner as the carbon uptake measurements except that incubation periods were 24 hours. The amount of ^{15}N incorporated into a sample was determined by mass spectrometry.

Inorganic nutrient concentrations of the water were measured using standard automated methods (Strickland and Parsons 1968). Hydrographic measurements of temperature and salinity were made using reversing thermometers and an induction salinometer on water samples collected with Nansen bottles.

Hydrography

In late winter and early spring the ice cover has reached its maximum extent, coinciding roughly with the seaward edge of the continental shelf (Fig. 21.1). Approaching the ice from the south there is a continuous decline in temperature and salinity from about 1.5 C and 32.3‰ near the Aleutian Islands to -1.5 C and 31.7‰ in the ice (Fig. 21.3).

The winter ocean is vertically well mixed. Generally water temperatures under the ice range from -1.5 to -1.8 C with less than 0.1 C difference in surface and bottom values. Salinity values follow a similar trend with somewhat more variation (Fig. 21.4). At the stations in the open water and in the ice front, there were nearly uniform salinity profiles in the upper 50 m. The profiles show a low-salinity zone around the ice front with increasing salinity both in the open water south of the ice and in the ice-covered regions farther north. This low-salinity zone at the ice front must develop in early spring as melting begins. In early April, although the ice was close to its maximum extent, the

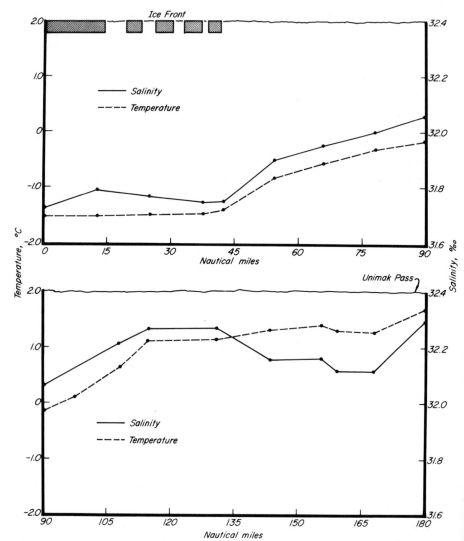

Fig. 21.3 Temperature and salinity transect at 10 m from the ice front to Unimak Pass, 20–21 April 1971.

low-salinity water at the ice margin was clearly evident. This zone of low-salinity water is expected to be more pronounced and probably forms or moves north with the receding ice front as spring develops.

Water column productivity
The standing stock and productivity of phytoplankton in the water under the ice and in the open water south of the ice attained the low values characteristic

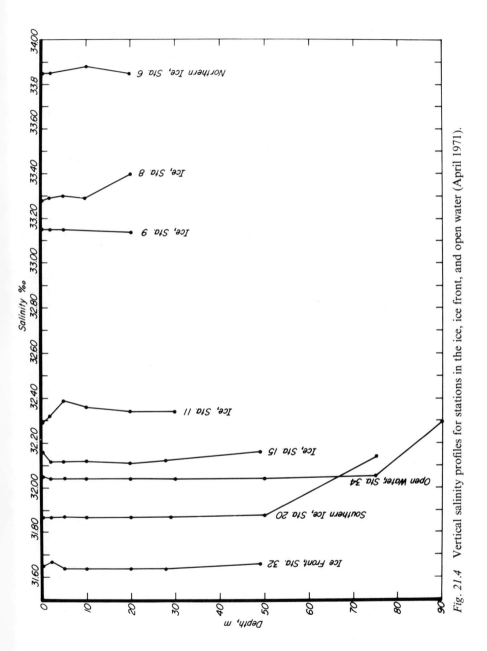

Fig. 21.4 Vertical salinity profiles for stations in the ice, ice front, and open water (April 1971).

of winter (Figs. 21.5 and 21.6). In contrast to this was the zone around the edge of the ice where much higher standing stocks and productivities occurred. Surface water south of the ice and under the ice had similar standing stocks that averaged about 0.2 to 0.9 mg chlorophyll a/m^3; at the ice front, this rose to 4.0 mg chlorophyll a/m^3 (Table 21.1). For the total water column, the water under the ice contained the least phytoplankton, averaging 3 to 9 mg chlorophyll a/m^2. In the one station in open water south of the ice, the concentration was low but the water column was deeper and resulted in a standing stock of 22.55 mg chlorophyll a/m^2. At the ice edge the water contained by far the highest amount of phytoplankton; the standing stock for three stations averaged 168 mg chlorophyll a/m^2. This high value was the result of increased concentration as well as relatively deep water.

The winter primary productivity data do not represent extensive measurements; however, a general view of rates is possible. The primary productivity of the surface water reflects a pattern similar to the standing stocks. The lowest rates occurred in the open sea south of the ice, with slightly higher values in the water under the ice, and the highest values were found at the ice front (Fig. 21.6 and Table 21.1). For the surface water under the ice, where there are enough stations to permit seasonal comparison, there seems to be little difference between the February, March, and April data, even though the measurements were made in different years and under varying light conditions. Apparently the water under the ice remains unproductive until ice break-up.

The primary productivity of the total water column was greatest at the ice front. Although we have only one station in this zone, the observed rate of 89.28 mg C/m^2-day is considerably greater than the 15 to 20 mg C/m^2-day in the water under the ice and probably equally greater than the open sea.

The geographical pattern of standing stocks and productivity in the ice-covered Bering Sea appears to be related to hydrographic conditions. The highest rates and standing stocks coincided with the zone of low-salinity at the ice front. Apparently this zone is where the spring increase in phytoplankton begins. This bloom is expected to follow the wake of the receding ice. Presumably a second bloom would occur in the ice-free waters in early summer after the development of the seasonal thermocline. This type of dual-component spring bloom has previously been described and theoretically accounted for by Marshall (1957) for the western Barents Sea. Using the relationships of critical depth and the depth of mixing (Sverdrup 1953), Marshall showed that the melting ice combined with the shallow water results in increased stability, and consequently the depth of the mixed layer is less than the critical depth, permitting an increase in phytoplankton. The phytoplankton bloom developed later in the more southern open sea because of the later development of stability; i.e., the mixed layer exceeded the critical depth until later in the year. This situation described for the western Barents Sea seems applicable to our observations in the Bering Sea as supported by hydrographic data and phytoplankton productivity and standing stock. In addition, the nutrient data presented in the next section support this scheme. A third component, discussed in a following section, is the production of the algae on the undersurface of the ice.

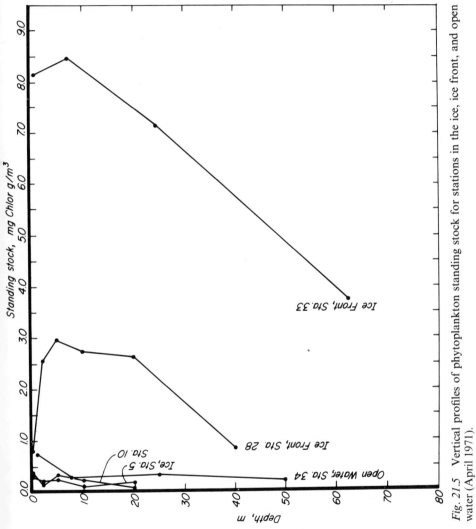

Fig. 21.5 Vertical profiles of phytoplankton standing stock for stations in the ice, ice front, and open water (April 1971).

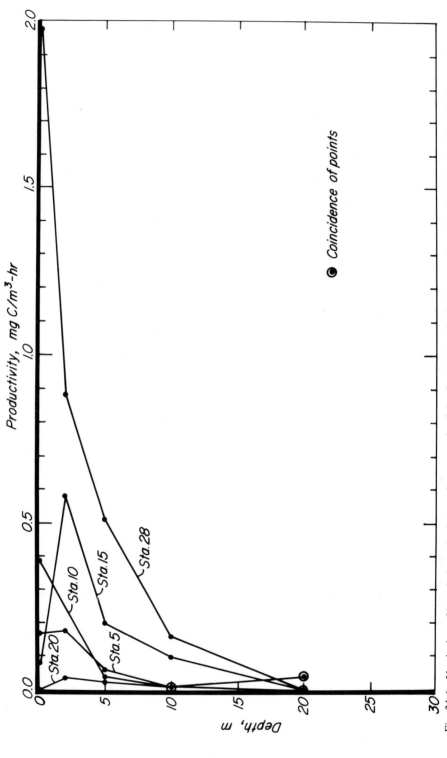

Fig. 21.6 Vertical profiles of phytoplankton primary productivity for stations in the ice, ice front, and open water (April 1971). Refer to Fig. 21.2 for location of stations.

TABLE 21.1 Primary productivity and standing stock of phytoplankton and sea-ice algae in the Bering Sea in winter (average values)

Region	Date	Primary productivity (carbon)					Standing stock (chlor a)		
		No. stations	Surface (0-m) (mg/m³)		Water column (mg/m²)		No. stations	Surface	Water column
			hourly	daily	hourly	daily		mg/m³	mg/m²
Open water south of ice	March 68	1	0.07	0.55			1	0.92	
	April 71	1					1	0.72	22.55
Water at ice front	April 71	1	1.98	23.76	7.44	89.28	3	4.03	167.60
Water under ice cover	February 70	8	0.26	1.54	3.48	20.86	8	0.44	8.75
	March 68	10	0.14	0.94			7	0.52	
	April 71	15	0.29	3.20	1.23	14.95	15	0.23	3.00
Sea ice	February 70	0	No apparent brown layer in ice				2	6.83	0.34
	March 68	2	5.55	44.40	0.28	2.22	5	59.49	2.97
	April 71	1	7.95	95.40*	0.40	4.77			

*Rate for 10% light intensity

Nitrogen uptake and under-ice productivity

Marine primary productivity can be measured by a number of techniques, the ^{14}C method of Steeman-Nielsen (1952) being perhaps the most popular due to its great sensitivity. Conventional measurements of primary production (e.g., the ^{14}C and chlorophyll a techniques) alone, however, do not reveal the capacity of a region to support production at higher levels in the food-chain. Productivity based on utilization of the different forms of inorganic nitrogen can add a new dimension to knowledge gained from the approach of carbon fixation. The ^{15}N technique of measuring productivity (Dugdale and Goering 1967) makes it possible to separate the fractions of primary productivity corresponding to new nitrogen (nitrogen advected in nitrate and molecular nitrogen) and regenerated nitrogen (ammonia) in the euphotic zone, thereby indicating the nutrient source responsible for phytoplankton growth.

A survey of ^{15}N-labeled ammonia and nitrate uptake in winter by micro-algae in sea ice and by phytoplankton in the surface waters of the Bering Sea was conducted in 1968 during phase A of the R/V *Alpha Helix* Bering Sea Expedition. Nitrogen uptake rates obtained from selected surface stations are given in Table 21.2. The percentage nitrate uptake in general was large, ranging from 22.2 to 90.9 for the phytoplankton. Sea ice algae likewise exhibited a high percentage nitrate uptake. Nitrate and ammonia utilization by phytoplankton was undetectable at station A-10, located in open water south of the ice front. The lowest rates of ^{14}C uptake also occurred in the open sea south of the ice, with slightly higher values in the water under the ice, and the highest values at the ice front. When present, nitrate appears to be the major source of inorganic nitrogen for winter algae growth in the Bering Sea.

Winter distribution of nitrate

In deep oceans, and in shallower seas during winter, nearly all of the combined inorganic nitrogen is present as nitrate, a generalization appearing to hold for the Bering Sea. Deepwater concentrations of nitrates are over 30 μg-atoms N/liter. Winter surface concentrations, in general, are somewhat lower (10–25 μg-atoms N/liter) and are drastically influenced by phytoplankton growth as discussed below. During summer, nitrate in some regions approaches unde-tectable levels, and nitrogen may limit phytoplankton productivity.

Winter nitrate concentrations in the water column of five eastern Bering Sea stations are depicted in Figure 21.7. Productivity and ice conditions at these stations were discussed previously. The extensive utilization of nitrate as a source of nitrogen for algae growth drastically alters the winter as well as the summer concentration of this nutrient in the euphotic zone. The use of nitrate to support the high productivity occurring at ice-front stations (28 and 33) is shown in the depth distribution of nitrate. Nitrate is much lower in the euphotic zone than in water below. In comparison, nitrate concentrations in the water column at stations in the ice (5 and 10) and in the open water (34), where primary productivity is low, are nearly uniform with depth. The profiles of nitrate, therefore, tend to support our ^{15}N uptake studies, which showed nitrate as the source of nitrogen for winter phytoplankton growth in the Bering Sea.

TABLE 21.2 Winter nitrogen uptake and productivity in the ice-covered Bering Sea at six stations in March 1968 (R/V *Alpha Helix* Bering Sea Expedition, phase A)

Station	Depth (m)	% Incident light	Nitrogen concentrations (μg-atoms/liter)		Nitrogen uptake (μg-atoms/liter-hr)		% Nitrate uptake*	Carbon uptake (μg-atoms/liter-hr)	Ratio C:N uptake*
			(NO_3^-)	(NH_4^+)	NO_3^-	NH_4^+			
A-10	0	100	11.4	2.4	0.0000	0.0000	—	0.006	—
A-11	0	100	10.3	2.4	0.0040	0.0040	50.0	0.013	16.3
A-16	0	100	0.9	0.3	0.0002	0.0001	66.7	0.011	36.7
A-17	0	100	1.2	2.6	0.0035	0.0099	26.1	0.016	1.2
A-18	0	100	2.2	2.9	0.0020	0.0007	22.2	0.002	2.2
A-21	0	100	11.8	0.7	0.0010	0.0001	90.9	0.025	22.7
Sea Ice 14	—	—	3.7	2.0	0.0042	0.0014	75.0	0.440	78.5

*% Nitrate uptake $= \dfrac{^{15}N - NO_3^- \text{ uptake}}{^{15}N - NH_4^+ \text{ uptake} + ^{15}N - NO_3^- \text{ uptake}} \times 100$

**NO_3^- uptake $+ NH_4^+$ uptake

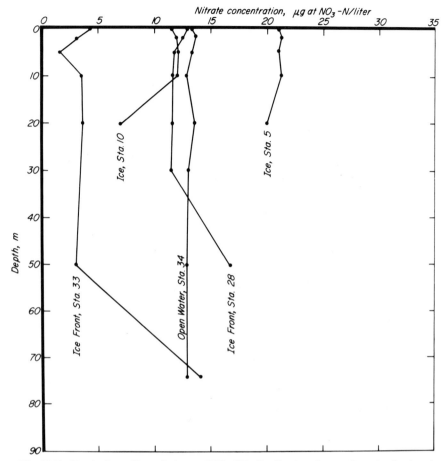

Fig. 21.7 Vertical profiles of nitrate (NO₃-N) distribution for stations in the ice, ice front, and open water (April 1971).

We have observed a rapid nitrate decline in the water near the bottom at a few ice-covered shallow shelf stations occupied during the late winter (e.g., stations 5 and 10, Figure 21.7), mostly near the Yukon-Kuskokwim Delta. Our observations are limited in number and geographic coverage, but it does appear that the sediments in this region are a sink for nitrate. One possible explanation is that reducing conditions exist in these sediments even when the overlying waters contain dissolved oxygen. Reducing conditions would allow bacterial nitrate reduction and denitrification to occur near the sediment-water interface. The rate of supply of nitrate to the sediments is probably controlled by vertical eddy diffusion. Sediments are known to be active sites of denitrification in other marine areas (Goering and Pamatmat 1971).

Peculiar oxygen conditions occur in the Bering Sea shelf regions during the winter when the sea is covered by ice (Lisitsyn 1969). The water in these regions is almost completely isolated from its two main sources of oxygen, the atmosphere and photosynthesis by phytoplankton. Correspondingly, the respiration of the benthos and the decay of organic matter in suspension and in the bottom sediments has a large oxygen demand. If replenishment of oxygen by circulation is restricted, then low oxygen conditions might prevail near the sediment-water interface. Lisitsyn (1969) reports that in late winter, oxygen concentrations in the water can fall to 20–30 percent of saturations in the ice-covered Yukon-Kuskokwim region. The high oxygen demand caused by the high organic load of these waters coupled with the limited supply of oxygen results in the observed low oxygen concentrations. In certain years some depressions can apparently even go anoxic, and hydrogen sulfide is produced.

Sea ice productivity

From late winter through break-up, the undersurface of the sea ice is covered by a dense community of micro-algae and other organisms. This community is generally restricted to the lower few centimeters of ice and exhibits a marked seasonal cycle. From our three cruises we have a fragmentary picture of the development of the ice community in late winter. In late January and February (1970), winter storms were still very much present and at this time the brown layer on the undersurface of the ice characterizing the ice algae community was not visible. No quantitative samples were taken. In March (1968) the lower ice surface was well colored over large portions of the ice-covered sea. Standing stock at this time averaged 6.83 mg chlorophyll a/m^3 or, on an areal basis, this is reduced to 0.34 mg chlorophyll a/m^2, since the community is restricted to the lower approximately 5 cm of ice (Table 21.1). During April (1971) the brown layer of ice was again distinct over wide areas, and the standing stock averaged 59.49 mg chlorophyll a/m^3 or 2.97 mg chlorophyll a/m^2, a quantity equal to the phytoplankton in the 30 m under-ice water column.

The productivity of the ice community is intense and strongly dependent on ambient light (Fig. 21.8). In late winter, when the sea ice is about 1 m thick and covered by a few to perhaps 50 centimeters of snow, the algal community is growing in light intensities ranging from less than 1 to 10 percent of surface intensity, depending on snow and ice thickness. With the decline of winter, solar radiation increases, the snow cover melts and the ice becomes more transparent to visible radiation; these conditions would increase the productivity of the ice community. From our observations, we estimate the productivity for the March–April period to range from 44.4 to 95.4 mg C/m^3 per day within the ice community layer or, on an areal basis, 2.2 to 4.8 mg C/m^2 per day (Table 21.1). These rates are 30 percent or less than those of the water column at the time.

Unfortunately, it has not yet been possible to follow completely the development of the sea-ice community from winter through spring, but evidently this community has a marked seasonal cycle and is not continuously present throughout winter. Rather, the ice community develops in late winter and increases until the ice disappears. The melting ice pack would promote the growth of the ice community with maximum standing stock being reached just prior to the disappearance of the ice. Supporting this, long strands of algae

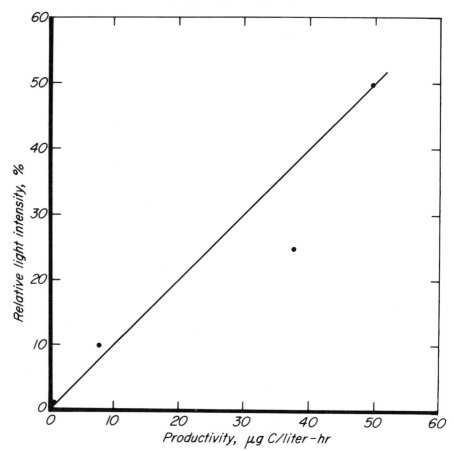

Fig. 21.8 Productivity of sea ice community in relation to relative light intensity.

were reported in May attached under the ice near St. Lawrence Island (C. Ray, personal communication).

Primary production cycles in the Bering Sea

The results of these studies reveal a complex productivity system that in part explains the high production of consumer trophic levels for which the Bering Sea is known. We realize that the data are not complete; nonetheless it is possible to outline the general scheme of primary production for the Bering Sea. It is tempting to place estimates of integrated rates of production on this scheme, but this is premature.

Unlike the North Pacific Ocean and Chukchi Sea, the annual increase in production in the Bering Sea begins in late February with the development of the algal community in the sea ice. The production of this community increases with the passing of winter and probably reaches a maximum just before the ice melts completely. The contribution of this community to the total annual production has not yet been assessed. The algae of the ice comprise the first spring bloom that occurs in the Bering Sea, preceding the bloom that occurs in the open water farther south.

In April, as the ice melts, a second spring bloom develops in the wake of the receding ice. This begins along the southern ice front, coinciding approximately with the edge of the continental shelf. This bloom is promoted by the stability associated with the low-density water around the melting ice.

The conventional spring bloom occurs in the open waters with the formation of the seasonal thermocline, first in the Aleutian Islands and then moving north with spring. In northern waters around the Bering Strait, the bloom occurs in summer and reaches the highest productivity known for the Bering Sea.

As a result of the seasonal ice cover, the annual primary production of the Bering Sea is actually increased. Furthermore, the annual spring increase in algal standing stock begins in the middle and northern Bering Sea rather than the expected southern waters.

Discussion

KASAHARA: In what manner are the algae on the undersurface of the ice utilized by algaevores?

McROY: From dives under the ice at the arctic ice station T-3, Dr. English of the University of Washington reported certain arthropods grazing directly on the undersurface of the ice. Also, animals have been observed just under the ice when a lead breaks open or other sudden exposure to the underside of the ice occurs.

WILIMOVSKY: While diving at the edge of the sea ice northwest of Point Barrow, I have observed two codfish species (*Boreogadus saida* and probably *Arctogadus glacialis*) feeding extensively on the underlayer of the ice. Similar findings for the first named species are reported by Bogorov.

HATTORI: Are species of ice algae quite different from planktonic varieties; what is their summer habitat?

McROY: Ice algae are generally not the same species found in the water column but are characterized by many pennate-type diatoms. Work by others at the Institute of Marine Science, University of Alaska, suggests that ice algae are benthic dwellers in the summer.

HATTORI: How do you explain the relatively high primary production along the edge of the ice?

McROY: The high ice-edge productivity is associated with low-salinity water formed by the melting ice. Such a condition was described for the western Barents Sea by Marshall (1958), who postulated that the stratification resulting from low-salinity water permits a bloom to develop at the ice edge. This case seems to apply to the Bering Sea.

Acknowledgments

Appreciation is expressed for icebreaker time provided by the U. S. Coast Guard and for research support by the Oceanography Section, National Science Foundation through NSF Grant GA 33387.

REFERENCES

APOLLONIO, S.
  1961 The chlorophyll content of arctic sea ice. *Arctic* 14: 197–199.

BUNT, J. S., and C. C. LEE
  1970 Seasonal primary production in antarctic sea ice at McMurdo Sound in 1967. *J. Mar. Res.* 23(8): 304–320.

BURKHOLDER, P. R., and E. F. MANDELLI
  1965 Productivity of micro-algae in antarctic sea ice. *Science* 149: 872–874.

BURNS, J. J.
  1970 Remarks on the distribution and natural history of pagophilic pinnipeds in the Bering and Chukchi Seas. *J. Mammal.* 51(3): 445–454.

DUGDALE, R. C., and J. J. GOERING
  1967 Uptake of new and regenerated forms of nitrogen in primary productivity. *Limnol. Oceanogr.* 12: 196–206.

GOERING, J. J., and M. PAMATMAT
  1971 Denitrification in sediments of the sea off Peru. *Investigacion Pesquera* 35(1): 233–242.

IRVING, L., C. P. McRoy, and J. J. BURNS
  1970 Birds observed during a cruise in the ice-covered Bering Sea in March 1968. *Condor* 72(1): 110–112.

LEONOV, A. G.
  1960 Regional oceanography, Part 1 [in Russian]. *Gidrometeoizdat*, Leningrad, 765 pp. (Transl. avail. Nat. Tech. Inf. Serv., Springfield, Va., AD 627508 and AD 689680.

LISITSYN, A. P.
  1969 *Recent sedimentation in the Bering Sea.* IPST Press, Jerusalem, 614 pp.

MARSHALL, P. T.
  1957 Primary production in the arctic. *J. Conseil.* 23(2): 173–177.

McRoy, C. P., J. J. GOERING, and W. S. SHIELS
  1972 Studies of primary productivity in the eastern Bering Sea. In *Biological oceanography of the northern North Pacific Ocean* [Motoda commemorative volume], edited by A. Y. Takenouti et al. Idemitsu Shoten, Tokyo, pp. 199–216.

MEGURO, H., K. ITO, and H. FUKUSHIMA
  1966   Diatoms and the ecological conditions of their growth in sea ice in the Arctic Ocean. *Science* 152: 1089–1090.
  1967   Ice flora (bottom type): a mechanism of primary production in polar seas and the growth of diatoms in sea ice. *Arctic* 20(2): 114–133.

NEESE, J. C., R. C. DUGDALE, V. A. DUGDALE, and J. J. GOERING
  1962   Nitrogen metabolism in lakes. 1. Measurement of nitrogen fixation with ^{15}N. *Limnol. Oceanogr.* 7: 163–169.

SEMINA, G. T.
  1955   The vertical distribution of phytoplankton in the Bering Sea [in Russian]. *Doklady Akad. Nauk SSSR* 101(5).

STEEMAN-NIELSEN, E.
  1952   The use of radioactive carbon (C^{14}) for measuring organic production in the sea. *J. Cons. Explor. Mer.* 18: 117–140.

STRICKLAND, J. D. H., and T. R. PARSONS
  1968   A practical handbook of seawater analysis. *Fish. Res. Bd. Can. Bull. 167*, 311 pp.

SVERDRUP, H.
  1953   On conditions for vernal blooming of phytoplankton. *J. Conseil.* 18(3): 287–295.

Part 6

METEOROLOGICAL PROCESSES

CHAPTER 22

The relationship between ice and weather conditions in the eastern Bering Sea

R. KONISHI *and* MITSURU SAITO

Nippon Suisan Kaisha, Ltd., Tokyo, Japan

Abstract

Results of a 12-year study during (1960–71) show that ice in the Bering Sea is influenced by changes occurring in weather conditions and sea currents in cycles of two years. Conversely, the distribution of ice exerts an influence on the course of atmospheric pressure and temperature, which were noted to follow a similar cyclic pattern.

The drift ice in 1971 was positioned farther south than in usual years and remained in the waters west of the Pribilof Islands until 10 June, an anomaly attributable to unusually persistent northerly winds during March to May. In contrast to this in 1967, there was a predominance of winds from the south in May, and the ice disappeared earlier than in usual years.

A stagnant weak front and weak low frequently exist along the ice edge. Low pressure flowing northward into the Bering Sea from south of the Aleutian Islands becomes stationary in the vicinity of drift ice and dissipates. On the other hand, a front moving southward with accompanying low pressure gains in force over the warm sea surface in the vicinity of the ice edge. As a result, a stagnant frontal zone is always found at the southern edge of the sea ice, and the low is seen to move along the frontal zone. Moreover, the occluding low will at times move westward. In this way, sea ice may also have the effect of blocking low pressure.

INTRODUCTION

Since the 1970–71 antarctic whaling season, the Meteorological Section of Nippon Suisan Kaisha, Ltd., has intercepted satellite pictures with an APT (Automatic Picture Taking) receiver equipped aboard their factory whaling ship and utilizes the pictures as supplemental charts for their weather analysis. During the period May to June 1971, the sea ice in the eastern Bering Sea was studied. Six comparatively clear pictures taken by NOAA-1 during a cloudless period 13–19 May 1971 are shown in Figures 22.1–22.3. Together with weather observation data gathered by ships and broadcast four times daily by teletype, daily sketches of sea ice were made from the satellite pictures. Three-day composite ice charts for this period were then compiled from the daily ice charts (Fig. 22.4). An ice crevice found in making the charts was treated as a void, and daily wind or current drift of the ice area was disregarded. By comparing the composite ice charts, one can trace the condition of melting sea ice.

Relationship between ice and atmospheric pressure

As shown in Figures 22.5 and 22.6, the Bering Sea drift ice in 1971 was positioned farther south than usual and remained in the waters west of the Pribilof Islands until around 10 June. This was attributable to unusually persistent northerly winds during March to May. This movement of the drift ice was due not only to the winds, however, but also to the water currents; moreover, the distribution of the ice influenced the course of the pressure depression.

Figure 22.7 gives a comparison of the locations of the ice edge over a three-day period in May 1971. The ice in the bay east of St. Paul Island moved from 17–18 May to expand in the east-west direction. On 19 May, however, the wind added its influence, and the ice around the bay was generally set adrift in a southwesterly direction. The ice on the south side of the bay had been thawing. The situation of the wind during this period was northeasterly 10–15 knots on 17 May and northwesterly 25–30 knots on 18 and 19 May. The thawing of the southern part of the ice edge and the formation of the bay were due presumably to the influence of the warm current that had flowed northward into the Bering Sea from south of the Aleutian Islands.

A stagnant weak front and low frequently exist along the edges of the ice. Low pressure flowing northward into the Bering Sea from south of the Aleutians becomes stationary in the vicinity of the drift ice and dissipates. Figure 22.8 indicates the average ice limits from 13–15 May 1971 and the tracks of the low center from 7–19 May 1971. Apparently the center of the low atmospheric pressure had hardly entered the area before ice appeared on the sea. The low then stagnated in the area south of St. Paul Island, where the sea-surface temperature was warm. In contrast to this, a front moving southward from the north with attendant low pressure gains in force over the warm sea surface along the ice edge. As a result, a stagnant frontal zone is always found at the southern edge of the sea ice, and the low is seen to move along the frontal zone and occasionally westward. In this way, the sea ice can exert the effect of also blocking *low* as well as high pressure.

Ice limits and sea-surface temperature

Figure 22.9 shows the deviation for five-day average sea-surface temperature (°C) from 11–15 May in 1971 and in 1970. A warm area is indicated on the sea southeast of St. Paul Island and a cold area on the sea northwest of the

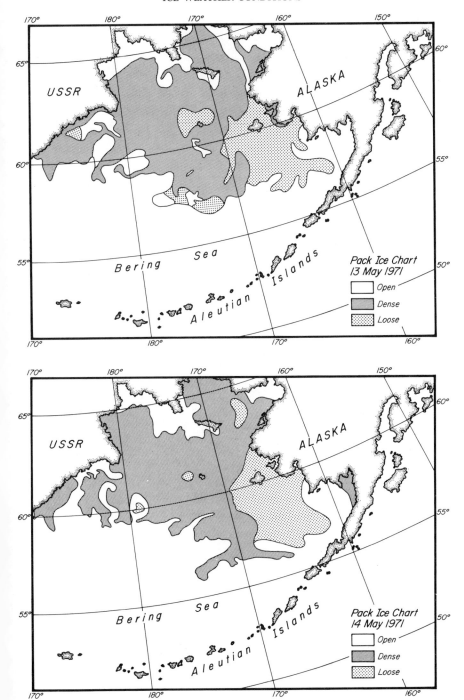

Fig. 22.1 Daily ice charts for the Bering Sea, 13 May (upper) and 14 May 1971 (lower).

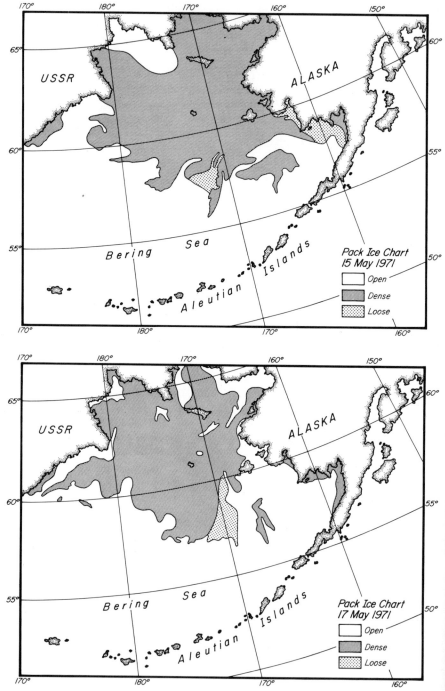

Fig. 22.2   Daily ice charts for the Bering Sea, 15 May (upper) and 17 May 1971 (lower).

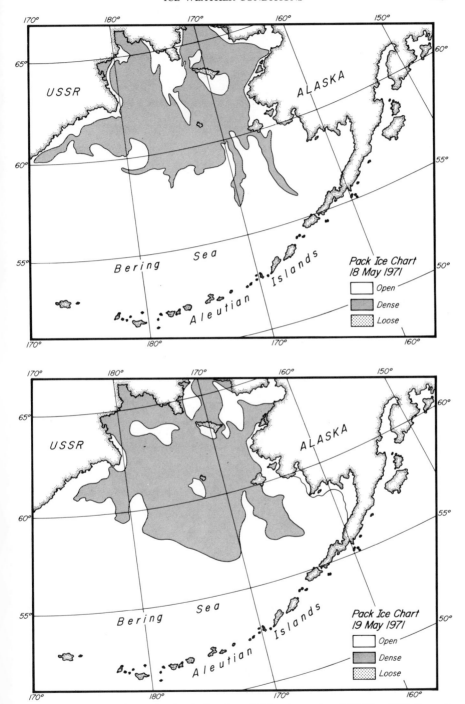

Fig. 22.3 Daily ice charts for the Bering Sea, 18 May (upper) and 19 May 1971 (lower).

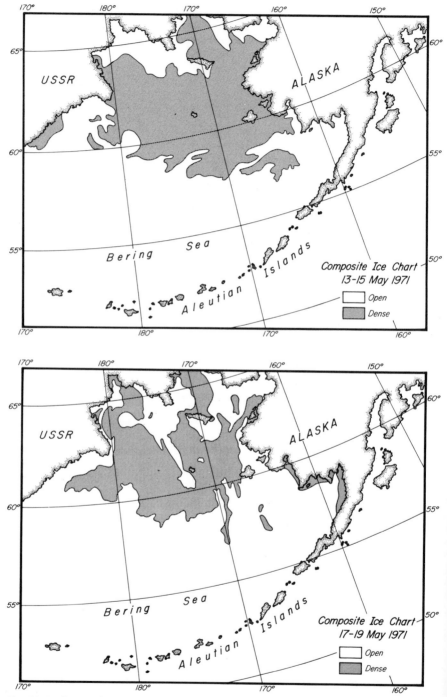

Fig. 22.4 Composite ice charts for the Bering Sea, 13–15 May (upper) and 17–19 May 1971 (lower).

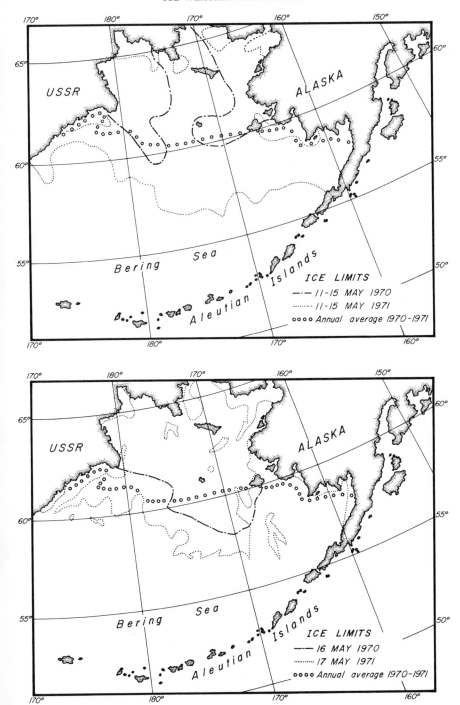

Fig. 22.5 Bering Sea ice limits for periods 11–15 May 1970 and 1971 (upper) and during 16 May 1970 and 17 May 1971 (lower), together with annual average for 1970–71.

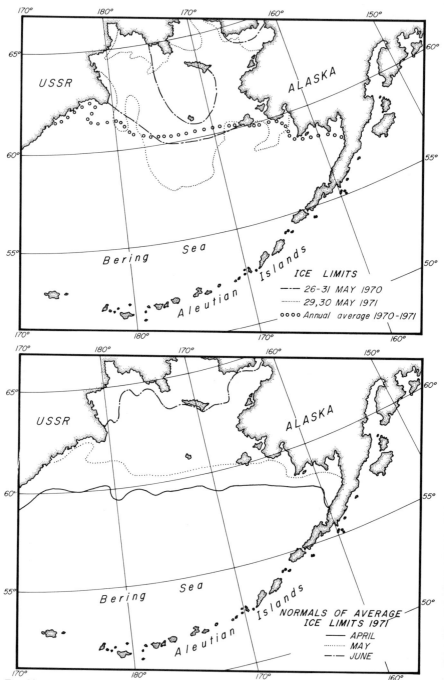

Fig. 22.6   Bering Sea ice limits for periods 26–31 May 1970 and 29, 30 May 1971 with annual average 1970–71 (upper); normals of average ice limits in April, May, and June 1971 (lower).

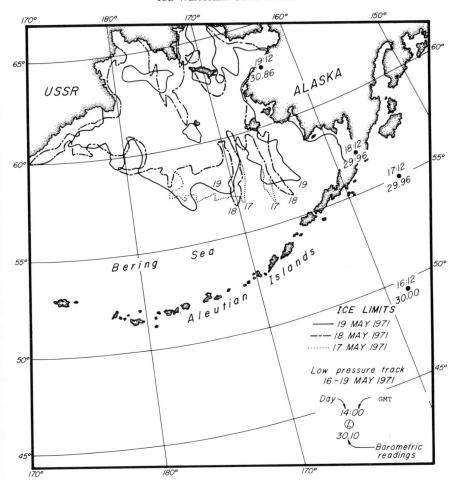

Fig. 22.7 Locations of ice edge on 17, 18, and 19 May 1971 and low pressure track during 16–19 May 1971.

island; the low pressure stagnated on the south side of the island and developed slightly.

According to the average sea-ice limits for 21–25 May and low center tracks for 18–28 May 1971 (Fig. 22.10), as compared with the first half of May, the frontal zone moved northward slightly, and the ice limits receded north accordingly. During this period, as well as in the first half of May, the low pressure did not enter the sea-ice area. Also, there was an area where the low pressure developed slightly and became stagnant in the central part of the Bering Sea. Here, the sea-surface temperature had become warm, presumably by influence of the ocean current.

In Figure 22.11 the difference is shown between the average sea-surface temperature during 16–20 May 1971 and the corresponding period of the previous year. It can be seen that in 1971 the surface temperature was low in

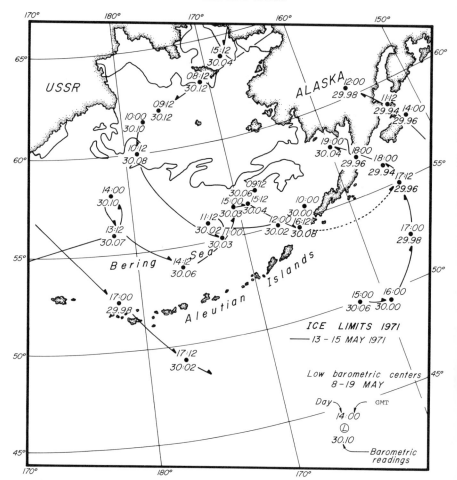

Fig. 22.8 Average ice limits in Bering Sea from 13–15 May 1971 and tracks of low barometric center during 8–19 May 1971.

the area west of St. Paul Island, high on its south side, and the adjacent low center developed more than was noted in 1970. As a result, the ice west of St. Paul Island tended to move south, while the ocean current to the east moved northward with great force, partly from the additional influence of the wind from the south.

Figure 22.12 indicates the difference in the average sea-surface temperature between 21–25 May 1971 and the corresponding period in 1970, together with the 1971 average ice limits and low-center tracks for 21–25 May. The sea-surface temperature was warm in certain areas south of St. Paul Island and in the western Bering Sea. In these areas the low pressure tended to turn in the counterclockwise direction and to develop slightly or become stagnant.

The average ice limits for 26–31 May 1971 and the tracks of the low center from 28 May–7 June 1971 are shown in Figure 22.13. The low center which had developed and moved north developed further in Bristol Bay, but it began to weaken upon reaching the leading part of the ice edge. Also, the low center which arrived from the north advanced westward along the south part of the ice where the surface temperature was warm. This may be considered to be a phenomenon forecasting the strong development of blocking high pressure in the east.

The tracks of the low center from 7–17 June 1971 are shown in Figure 22.14. The Bering Sea was overcast by the ridge of the atmospheric pressure, and an independent high pressure appeared. The ice remained south in the waters west of St. Paul Island until about 10 June. The frontal zone shifted further north than it had stood at the end of May. In mid-June, therefore,

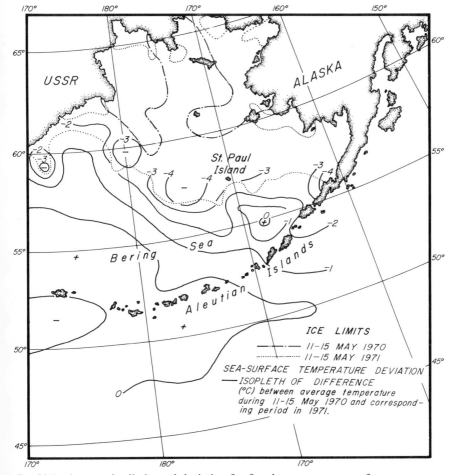

Fig. 22.9 Average ice limits and deviation for five-day average sea-surface temperature (°C) for 11–15 May 1971 and for corresponding period in 1970.

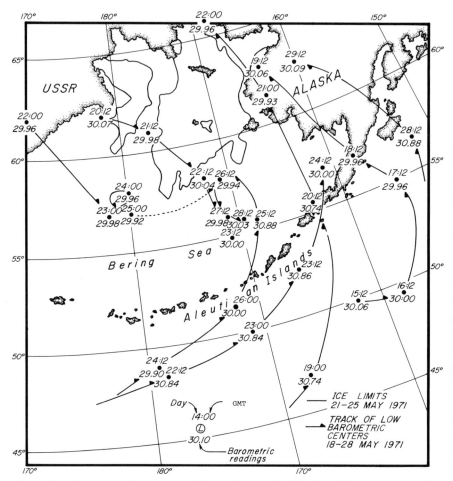

Fig. 22.10 Average ice limits in the Bering Sea for 21–25 May 1971 and track of low barometric centers for 18–28 May 1971.

the low center advanced eastward over the Bering Sea but deteriorated north of Bristol Bay. Apparently ice remained predominant north of this low center track, but detailed information is unavailable due to cloud interception with the frontal zone.

Phenomena of abnormally low water temperature and delay in sea ice melting time during 1971

In June 1971, the Bering Sea began to show a tendency to blocking, and an independent high pressure system was formed by a ridge which advanced westward from the direction of Alaska. The date of appearance and the intensity of this high pressure vary from year to year and are connected with the delay or advance in seasonal advent.

According to the charts of the mean monthly surface pressure from January to April of 1971, the condition was similar to that in 1959 and the polar high atmospheric pressure so unusually strong that its spreading portion covered the northern and western Bering Sea. Meanwhile, the low pressure area which had overlain the Gulf of Alaska advanced westward gradually with the oncoming season and entered Bristol Bay in May. Consequently, the track which the low pressure pattern followed on the daily chart advanced northward and moved gradually west. The location of the average low for each month was generally typical of past years, although there was more active movement than usual that resulted in a prevailing northerly wind east of the Bering Sea from March to May.

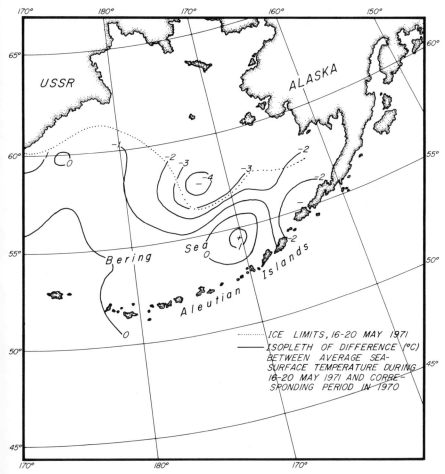

Fig. 22.11 Average ice limits in the Bering Sea for 16–20 May 1971 and difference between average sea-surface temperature during that period and the corresponding period of the previous year.

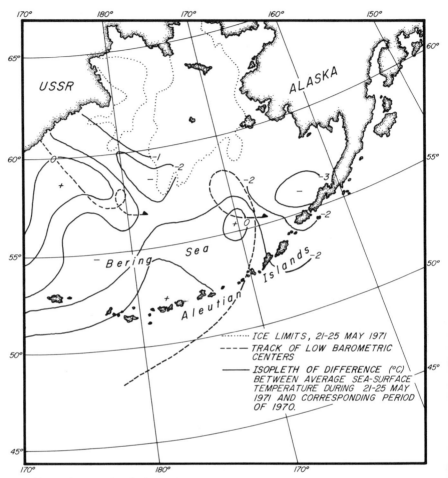

Fig. 22.12 Average ice limits in the Bering Sea and track of low barometric centers for 21–25 May 1971; difference in average sea-surface temperature between that period and the corresponding period of the previous year.

In June the Aleutian low pressure system moved to the western Bering Sea and became weakened. Meanwhile, a ridge extended southward from Alaska, merged with the high pressure area over the northern Pacific, gradually grew in intensity, and advanced slowly westward. Thus, in the eastern Bering Sea the southerly wind began to increase from the middle of the month, the ambient and surface temperatures rose gradually, and the ice west of St. Paul Island vanished.

Comparative pattern of ice and weather conditions during 1967–71
During 1967 to 1971 in the Bering Sea, the ice lay most southerly in 1971 and most northerly in 1967. In the following order, the ice limit extended increasingly southward: 1967, 1969, 1968, 1970, and 1971. From 1967 through 1970

the ice extension and recession tendencies alternated every two years, but in 1971 the ice lay anomalously south.

A comparison of the mean monthly surface chart for May 1971 (Fig. 22.15) with that for the same month in 1967 (Fig. 22.16) reveals that the low pressure site that was evident in the Gulf of Alaska during March and April 1971 moved westward to Bristol Bay. This shift caused a northerly wind to prevail from April to May 1971. In May 1967 the Aleutian low originating in the western Bering Sea in March and April moved more easterly than in usual years, extended south to Kodiak and was weaker than normal. Also, the ridge was situated more easterly than usual and covered the Bering Sea, resulting in the prevalence of a southerly wind from March to April 1967. Both the ambient and surface temperatures were high, and the sea ice extended

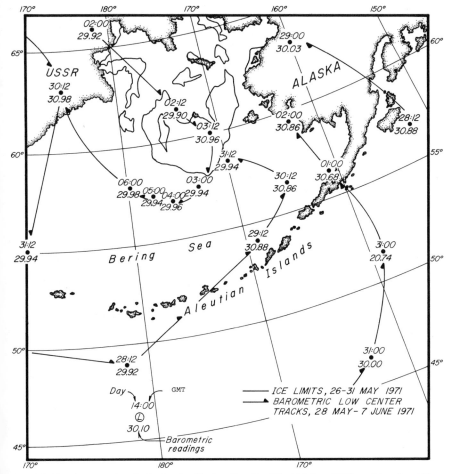

Fig. 22.13 Average ice limits in the Bering Sea for 26–31 May 1971 and tracks of low barometric centers for 28 May–7 June 1971.

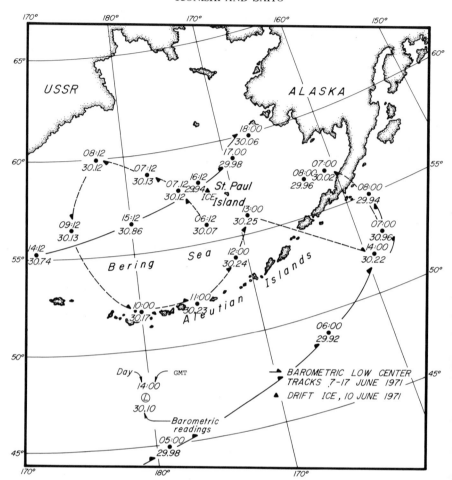

Fig. 22.14 Tracks of low barometric centers in the Bering Sea during 7–17 June 1971. Ice remained south in the waters west of St. Paul Island until about 10 June.

farther north than in more typical years, receding to St. Matthew Island and vicinity at the end of March. In comparing 500-mb five-day mean charts for the period 11–15 May 1971 (Fig. 22.17) with the same period in 1967 (Fig. 22.18), it is shown that the Bering Sea was covered by a ridge in 1967, as opposed to a trough in 1971, demonstrating a means by which cold air moves southward on the Bering Sea.

The annual May ambient temperature means on St. Paul Island for 12 years (1960–71) are shown in Figure 22.19. The mean value for 1971 was −2.3 C, compared with a maximum for the period of 3.4 C that occurred in 1967. Except for the 1965 and 1971 data, a two-year cycle was observed in the temperature variation.

In May 1968 the Aleutian low was a little more easterly than normal, as

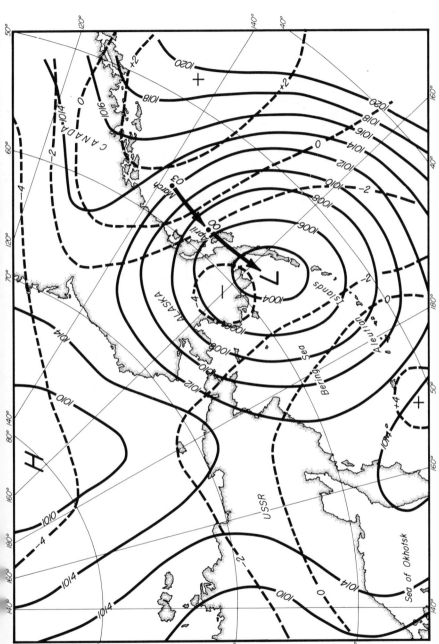

Fig. 22.15   Mean monthly surface chart of atmospheric pressure for May 1971, showing low center over the Bering Sea.

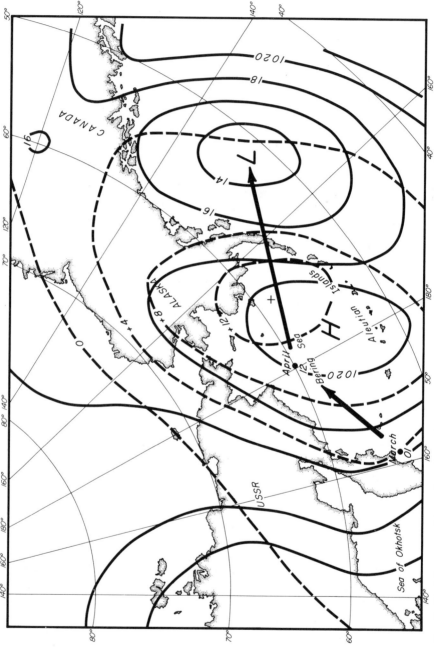

Fig. 22.16 Mean monthly surface chart of atmospheric pressure for May 1967 over the Bering Sea and Gulf of Alaska.

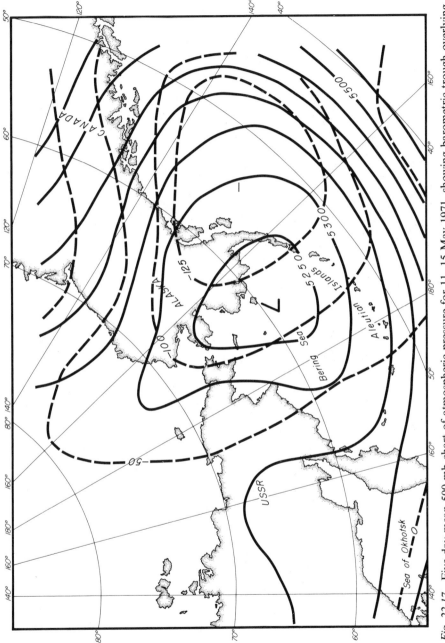

Fig. 22.17 Five-day mean 500-mb chart of atmospheric pressure for 11–15 May 1971, showing barometric trough overlying the Bering Sea.

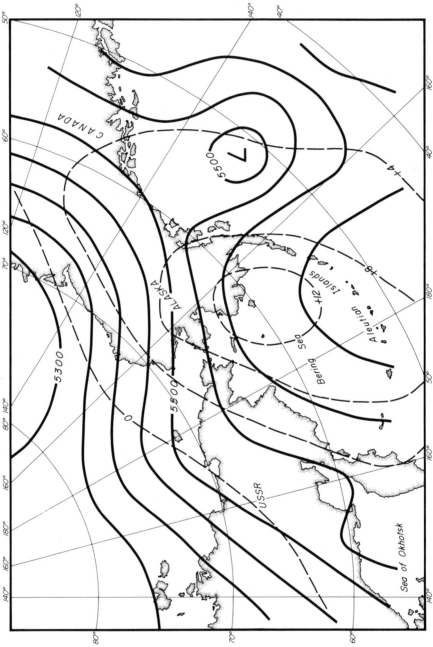

Fig. 22.18 Five-day mean 500-mb chart of atmospheric pressure for 11–15 May 1967, showing barometric ridge covering the Bering Sea

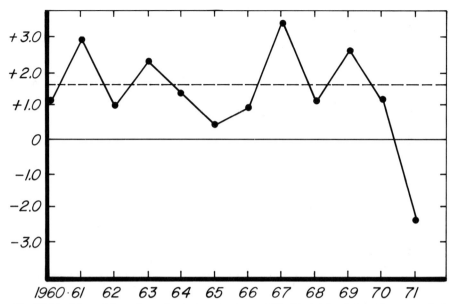

Fig. 22.19 Mean May atmospheric temperature at St. Paul Island for 12-year period 1960–71. A two-year cycle is seen in the variation of temperature except for 1965 and 1971.

was the case in 1967, but it reverted slightly westward from where it lay in the previous year. The ridge moved further west than in the previous year and covered the western Bering Sea. This situation brought on a tendency toward a more frequent northerly wind during May 1968 (Fig. 22.20) than in 1967, with the result that the air and surface temperatures of the Bering Sea were lower than those in the previous year. Although no ice data were available for 1968, the ice probably lay further south than in 1967.

In May 1969 the Aleutian low and the North Pacific high were very westerly, and a ridge was formed from Alaska to eastern Siberia. As a result there was frequently southerly wind on the Bering Sea, and both the air and surface temperatures were high. The ice around Bristol Bay receded at the end of April to north of St. Matthew Island.

In May 1970 a strong North Pacific high spread westward and covered the western Bering Sea. The Aleutian low, in contrast to its location in 1969, was slightly easterly and in the vicinity of Cold Bay. Thus, in the eastern Bering Sea, a pressure pattern developed in which a northerly wind was evident, although weaker than in 1968, and the ice lay slightly farther south than in 1969.

In this manner, the changes seen in the wind and ice in the years 1967 through 1970 followed a two-year cycle pattern. In May 1971, however, an exception occurred in the cycle, with northerly winds prevailing as in the preceding year. The atmospheric temperature declined during 12–14 May

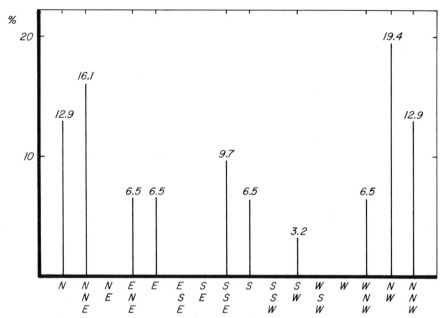

Fig. 22.20 Frequency distribution of wind direction at St. Paul Island during May 1968, indicating prevalence of northerly winds.

1971 (Fig. 22.21). Despite the northerly wind in effect from 17–19 May, the temperature drop was slight. The ice at this time was tending to drift on a northbound current.

The same cycle can also be observed with respect to the position and magnitude of the pressure distribution, which cycle has continued since 1960.

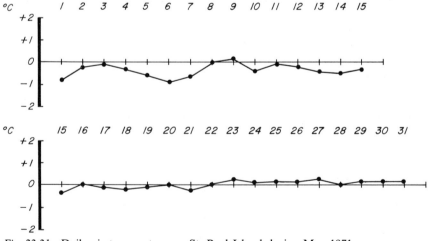

Fig. 22.21 Daily air temperature on St. Paul Island during May 1971.

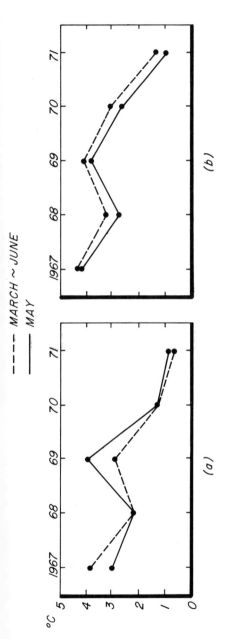

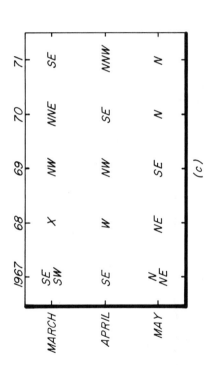

Fig. 22.22 Annual environmental observations made from ship (*Shikishima Maru*) in the eastern Bering Sea from March to June between 1967 and 1971: (a) monthly mean air temperature (°C); (b) monthly mean sea-surface temperature (°C); (c) direction of most frequent winds.

The mean atmospheric and water temperature and most frequent wind direction were observed by ship during an annual period from March to June between 1967 and 1971 (Fig. 22.22). A cyclic change with intervals of two years was noted in every month after 1967, with the exception of 1971, during which temperatures were abnormally low continuously from the previous year.

On the other hand, the expansion of the arctic high over the Bering Sea was pronounced in 1959, 1960, and 1971, while in 1965 it was strong only in May. The two-year cycle was observed in the pressure pattern as well, with the exception of 1959 and 1960. Figure 22.23 shows the changes in the difference of the mean monthly surface pressure between 60°N 180°W and 60°N 160°W in an annual period from March to May 1967 through 1971. The atmospheric pressure was higher in the west during 1968, 1970, and 1971, resulting in a prevailing northerly wind. A southerly wind prevailed in March and April of 1967 but only in May 1969. The two-year pattern appeared between 1967 and 1970, except for 1965. Three-year periods in 500-mb mean monthly surface pressure (Fig. 22.24) are seen for May 1963–69 and for April 1961–67. The same period is also reflected in the sea-surface temperature of the North Pacific Ocean.

As described above, the ice in the Bering Sea is influenced by changes occurring in meteorological and oceanographic conditions, which therefore form an essential element in forecasting ice conditions.

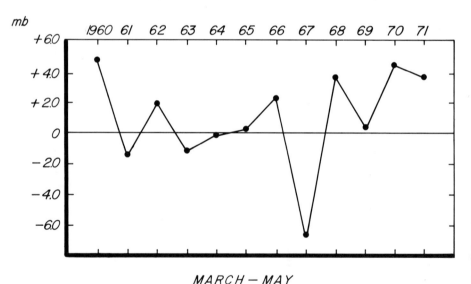

MARCH — MAY

Fig. 22.23 Trend of monthly meridional indices indicated by surface pressure differences between 60°N 180°E and 60°N 160°W in a period from March to May between 1967 and 1971 (+ west area of higher atmospheric pressure than east, northerly winds prevailing; − west area of lower pressure than east with prevailing southerly winds). Two-year cycle is apparent except for 1965.

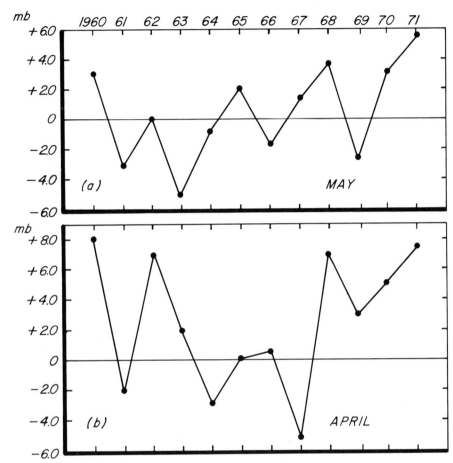

Fig. 22.24 Trend of monthly meridional indices indicated by surface pressure differences between 60°N 180°E and 60°N 160°W between 1960 and 1971 for (a) May, with three-year period appearing in 1963–1969, and (b) April, three-year period appearing in 1961–67.

Discussion

TAKENOUTI: The May 1971 monthly mean temperature was approximately 5 C lower than average. Is such a difference common at higher latitudes comparable to that of St. Paul Island?

SAITO: This was an exceptional deviation from average for the 12-year period studied (1960–71). A total dispersion of 5.7 C is observed between the maximum monthly mean temperature of +3.4 C for May 1967 and the minimum May mean of −2.3 C noted in 1971.

WADA: The abnormally low temperature of the Bering Sea in May 1971 may be closely related to the cool summer in northern Japan that year which resulted in a poor rice harvest.

FAVORITE: Your sea surface temperature distribution shown for May 1971 implies that considerable data were obtained over a large area in the eastern Bering Sea. What was the name of the vessel and in what activity was she engaged?

SAITO: The data were obtained by numerous fishing boats.

McCLAIN: With the use of satellite pictures, Mr. Saito has demonstrated a meteorologically-induced abnormality in ice conditions over the Bering Sea in 1971 which in turn affected the weather systems.

CHAPTER 23

Intensification of the Okhotsk high due to cold sea surface

Takashi Ōkawa

Sapporo District Meteorological Observatory, Sapporo, Japan

Abstract

Although Siberia and Alaska are heated by insolation throughout early summer and heat gained is transported to the upper layer by convection and diffusion, the air over the Bering Sea remains cool because of the cold sea surface. As a result, in the upper troposphere a stationary ridge is formed over the eastern part of Siberia and a trough over the Bering Sea. This pattern tends to create a convergence field over the Sea of Okhotsk and stimulates formation of a low level anticyclone, the Okhotsk high, which always brings cool and cloudy weather to northern Japan.

The pressure rise due to cooling of the lower layer of the atmosphere above the cold sea surface is estimated to be 2 mb/12 hours. The atmosphere in the close proximity of the cold sea surface is cooled and tends to remain in place because of the increase in its density and the surface friction. As a result, this cold air forms a distinct shielding layer about 50–100 m in height separated from the Ekman layer. When the shielding layer is formed, the descending anticyclonic current receives less friction because it does not reach the sea surface, and it tends to decrease its deviating angle from the isobar. This means a decrease in the amount of outflow from the anticyclone domain and an additional increase in pressure as a resulting effect of the shielding layer.

INTRODUCTION

The anticyclone which appears over the Sea of Okhotsk in the early summer has been discussed in Japan since the beginning of the century because of its direct association with the seasonal *baiu* phenomenon of persistent raininess. Okada (1910) suggested that this anticyclone, referred to as the *Okhotsk high*, is formed by cooling of the air column due to the cold surface of the Sea of Okhotsk. After World War II, Okuda (1951) pursued study of the Okhotsk high by use of weather charts and explained its development from the perspective of a *blocking* high pressure system.

It has been further confirmed (Ōkawa 1973) that a main factor affecting the generation and development of the high might be found in the flow pattern of the upper troposphere; however, two secondary factors are attributed to the cold sea surface. First considered is a pressure rise caused by cooling of the air column, which Takahashi (1968) estimated as 1 mb/day and for which Yagi (1969) showed a value of 2–3 mb/36 hours in his study. The other factor is a *shielding layer effect* (Wexler 1951), which is evaluated in this paper.

Formation and development of the Okhotsk high

During early summer (June), when Siberia and Alaska are heated by insolation, the heat gained is transported to the upper layer by processes of convection and diffusion; the air over the Bering Sea remains cool, however, because of the cold sea surface. As a result, a stationary ridge is formed in the upper troposphere over eastern Siberia, and a trough develops over the Bering Sea. These formation processes were recently confirmed by a numerical simulation method (A. Katayama, A. Arakawa, and Y. Mintz 1971, unpublished; discussed by Asakura in Chapter 24, this volume).

A convergence field over the Sea of Okhotsk (Sutcliff 1947) results from this pattern and stimulates the formation of a low level anticyclone, or the Okhotsk high. This phenomenon is particularly important in Japan because it is consistently associated with cool and cloudy weather in the northern part of the country.

In addition to the Okhotsk high described by the above pattern and here designated type E, there are two other forms of the Okhotsk high produced by upper level patterns N and R, respectively. Type N appears below the upper level cold low which moves southward from the Arctic Ocean to the Sea of Okhotsk, crossing the eastern part of Siberia. Type R appears below the upper level ridge situated over the northern part of the Sea of Okhotsk. All three Okhotsk highs were evident during the baiu period from 31 May to 14 July over 21 years (1950–1970) and were classified according to surface and 500-mb weather chart data recorded at 00Z and 12Z. More than 50 percent of the patterns were identified as type E, about 35 percent were type R, and 12.5 percent were type N (Fig. 23.1).

A variance analysis of the monthly mean 500-mb height field has been computed (Ōkawa 1964) from 5-day mean height anomalies by use of the F-distribution hypothesis and distribution of F_0 calculated by the following equation (1):

$$F_0 = \frac{N - k}{k - 1} \frac{\sum\limits_{i=1}^{k} n(\bar{x}_i - \bar{x})^2}{\sum\limits_{i=1}^{k} \sum\limits_{j=1}^{n} (x_{ij} - \bar{x}_i)^2} \tag{1}$$

where

x_{ij} : anomaly of 5-day mean 500-mb height at a given mesh point,

$\bar{x}_i$: anomaly of monthly mean 500-mb height at a given mesh
point (mean value of x_{ij} in each year),
i : years, from 1946 to 1961,
j : 5-day periods in a month,

Therefore,

$k = 16$ (1946–61)
$n = 6$ (for one month consisting of six 5-day periods), and
$N = n \times k$.

The large value of F_0 indicates that the anomalies of monthly mean height vary considerably from year to year; consequently, in the domain of large F_0, the monthly mean height indicates the characteristics of an individual year. Conversely, in the domain of small F_0, the monthly mean heights vary only slightly from year to year. In the latter case, it can be concluded that a trough and a ridge in an average chart are stable and appear every year; i.e., if a trough or ridge appears over the area with small F_0 in the monthly normal chart, such a trough or ridge can be expected to occur in that month every year (Fig. 23.2).

In the case of the distribution of F_0 in June (Fig. 23.3), small values (< 1.0) in the Bering Sea area and large ones in the area near the north coast of the Sea of Okhotsk (3.0) and near the west coast of Canada (4.0) mean that the Bering Sea trough will persist in the baiu season of June and July each year. There is, however, a small east or west fluctuation from its normal position from year to year. Also, a trough formed in the baiu season each year, having

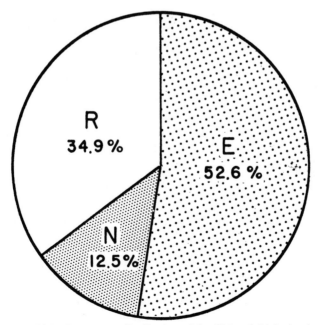

Fig. 23.1   Percentage distribution of the Okhotsk high classification by type.

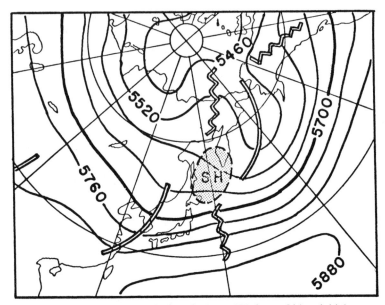

Fig. 23.2   The normal 500-mb chart in June. SH shows Okhotsk high.

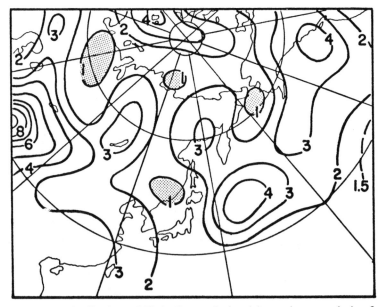

Fig. 23.3   Distribution of F_0 in June calculated by variance analysis of anomalies of monthly mean 500-mb height.

an F_0 value of about 1.0, is observed over Korea and is a main factor linking development of the baiu phenomenon with the Bering Sea trough.

Figure 23.4 shows how closely the appearance of the Okhotsk high is related to the flow pattern between the Siberian ridge and the Bering Sea trough as defined in the preceding section. Here, $(H_R - H_T)$ is an index showing the degree of 500-mb level northeasterly flow over the Sea of Okhotsk and is calculated from a height difference between H_R and H_T, where H_R is an index representing the strength of the ridge over eastern Siberia and H_T that of the Bering Sea trough, computed as follows:

$$H_R = \frac{1}{4}\left[\left(\begin{array}{c}60°N\\130°E\end{array}\right) + 2 \times \left(\begin{array}{c}60°N\\140°E\end{array}\right) + \left(\begin{array}{c}60°N\\150°E\end{array}\right)\right]$$

$$H_T = \frac{1}{4}\left[\left(\begin{array}{c}50°N\\160°E\end{array}\right) + 2 \times \left(\begin{array}{c}50°N\\170°E\end{array}\right) + \left(\begin{array}{c}50°N\\180°W\end{array}\right)\right]$$

where the latitudes and longitudes in parentheses indicate a mean 500-mb height in the baiu period. For every year, during the baiu season, there are 90 sheets of surface weather charts. The percentage of sheets in which an Okhotsk high of type E, N, or R is found is calculated against this total number of sheets for that type over composite years and are classified as E, N, or R, respectively. The E, N, and R are defined as a frequency rate of the Okhotsk high for each type. Then $E + R - N$ are calculated (Fig. 23.4). The reason N is subtracted from the sum of E and R is because of its negative relationship to $(H_R - H_T)$. It is seen that there exists a distinct correlation between $E + R - N$ and $(H_R - H_T)$ in terms of a coefficient of $+0.799$, although the number of data is rather small ($n = 21$).

Shielding layer hypothesis for development of the Okhotsk high

As it has been stated earlier, the main factor contributing to the development of the Okhotsk high exists in the upper flow pattern. As secondary factors, however, two effects of the cold sea surface are considered. One is a pressure rise caused by cooling of the air column due the cold sea surface, and the other is by action of the shielding layer described in this section. The mechanics

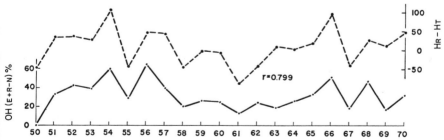

Fig. 23.4 Relation between the frequency of the Okhotsk high OH(E+R−N)% shown by lines and (H_R−H_T) indicated by broken lines which show degrees of 500-mb level northeasterly flow over the Sea of Okhotsk.

stated here are applicable not only to the Okhotsk high but also to any anticyclone formed over a cold sea such as the Bering Sea or Arctic Ocean.

The atmosphere in the close proximity of a cold sea surface is cooled and
tends to remain there because of increased density and due also to surface
friction. As a result, this cold air may form a distinct shielding layer 50–100 m
in height and separated from the Ekman layer (Fig. 23.5).

When a shielding layer is formed, the surface friction acting upon a
descending anticyclonic current is reduced because the current does not reach
the sea surface; this tends to decrease the angle of deviation to the isobar, thus
decreasing in turn the amount of outflow from the domain of the anticyclone;
consequently, the pressure gradually rises. The amount of air flowing out of
the anticyclone is estimated here according to Taylor (1915), whose early data
on the Scotia Expedition are also used here for lack of other data for the cold
sea surface.

Taylor's formulas. The equations of motion of an incompressible viscous
fluid are

$$\left.\begin{array}{c}
\dfrac{du}{dt} + 2\omega v \sin\theta = -\dfrac{1}{\rho}\dfrac{\partial p}{\partial x} + \dfrac{\mu}{\rho}\nabla^2 u \\[2ex]
\dfrac{dv}{dt} - 2\omega u \sin\theta = -\dfrac{1}{\rho}\dfrac{\partial p}{\partial y} + \dfrac{\mu}{\rho}\nabla^2 v \\[2ex]
\dfrac{dw}{dt} = -g - \dfrac{1}{\rho}\dfrac{\partial p}{\partial z} + \dfrac{\mu}{\rho}\nabla^2 w
\end{array}\right\} \quad (2)$$

where u, v, and w are components of the wind velocity parallel to the
coordinates x, y, and z (positive directions north, east, and upward,

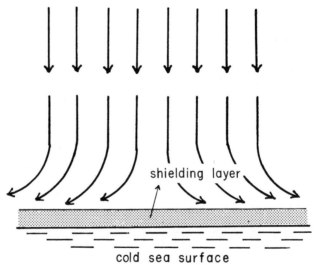

Fig. 23.5 Illustration of the descending anticyclonic currents
and shielding layer over the cold sea surface.

respectively); ω is the angular velocity of the earth's rotation; and θ is the latitude.

If the motion is assumed to be horizontal, steady, and uniform, and the pressure is represented as

$$p = \text{constant} - g\rho z + Gy,$$

equations (2) become

$$0 = -2\omega v \sin\theta + \frac{\mu}{\rho}\frac{d^2u}{dz^2}$$

$$0 = 2\omega u \sin\theta - \frac{G}{\rho} + \frac{\mu}{\rho}\frac{d^2v}{dz^2}$$

$$\left.\right\} \quad (3)$$

Considering that u and v remain finite for infinite values of z, the solutions of equations (3) are:

$$u = A_2 e^{-BZ} \cos BZ - A_4 e^{-BZ} \sin BZ + Q_g \qquad (4)$$

$$v = A_2 e^{-BZ} \sin BZ + A_4 e^{-BZ} \cos BZ \qquad (5)$$

where

$$B = \sqrt{\frac{\omega \sin\theta}{K}}$$

$$K = \frac{\mu}{\rho}; \text{ coefficient of eddy viscosity and}$$

$$Q_g = \frac{G}{2\mu B^2}; \text{ velocity of geostrophic wind.}$$

Supposing that there is slippage at the surface, we take one boundary condition

$$\left[\frac{du/dz}{u}\right]_{z=0} = \left[\frac{dv/dz}{v}\right]_{z=0},$$

and let the surface wind deviate an angle α from the geostrophic wind. Then

$$\tan\alpha = -\left[\frac{v}{u}\right]_{z=0} = \frac{A_4}{A_2 + Q_g} \qquad (6)$$

and

$$A_2 = \frac{-\tan\alpha(1 + \tan\alpha)}{1 + \tan^2\alpha}Q_g$$

$$A_4 = \frac{-\tan\alpha(1 - \tan\alpha)}{1 + \tan^2\alpha}Q_g.$$

$$\left.\right\} \quad (7)$$

Knowing the values of Q_g, K, and α, the velocity components u, v at a certain height z can consequently be computed by equations (4) and (5).

Computed rise in pressure. Applying Taylor's formulas to the case of a circular anticyclone as shown in Figure 23.6, the outflow of an amount of air Q is computed by

$$Q = 2\pi r \bar{\rho} \int_0^\infty (-v)dz \qquad (8)$$

where

r = radius of the anticyclone

$\bar{\rho}$ = mean density of air, 1.25×10^{-3} C.G.S. (at 990 mb, virtual temperature 3 C).

Hereafter, supposing a circular anticyclone with a radius of 500 km located at about 50°N, we estimate the surface pressure increment (ΔP_n) due to Q, which is assumed to be accumulated over the area of the circle, then

$$\Delta P_n = \frac{Q}{\pi r^2}$$

$$= 0.5665 \frac{\sqrt{K}\tan\alpha}{1 + \tan^2\alpha} Q_g \text{ mb/day} \qquad (9)$$

where the unit of Q_g is m/sec, and that of K is cgs units $\times 10^{-2}$. Table 23.1 shows values of ΔP_n corresponding to various K and α with 1 m/sec of Q_g.

Then, the pressure rise after the formation of a shielding layer (ΔP) is

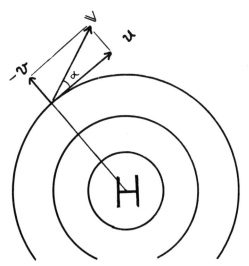

Fig. 23.6 Illustration of wind components and deviating angle of surface wind from the geostrophic current.

computed from a difference between the pressure increment without a shielding layer (ΔP_1) and that with a shielding layer (ΔP_2); that is

$$\Delta P = \Delta P_1 - \Delta P_2,$$

where the top height of the shielding layer is taken as the lower limit of integral in equation (8) for the calculation of ΔP_2.

From observations made by means of kites on board the ice patrol ship *Scotia* in the North Atlantic in summer 1913, Taylor (1915) concluded that K lies between 0.77×10^3 and 6.9×10^3, and α lies between 10 and 20 degrees of arc. From these values the pressure rise is estimated as follows:

without a
shielding layer assuming $\begin{cases} K = 6.9 \times 10^3 \text{ cgs} \\ \alpha = 15° \\ Q_g = 6 \text{ m/sec} \end{cases}$ $\Delta P_1 = 7.20 \text{ mb/day}$

with a
shielding layer assuming $\begin{cases} K = 0.9 \times 10^3 \text{ cgs} \\ \alpha = 5° \\ Q_g = 6 \text{ m/sec} \end{cases}$ $\Delta P_2 = 0.89 \text{ mb/day}$

$$\Delta P = \Delta P_1 - \Delta P_2 = 6.31 \text{ mb/day} \doteqdot 3 \text{ mb/12 hours.}$$

The value 3 mb/12 hours seems, however, to be a slight over-estimation; taking into account the values in Table 23.1, therefore, the pressure rise is re-calculated:

without a
shielding layer assuming $\begin{cases} K = 5.5 \times 10^3 \text{ cgs} \\ \alpha = 17.5° \\ Q_g = 5 \text{ m/sec} \end{cases}$ $\Delta P_1 = 6.03 \text{ mb/day}$

with a
shielding layer assuming $\begin{cases} K = 3.0 \times 10^3 \text{ cgs} \\ \alpha = 7.5° \\ Q_g = 5 \text{ m/sec} \end{cases}$ $\Delta P_2 = 2.01 \text{ mb/day}$

$$\Delta P = \Delta P_1 - \Delta P_2 = 4.02 \text{ mb/day} \doteqdot 2 \text{ mb/12 hours.}$$

This value 2 mb/12 hours seems acceptable as an effect of the shielding layer in practice because rather modest values of K and α are used. In this case, the pressure rise during 12 hours is considered important, for the following reason. Looking at the Okhotsk high development processes in daily weather charts, it is noted that frequently the high develops rapidly at a rate of 4 mb/12 hours during half a day in the initial stage and then tends to reach a steady state during which the rate of development decreases.

An example of such a case is shown in Figure 23.7. At 12Z on 7 June 1966, only a moderate pressure saddle (1017–1019 mb) in the southern part of the Sea of Okhotsk was found except for an anticyclone (1020 mb) near the northern sea coast. At 00Z of the following day, 12 hours later, the anticyclone near the northern coast was moving westward, weakening in intensity, and a pressure

TABLE 23.1   Values of surface pressure increment ΔP_n (mb/day) for 1 m/sec of out-flowing air Q_g

$\times 10^3$ K ╲ $\alpha°$	2.5	5.0	7.5	10.0	12.5	15.0	17.5	20.0
0.5	0.055	0.110	0.164	0.217	0.268	0.317	0.363	0.407
1.0	0.078	0.156	0.232	0.306	0.379	0.448	0.514	0.576
1.5	0.096	0.191	0.284	0.375	0.464	0.549	0.629	0.705
2.0	0.110	0.220	0.328	0.433	0.535	0.633	0.727	0.814
2.5	0.123	0.246	0.367	0.484	0.599	0.708	0.812	0.910
3.0	0.135	0.269	0.402	0.531	0.656	0.776	0.890	0.997
3.5	0.146	0.291	0.434	0.573	0.708	0.838	0.961	1.077
4.0	0.156	0.311	0.464	0.612	0.757	0.896	1.028	1.152
4.5	0.166	0.330	0.492	0.650	0.803	0.950	1.090	1.221
5.0	0.175	0.348	0.518	0.685	0.846	1.001	1.149	1.287
5.5	0.183	0.365	0.544	0.718	0.888	1.050	1.205	1.350
6.0	0.191	0.381	0.568	0.750	0.927	1.097	1.259	1.410
6.5	0.199	0.397	0.591	0.781	0.965	1.142	1.310	1.468
7.0	0.207	0.412	0.613	0.811	1.002	1.185	1.359	1.523

rise of about 3–5 mb occurred over the entire Sea of Okhotsk, with the center of the high (1024 mb) located southwest of Kamchatka. At 12Z of the same day, another 12 hours later, no such pressure rise was observed as that seen at 00Z 8th, although the area enclosed by the 1024-mb isobar was wider than that at 00Z.

Remarks and suggestions for further research
It has been pointed out that the Ōkhotsk high reaches a steady state after the initial development stage of about 12 hours. From this phenomenon, the author infers that the amount of air converging in the upper level of the troposphere must balance with a divergence in a lower layer, which has a strong peak just above the Ekman layer (about 850-mb level). It can be considered, therefore, that the amount of divergence at all levels from the top of the troposphere down to the sea surface must be calculated both in initial and steady-state stages, in order to estimate the effect of the shielding layer on the development of the high.
    First of all, however, some observational results must be obtained to prove

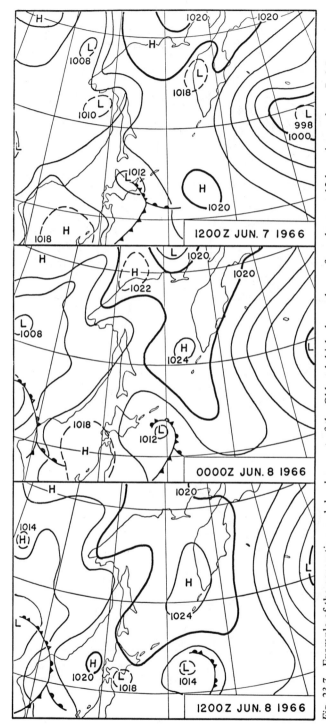

Fig. 23.7 Example of the generation and development of the Okhotsk high shown by surface charts at 12-hour intervals on 7–8 June 1966.

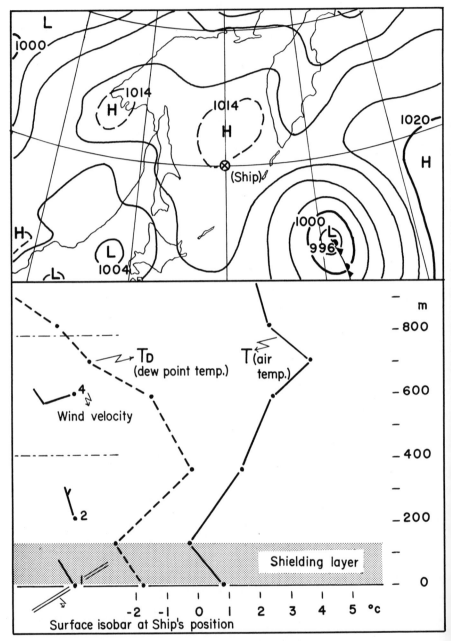

Fig. 23.8 An example of sounding at 50°N, 150°E over the Sea of Okhotsk at OOZ 21 May 1971 (lower part) and weather chart at the time of measurement (upper part).

the existence of the shielding layer. In early summer from 20 May to early June 1971, soundings were conducted from aboard the R/V *Keifu Maru* of the Japan Meteorological Agency by use of radiosondes and rawins (radio winds-aloft observations) in the central region of the Sea of Okhotsk (50°N, 150°E). These data show the existence of a shielding layer, as illustrated in Figure 23.8 together with a local weather chart at the time of measurement. In the upper chart of Figure 23.8 (00Z, 21 May 1971), a high with a center of 1014 mb is shown in the middle of the sea with the ship located south of the center. The lower portion of the figure shows the results of the sounding, where T is an ascent curve of air temperature, T_D is dew point temperature, and the direction of the surface isobar at the ship's position is indicated. The mean wind, averaged over a thickness of about 400 m, is illustrated in the figure with dashes and broken lines, except for a surface wind. For lack of a detailed wind sounding, the shielding layer is assumed from the temperature curve and is shown by a shaded layer in the figure.

Considering the results of the wind sounding: NW(330°) 1 m/sec, at the sea surface, NNW(345°) 2 m/sec, as a mean for height interval 0–413 m and WSW(253°) 4 m/sec, for 413–784 m, the wind at about 200–300 m was northeast, which coincides closely with the direction of the surface isobar. This suggests the existence of the shielding layer as shown in Figure 23.8.

Still it is impossible to clearly determine a shielding layer by observations with routine soundings of radiosondes and rawin with counting at rather rough intervals, and detailed sounding by a tethered balloon will have to be done in the future for certain detection of this layer.

Precise and detailed measurements of the upper level wind would permit determination of heights H_1 and H_2, at which points wind direction and speed, respectively, attain those of the geostrophic wind. Then it would be easy to estimate an eddy viscosity coefficient K (assuming a constant throughout the layer in question) and a deviation angle α by Taylor's formulas. It should be noted that in sea areas the direct determination of α with weather charts is more erroneous than often supposed, because of errors in drawing the isobar due to lack of sufficient observations. Consequently, detailed soundings with a tethered balloon are a prerequisite to further study of the effect of the shielding layer.

Discussion

McCLAIN: Is there a tendency for high pressure to intensify over the Sea of Okhotsk also in the winter when ice covers the sea?

ŌKAWA: My studies have been limited to the period from latter May to mid-July. Under winter conditions, however, I would suppose that a shielding layer would not form easily because of the severe and persistent northwesterly monsoons which disturb the air layer just above the surface. Such a shielding layer might be formed in the clear and calm winter night conditions, but the magnitude of this effect would be comparable to that of the pressure increment caused by cooling of the air column by the snow covering the Sea of Okhotsk.

McCLAIN: Dr. Tabata showed in his film how ice forms on the Sea of Okhotsk because of the lack of deep water convection that confines cooling to the shallow top layer. It is possible that this surface ice formation generates a feedback mechanism returning heat to the atmosphere through radiation while intensifying high pressure over the area; the high pressure and clear skies would in turn promote even further radiation loss of heat from the ice.

Acknowledgments

The author expresses gratitude to Dr. K. Suda of the Japan Meteorological Agency for suggesting this problem and offering his helpful advice. Appreciation is extended to Prof. A. Y. Takenouti, Faculty of Fisheries, Hokkaido University, for symposium arrangements and manuscript review, and to Dr. H. Wada, Chief of the Long-Range Forecasting Section, JMA, and convener of the meteorological section of this symposium, for his guidance and encouragement.

Acknowledging personnel of the Sapporo District Meteorological Observatory, the author thanks Dr. K. Mohri, Director, for the opportunity to participate in the symposium, and Messrs. A. Fujinori, Head of the Technical Division, and T. Sugimoto, Chief of the Forecasting Section, for their continuing assistance.

REFERENCES

OKADA, T.
  1910   Baiu, or rainy season in Japan. *Bull. Cent. Meteor. Obs. Tokio* 1(5): 1–82.

ŌKAWA, T.
  1964   Variance analyses of 500-mb height in the Northern Hemisphere. *J. Meteor. Res. JMA* 16: 367–375.
  1973   Growth mechanism of the Okhotsk high. *J. Meteor. Res. JMA* 25: 65–77.

OKUDA, Y.
  1951   Baiu. *Research Note for Weather Forecasting* 2(1): 27–65.

SUTCLIFF, R. C.
  1947   A contribution to the problem of development. *Quart. Journ.* 73: 370–383.

TAKAHASHI, K.
  1968   Air-sea interaction, *Synoptic Meteorology.* Iwanami-Shoten, Tokyo, pp. 336–341.

TAYLOR, G. I.
  1915   Eddy motion in the atmosphere. *Roy. Soc. Phil. Trans., Series A*, 215(1).

WEXLER, H.
   1951   Anticyclone, *Compendium of Meteorology*. Amer. Meteorol. Soc., Boston, Mass., pp. 621–629.

YAGI, S.
   1969   Drawing of prognostic surface chart by use of numerical prediction data and its revising due to the surface condition. *Bull. Meteorol. Res. Sapporo District Meteorol. Obs.*

CHAPTER 24

The role of the Bering Sea in long-range weather forecasting in Japan

Tadashi Asakura

Forecast Division, Japan Meteorological Association, Tokyo, Japan

Abstract

The sea condition and pressure pattern over the Bering Sea are important factors in seasonal weather forecasting in Japan. The cold air pool over the Bering Sea in early summer is formed by cooling as a result of air-sea interaction and by a marked contrast of heating between land and sea. The role of the cold source near the Bering Sea is enhanced by the effect of the heat source near India.

Effect of Bering Sea on rice harvest in northern Japan

The study of seasonal weather forecasting in Japan started early in this century, when farmers were distressed by frequently bad rice harvests. The cause of the bad harvest was ascribed to the cold sea current called *Oyashio* in accord with the familiar saying, "A bad harvest comes from the sea."

When the Aleutian low develops in the winter and the North Pacific anticyclone moves to the south in the spring, a persistent and strong northeasterly wind flows over the Bering Sea during both winter and spring. This results in an acceleration of the Oyashio Current, which cools the sea surface off the eastern coast of Japan.

The cool summer of 1971 coincided with a bad rice harvest in northern Japan, and the sea-surface temperature was below normal off the eastern coast (Fig. 24.1). Synoptic processes at the 500-mb level between the previous winter and the cool summer of 1971 showed characteristics similar to those long recognized.

The Aleutian low develops in the southeastern Bering Sea, the anticyclone develops north of Kamchatka during the winter, and the Pacific anticyclone is displaced southward during the spring. These pressure patterns tend to strengthen the Oyashio Current and bring a cool summer to northern Japan. It is apparent, then, that the Aleutian low during the winter is a very important factor in the seasonal forecasting of the summer season, but the prediction of the Aleutian low itself is a very difficult problem.

In studying the behavior of the Aleutian low, a relationship was found between heavy precipitation at Canton Island and a negative height anomaly that may be a cyclonic development (Fig. 24.2). Bjerknes (1966, 1969) has shown that a large positive sea-surface temperature anomaly in the equatorial sea brings a heavy intensification of the Hadley Circulation, resulting in the development of western flow and of a low in the Gulf of Alaska. This may suggest that air-sea interaction is very important in predicting the Aleutian low not only in a localized area such as the Bering Sea but in a vast and remote sea as well.

For seasonal weather forecasting, special attention has been paid to the behavior of atmospheric disturbances over the Bering Sea, including the Aleutian high in the stratosphere.

Cold air pool in the Bering Sea during early summer

Another synoptic process operating over the Bering Sea from winter to summer is the formation of a cold air pool in May and June. Cold air is steered from this pool to northern Japan by the northeast monsoon. The geographical feature of monthly mean temperature change in the lower (500–850 mb) and

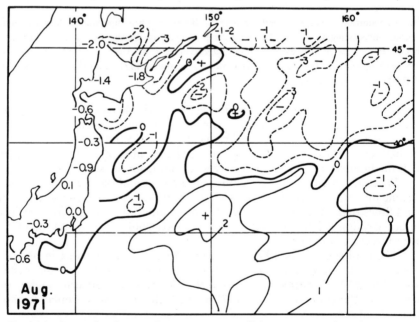

Fig. 24.1 Anomaly of sea-surface temperature off the east coast of northern Japan in August 1971.

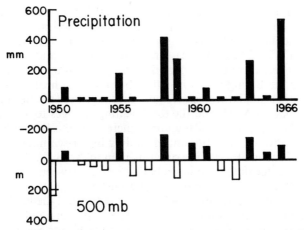

Fig. 24.2 Precipitation at Canton Island (J. Bjerknes 1969) and 500-mb height anomaly at selected point (60°N, 180°E) in the Bering Sea for January 1950–1966.

upper (100–500 mb) troposhere are shown in Figure 24.3. In the lower troposphere, warming areas are found over the continents of Asia and North America, and cooling is noted near the Bering Sea. Likewise, in the upper troposphere, warming is found over Asia, centered near the Tibetan Plateau, and cooling occurs over the Pacific Ocean near the Bering Sea. Comparing the above observations with the atmospheric heat distribution, the warming corresponds to the heat source and the cooling to the cold source (Asakura and Katayama 1964; Asakura 1968). Heating or cooling, then, is an important factor in the formation of a cold pool.

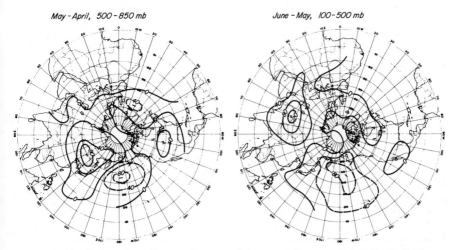

Fig. 24.3 Deviations from the zonal mean of the monthly mean 500–850-mb thickness change for May–April (left) and 100–500-mb thickness change for June–May (right).

A. Katayama, Y. Mintz, and A. Arakawa (1971, unpublished) studied the seasonal variation of the general circulation in a numerical simulation of the atmosphere and successfully reproduced the formation of a cold air pool over the Bering Sea (Fig. 24.4). The sea-surface temperature and sea ice were held constant, however, through the period of this simulation. In April, the heat exchange over the Bering Sea is negative, but the total heating is positive. In June, on the other hand, the negative heat exchange over the Bering Sea covers a wider area than in April, and the total heating changes to a negative value. In other words, the cooling of the atmosphere over the Bering Sea is taking place in June; however, the amount of cooling by air-sea interaction over the Bering Sea in June is of the same order as than in April.

A much greater difference in diabatic heating between April and June is found over both continents than over the Bering Sea. A great contrast is formed in heating or cooling distribution between the ocean and the land surfaces. An especially intensive heat source is found over Asia centered at the Tibetan Plateau. The formation of a cold air pool over the Bering Sea is then effected partly by cooling due to air-sea interaction and partly by the contrast of heat distribution between the Asian continent and the Bering Sea. The ice conditions would thus be an important factor in predicting a cold pool.

Role of heat source near the Tibetan Plateau and cold source near the Bering Sea
The importance of the cold source has been stressed in development of the Okhotsk high during the baiu (rainy) season of June to July in Japan. On the other hand, the heat source near the Tibetan Plateau plays an important role in the southwest monsoon in India. Also, the baiu in East Asia is closely related to the southwest monsoon in India. This fact can be explained only by the cold source near the Bering Sea.

The role of the cold and heat sources for the atmosphere is discussed on the basis of thermodynamic and vorticity equations, considering the effects of the normal heat source near India and the normal cold source near the Bering Sea. Calculations are carried out for 500-, 700-, and 1000-mb levels in the northern hemisphere, assuming a geostrophic approximation, grid size of 300 km, and a time interval of 90 min.

The normal distribution of atmospheric heating in the layer between

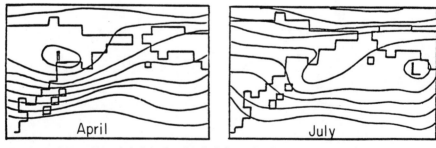

Fig. 24.4 Mean 400-mb height for April (left) and July (right), averaged for years 1, 2, and 3 of numerical simulation. Height contours are at intervals of 120 m (left) and 60 m (right). (K. Katayama, Y. Mintz, and A. Arakawa 1971, unpublished).

1000 and 500 mb is shown in Figure 24.5. A large heat source dominates in the low latitudes with a center in the vicinity of India, and a cold source is found over the ocean south of Kamchatka. Japan is situated just between the heat source and the cold source. To study the role of the heat and cold sources, flow patterns at the 700-mb level are calculated by operating the heat source and cold source on a circular pattern (Fig. 24.6).

As revealed on the tenth day of the numerical experiment, the result (Fig. 24.6) is that a blocking flow pattern is formed in East Asia with a trough near Kyushu and a ridge over the Sea of Okhotsk. In India, a monsoon trough is formed with a ridge over the Tibetan Plateau. In a recalculation, this time neglecting the heat source near India and considering only the cold source near the Bering Sea, the results show that the blocking flow pattern could not be formed—that is, the role of the cold source near the Bering Sea is enhanced in combination with the heat source near India.

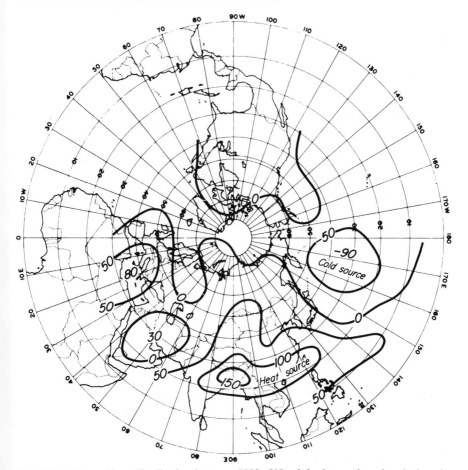

Fig. 24.5 Normal heat distribution between 1000–500-mb in the northern hemisphere in June.

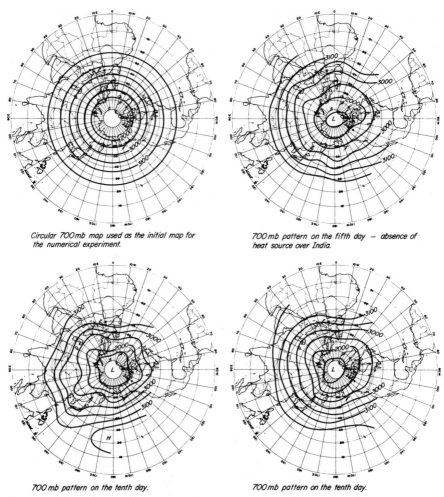

Circular 700mb map used as the initial map for the numerical experiment.

700mb pattern on the fifth day — absence of heat source over India.

700mb pattern on the tenth day.

700mb pattern on the tenth day.

Fig. 24.6 700-mb height contour. Left: Initial circular pattern and blocking flow pattern on tenth day. Right: 700-mb pattern on fifth and tenth day, excluding heat source over India.

Discussion

CAMERON: Please clarify the assumptions you made in deriving the model you used to calculate the cold pool development following cyclonic circulation centered on the pole.

ASAKURA: Quasi-geostrophic wind approximation is applied in this most recent model, and the heat or cold source is held constant throughout the simulation.

REFERENCES

ASAKURA, T.
   1968   Dynamic climatology of the atmospheric circulation over East Asia centered in Japan. *Meteorol. Geophys.* 19: 1–69.

ASAKURA, T., and A. KATAYAMA
   1964   On the normal distribution of heat sources and sinks in the lower troposphere over the Northern Hemisphere. *J. Meteorol. Soc. Japan*, Series 2, 42(4): 209–244.

BJERKNES, J.
   1966   A possible response of the atmospheric Hadley Circulation to equatorial anomalies of ocean temperature. *Tellus* 18: 820–829.
   1969   Atmospheric teleconections from the equatorial Pacific. *Monthly Weather Review* 7: 163–172.

CHAPTER 25

Many-yearly variations of the atmospheric circulation and long-term trends in the change of hydrometeorological conditions in the Bering Sea area

A. A. GIRS

Arctic and Antarctic Scientific Research Institute, Leningrad, USSR

Abstract

The patterns of atmospheric circulation of the northern hemisphere and their long-term variations are discussed. The existence of long-term trends of the same sign in the changes of the water salinity in the Bering Strait has been revealed and their association with the epochs of long-term atmospheric circulation is indicated. Prognostic suggestions on the tendency of salinity variations in the Bering Strait in the coming epoch are given.

CLASSIFICATION OF ATMOSPHERIC MACROPROCESSES

The atmospheric circulation of the northern hemisphere has been classified by Vangengeim (1941) into two meridional patterns and one zonal type (designated E, C, and W, respectively, in the Atlantic-Eurasian sector of the hemisphere; M_1, M_2, and Z in the Pacific-American sector). The distinguishing features of each of these patterns and their nine combinations are based upon certain properties of long thermobaric waves within the troposphere, the general scheme of which is illustrated in Figures 25.1–25.3.

In order to evaluate the surface fields of the meteorological elements peculiar to these nine patterns (Girs 1959, 1960, 1966), it is necessary to consider the following conditions that prevail at the earth's surface (Fleagle 1945): below *eastern* upper-level ridges (western upper-level troughs) there usually occur areas of negative temperature anomalies, positive pressure anomalies, and precipitation deficits; associated below *western* upper-level ridges (eastern upper-level troughs), conversely, are positive temperature anomalies, negative pressure anomalies, and areas of excessive precipitation. Maps of the surface meteorological fields of the W, C, E, Z, M_1, and M_2 patterns have been published (Girs 1959, 1960, 1966), showing these features and their corresponding epochal histories.

Impact on hydrosphere
Patterns of circulation in the atmosphere are reflected also in characteristics of the hydrosphere. For example, it is noted (Girs 1971) that during the processes of the E pattern, the level of the Caspian Sea decreases, the level of Baikal Lake increases, the extent of the Barents Sea ice cover is reduced, and the water temperature in the north Atlantic Ocean increases. An opposite picture occurs during processes of the W pattern (Girs 1971).

Hydrometeorological system
In the Pacific-American sector the processes of the M_1, M_2, and Z patterns are likewise distinguished in characteristics both of the atmosphere and hydrosphere: an upper-level *trough* develops at the meridians of the Aleutian Islands when M_2 processes are operating (Figs. 25.1–25.3), during which period the Bering Sea is characterized by the predominance of northern winds (the trailing edge of the Aleutian low), negative air temperature anomalies, surging of cold arctic water (Girs 1956, 1959, 1960), and a decrease of salinity in the Bering Strait. During M_1 processes the picture is opposite: an upper-level

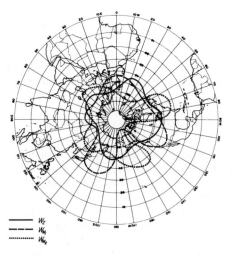

Fig. 25.1 Location scheme of main upper-level (500 mb) ridges and troughs during W pattern of atmospheric macroprocesses.

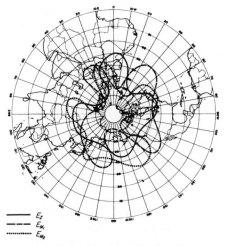

Fig. 25.2 Location scheme of main upper-level (500 mb) ridges and troughs during E pattern of atmospheric macroprocesses.

ridge develops at the meridians of the Aleutian Islands, extending the ridge of the Hawaiian high in this location; southwestern winds prevail in the western part of this ridge, warm water surges, warm air is advected into higher latitudes, and salinity increases in the Bering Sea and Bering Strait.

During zonal processes in the Bering Sea, conditions similar to those of the M_1 processes are developed. At this time there is usually an eastward shift of lows across the Aleutian Islands with a subsequent plunge into western America. The *leading* part of these lows is characterized by prevailing southern winds

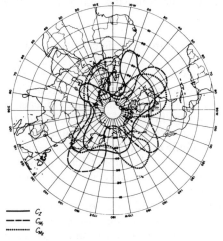

Fig. 25.3 Location scheme of main upper-level (500 mb) ridges and troughs during C pattern of atmospheric macroprocesses.

that contribute to the northern advection of warm air, surging of warm water, and an increase in salinity. When the lows reach the meridians of the Aleutian Islands, however, their trailing portion becomes associated with the predominance of northern winds, surging of cold water from more northern latitudes, advection of cold air, and a decrease in salinity: that is, conditions now resemble those observed during the M_2 processes.

Epochal patterns

The study of many-yearly variations in the occurrence of macroprocesses W, C, E, Z, M_1, and M_2 from 1900 to 1971 has shown (Girs 1966, 1967, 1971) that the annual frequency of these processes changes distinctly with respect to time, on the basis of which an epochal classification can be made. For example, during the epoch designated W_{Z+M_1} from 1900 to 1928, the frequency of W processes in the Atlantic-Eurasian sector and that of the Z and M_1 processes in the Pacific-American sector were systematically above the normal. Three further epochs have been recognized to date, according to the combination of atmospheric patterns defining them: E_{M_2} (1929 to 1939), C_{M_2} (1940 to 1948), and $(E + C)_{M_1+Z}$ (1949 to apparently 1973).

These epochs caused long-term trends of one sign in the change of a number of characteristics of the atmosphere and hydrosphere of both the northern and southern hemispheres, as analyzed in recent literature (Girs 1971). For example, the curve of many-yearly variations in water temperature in the north Atlantic Ocean (Fig. 25.4) implies that during the W_{Z+M_1} epoch (1900 to 1928) water temperature was generally below the normal and caused the integral curve to fall. The epoch of pattern E (1929 to 1939) was typified by an increase in water temperature, mainly in the area between Iceland and Europe; also during epoch C (1940 to 1948) an increase was noted but mainly in the area west of Iceland. Figure 25.4 shows the curve of sea level variation in the East Pacific Ocean near San Francisco. During epoch W_{Z+M_1} (1900 to 1928) the sea level tended to fall, recovering during the following epoch E_{M_2} (1929 to 1939).

Salinity trends

To characterize the effect of atmospheric circulation on the hydrologic regime of the Bering Sea, we analyzed salinity data for the Bering Strait along the cross-section of Cape Peyek-Cape Prince of Wales from 1941 to 1965. We calculated (Fedorova 1968) the salinity normal from these data for the given years, determined the departures from the normal for each year, and constructed the integral curve of salinity shown in Figure 25.5 along with frequency curves of the Z, M_1, and M_2 macroprocesses for the same years. Comparison of these curves enabled us to conclude that salinity in the area of the Bering Strait varies significantly with respect to time, and we have identified long-term periods during which salinity was systematically above or below the normal (the rise or fall of the integral curve).

Long-term hydrological trends are commensurate in duration and time with abrupt changes in the trends for development of atmospheric circulation. For instance, a general drop in the salinity curve was noted from 1946 to 1953, during the same years an abnormal development was observed in the M_2

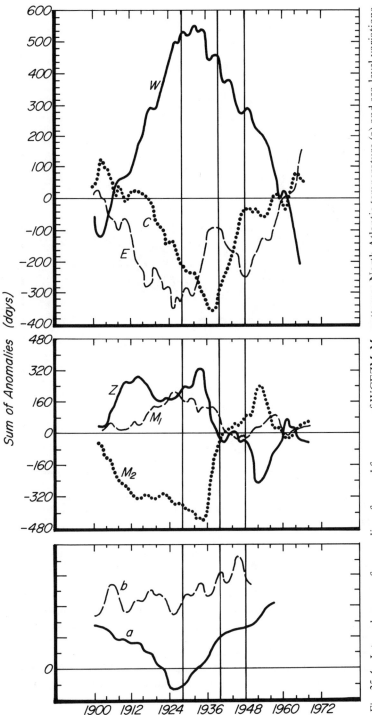

Fig. 25.4 Integral curves of anomalies of annual frequency of WCEZM₁,M₂ patterns; North Atlantic temperature (a) and sea-level variations near San Francisco (b).

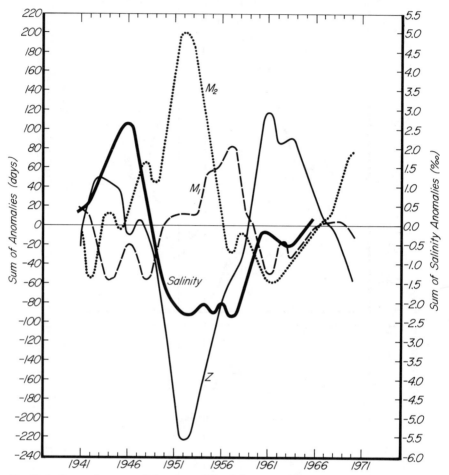

Fig. 25.5 Integral curves of the anomalies of the annual frequency of Z, M_1, and M_2 atmospheric processes and water salinity in the Bering Strait (Cape Peyek-Cape Prince of Wales profile).

processes to contribute to the above-mentioned decrease in salinity. An abrupt change in the course of both the salinity and M_2 curves took place simultaneously in 1953. From 1954 to 1957 slight variations were noted in the salinity pattern at the same time that M_2 processes weakened and Z and M_1 processes developed. Again, the M_1 processes contributed to an increase in salinity, and the Z processes could cause either a decrease or increase (Fig. 25.5). In the years that M_1 processes developed in combination with Z, salinity decreased somewhat—just as in the years during which M_1 weakened and only Z developed. An abrupt change toward a tendency of the integral salinity curve to *rise* took place in 1957 concomitant with an abrupt change toward falling in the M_1 curve.

The salinity curve rose appreciably from 1958 to 1961, corresponding to a rise in the integral curve for frequency of Z processes. An abrupt change in the course of both curves occurred in 1961. During the years in which M_1 and M_2 processes weakened and Z processes became manifest in such characteristics as a prevailing southern wind over the Bering Sea, there was an increase in salinity. The frequency of Z processes fell systematically below normal after 1961, opposed by a rise in the integral curve for M_2 and, to a lesser extent, for M_1 processes. An explanation was offered by this for the delay in the rising of the salinity curve after 1961 and transition to its slight variations. Although there are no salinity data available for the period after 1965, we suggest that the salinity curve tended to fall during the ensuing years due to relatively intensive development of M_2 processes.

In considering briefly a part of the salinity curve for 1942 to 1946, it can be seen (Fig. 25.5) that in those years the salinity was above the normal and coincided during 1942–1944 with development of zonal and M_2 processes. This fact is discrepant with the earlier conclusion that M_2 processes are typically associated with a salinity decrease. In this case, however, zonal processes of simultaneous development apparently interacted sufficiently to modify the character of the M_2 processes in such a manner that an increase in salinity could occur. During the following period 1945 to 1946 the rise of the salinity curve was evidently conditioned by the development of M_1 processes.

Future research

When it becomes possible to predict the pattern of atmospheric circulation over many years, one can then forecast probable long-term trends in the change of salinity in the Bering Sea and Bering Strait. Further, since salinity changes are accompanied by changes in other hydrological characteristics of the sea, the forecast of epochal transformations of atmospheric circulation patterns might well lead both to very long-term projections of hydrometeorological characteristics of the northern hemisphere in general and of the Bering Sea specifically. Development of so complex a capability as long-term forecasting of atmospheric circulation is at present under only preliminary study; the solution of this magnitude a problem entails an approach to the underlying question of why circulation epochs even occur (Girs 1971): the general explanation is that they appear primarily under the influence of a cosmogeophysical combination of factors such as solar activity, polar nutation, change of velocity in the earth's rotation, and tidal phenomena in the world ocean. These factors are considered in predicting that the current epoch of combined $E+C$ circulation will last apparently until 1973–1974 (Girs 1971), after which an epoch will develop of western (W) circulation together with M_1 and Z processes in the Pacific-American sector of the hemisphere—that is, processes which can be expected to contribute to an increase in Bering Sea salinity.

The hydrological and meteorological trends occurring in the Bering Sea are of great value in very long-term forecasting of sea and weather characteristics. Not to be neglected is the effect of these trends on short-term forecasts. Methods for applying longer-term trends to short-term forecasts are considered in detail in the literature (Girs 1971, Chapter 6).

Discussion

COACHMAN: On what data is the long-term salinity curve for the Bering Strait based?

GIRS: These Bering Sea station data are published in the Soviet journal *Okeanologia*, Vol. 8, No. 1, 1968.

REFERENCES

FEDOROVA, Z. P.
  1968   Transport of salts through the Bering Strait into the Chukchi Sea. *Okeanologiya* 8(1).

FLEAGLE, R. G.
  1945   The field of vertical motion in selected weather situations. *J. Amer. Geophys. Un.* 26: 359–366.

GIRS, A. A.
  1956   Interrelation of processes in the atmosphere and hydrosphere. U. S. Weather Bureau (OP-03A3) OPNAV-PO3–31.
  1959   Standard characteristics of the main varieties of patterns of atmospheric circulation during the cold season of the year. *Problemy Arktiki*, No. 7, *Morskoy Transport*, Leningrad.
  1960   Standard characteristics of the main varieties of patterns of atmospheric circulation during the warm season of the year. *Problemy Arktiki i Antarktiki*, No. 2, *Morskoy Transport*, Leningrad.
  1967   On the peculiarities of the arctic meteorological regime in different stages of the circulation epoch of 1949–1964. In *Proceedings of Symposium on Polar Meteorology*. [5–9 September 1966, Geneva, Switzerland]. WMO-No. 211 TP 111.
  1966   Heat regime of the Soviet Arctic related to the main patterns of the atmospheric circulation and their many-yearly variations. In *Proceedings of Symposium on the Arctic Heat Budget and Atmospheric Circulation*. RAND Corp., Santa Monica, Calif., 1966.
  1971   *Many-yearly variations of atmospheric circulation and long-term hydrometeorological forecasts.* Gidrometeoizdat (Leningrad).

VANGENGEIM, G. YA.
  1941   The prediction of seasonal distribution of meteorological elements. *Izv. Akad. Nauk SSSR*, Geophys. and Geogr. Series, No. 3.

Part 7

GEOLOGICAL PROCESSES

CHAPTER 26

Cenozoic sedimentary and tectonic history of the Bering Sea

C. HANS NELSON, DAVID M. HOPKINS *and* DAVID W. SCHOLL

U. S. Geological Survey, Menlo Park, California

Abstract

The Bering Sea consists of an abyssal basin that became isolated from the Pacific Ocean by the development of the Aleutian Ridge near the end of Cretaceous time and by formation of a large epicontinental shelf area that first became submerged near the middle of the Tertiary Period. We postulate that the sediment eroded from Alaska and from Siberia during Cenozoic time has been trapped in subsiding basins on the Bering shelf and in abyssal basins during the Tertiary, collected in continental rise and abyssal plain deposits of the Bering Sea during Pleistocene periods of low sea level, and has been transported generally northward from the Bering shelf through the Bering Strait into the Arctic Ocean during periods of high sea level in Pleistocene and Holocene time.

Filling of subsiding basins on the shelf was dominated by continental sedimentation in the early Tertiary and by marine deposition in the later Tertiary. River diversions caused by Miocene uplift of the Alaska Range increased the drainage area of the Yukon River twofold or more. This change established the Yukon as the dominant source of river sediments (90 percent) reaching the Bering Sea and greatly accelerated sedimentation in the basins.

The Quaternary Period has been presumably a time of alteration between two modes of sedimentation. When sea level was glacioeustatically lowered, the Yukon and other rivers extended their courses across the continental shelf and delivered most of their sediments to the abyssal basin; an incidental result was the cutting of some of the world's largest submarine canyons. While sea level was high, much sediment from the Yukon River was swept northward by current action through the Bering Strait and deposited on the continental shelf to the north.

INTRODUCTION

The Bering Sea, its resources, and its environmental setting are of intense interest to planners and scientists of several nations. Its biological resources are exploited by Russian, Japanese, American, and Korean fishing fleets. Parts of the floor of the Bering Sea contain deposits of valuable metals such as gold, tin, and platinum; other parts probably hold major accumulations of oil and gas. The minimal industrialization and urbanization of the Alaskan and Siberian continental margins and hinterlands have left these areas nearly unaltered. Future coastal development, mineral exploitation on land, on the sea floor, and in the sea, overfishing, and possible aquacultural developments will affect this pristine environment. Consequently, baseline environmental studies are needed to inventory the natural resources, to provide information for future quality-standards monitoring, and to establish natural-condition criteria for assessing polluted environments elsewhere.

The geology of the Bering Sea has been studied since the middle of the nineteenth century; an excellent review of earlier studies was prepared by Lisitsyn (1966). The state of knowledge resulting from Soviet and American studies of the marine geology, shoreline geology, and paleontology of the Bering Sea region between 1948 and 1965 is summarized in *The Bering land bridge* (Hopkins 1967b).

The modern era of marine geologic study began in the late 1940s with a series of Navy Electronics Laboratory cruises, during which data were gathered for the first published maps of the bottom sediments (Dietz et al. 1964) and the first seismic reflection profiles were obtained (Moore 1964). Cruises during the 1950s by scientists of the Institute of Oceanology, Academy of Science, USSR, resulted in a general discussion of the bottom morphology of the Bering Sea (Udintsev et al. 1959) and in a monographic treatment of the sediments and sedimentary processes of the western Bering Sea (Lisitsyn 1966). Cruises during the late 1950s and early 1960s by the Institute of Fisheries and Oceanography (VNIRO, Ministry of Fisheries, USSR) resulted in much new data on the sediments of the eastern Bering Sea (Gershanovich 1967, 1968). In the 1950s and 1960s, the sediments and topography of the Chirikov Basin and Norton Sound areas (Fig. 26.1) of the northern Bering Sea were investigated by the Department of Oceanography, University of Washington, in cooperation with the U. S. Coast Guard and the U. S. Geological Survey (Creager and McManus 1967; Creager et al. 1970; McManus and Smyth 1970; Grim and McManus 1970). Sediment and seismic profiling studies by the University of Washington were recently extended to the southcentral and southwestern continental shelf of the Bering Sea (Knebel and Creager 1970; Kummer and Creager 1971). Recent Soviet geophysical studies of the oil and gas potential of the Gulf of Anadyr have been reviewed by Verba et al. (1971). Bottom sediments of the Bristol Bay area were studied by Sharma (1970) of the University of Alaska. Studies of the deep structure of the abyssal basin of the Bering Sea were pioneered by Shor (1964). The Bowers Bank-Shirshov Ridge system has been studied by Lamont-Doherty Geological Observatory (Ewing et al. 1965; Ludwig, Houtz and Ewing 1971; Kienle 1970) in cooperation with Japanese investigators (Ludwig, Murauchi and Den 1971b).

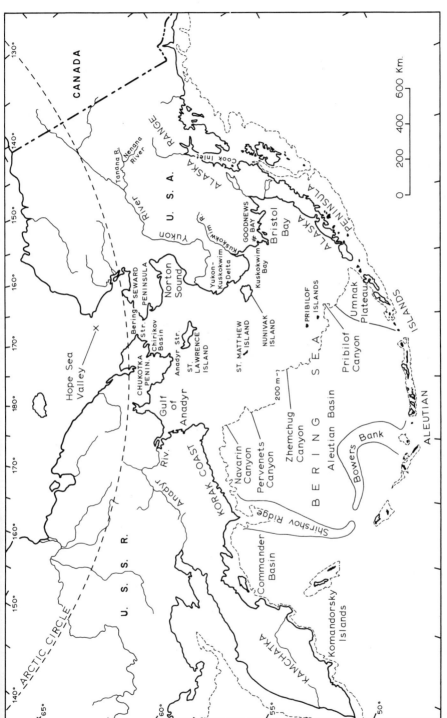

Fig. 26.1 Submarine and continental physiographic features in the Bering Sea region.

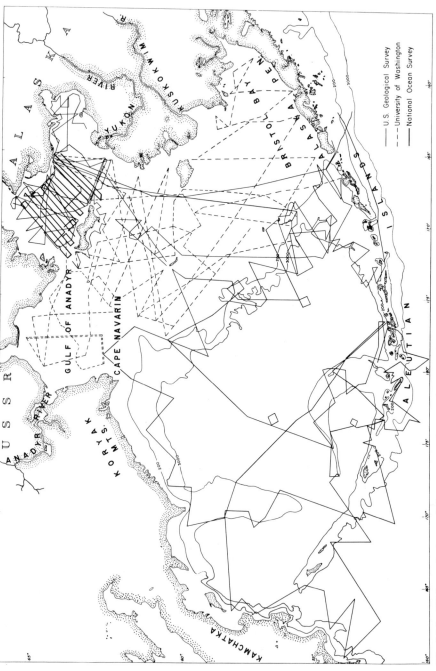

Fig. 26.2 High resolution and deep penetration continuous seismic reflection track lines completed in the Bering Sea by the University of Washington, National Ocean Survey and the U.S. Geological Survey.

Knowledge of the Bering Sea has been further advanced by offshore research conducted by the U. S. Geological Survey in cooperation with the U. S. Navy Undersea Center, the U. S. National Oceanographic and Atmospheric Agency, and the University of Washington. The U. S. Geological Survey studies have resulted in the acquisition of records for 35,000 km of deep-penetration seismic-reflection profile (SRP) records (Fig. 26.2), with shipborne magnetometer and bathymetric data for most of this distance; high-resolution SRP records for 4200 km; 35 dredge samples from the area of the continental slope; about 1000 large grab samples and 200 box cores from the continental shelf (Figs. 26.3 and 26.4).

Bering Sea research by the U. S. Geological Survey helped to describe the structure and origin of the abyssal basin and major topographic features in the deep Bering Sea, including the Umnak Plateau, the Bowers Bank, and Shirshov Ridge (Scholl et al. 1974) (Fig. 26.1); the nature and history of the continental margin (Scholl et al. 1968; Hopkins et al. 1969; Scholl, Buffington et al. 1970; Scholl and Marlow 1970a); the Tertiary tectonic pattern and the distribution, configuration, thickness, and probable age of Tertiary sedimentary basins on the shelf (Scholl and Hopkins 1969; Walton et al. 1969; Scholl and Marlow 1970b; Hopkins and Scholl 1970); the general distribution and nature of the unconsolidated sediments and the resource potential of detrital minerals in the northern Bering Sea and the Goodnews Bay area (Silberman 1969; Greene 1970; Moll 1970; Tagg and Greene 1973; Nelson 1971a; Venkatarathnam 1971; Sheth 1971; Nelson and Hopkins 1972); and the broad outlines of the Quaternary paleogeography in the northern Bering Sea (Hopkins 1967a, 1972; Hopkins et al. 1972). Studies have also been conducted in the adjacent Chukchi Sea to the north (Grantz et al. 1970a,b) and in the Aleutian Trench-Gulf of Alaska region to the south (von Huene and Shor 1969; Marlow et al. 1970; von Huene et al. 1970).

A new phase of research began with the first deep-sea drilling in the Bering Sea during the cruise of the R/V *Glomar Challenger* in 1971, the results of which are available in a preliminary account (Scholl, Creager et al. 1971).

TERTIARY GEOLOGIC HISTORY

Tectonic framework and plate motions

The continental shelf of the Bering Sea has long been known to be an area of continental crust that connects Asia and North America (Hopkins 1959); the deep part of the sea is a small abyssal ocean basin, floored by oceanic crust, and overlain by a thick sequence of undeformed sediments (Shor 1964; Menard 1967; Ludwig, Houtz and Ewing 1971). Recent studies have suggested that the continental shelf of the Bering Sea is a submerged part of a large continental plate that includes all of North America and northeastern Siberia (Churkin 1972), and the abyssal basin of the Bering Sea is thought to be a northern embayment of the Pacific Ocean that became isolated by the development of the Aleutian Ridge prior to late Eocene time but no earlier than late Cretaceous time (Scholl, Greene and Marlow 1970; Scholl and Buffington 1970; Scholl et al.

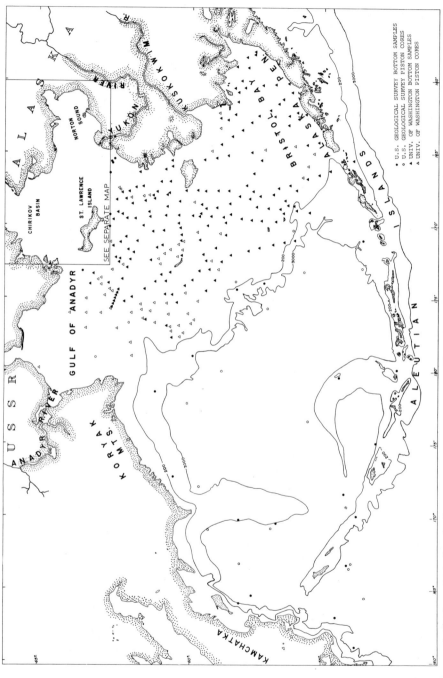

Fig. 26.3 Location of geologic samplings in the Bering Sea, excluding northeastern portion.

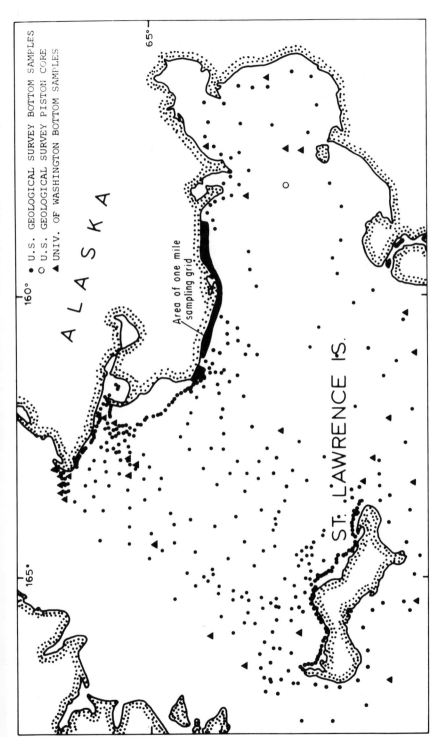

Fig. 26.4 Location of geologic samplings in the northeastern Bering Sea.

Legend (rotated):

- U.S. GEOLOGICAL SURVEY BOTTOM SAMPLES
- ○ U.S. GEOLOGICAL SURVEY PISTON CORE
- ▲ UNIV. OF WASHINGTON BOTTOM SAMPLES

ALASKA

Area of one mile sampling grid

ST. LAWRENCE IS.

160° 165° 65°

1974). The interaction of oceanic and continental crust was localized at the base of the continental slope of the Bering Sea during most of Mesozoic time. A shift in the zone of subduction southward away from Beringia [1] to the present site of the Aleutian trench-arc system then took place near the beginning of the Tertiary Period.

The tectonic history of the Bering Sea is complicated by the fact that prior to Miocene time the Pacific Plate did not impinge directly upon the North American-Siberian Plate but was separated from it by the Kula Spreading Center and the Kula Plate; the Kula Center and much of the Kula Plate were destroyed by subduction near the middle of the Tertiary Period (Grow and Atwater 1970). The floor of the abyssal basin of the Bering Sea is probably a remnant of the oceanic Kula Plate that has been connected to the Beringian segment of the continental North American-Asian Plate since the beginning of Tertiary time.

Much of the deeper Bering Sea is occupied by two abyssal plains, the Aleutian Basin and the Commander Basin (Fig. 26.1), which grade at their northern edges into a continental rise consisting of a series of coalescing deep-sea fans (Scholl et al. 1968). Seismic reflection profiles across the continental margin indicate that the continental basement is down-flexed and extends some distance seaward into the Aleutian Basin; this may in part reflect foundering of the continental margin as an isostatic response to loading by terrigenous debris deposited on the continental rise. The Umnak Plateau, a large area of intermediate depth occupying the southeastern corner of the abyssal basin (Fig. 26.1), may also be a mass of foundered continental crust. A perplexing feature of the abyssal portion of the Bering Sea is the Shirshov Ridge-Bowers Bank complex (Fig. 26.1), which consists of two large, nearly-connecting aseismic ridges extending in an S shape southward from the Siberian coast to the Aleutian Ridge and isolating the Commander Basin from the Aleutian Basin.

The tectonic outline of continental Beringia is dominated by a gigantic structure composed of the Alaskan orocline on the east and the Chukotkan orocline on the west, two sharp flexures concave toward the Pacific Ocean. The broad intervening faulted flexure is submerged on the continental shelf and concave toward the Arctic Ocean (Fig. 26.5). Hopkins and Scholl (1970) believe that large transcurrent faults, developed mostly during late Cretaceous and Paleocene time, are the strongest expression of the flexure system that probably reflects the bending of Siberia relative to North America as a consequence of rifting in the Atlantic and Arctic basins. The southward bulge of the Alaskan part of the North American continental margin may have resulted from this oroclinal folding. The shift in the site of the subduction zone from the continental margin in the Bering Sea to the Aleutian Trench may have been a response to bending of the continental margin.

Although the major oroclinal folding appears to have been complete before Oligocene time, the axial trends of basins and upwarps involving middle and

---

[1] Beringia includes the present area of western Alaska, northeastern Siberia, and shallow parts of the Bering and Chukchi seas.

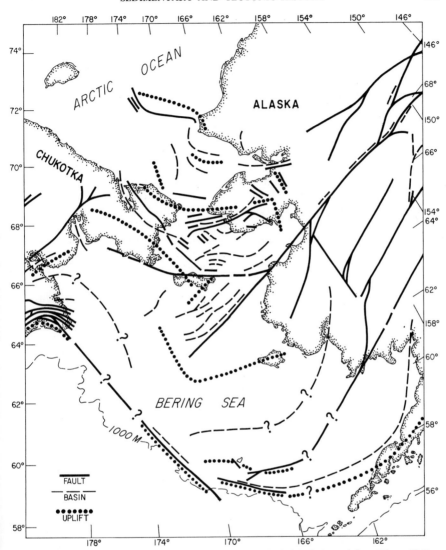

Fig. 26.5 Cenozoic structures of the Bering and Chukchi shelves (after King, 1969, and sources cited for Figure 26.7).

late Tertiary rocks generally conform to the configuration of the oroclines; many of the transcurrent faults remain active, displacing upper Tertiary and Quaternary sediments (Grim and McManus 1970). Movement along transcurrent faults just north and south of Bering Strait has probably played an important role in the development and destruction of intercontinental land connections in the area of Bering Strait.

Bering shelf acoustic and geologic units

Three broad groups of rocks and sediments have been distinguished on the continental shelf in the Bering Sea: folded rocks below a strongly reflecting horizon, termed the acoustic basement by Scholl et al. (1968); the main layered sequence (MLS), a unit of gently deformed sediments of Tertiary age above the acoustic basement; and the generally flat lying sediments of Quaternary age at or near the sea floor.

Acoustic basement

The acoustic basement is an erosion surface that bevels all rocks underlying the MLS. The petrology and age of the rocks below the acoustic basement are not well known; however, speculations based on magnetic mapping, extrapolation from rocks exposed on the mainland and insular areas, deep profiling, and dredging at the continental margin suggest that three distinctive tectonic units may occupy successively more southern belts (Fig. 26.6).

The basement structure beneath the Chirikov Basin and much of Norton Sound consists probably of strongly lithified rocks of Paleozoic age, metamorphosed in many places. Similar rocks are exposed on northern Chukotka Peninsula, on the Seward Peninsula, and on St. Lawrence Island.

The Okhotsk Volcanic Belt of late Mesozoic and early Cenozoic volcanogenic rocks (Tilman et al. 1969) extends across southern Chukotka Peninsula and beneath Anadyr Strait to southwestern St. Lawrence Island, thence south of the central and eastern part of the island (Patton and Csejtey 1970; Patton 1970). The southern limit of the Okhotsk Volcanic Belt lies within 100 km south of St. Matthew Island according to the silicic volcanic rocks exposed on the island and the presence of many steep magnetic anomalies in the surrounding area (Scholl and Marlow, unpublished data). Broadly undulating reflectors detectable beneath the acoustic basement in this region (Scholl and Marlow 1970b) are suggestive of the open folding that characterizes the Okhotsk Volcanic Belt (Tilman et al. 1969). Patton (1970) suggests that the Okhotsk Volcanic Belt swings northeastward from St. Matthew Island, passing beneath Norton Sound to the Alaskan mainland (Fig. 26.6).

The southernmost tectonic unit (Fig. 26.6) seems to consist of intensely deformed Cretaceous flysch sediments intruded by serpentinite. A similar complex is exposed extensively in mainland areas of Alaska and Siberia (Scholl et al. 1966, 1968, 1974); lithified turbidites of late Cretaceous (Campanian) age have been recovered from below the acoustic basement in the walls of the Pribilof Canyon (Hopkins et al. 1969); serpentinite lies beneath the Quaternary lavas that make up most of St. George Island (Barth 1956).

Since its formation in early Tertiary time, the erosion surface represented by the acoustic basement has been considerably deformed: it has been flexed into upwarps in insular and peninsular areas, where rocks of Paleocene and older age are now exposed, and into downwarps that occur as numerous coalescent sedimentary basins on the continental shelf; these basins are now filled with strata of the main layered sequence (Fig. 26.7). The acoustic basement is sharply downflexed at the continental margin and can be traced a considerable distance seaward beneath the continental rise (Scholl et al. 1966, 1968).

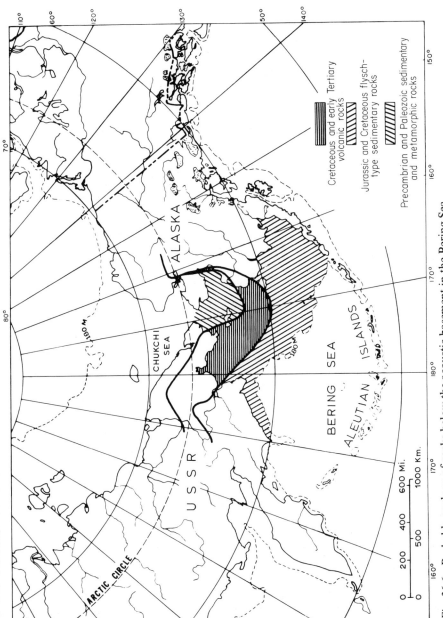

Fig. 26.6 Probable geology of rocks below the acoustic basement in the Bering Sea.

Cretaceous and early Tertiary volcanic rocks

Jurassic and Cretaceous flysch-type sedimentary rocks

Precambrian and Paleozoic sedimentary and metamorphic rocks

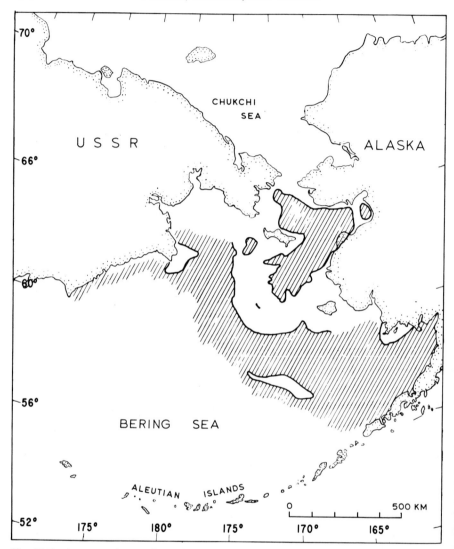

Fig. 26.7 Areas on the continental shelf of Bering Sea that are underlain by 500 m or more of main layered sequence (data from Scholl et al. 1968; Scholl and Hopkins 1969; Grantz et al. 1970a, 1970b; Scholl and Marlow 1970b; Verba et al. 1971; Kummer and Creager 1971).

From these limited data a late Mesozoic-early Cenozoic history of the Bering Sea area can be postulated. An episode of volcanism that began during Cretaceous time and continued into the Palocene Epoch in the Okhotsk Volcanic Belt extended through the Bering Sea to the Alaska mainland. Clastic and volcanic sediments were deposited at about the same time in several small, narrow, fault-bounded basins on the Alaskan mainland—for example, those filled with Chickaloon and Cantwell formations (Wolfe and Wahrhaftig 1966, 1970). Soon uplift processes ended sedimentation, however, and a long period of erosion began throughout Alaska, the Chukotka Peninsula, and the continental shelves of the Bering and Chukchi seas. Near the end of Oligocene time most of Beringia was a surface of marine and subaerial planation. This surface was later warped and the basins filled with Tertiary sediments of the main layered sequence.

Main layered sequence

The MLS is an acoustic unit that represents gently deformed sedimentary strata underlying most of the floor of the continental shelf of the Bering Sea. The unit is over 500 m thick in about two-thirds of the continental shelf area (Fig. 26.7); locally it is about 4 km thick (Scholl and Marlow 1970a; Scholl et al. 1974).

In a few places the MLS grades upward into Quaternary sediments with no visible unconformity. More commonly, however, the gently dipping strata of the MLS are truncated by a nearly horizontal erosion surface a short distance below the sea floor (Scholl and Hopkins 1969; Grim and McManus 1970). Outcrops of gently dipping beds within the MLS form cuestas on the sea floor south and east of St. Lawrence Island and probably elsewhere. The MLS forms a thick prograded sequence at the continental margin off Bristol Bay and the Gulf of Anadyr, and apparently a thinner mass of prograded MLS originally lay on the downwarped acoustic basement at the continental margin in central Bering Sea. The prograded MLS is modified by submarine slumps off Bristol Bay and the Gulf of Anadyr; on the central continental slope, the MLS has been intensely dissected by submarine canyons and gullies that cut into and below the acoustic basement (Scholl et al. 1968; Scholl, Buffington et al. 1970).

The MLS is an offshore analog to gently folded nonmarine sediments of middle and late Tertiary age that fill basins in mainland areas. Examples include the Kenai Formation in the Cook Inlet Basin of southern Alaska (Wolfe and Wahrhaftig 1966), the Tertiary Coal-bearing Formation of the central Alaska Range (Wahrhaftig et al. 1969), the Kougarok Gravel of the Seward Peninsula (Hopkins 1963), and the Koinatkhun Suite of the Chukotka Peninsula (Baranova et al. 1968; Biske 1973).

The MLS has been sampled at many locations near the shoreline and near the continental margin. Samples dredged from the continental margin consist of shallow-water detrital marine sediments rich in diatoms of Oligocene to late Pliocene or early Pleistocene age (Hopkins et al. 1969; Scholl et al. 1974). Several exploratory drill holes off Nome terminated in marine near-shore sands and clayey silts of early Pliocene age (D. M. Hopkins and C. H. Nelson, unpublished data). Marine detrital sediments of Oligocene age, nonmarine sediments of late Oligocene or early Miocene age, and marine sediments of late Miocene age are exposed near the west coast of the Gulf of Anadyr (Petrov 1966 and unpublished data; Baranova et al. 1968). Nonmarine sediments of late Oligocene age are exposed on northwestern St. Lawrence Island (Patton and Csejtey 1970) late Miocene to early Pliocene and marine limestone containing diatoms (I. Koizumi, Osaka Univ., written communication, 1971) has

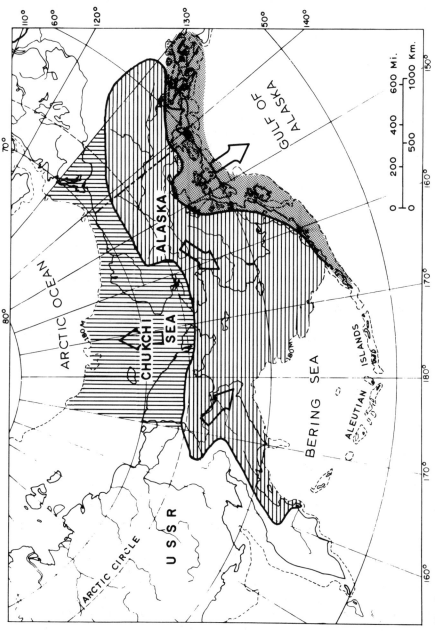

Fig. 26.8 Areas tributary during Early Tertiary time to the Bering Sea, the Gulf of Alaska and the Arctic Ocean.

been dredged from the sea bottom about 30 km south of St. Lawrence Island.

Stratigraphic studies of middle and late Tertiary beds in the Nenana and Cook Inlet Basins of mainland Alaska (Wahrhaftig et al. 1969; Kirschner and Lyon 1973) indicate major modification of the drainage pattern in Alaska that must have had pronounced effects upon sedimentation in the Bering Sea. Prior to late Miocene time much of central Alaska was drained by the tributaries of a trunk stream that flowed southward across the present site of the Alaska Range to the area of Cook Inlet and thence to the Gulf of Alaska (Fig. 26.8). Uplift of the Alaska Range during late Miocene time defeated this trunk stream, diverting the drainage from central Alaska westward to the Bering Sea by way of the Yukon River and its tributaries. This drainage diversion must have resulted in an enormous increase in the volume of sediment reaching the Bering Sea and a change in its character; the event is probably recorded by changes in the petrology of the sediments making up the MLS.

The geologic history of the Main Layered Sequence in the Bering Sea begins, then, after a prolonged period of denudation. The erosional carving of the acoustic basement took place probably from Eocene into Oligocene time. The erosional detritus from the present Bering continental shelf, as well as from southwestern Alaska and the southeastern Chukotka Peninsula, presumably was deposited in the Aleutian and Commander basins of the Bering Sea (Fig. 26.8). Sediment from parts of Beringia north of the Arctic Circle was deposited in the Canadian Basin of the Arctic Ocean; sediment from central, southern, and southeastern Alaska, and the southern Yukon Territory of Canada emptied into the Gulf of Alaska (Fig. 26.8). By late Oligocene time, however, much of the sediment was being trapped in small basins on the Alaskan and Chukotka peninsula mainlands and in a series of large shelf basins. Marine sediments were deposited on the actively subsiding continental margin and intermittently in the Anadyr Basin, but areas farther north were still emergent and receiving continental sediments.

During late Miocene time, uplift of the Alaska Range blocked at least one and possibly several drainage-ways that had been tributary to the Gulf of Alaska. These drainage-ways became integrated to form the present Yukon River, resulting in nearly 50 percent increase in the land area shedding sediment to the Bering Sea (Fig. 26.9) and probably in a significant increase in sediment deposited in its offshore basins. Late in the Miocene Epoch, owing either to eustatic rise in sea level or to structural downwarping, the sea gradually invaded basins successively northward on the continental shelf. Biogeographical evidence summarized by Hopkins (1967b) indicates that Bering Strait was inundated and finally formed a continuous marine connection between the Pacific and Arctic oceans by the end of the Miocene Epoch (Fig. 26.10).

Most of Miocene and Pliocene time was a period of progradational sedimentation at the continental margin, and deposition of the MLS built the shelf outward a considerable distance off Bristol Bay and off the Gulf of Anadyr; the continental margin prograded somewhat less in the central part of the Bering Sea area. Indirect biogeographical evidence indicates that the Bering Strait seaway was blocked briefly about five million years ago, forming a narrow Siberian-Alaska land bridge (Fig. 26.11); the seaway was re-established about

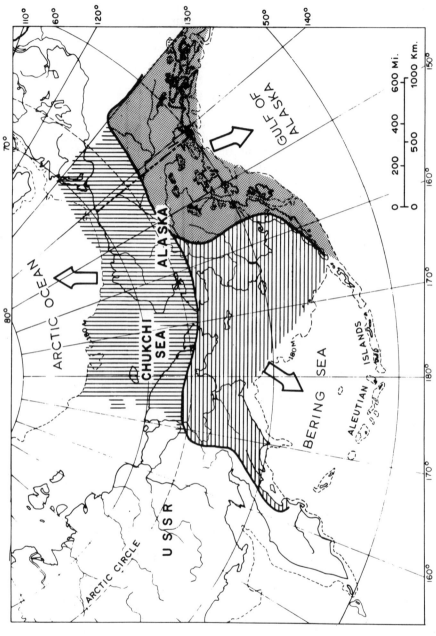

Fig. 26.9 Areas tributary from Late Miocene through Early Pleistocene time to the Bering Sea, the Gulf of Alaska and the Arctic Ocean.

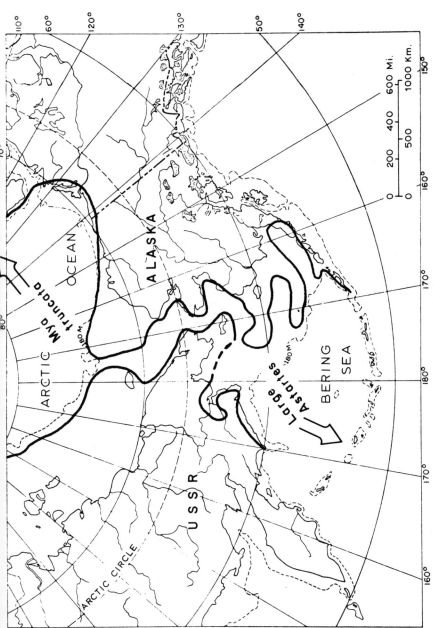

Fig. 26.10 Probable configuration of the Bering Seaway ten million years ago.

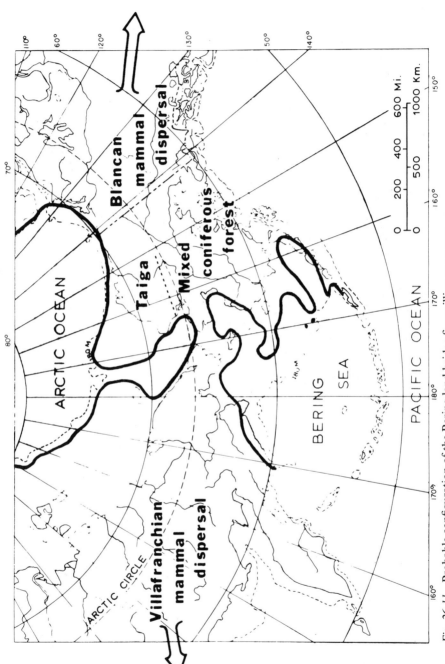

Fig. 26.11 Probable configuration of the Bering land bridge five million years ago.

three million years ago, and the Bering Sea assumed approximately its present configuration (Hopkins 1967a). In late Pliocene time, clastic progradation ceased (Hopkins et al. 1969).

QUATERNARY GEOLOGIC HISTORY

Structural and stratigraphic framework

High-resolution seismic profiling on the continental shelf of the Bering Sea indicates that the Tertiary Main Layered Sequence is overlain, generally unconformably, by a more complex suite of deposits of probable Quaternary age (Walton et al. 1969; Grim and McManus 1970; Nelson 1970; Kummer and Creager 1971). The Quaternary sediments on the continental shelf are rarely thicker than 100 m and in many places are only a few meters thick; they are absent entirely in some near-shore and insular areas, and pre-Quaternary bedrock is exposed on the sea floor.

Quaternary sediments nearly 1 km thick mantle most of the deep basin of the Bering Sea; thicker deposits may underlie the continental rise (Scholl et al. 1968; Scholl et al. 1971). A slope-mantling unit—found only on the outer shelf, continental slope, and on part of the Umnak Plateau in the southeastern sector—seems to consist of hemipelagic sediment draped over an undulating surface in areas of maximal organic productivity. A continental-rise unit is characterized by acoustic and topographic features, indicating that it probably consists of turbidites from a complex system of coalescing fans radiating beyond mouths of submarine canyons. Yet another Quaternary unit, the abyssal-plain (Scholl et al. 1968), contains a large surface component of diatom tests (Gershanovich 1967); the major part is detrital sediment shed from Beringia during Quaternary time, carried through submarine canyons, and deposited in the deepest parts of the Bering Sea (Scholl, Buffington et al. 1970; Scholl et al. 1971).

Tectonically the Quaternary deposits are relatively undisturbed; however, fault scarps break the sea floor in the northern Bering Sea (McManus and Grim 1970), near the Pribilof Islands (D. M. Hopkins, unpublished data), and perhaps elsewhere. A large closed depression north of the middle part of St. Lawrence Island, shown on a new bathymetric map of the region (U. S. Coast and Geodetic Survey 1969) is evidently of tectonic origin. Near-shore areas of Goodnews Bay seem to have undergone tectonic unwarping in late Quaternary time (Hopkins 1959; P. Barnes and A. R. Tagg, unpublished data), and warped terraces indicate considerable tectonic activity along the coast of the Seward Peninsula (Sainsbury 1967; D. M. Hopkins, unpublished data). Our unpublished studies show that lava flows and plugs of Quaternary age are present on the sea floor north of the central part of St. Lawrence Island, near the Pribilof Islands, and on the continental slope along the Aleutian Ridge.

Pleistocene glacial and fluvial deposition

The Pleistocene history of the Bering continental shelf is characterized by repeated episodes of subaerial exposure, fluvial sedimentation, and massive glacial encroachment, as well as deposition. Evidence of Siberian glaciers on the continental shelf is shown in highly deformed deposits observed in seismic profiles in the northern Gulf of Anadyr (Kummer and Creager 1971) and western Chirikov Basin (Grim and McManus 1970). From seismic records three distinct episodes of glaciation are recognized. The Siberian glaciers pushed as far as 150 km beyond the shoreline of the Chukotka Peninsula (Fig. 26.1). They encroached on St. Lawrence Island and deposited a series of morainal ridges on the continental shelf that are now expressed as gravel bars extending northward from the island (Fig. 26.12) (Nelson and Hopkins 1972; Hopkins et al. 1972). Other early-to-middle Pleistocene valley glaciers of Seward Peninsula pushed debris a few kilometers seaward from the present coast. The distribution of glacial deposits in southwestern Alaska (Coulter et al. 1965) indicates also that glaciers extended far beyond the shoreline into Bristol Bay.

Extensive migration of the Yukon River mouth and channels during Pleistocene periods of low sea level is indicated by reconnaissance seismic profiling. At least two major bodies of Yukon River sediment on the Bering shelf are apparent in high resolution profiles, and presumably there are others. In one such deposit of northern Norton Sound, prograded beds and distributary channels (Moore 1964) appear to extend as a buried surface far southwestward into Chirikov Basin (Grim and McManus 1970). Another younger deposit, discovered by Knebel and Creager (1973), lies between St. Lawrence and St. Matthew islands and probably formed during late Wisconsin time about 11,000 to 16,000 years ago. The presence of large, partially buried channels suggests that just prior to 16,000 years ago, the main Yukon River drained to the southwest of St. Lawrence Island and through the central shelf. The Pribilof and Zhemchug submarine canyons were probably carved by turbidity currents when the Yukon River reached the continental margin during such different times of lowered sea level (Scholl, Buffington et al. 1970).

Other trunk streams as well can be traced across the continental shelf. The Kuskokwim River evidently flowed south to central Bristol Bay and thence southwestward toward the abyssal Bering Sea (Hopkins 1972); turbidites from an extended Kuskokwim River may have carved the Bering Submarine Canyon (Scholl, Buffington et al. 1970). It appears the Anadyr River flowed southward through the Gulf of Anadyr to the Pervenets Canyon. Areas north of St. Lawrence Island were drained by a north-flowing trunk stream that passed through the Bering Strait to join the Hope Sea Valley (Creager and McManus 1967; Hopkins 1972). Faulting created two large depressions: one north of central St. Lawrence Island and the other south of Bering Strait. These depressions seem to have contained large lakes when the continental shelf was last exposed.

Holocene marine deposition

In Holocene time the erosional and depositional features of the Pleistocene have been masked by sediments in parts of the Bering Sea. In the southern Bering Sea, offshore from local mud-filled embayments along the shoreline, the Holocene deposits form a classic gradational sequence ranging from nearshore coarse sands to muds at the shelf edge (Sharma 1970). Farther north, however, Holocene sediments on the Bering shelf do not form as nearly complete a cover or show such a pattern of gradation. Rather, beginning with the northern and

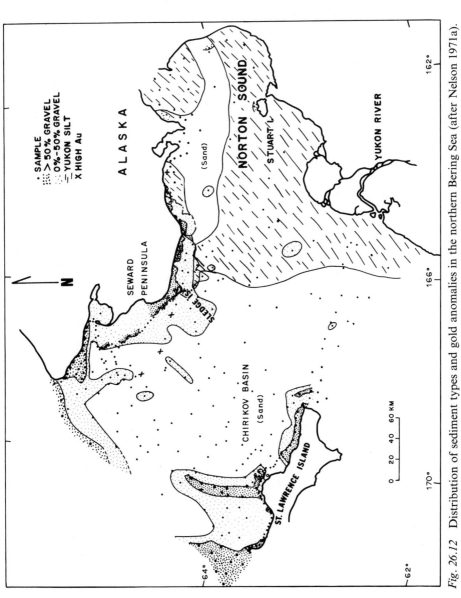

Fig. 26.12 Distribution of sediment types and gold anomalies in the northern Bering Sea (after Nelson 1971a).

western margins of the central shelf area (Lisitsyn 1966), the sediments are extremely heterogeneous. Here the distribution of relict and modern surface sediments is patchy and dependent upon positions of bedrock and glacial debris outcrops on the sea floor, locations of river sediment inflow, and water current velocity and patterns.

During high sea level of the Holocene, strong currents have affected the shape of the sea floor and distribution of sediment. In the Bering and Anadyr Straits, relict gravels and hummocky topography of apparent glacial origin remain exposed (Grim and McManus 1970; Nelson and Hopkins 1972); on the lee side of such current-swept channels, weaker currents have deposited sediments to form shoals like those north of Cape Prince of Wales (Creager et al. 1970) and Northeast Cape (Nelson 1970). Most of Chirikov Basin retains a thin cover of relict fine marine sands from the Holocene transgression except for the Siberian morainal ridges, near-shore areas, and straits, where current scour preserves the glacial and bedrock gravels (Fig. 26.12).

In the northern part of the Bering Sea, limited Holocene deposits and surface topographical features such as former beach ridges, outwash fans, stream valleys, and extensive relict sediments (Nelson and Hopkins 1972) suggest that sediments from the Yukon River and other streams have been swept from most of the northern Bering Sea into the Arctic Ocean by strong northward-moving currents during high sea level (Fleming and Heggarty 1966; Husby 1969, 1971). The present current regime and movement of Yukon-Kuskokwim sediment may explain the general lack of modern sediments in the northern Bering Sea and the presence of Holocene deposits on the epicontinental shelf to the north (Creager et al. 1970; McManus et al. 1969) and south (Sharma 1970).

Although nearly 90 million metric tons and 90 percent of the river sediment supplied to the Bering Sea are derived from the Yukon-Kuskokwim system (Lisitsyn 1966), a layer less than 20 cm thick of Holocene Yukon silt covers Norton Sound seaward of the river delta (Fig. 26.12) (C. H. Nelson, unpublished data). A recent radiocarbon date of 10,500 B.C. (C. H. Nelson and M. Rubin, unpublished data) for well-defined subaerial peat layers immediately below the Yukon silt indicates rates of Holocene sedimentation as low as 2 cm/100 years. Correlative paleontologic analyses in central and southern Norton Sound reveal the same low accumulation; yet the Yukon River input suggests that deposits here should be to the order of meters in thickness if all material of the past 10,500 years sedimented in Bering Sea.

The low rate of Holocene sedimentation in Norton Sound, the almost complete lack of Holocene sediment in the Chirikov Basin, but a relatively high rate in the Chukchi Sea suggests a major displacement and bypassing of Yukon-Kuskokwim sediment over the Bering epicontinental shelf. The several meters of Holocene silt in the Chukchi Sea (McManus et al. 1969) and high concentrations (26 g/m^3) of suspended sediment in the easternmost Bering Sea (McManus and Smyth 1970) indicates that major portions of Yukon sediment are being flushed out of Norton Sound and Bering Sea into the Chukchi Sea. This flushing apparently has occurred throughout Holocene time and could have taken place during other periods when the Bering Strait was submerged.

In contrast to the few measurements in eastern Bering Sea, studies of suspended sediment transport as well as other modern sedimentary processes have been well documented in western Bering Sea (Lisitsyn 1966). Shoreline erosion, ice rafting and organic production appear to be more important in the western Bering Sea, and measurements of suspended sediment indicate transport rates to be several orders of magnitude lower in the western Bering Sea (2–5 g/m^3; Lisitsyn 1969) than in the eastern part (5–25 g/m^3; McManus and Smyth 1970) of the Bering Sea. Lisitsyn (1966) suggests that pack ice and shorefast ice movements may be rafting pebbles as far as 100 km offshore. Gershanovich (1967) does not find evidence to support this in the eastern Bering Sea, nor do we (Fig. 26.12). Our data indicate that present sea floor gravel patches (at least 50 percent gravel) originated when the Holocene or earlier transgressions reworked glacial drift of the sea floor surface. Grounding ice now plucks up pebbles from these exposed gravel patches but drops scattered pebbles into finer sized sediments within 10–20 km of the gravel-rich regions (Fig. 26.12).

Sedimentary history and heavy metal concentration

The Bering Sea continental shelf is a mosaic of modern sediments (Sharma 1970) and of relict deposits formed in shallow water, at the strand, or in subaerial environments at times when sea level was lower than at present (McManus et al. 1969; Nelson 1970) (Fig. 26.12). Glacial deposits were formed on the Quaternary subaerial shelf far seaward of the present shoreline; subaerial drainage systems and glacial scouring dissected previous sea floors; and shoreline transgressions, regressions and late Quaternary stillstands at about −12, −17, −21, and −38 m (Tagg and Greene 1973; Nelson and Hopkins 1972) built shoals, partially filled in old stream valleys, and smoothed over the older topographic features. In other areas, shoreline processes and current action reworked, winnowed, and prevented deposition through sediment bypassing (Knebel and Creager 1970; Nelson 1970).

In places near mineralized land areas, the complex subaerial and marine processes have resulted in offshore placer deposits. Gold- and platinum-bearing glacial debris has been reworked by oscillations of the shoreline to form nearshore placers in surface relict gravels (Nelson and Hopkins 1972; A. R. Tagg and H. Barnes, unpublished data) (Fig. 26.12). Beach processes and strong current action have transported, winnowed, and concentrated the tin (as cassiterite), although tin placers of ore grade have not yet been recognized (Nelson 1971a).

Contemporary processes concentrate non-economic quantities of toxic heavy metals in near-shore regions, particularly along beaches of southern Seward Peninsula (Nelson 1971b; Nelson et al. 1972) and in the Kuskokwim River (Clark et al. 1971) and Goodnews Bay regions (H. Barnes, unpublished data). Limited reconnaissance indicates that modern beaches of the southern Seward Peninsula contain anomalies as high as 1.3 ppm Hg, 40 ppm Cr, and 75 ppm Cu. Even though data are not yet available for most toxic metals, general distribution trends are suggested by numerous gold (Nelson and Hopkins 1972) and mercury values in the northern Bering Sea (Figs. 26.12 and 26.13); their mean value content is highest in modern beaches (0.22 ppm Hg, 0.155 ppm

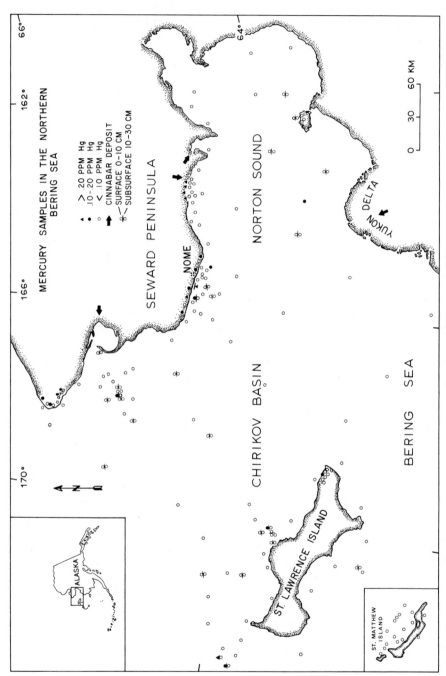

Fig. 26.13 Mercury distribution in the northern Bering Sea (after Nelson et al. 1972).

Au at Nome) and relict near-shore gravels (0.1 ppm Hg; 0.556 ppm Au at Nome) and lowest in offshore sands (0.03 ppm Hg; 0.001 ppm Hg). Although gold mining may have been partially responsible for mercury anomalies near Nome, Alaska, much greater mercury concentrations in Pliocene sediments immediately offshore (0.6 ppm) and in modern beach sediments at Bluff (0.45–1.3 ppm) indicate that the contamination effects of mining may have been insignificant. Normal background content of mercury (0.03 ppm) throughout the central Bering Sea areas and low mean values (0.03–0.05 ppm) immediately offshore from mercury-rich beaches suggests that effects of sediment dispersal of naturally and artificially introduced mercury has been minimal offshore in bottom sediments.

FUTURE RESEARCH NEEDS

The extensive background studies conducted thus far in the Bering Sea set the stage for topical geologic research of significant social value as well as high scientific interest. Future investigations should be focused upon the biological mineral, and fuel resources of the Bering Sea; upon geologic processes that can contribute to the understanding, use, and maintenance of these resources; and upon human activities that threaten the wise use of the resources. Because the biological resources of the Bering Sea are extensively used by several Pacific nations and because the contiguous mineral resources of the continental shelf are owned both by the United States and the Soviet Union, any effective research effort must be international in scope. It is proposed that geological investigations be planned in several subject areas:

Studies, including shallow bore-holes, should be made of the fuel and mineral resource potential, sedimentology, age, and tectonics of the Main Layered Sequence and of rocks below the Acoustic Basement on the continental shelf. This knowledge would facilitate the proposed joint Soviet-American research project to compile a 1 : 1,000,000 scale geologic map of western Alaska, northeastern Siberia, and the intervening seas;

There is a need for studies to define the hydrocarbon resource potential, age, and dispersal history of the sediments that comprise the deep-sea fans at the base of the continental slope. This study should also delineate potential mineral (Au, Pt, Sn) and fuel resources and assess potentially hazardous quantities of trace metals (Hg, Cu, Cr) that may be widely distributed in the sediment of the Bering Sea floor;

Studies with bore-holes should be conducted in submerged and buried beaches and in alluvial channels and deltas, in order to outline placer and toxic metal concentrations, sources, dispersal, and geologic history;

Paleoecologic studies are needed to gather pre-historic baseline information on natural concentration and cycling of trace metals, climatic variability, distribution changes of organisms, and normal ranges of oceanographic conditions. These data will aid in the evaluation of actual and potential man-made changes (for example, damming arctic rivers and the Bering Strait) in the world-wide atmospheric, hydrologic, and chemical systems in the particularly sensitive polar regions;

Geologic studies should coincide with other biologic investigations to monitor modern offshore sediment and organism concentrations and dispersal of toxic metals from natural onshore deposits along the southern Seward Peninsula, Kuskokwim Bay, and Goodnews Bay;

Synoptic studies of modern geologic processes should include major sediment and ice movement processes to assess effects of projected offshore mining, harbor facilities, and mineral transport systems.

Discussion

NESHYBA: Have there been any measurements of heat flow through the Bering Sea floor?

NELSON: Possibly Lamont-Doherty Geological Observatory might have made some measurements on the deep floor near the trench. A heat flow research program has been started in coastal areas of the Beaufort Sea to study permafrost regions underlying the shelf.

McROY: Are there any sediment data for the winter, especially in the Yukon Delta region?

NELSON: No winter data are available to my knowledge; all of our sampling has been done in the summertime. The Yukon winter system would be interesting to study, however, because our summer sampling has indicated that the sediment is changing very drastically. These changes could quite possibly be seasonal, as well as coincident with the storm pattern, and we would certainly need to know this in determining the Yukon flushing system.

McROY: It was reported by the U. S. Fish and Wildlife Service about 1949 that the Eskimos on St. Lawrence Island know good fishing will be found over areas of "grass bottom". This organic material may be seasonal but suggests that sediment studies throughout the year could be important to the local economy.

NELSON: Yes, I agree, and it might be speculated that a reduced current regime could allow accumulation of organic material in winter and washing away of the grasses in stronger summer currents.

REFERENCES

BARANOVA, YU. P., S. F. BISKE, V. F. GONCHAROV, I. S. KAL'KOVA, and A. D. TITKOV
    1968   Cenozoic of the northeast of the U.S.S.R. [in Russian]. *Akad. Sci., USSR, Siberian Div., Inst. Geol. Geophys.,* Tr. 128, 125 pp. (English trans. available from D. M. Hopkins, USGS, Menlo Park, Calif.).

BARTH, T. F. W.

1956 Geology and petrology of the Pribilof Islands, Alaska. *U. S. Geol. Survey Bull.* 1028-F: 101–160.

BISKE, S. F.

1973 Correlation of Tertiary nonmarine sediments in Alaska and northeastern Siberia. *In Proceedings of Second International Symposium* [Feb. 1971] *on Arctic Geology*, edited by M. G. Pitcher. Amer. Ass. Petrol. Geol. Mem. 19: 239–245.

CHURKIN, M.

1972 Western boundary of the North American continental plate in Asia. *Geol. Soc. Amer. Bull.* 83: 1027–1036.

CLARK, A. L., W. H. CONDON, J. M. HOARE, and D. H. SORG

1971 Analysis of stream-sediment samples from Taylor Mountains D-8 Quadrangle. *U. S. Geol. Survey Open File Rep.*, 60 pp.

COULTER, H. W., ET AL.

1965 Map showing extent of glaciation in Alaska. *U. S. Geol. Survey Miscellaneous Geologic Investigations Map* I-415.

CREAGER, J. S., R. J. ECHOLS, M. L. HOMES, and D. A. MCMANUS

1970 Chukchi Sea continental shelf sedimentation [Abstract]. *Amer. Ass. Petrol. Geol. Bull.* 54(12): 2475.

CREAGER, J. S., and D. A. MCMANUS

1967 Geology of the floor of Bering and Chukchi Seas—American studies. In *The Bering land bridge*, edited by D. M. Hopkins. Stanford Univ. Press, Stanford, Calif., pp. 32–46.

DIETZ, R. S., A. J. CARSOLA, E. C. BUFFINGTON, and C. J. SHIPEK

1964 Sediments and topography of the Alaska shelves. In *Papers in marine geology*, edited by R. L. Miller. MacMillan, New York, pp. 241–256.

EWING, M., W. J. LUDWIG, and J. EWING

1965 Oceanic structural history of the Bering Sea. *J. Geophys. Res.* 70: 4593–4600.

FLEMING, R. H., and D. HEGGARTY

1966 Oceanography of the southeastern Chukchi Sea. In *Environment of the Cape Thompson region, Alaska*, edited by N. J. Wilimovsky, and J. N. Wolfe. U. S. Atomic Energy Comm., pp. 697–754.

GERSHANOVICH, D. E.

1967 Late Quaternary sediments of Bering Sea and the Gulf of Alaska. In *The Bering land bridge*, edited by D. M. Hopkins. Stanford Univ. Press, Stanford, Calif., pp. 32–46.

1968 New data on geomorphology and recent sediments in the Bering Sea and the Gulf of Alaska. *Mar. Geol.* 6: 281–296.

GRANTZ, A., S. L. WOLF, L. BRESLAU, T. C. JOHNSON, and W. F. HANNA

1970a Chukchi Sea, seismic reflection and magnetic profiles—1969, between northern Alaska and International Date Line. *U. S. Geol. Survey Open File Rep.*, 26 pp.

1970b Reconnaissance geology of the Chukchi Sea as determined by acoustic and magnetic profiling. In *Proceedings of Geologic Seminar on North Slope of Alaska*, edited by W. L. Adkinson and W. W. Brosge. Amer. Ass. Petrol Geol., Pac. Section, Los Angeles, Calif., F1-F28.

GREENE, H. G.

1970   A portable refraction seismography survey of gold placer areas near Nome, Alaska. *U. S. Geol. Survey Bull.* 1312-B, 29 pp.

GRIM, M. S., and D. A. MCMANUS

1970   A shallow-water seismic-profiling survey of the northern Bering Sea. *Mar. Geol.* 8: 293–320.

GROW, J. A., and T. ATWATER

1970   Mid-Tertiary tectonic transition in the Aleutian Arc. *Geol. Soc. Amer. Bull.* 81: 3715–3722.

HOPKINS, D. M.

1959   Cenozoic history of the Bering Land Bridge. *Science* 129: 1519–1528.
1963   Geology of the Imuruk Lake area, Seward Peninsula, Alaska. *U. S. Geol. Survey Bull.* 1141-C, 101 pp.
1967a The Cenozoic history of Beringia—a synthesis. In his (ed.) *The Bering land bridge*, Stanford Univ. Press, Stanford, Calif., pp. 451–481.
1967b (Ed.). *The Bering land bridge*. Stanford Univ. Press, Stanford, Calif., 495 pp.
1972   The paleogeography and climatic history of Beringia during late Cenozoic time. *Internord* 12: 121–150.

HOPKINS, D. M., R. W. ROWLAND, and W. W. PATTON

1972   Middle Pleistocene mollusks from St. Lawrence Island and their significance for the paleo-oceanography of the Bering Sea. *Quaternary Res.* 2: 119–134.

HOPKINS, D. M., and D. W. SCHOLL

1970   Tectonic development of Beringia, late Mesozoic to Holocene [Abstract], *Amer. Ass. Petrol. Geol. Bull* 54(12), p. 2486.

HOPKINS, D. M., D. W. SCHOLL, W. O. ADDICOTT, R. L. PIERCE, P. B. SMITH, J. A. WOLFE, D. GERSHANOVICH, B. KOTENEV, K. E. LOHMAN, J. E. LIPPS, and J. OBRADOVICH

1969   Cretaceous, Tertiary, and Early Pleistocene rocks from the continental margin in the Bering Sea. *Geol. Soc. Amer. Bull.* 80: 1471–1480.

HUSBY, D. M.

1969   Report oceanographic Cruise USCG Northwind, northern Bering Sea–Bering Strait–Chukchi Sea, 1967. *U. S. Coast Guard Oceanogr. Rep.* 24, Washington, D. C., 75 pp.
1971   Oceanographic investigations in the northern Bering Sea and Bering Strait, June–July, 1968. *U. S. Coast Guard Oceanogr. Rep.* 40, Washington, D. C., 50 pp.

KIENLE, J.
1970 Gravity and magnetic measurements over Bowers Ridge and Shirshov Ridge, Bering Sea. *J. Geophys. Res.* 76: 7138–7153.

KING, P. B., JR.
1969 Tectonic map of North America, *U. S. Geol. Survey Map.*

KIRSCHNER, C. E., and C. A. LYON
1973 Stratigraphic and tectonic development of Cook Inlet petroleum province. In *Proceedings of Second International Symposium* [Feb. 1971] *on Arctic Geology,* edited by M. G. Pitcher. Amer. Ass. Petrol. Geol. Mem. 19: 396–407.

KNEBEL, H. J., and J. S. CREAGER
1970 Holocene sedimentary framework of east-central Bering Shelf: Preliminary results [Abstract]. *Amer. Ass. Petrol. Geol. Bull.* 54(12): 2491.
1973 Yukon River: evidence for extensive migration during the Holocene Transgression. *Science* 79: 1230–1231.

KUMMER, J. T., and J. S. CREAGER
1971 Marine geology and Cenozoic history of Gulf of Anadyr. *Mar. Geol.* 10(4): 257–280.

LISITSYN, A. P.
1966 Recent sedimentation in the Bering Sea [in Russian]. *Inst. Okeanol. Akad. Nauk USSR.* (Transl. by Israel Program for Scientific Translations, available from U. S. Dept. Commerce, Clearinghouse for Fed. Sci. and Tech. Info., 1969, 614 pp.).

LUDWIG, W. J., R. E. HOUTZ, and M. EWING
1971 Sediment distribution in the Bering Sea: Bowers Ridge, Shirshov Ridge, and enclosed basins. *J. Geophys. Res.* 76: 6367–6375.

LUDWIG, W. J., S. MURAUCHI, N. DEN ET AL.
1971 Structure of Bowers Ridge, Bering Sea. *J. Geophys. Res.* 76: 6350–6366.

MARLOW, M. S., D. W. SCHOLL, E. C. BUFFINGTON, R. E. BOYCE, T. R. ALPHA, P. J. SMITH, and C. J. SHIPEK
1970 Buldir depression—A late Tertiary graben on the Aleutian Ridge, Alaska. *Mar. Geol.* 8: 85–108.

MCMANUS, D. A., J. C. KELLEY, and J. S. CREAGER
1969 Continental shelf sedimentation in an arctic environment. *Geol. Soc. Amer. Bull.* 80: 1961–1984.

MCMANUS, D. A., and C. S. SMYTH
1970 Turbid bottom water on the continental shelf of northern Bering Sea. *J. Sed. Petrol.* 40(3): 869–877.

MENARD, H. W.
1967 Transitional types of crust under small ocean basins. *J. Geophys. Res.* 72: 3061–3073.

MOLL, R. F.
   1970   Clay mineralogy of the north Bering Sea shallows. M. S. Thesis, Univ. So. Calif., 101 pp.

MOORE, D. G.
   1964   Acoustic reflection reconnaissance of continental shelves: eastern Bering and Chukchi Seas. In *Papers in marine geology*, edited by R. L. Moore. MacMillan Co., N. Y., pp. 319–362.

NELSON, C. H.
   1970   Late Cenozoic history of deposition of northern Bering shelf [Abstract]. *Amer. Ass. Petrol. Geol.* 54(12): 2498.
   1971a Northern Bering Sea, a model for depositional history of Arctic Shelf placers; ecologic impact of placer development. In *Proceedings of the First International Conference on Port and Ocean Engineering under Arctic Conditions*, Tech. Univ. of Norway, Trondheim, Vol. 1 pp. 246–254.
   1971b Trace metal content of surface relict sediments and displacement of northern Bering Sea Holocene sediments [Abstract]. *Abstract Volume, Second National Coastal and Shallow Water Research Conference*, Univ. So. Calif. Press, p. 269.

NELSON, C. H., and D. M. HOPKINS
   1972   Sedimentary processes and distribution of particulate gold in northern Bering Sea. *U. S. Geol. Survey Prof. Paper* 689, 27 pp.

NELSON, C. H., D. E. PIERCE, K. W. LEONG, and F. F. H. WANG
   1972   Mercury distribution in ancient and modern sediment of northeastern Bering Sea. *U. S. Geol. Survey Open File Rep.* 533, 25 pp.

PATTON, W. W., JR.
   1970   Mesozoic tectonics and correlations in Yukon-Koyukuk Province, west-central Alaska [Abstract]. *Amer. Ass. Petrol. Geol. Bull.* 54(12): 2500.

PATTON, W. W., JR., and B. CSEJTEY, JR.
   1970   Preliminary geologic investigations of western St. Lawrence Island, Alaska. *U. S. Geol. Survey Open File Rep.*, 29 pp.

PETROV, O. M.
   1966   Stratigraphy and fauna of marine mollusks of Quaternary deposits of the Chukotka Peninsula [in Russian]. *Akad. Nauk SSSR Geol. Inst.*, Tr. 155: 257 pp. (Partial transl. available from D. M. Hopkins).

SAINSBURY, C. L.
   1967   Quaternary geology of western Seward Peninsula, Alaska. In *The Bering land bridge*, edited by D. M. Hopkins. Stanford Univ. Press, Stanford, Calif., pp. 121–143.

SCHOLL, D. W., E. C. BUFFINGTON, and D. M. HOPKINS
   1966   Exposure of basement rock on the continental slope of Bering Sea. *Science* 153: 992–994.
   1968   Geologic history of the continental margin of North America in Bering Sea. *Mar. Geol.* 6: 297–330.

SCHOLL, D. W., and D. M. HOPKINS

1969 Newly discovered Cenozoic basins, Bering shelf, Alaska. *Amer. Ass. Petrol. Geol. Bull.* 53: 2067–2078.

SCHOLL, D. W., and E. C. BUFFINGTON

1970 Structural evolution of Bering continental margin: Cretaceous to Eocene [Abstract]. *Amer. Ass. Petrol. Geol. Bull.* 54(12): 2503.

SCHOLL, D. W., E. C. BUFFINGTON, D. M. HOPKINS, and T. R. ALPHA

1970 The structure and origin of the large submarine canyons of the Bering Sea. *Mar. Geol.* 8: 187–210.

SCHOLL, D. W., H. G. GREENE, and M. S. MARLOW

1970 The Eocene age of the Adak "Paleozoic?" locality, Aleutian Islands, Alaska. *Geol. Soc. Amer. Bull.* 81: 3583–3592.

SCHOLL, D. W., and M. S. MARLOW

1970a Diapirlike structures in southeastern Bering Sea. *Amer. Ass. Petrol. Geol. Bull.* 54: 1644–1650.

1970b Bering Sea seismic profiles, 1969. *U. S. Geol. Survey Open File Rep.*

SCHOLL, D. W., J. S. CREAGER ET AL.

1971 Deep sea drilling project Leg 19. *Geotimes* 16(11): 12–15.

SCHOLL, D. W., E. C. BUFFINGTON, and M. S. MARLOW

1974 Plate tectonics and the structural evolution of the Aleutian-Bering Sea region. In *The geophysics and geology of the Bering Sea region*, edited by R. B. Forbes. Geol. Soc. Amer. Mem. 151.

SHARMA, G. D.

1970 Recent sedimentation on southern Bering shelf [Abstract]. *Amer. Ass. Petrol. Geol. Bull.* 54(12): 2503.

SHETH, M.

1971 A heavy mineral study of Pleistocene and Holocene sediments near Nome, Alaska. *U. S. Geol. Survey Open File Rep.*, 83 pp.

SHOR, G. G., JR.

1964 Structure of the Bering Sea and the Aleutian Ridge. *Mar. Geol.* 11: 213–219.

SILBERMAN, M. L.

1969 Preliminary report on electron microscopic examination of surface texture of quartz sand grains from the Bering shelf. *U. S. Geol. Survey Prof. Paper* 650-C, pp. C33–C37.

TAGG, A. R., and H. G. GREENE

1973 High-resolution seismic survey of a near-shore area. Nome, Alaska *U. S. Geol. Survey Prof. Paper* 759-A: A1–A23.

TILMAN, S. M., V. F. BELYI, A. A. NIKOLAEVSKII, and N. A. SHILO

1969 Tectonics of northeastern USSR [in Russian]. *Akad. Nauk SSSR, Nauchno-Issled. Inst.*, Tr. 33, 78 pp. (Transl. available from Nat. Transl. Center, Chicago, Ill.).

Udintsev, G. B., I. G. Biochenko, and V. F. Kanaev

  1959   Bottom relief of the Bering. In *Geographical description of the Bering Sea*, edited by P. L. Bezrukov. *Inst. Okeanol. Akad. Nauk SSSR*, Tr. 29, pp. 14–16. (Transl. by Israel Program for Scientific Transl., 1964).

U. S. Coast, and Geodetic Survey

  1969   St. Lawrence Island to Port Clarence, 1 : 250,000. Environ. Sci. Serv. Admin. Bathymetric Map (Prelim.), C&GS PBM-1.

Venkatarathnam, K.

  1971   Heavy minerals on the continental shelf of the northern Bering Sea. *U. S. Geol. Survey Open File Rep.*, 93 pp.

Verba, M. L., G. I. Gaponenko, S. S. Ivanov, A. N. Orlov, V. I. Timofeev, and Iu. F. Chernenkov

  1971   Deep structure and oil and gas prospects of the northwestern part of Bering Sea [in Russian]. *Nauchno-Issled. Inst. Geol. Arktiki (NIIGA), Geofiz. Metody Razvedki v Arktike* 6: 70–74. (English transl. available from Nat. Transl. Center, Chicago, Ill.).

von Huene, R., and G. G. Shor

  1969   The structure and tectonic history of the eastern Aleutian Trench. *Geol. Soc. Amer. Bull.* 80(10): 1889–1902.

von Huene, R., et al.

  1970   Marine geophysical study around Amchitka Island, western Aleutian Islands, Alaska. *U. S. Geol. Survey Rep.* USGS-474-74, 25 pp.

Wahrhaftig, C., J. A. Wolfe, E. B. Leopold, and M. A. Lamphere

  1969   The coal-bearing group in the Nenana coal field Alaska. *U. S. Geol. Survey Bull.* 1274-D, 30 pp.

Walton, F. W., R. B. Perry, and H. G. Greene

  1969   Seismic reflection profiles northern Bering Sea. U. S. Dept. Commerce, ESSA, Operational Data Rep. C&GS DR-8.

Wolfe, J. A., and C. Wahrhaftig

  1966   Tertiary stratigraphy and paleobotany of the Cook Inlet region, Alaska. *U. S. Geol. Survey Prof. Paper* 398-A, 29 pp.

  1970   The Cantwell Formation of the central Alaska Range. *U. S. Geol. Survey Bull.* 1294-A, pp. 41–46.

CHAPTER 27

Contemporary depositional environment of the eastern Bering Sea

Part 1. Contemporary sedimentary regimes of the eastern Bering Sea

GHANSHYAM D. SHARMA

Institute of Marine Science, University of Alaska, Fairbanks, Alaska

Abstract

A shallow shelf constitutes the greater part of the eastern and northeastern Bering Sea. The extensive shelf is smooth with an extremely gentle gradient and is generally bordered by a steep continental margin scarred with canyons. Most detritus deposited in the Bering Sea originates along the eastern coast. The coarse detritus is retained on the shallow shelf; the fines are carried to the shelf edge, resulting in seaward extension of the shelf. In the northeastern Bering Sea, however, the sediments are carried northward by predominant currents, flushed through the Bering Strait and finally deposited in the Chukchi Sea. Relatively meager sediment deposition and submarine erosion in this part of Bering Sea has left the relict sediment uncovered.

The eastern Bering Sea receives sediments from various sources. The Yukon and Kuskokwim rivers and the relatively young, rugged coastline account for the continental sediment contribution. Appreciable amounts of biogenous sediments and large amounts of suspended sediments brought by incoming North Pacific water are also deposited in the Bering Sea. Locally, volcanic ash transported by the wind has been reported by some investigators. In the shallow shelf area, the sediments are dispersed in the water column and transported and graded by frequent storms. The coastal and shelf area in the Bering Sea represent a high-energy depositional environment. The water movement is the major control for the sediment transport and deposition in the eastern Bering Sea.

During winter months, the northern and most of the eastern Bering Sea are covered with ice. A significant amount of sediments (silt and clay) is incorporated in sea ice and carried southward. The similarity of clay minerals in sea-ice sediments collected from various parts of the Bering Sea suggests a common source for these sediments. It is postulated that most of these sediments were in suspension during freezing of seawater and aggradation of sea ice.

INTRODUCTION

The Bering Sea is a unique subarctic body of water lying between 52°–66°N and 162°E–157°W, covering an area 2.25 × 10⁶ km². Almost half of the Bering Sea floor constitutes a gentle, uniformly sloped continental shelf. Approximately 80 percent of the total shelf lies in the eastern Bering Sea. Most of the shelf has been recently submerged and therefore provides a unique opportunity to study sediment transport and dispersion and the evolution of the contemporary continental shelf. Three major rivers—the Yukon, Anadyr, and Kuskokwim— and a relatively young coast provide plentiful sediments deposited in the Bering and Chukchi Seas. In the southeastern Bering Sea shelf, the thickness of contemporary sediments varies from 1.5 to 6 m, while the northern shelf is blanketed by a thin veneer of recent Yukon and Anadyr river sediments. Relict glacial sediments also cover a large area of the northern region. This paper describes the nature and sources for sediments deposited in the eastern Bering shelf and the modes of sediment transport and dispersal in the eastern Bering Sea.

OCEANOGRAPHIC SETTING

The water movement and structure in the Bering Sea are intrinsically controlled by meteorological conditions. The atmospheric patterns prevailing over the region cause Pacific Ocean water to move northward into the Bering Sea through various passes along the Aleutian arc. It is estimated that approximately 1.5 × 10¹⁴ m³ of Pacific water enter the Bering Sea annually. Most of it returns, however, leaving only about 2.1 × 10¹³ m³, or 14 percent of the total water in the Bering Sea. The added Pacific water and the surface flow contributed by various rivers then flow northward, spilling over the Bering Strait into the Chukchi Sea. The amount of water moving through the Bering Sea varies seasonally as well as annually.

Since most of the Bering Sea is located in the subarctic climate, cyclonic circulation predominates over the region, and renders the eastern half warmer than the western half. The annual atmospheric circulation is controlled by the Honolulu, Arctic, and Siberian highs; seasonally, the weather in the region is influenced by the Aleutian low and Asiatic depression. The summer shift of the Honolulu anticyclone west and northwestward leads to an intensification of the cyclonic circulation and an increase of the frequency of south winds. The winter eastward shift of the Honolulu anticyclone to the east results in a simultaneous advance of the Arctic anticyclone to the south and thus to an intensification of north wind frequency. The direction, intensity, and duration of winds influence the Bering Sea water exchange between the Pacific Ocean to the south and the Arctic Ocean to the north.

The Bering Sea lies in the path of both extratropical cyclonic and Asiatic anticyclonic storms. Cyclonic storms occur in the north extratropical region so frequently that several are usually present. The general track of these migratory cyclonic storms is along the Japanese and Kurile islands, thence north and east over the Bering Sea. The energy imparted by these frequent storms to the sea waves mostly dissipates on the shallow shelf, causing destruction of water structure on the shelf and mass sediment movement away from shore.

Another unusual feature of the eastern Bering Sea is the sea ice cover during winter months. Pack ice covers the northern shelf area and often extends as far as the Pribilofs. The inflow of warm Pacific water along the Alaska Peninsula generally inhibits formation of sea ice in this region. Generally shore ice starts to form in December and often continues until late April. During early May the ice break-up begins along the eastern coast, and in late June most of the Bering Sea, including Bering Strait, is ice-free.

GEOLOGY

Regional geology, tectonics and analyses of reflection profiles from the eastern Bering Shelf have been described by Nelson et al. (Chapter 26, this volume). The morphology of the shelf floor is featureless and quite level except for a few shallow depressions. The average slope is approximately 1 m/3 km. The shelf displays a degree of leveling and slope uniformity that are extremely rare in other parts of the world oceans. As a result of smooth topography, sediments deposited on the shelf are independent of the usual topographic complications. The distribution of contemporary sediments is controlled primarily by the water dynamics, and the mineralogical character is related to various provenance. On the basis of these considerations, the eastern Bering Shelf can be divided into five regions, each with its own sediment characteristics and sediment distribution. The bottom sediments described are from: the Southeastern Shelf, including Bristol Bay; the Central Shelf, a broad region lying between St. Matthews and Nunivak islands; the Northeastern Shelf, the region bounded by Seward Peninsula, the Yukon Delta, St. Lawrence Island and the Russian coast in the west, generally known as Chirikov Basin and Norton Basin; the Northwestern Shelf, essentially covering the Gulf of Anadyr; and finally, the Outer Shelf, a north-south oriented area running parallel to the continental margin (Fig. 27.1).

Southeastern Bering Shelf

The Southeastern Bering Shelf is a triangular embayment, commonly known as Bristol Bay (Fig. 27.2). The area is bounded on the north and the east by the southern spurs of the Kilbuk Mountains and along the south by the Alaska Peninsula. From the apex the shelf extends approximately 450 km westward to a width of about 380 km. The region receives the drainage of numerous rivers and lakes, notably the Nushagak River from the north and the Kvichak River from the east, which form modest deltas at their mouths. The bottom

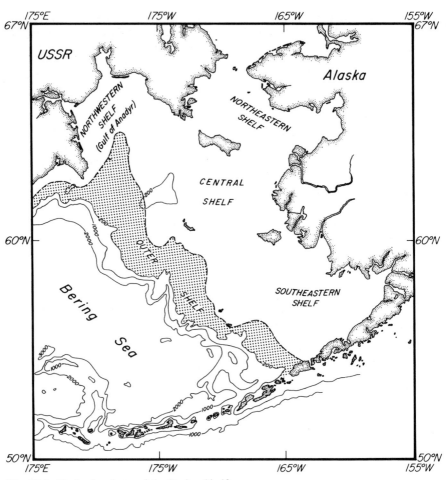

Fig. 27.1   Basinal regimes of the Bering Shelf.

morphology features a series of banks in the north and shallow depressions along the Alaska Peninsula. The depth increases gradually from the head of the bay to 100 m over a distance of 450 km with an average gradient of only 2.2×10^{-2} percent.

The sediments from the Southeastern Bering Shelf display a varying sediment texture. Near-shore sediments consist of gravel and coarse sand, while a greater part of the shelf is covered with fine to medium sands, and further offshore sediments become progressively finer (Fig. 27.3). The sediment mean size decreases with increasing water depth (Fig. 27.4). The sorting in sediments on the Southeastern Bering Shelf varies over a wide range and is a function of the sediment mean size (Fig. 27.5). Nearshore coarse sediments are extremely poorly sorted, medium and fine sands deposited on the mid-shelf are moderately well sorted, and offshore the sorting deteriorates with increasing silt and clay components. The nearshore sediments are strongly coarse-skewed, and the size distribution of mid-shelf sand is nearly symmetrical. Most sediments are leptokurtic to extremely leptokurtic.

Water samples from various depths for suspended load measurements in the Southeastern Bering Shelf were collected during summer under calm sea conditions. The suspended load consisted both of biogenic matter (mostly pelagic diatoms) and clay-sized sediments. Sediments in suspension varied from 1.6–12.0 mg/liter; however, most measurements were between 7.0–9.0 mg/liter. The lower euphotic zone at 10–20 m generally had maximum suspended load (Fig. 27.6) and is attributed to increased biogenic contributions. The surface sediment concentration, with the exception of local influence such as that near Port Moller, generally decreased with increased distance from shore.

The percent weight distribution of organic carbon in the surface sediments of the Southeastern Bering Shelf is shown in Figure 27.7. Higher concentrations of organic carbon occur near Togiak Bay and the Outer Shelf. The content of organic carbon increases with increasing clay content in the sediments (Fig. 27.8). The abundance of organic carbon in marine sediments is controlled

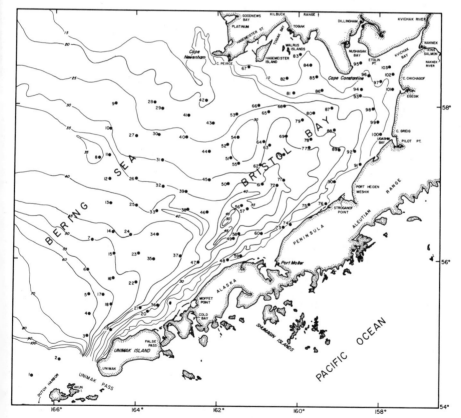

Fig. 27.2 Bathymetry and sampling locations on Southeastern Bering Shelf.

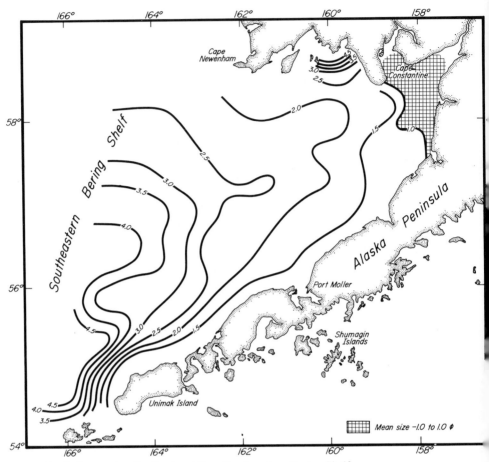

Fig. 27.3   Sediment size distribution (φ) on Southeastern Bering Shelf.

by production in the water column, transportation and destruction by both biological and non-biological processes. High productivity due to upwelling and wind-induced mixing near Unimak Pass and the Outer Shelf has been observed by J. J. Goering (personal communication, 1972) and may account for the high values in these regions. Although no data are available on the amount of organic matter contributed by local rivers, it is commonly known that rivers carry significant amounts of organic carbon to the oceans. This source may account for high amounts of organic carbon in Togiak Bay. The increasing organic carbon with increasing clays in the sediments (Fig. 27.8) strongly suggests that organic matter is generally transported and deposited with fine-grained or clay sized sediments. This perhaps results from adsorption of organics on the clay particles or by the current regimes which control distribution and deposition of materials carried in suspension.

The important sources of detrital sediments in the Southeastern Bering

Shelf are the drainage from the east and north and the contemporary volcanism along the Alaska Peninsula in the south. The common occurrence of euhedral and relatively unaltered hypersthene, magnetite, and ilmenite indicate a nearly basic and ultrabasic source. Reddish-brown pleochroic hornblende, a characteristic constituent of basic and ultrabasic rocks exposed in the Alaska Peninsula, was observed in most sediments and suggested that a significant source of sediments is in the south. High grade metamorphic minerals

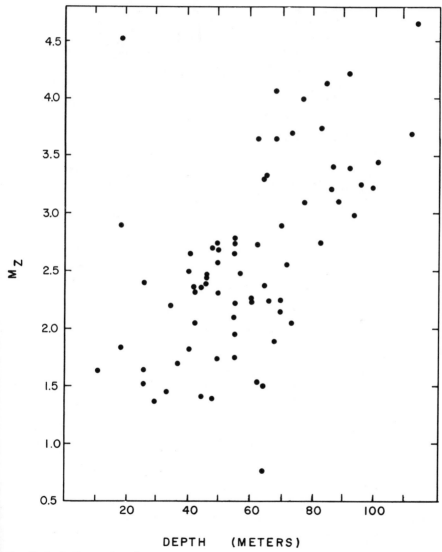

Fig. 27.4 Sediment size (φ) versus depth (m) on Southeastern Bering Shelf.

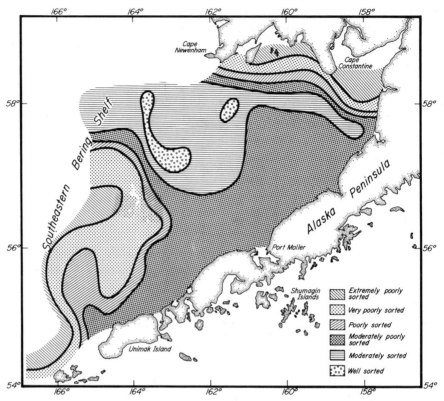

Fig. 27.5 Sediment sorting on Southeastern Bering Shelf.

sillimanite, garnet, staurolite, and epidote in the sediments link their source in the metamorphic province of the Alaska mainland.

Central Bering Shelf

The Central Bering Shelf extends from the Kuskokwim Delta north to the southern end of the Yukon Delta and shores of St. Lawrence Island. The shelf has two large islands—Nunivak and St. Lawrence. The average depth of the shelf is 50 m.

The sediments from the Central Bering Shelf have been studied by the University of Washington. The shelf is covered by sand (50–90 percent) and silt. The sediment mean size decreases with increasing depth and increasing distance from the shore (3–6 ϕ). Locally the Kuskokwim River, which carries large amounts of silt, controls the sediment distribution. In contrast to the unequivocal graded sand on the Southeastern Bering Shelf, the silt and sand composition in the Central Bering Shelf sediments varies throughout. The sediments are strongly fine-skewed and moderately sorted in the mid-shelf, while sorting becomes poorer towards the shore and offshore.

Sediments in suspension from a southeast transect between St. Lawrence Island and the Alaskan coast have been measured by McManus and Smyth (1970). The suspended load varied between 5 mg/liter in the mid-shelf to 10 mg/liter near the Alaska mainland and increased with increasing depth. The material in suspension consisted of more than 85 percent mineral grains with a mode of 20 μ(5.5 ϕ).

The sources for sediments on the Central Shelf are Quaternary alluvial, palustrine, lacustrine and glacial deposits and Cenozoic andesites, basalts and intrusives exposed along the coast; Tertiary shales, sandstones, clays, lignite and conglomerates drained by the Kuskokwim River; and the biogenic processes.

Northeastern Bering Shelf

The Northeastern Bering Shelf is defined by the Bering Strait to the north, by a 46-m deep sill separating the Gulf of Anadyr to the west, and by a 32-m deep sill across the Alaska Mainland and St. Lawrence Island to the south. The Northeastern Shelf is a shallow basin almost surrounded by the land mass. The

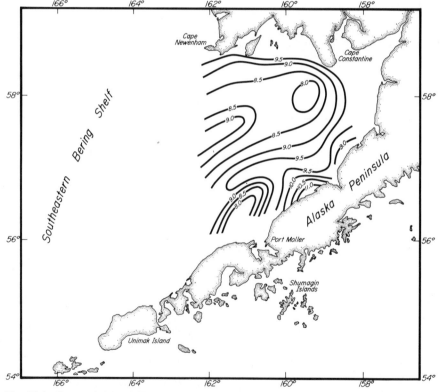

Fig. 27.6 Suspended load distribution at 15 m in the southeastern Bering Sea.

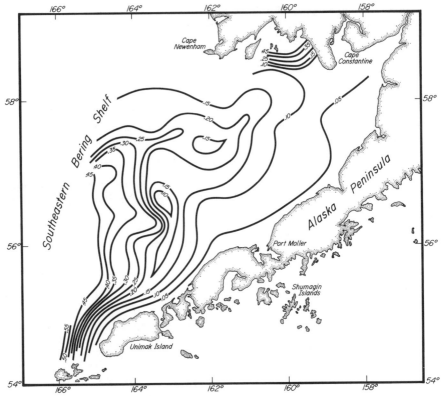

Fig. 27.7   Organic carbon (weight percent) in Southeastern Bering Shelf sediments.

eastern part, Norton Sound, is less than 30 m deep and receives the drainage of the Yukon River from the southeast. The western part, Chirikov Basin, is slightly deeper with a few channels.

The sediments from the Northwestern Bering Shelf have been described by Nelson and Hopkins (1972), Creager and McManus (1967) and Venkata-rathnam (1969). The sediment distribution in this region is complex because the basin is covered by both glacially derived relict and marine contemporary deposits. Submarine exposures of sandy gravel, reworked relict glacial morainal deposits, cover significant areas. Gravel sediments are common off the coast of Seward Peninsula and St. Lawrence Island, in offshore areas of the west-central shelf and in the Bering Strait. In the east, the relict morainal and outwash deposits extend close to the present Alaskan shoreline. The Bering Strait gravel which extends southwest towards St. Lawrence Island is probably a morainal ridge deposited by glaciers that originated in Siberia.

The marine sediments consist of silt and sand that cover much of the shelf. In the southeastern basin, the Yukon River has built a thin delta of clayey silt which overlies the silty sand (Fig. 27.9).

Nine heavy-mineral assemblages are recognized in the Northeastern Shelf

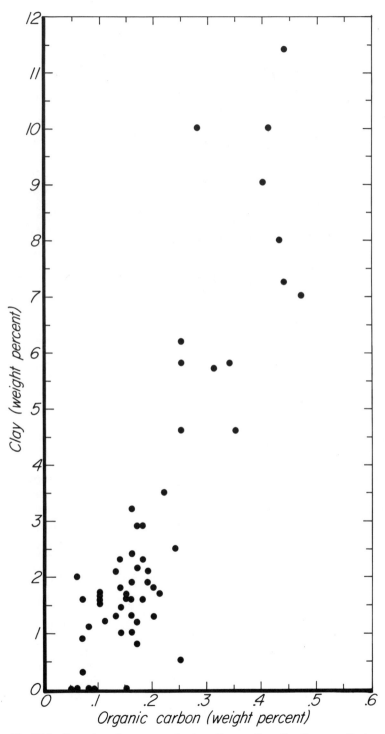

Fig. 27.8 Organic carbon versus clay in sediments from Southeastern Bering Shelf.

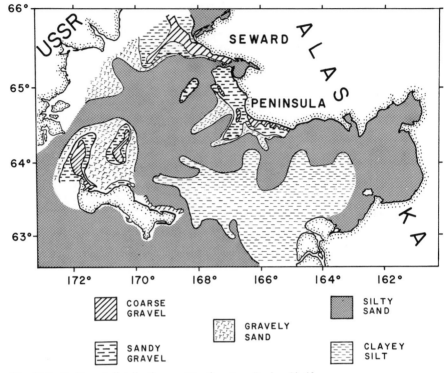

Fig. 27.9   Sediment distribution on Northeastern Bering Shelf.

by Venkatarathnam (1969). The entire shelf, however, can be broadly divided into three regions on the basis of heavy-mineral distribution. An area of low heavy-mineral content extends northwest-southeast along the broad shallow valley in the central part of the shelf, separating areas of relatively high concentrations on either side. The heavy mineral distribution and sediment texture suggest that much of the silty sand was deposited during lower sea levels, whereas the deposition of the clayey silt in Norton Sound is controlled by contemporary environment.

Source areas for detrital sediments of the Northeastern Shelf are Yukon River-clayey silt characterized by a high content of clinopyroxene, hornblende and hypersthene; the Siberian coast contribution of relict glacial deposits characterized by hornblende, opaque minerals with sphene, zircon and metallic copper; and the Central and Northwestern Shelf. The source for silty sand deposited during lower sea level in the western parts is not known.

Northwestern Bering Shelf

This part of the Bering Shelf, commonly known as the Gulf of Anadyr, is deeper than the other shelf areas. Much of the gulf shelf lies between 50–100 m isobath. The gentle slope is cut by numerous valleys with an average relief of 20 m. The shelf displays broad ridges measuring 20 to 30 km at the base and up to 20 m high. These bottom features are the manifestation of buried bedrock relief. The thickness of unconsolidated sediments in the Gulf of Anadyr varies between 15–20 m, and the upper 5 m is entirely of post-glacial origin (Lisitsyn 1966). The major Anadyr River empties into the Gulf of Anadyr.

Sediments in the Northwestern Bering Shelf are described in detail by Lisitsyn (1966). The nearshore sediments in the Gulf of Anadyr consist of gravels which form a broad, 50–60 km wide belt extending from shore to depths of 30–40 m. Adjacent to the gravels, coarse sands cover the gulf and generally are found at depths down to 50 m. Much of the gulf is covered by dark gray, well-sorted sands. These sands are predominant in deeper portions of the gulf, where they form a carpet as wide as 300 km. Generally, sands are separated from the coast by gravels; however, in the western part of the gulf the sand extends to the shore. Often fine sand contains large proportions of biogenic fragments. Fine sands with good sorting and rich in heavy minerals are widespread at depths varying from 50 to 90 m and cover extensive areas ranging from 150–170 km in width.

The suspended-matter concentrations over the Northwestern Shelf vary seasonally as well as vertically and laterally in the water column. Concentrations in surface water in the Gulf of Anadyr reach a peak in spring. Values up to 13 mg/liter have been observed during high flow of the Anadyr River. Normally the concentrations vary between 5 to 7 mg/liter in the Gulf and 3 to 5 mg/liter on the adjacent shelf. The suspended load seldom decreases with increasing depth. Bottom turbid layers with high quantities of suspended material have often been observed. The suspended material consists of mineral and biogenous (diatomaceous) constituents.

The Gulf of Anadyr sediments are predominantly sand, consisting of significant amounts of rock fragments and muscovite. The light minerals (feldspars and quartz) display reddish brown iron stain coating. The sands are well polished and show a fairly high degree of roundness. The heavy minerals include pyroxene, hornblende and epidote. The magnetic content in the sediments is low, while ilmenite is fairly common. The Gulf of Anadyr sands can be divided into two mineralogical provinces: the eastern and western. The sediments in the eastern part of the Gulf are characterized by a high content of K-feldspar, quartz, and hornblende fibrous amphiboles, which are related to granitoids of the Chukchi Peninsula. The western sector sediments transported by the Anadyr River consist of abundant plagioclase and ilmenite and low amounts of quartz, hornblende, and pyroxene. The typical minerals of this region are black ore minerals, hypersthene, augite-diopside, zircon, tourmaline, and rock fragments. The heavy mineral content in fine sand from this part of the Gulf exceeds 40 percent. The highest heavy-mineral concentration occurs at the Anadyr River mouth and decreases offshore.

The distribution of clays (chlorite, illite, and montmorillonite) and heavy minerals in the Gulf of Anadyr bear out that most sediments in the Northwestern Bering Shelf are brought by the Anadyr River, which drains the Anadyr Basin. The coarse material originates in the shore and is generally transported along the coastline by ice. The northern coast of the Gulf of Anadyr is characterized by Mesozoic sedimentary and igneous rocks. Thick Quaternary continental and marine deposits cover most of the western coast. The lacustrine and paludal deposits consist of silt. Marine Quaternary deposits are silty clays and sand with pebbles and shells. Isolated exposures of volcanics are found throughout the coast.

Outer Bering Shelf

The Outer Bering Shelf, a narrow belt parallel to the continental margin, lies between 100–170 m isobaths. At its southern proximity, north of Unimak Island, the shelf is approximately 120 km wide; to the north it becomes narrow near the Pribilof Islands but widens again and reaches a maximum width of about 350 km near the Gulf of Anadyr. The gradient of the Outer Shelf varies by about 10^{-1} to 10^{-2}, being steepest in the vicinity of the Pribilof Islands and very gentle in the north.

Outer Shelf sediments consist of varying mixtures of sand, silt, and clays. Sands are predominant in the lesser depths, while silt, clay, and biogenous content increase in the deeper parts. The silt content at the outer edge may reach over 50 percent by weight. The sediments are generally fine-skewed, poorly-sorted, and leptokurtic.

The suspended load in waters of the Outer Shelf varies seasonally. During sea ice cover, the suspended load in the surface water is low but may become as high as 3.0 mg/liter during phytoplankton bloom. The suspended load generally increases with depth. A significant increase in the suspended load occurs during and after severe storms, a condition that is attributed to roiling of the bottom sediments by storm waves. Concentrations up to 20 mg/liter in near-bottom waters have been observed by various investigators (Lisitsyn 1966).

The mineral distribution in the Outer Shelf is complex because of the multiple sources and intermixing of the sediments from these sources. Sediments in the southern portion are dominated by detritus originating in Bristol Bay and Kuskokwim River drainage. The central portion is covered by sediments originating in the adjacent coast, islands, and Kuskokwim River. The sediments in the Northwestern Outer Shelf are dominated by the influx of Anadyr River. The biogenic constituent in the Outer Shelf sediments is significantly higher than in sediments from the Inner Shelf.

HYDRODYNAMICS AND SEDIMENT TRANSPORT

Sediment transport and distribution on the Bering Shelf are controlled primarily by dominant currents and the wave field. During winter, considerable amounts of sediments are transported by ice. The dominant water movement of the Eastern Bering Shelf is caused by Pacific water entering near Unimak Island, moving northward towards St. Matthew Island and eastward towards the head of Bristol Bay. The northbound stream bifurcates near St. Matthew Island into the Lawrence and Anadyr currents, both merging prior to passing through the Bering Strait (Fig. 27.10). The Lawrence Current flows between the Alaska mainland and St. Lawrence Island. The Anadyr current is the portion of the main stream flowing northwest which is deflected eastward at the mouth of the

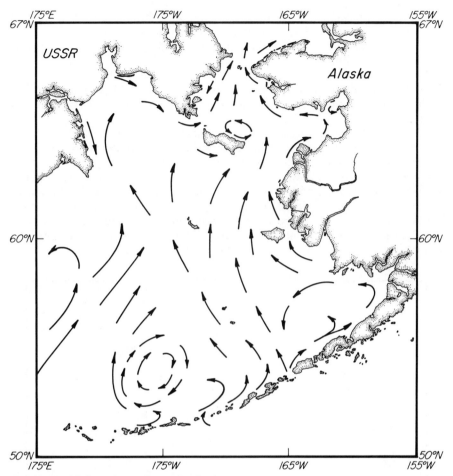

Fig. 27.10 Major currents in eastern Bering Sea.

Gulf of Anadyr to join the Lawrence Current south of the Bering Strait. The speed of these currents ranges up to 40 cm/sec. The movement of the water may occasionally reverse, although the net movement is to the north.

The water moving eastward along the Alaska Peninsula reaches the head of Bristol Bay and is deflected westward by the waters of the Kvichak and Nushagak rivers. Near the mouth of Kuskokwim Bay, the westward flowing water is mixed with the Kuskokwim River flow and directed towards the south to form a cyclonic gyre. Part of the Kuskokwim flow is carried northwards.

The movement of water generally affects the sediment distribution. The extent of sediment transport and deposition by these currents is best illustrated by heavy-mineral distribution in various regions of the shelf (Sharma et al. 1972; Sharma 1972). On the southern shelf the heavy-mineral distribution suggests a current moving along the Alaska Peninsula and forming a cyclonic gyre in Bristol Bay (Fig. 27.11). On the Central Shelf the major water movement is to the north and carries fraction of Kuskokwim River sediments. The sediments in the Northeastern Shelf are influenced by the northward moving waters and the influx of the Yukon River. Exposures of submarine relict sediments suggest areas of nondeposition. Most sediments brought by the Yukon River are carried into the Chukchi Sea by currents. The Gulf of Anadyr sediment distribution results from southward moving currents along the western coast.

The graded spatial variations in sediments from the Southern, Central, and Northeastern shelves indicate that the sediments from the coast are carried to the deeper parts of the shelf and that silt and clay from these sediments are ultimately deposited on the Outer Shelf, the slope, and deep basins. The sediment transport across the shelf is brought about by long waves generated by frequent severe storms in the Bering Sea. Computations from synoptic surface-wind data for severe annual storms in the Bering Sea reveal that these storms do indeed generate long waves capable of setting sand particles in motion at depths up to 94 m (Sharma et al. 1972; Sharma 1972).

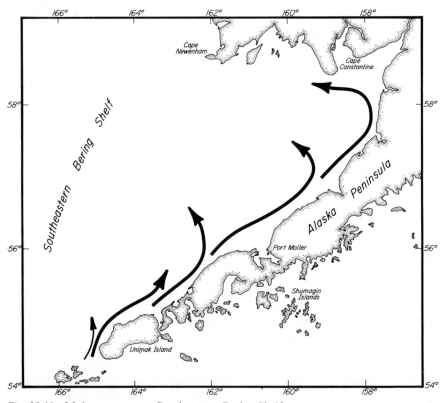

Fig. 27.11 Major currents on Southeastern Bering Shelf.

The sediment transport by wave action on the shelf is also indicated by C–M diagrams (Sharma et al. 1972; Sharma 1972). The sediments deposited are characteristic of particles transported in graded suspension and as bedload. Near-bottom turbid layers and increasing concentration of sand-silt sediments with increasing depth reported by Lisitsyn (1966) and McManus and Smyth (1970) further indicate a high-energy environment at the depositional interface. Roiling of sediments by storms on the shelf has been observed by Lisitsyn (1966), who suggests that such action occurs frequently at depths of 80 to 90 m and results in winnowing of fines and sorting of sands. It is suggested that sands are winnowed from the near-shore area, where significant interaction of the long waves with the bottom occurs. Offshore, as water depth increases, the sands begin to fall out but are resuspended and sorted during every severe storm. Beyond the 100-m isobath on the Outer Shelf, the effect of waves on the bottom sediments diminishes rapidly with increasing water depths—thus preventing elutriation of fine sediments. Silt and clays, therefore, are deposited only on the Outer Shelf.

During the winter most of the Bering Shelf has an ice cover, a significant mechanism of sediment transport. It is difficult, however, to estimate the amount of sediments transported by ice in this region. It was observed that the amount of sediments in sea ice varies regionally as well as temporally (Fig. 27.12). The sea ice samples collected during January–February 1970 provided information concerning the nature of sea ice and sediment therein (Fig. 27.13). The sea ice samples collected from the Bering Shelf were generally away from shore.

Bering shelf ice throughout the region displayed a consistent crystallographic structure. Sea ice was layered, with a bubbly, milky, fine-grained layer on the top and a layer of clear, dense, bubble-free ice near the bottom. The ice near the top contained a layer with vertically oriented crystals (Fig. 27.14). This layer was underlain by a thick layer of clear, dense ice consisting of long, tapered crystals with a horizontal C-axis. This particular crystal fabric is quite common in lake ice and streams ice. A layer of clear, very fine-grained, dense, bubbly, milky ice was observed at the bottom (Sharma et al. 1971).

Besides crystal orientation, the vertical layering of Bering Shelf ice can also be differentiated on the basis of crystallinity. A maximum of five layers were observed in some samples (Fig. 27.15). These layers were characterized primarily by the crystal orientation, size and shape of individual crystal, air bubbles, salt pockets and sediments or phytoplankton content in the ice.

Sediments in Bering Shelf ice were mostly found at the lower boundary of the middle ice layer with horizontal C-axis. The amount of sediment in sea ice varied significantly. These sediments consisted of sand, silt, and clays. The sand varied from 1 to 8 percent, silt from 35 to 80 percent, and clays from 12 to 60 percent of the total weight (Fig. 27.16). The bottom samples in the same area were dominantly medium to fine sands (Fig. 27.17).

The clay mineralogy of these sediments, determined by X-ray diffraction, was found to be uniform. The uniformity of clay minerals and texture of the sediments (grain size distribution) suggest that the sediment incorporated in

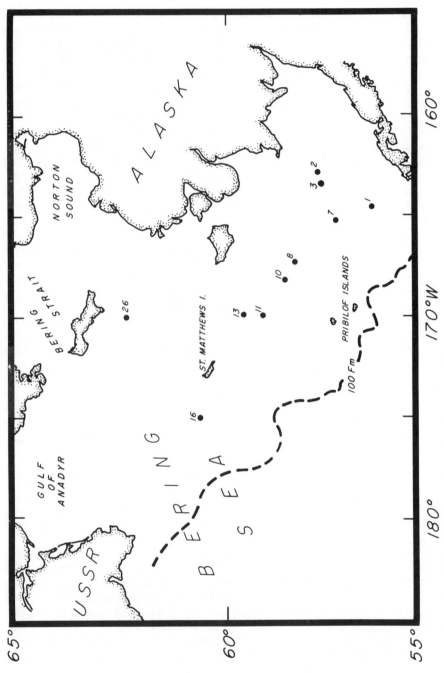

Fig. 27.12   Station locations for sea ice and sediment sampling.

Fig. 27.13   Distribution of sediments in sea ice.

the ice was in suspension during freezing of seawater and aggradation of sea ice. Fine sand and coarse silt found in sea ice can be brought in suspension by storms. It is not clear, however, how the energy for suspension is supplied during the ice cover.

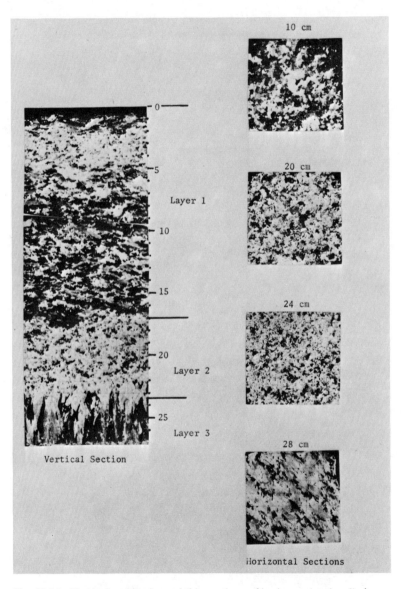

Fig. 27.14 Vertical and horizontal thin sections of ice layers (station 8) shown under polarized light.

The amount of sediments carried by ice varies from year to year, and therefore no quantitative estimates can be given. Nevertheless, during the period of study it was observed that large stretches of sea ice contained significant amounts of sediment. The final deposition of these sediments is controlled by ice movement and melting.

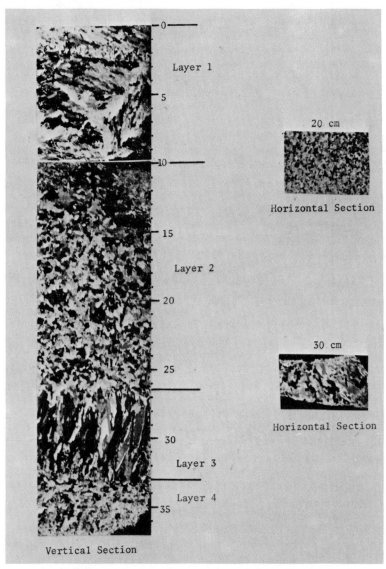

Fig. 27.15 Vertical and horizontal thin sections of ice layers (station 5) shown under polarized light.

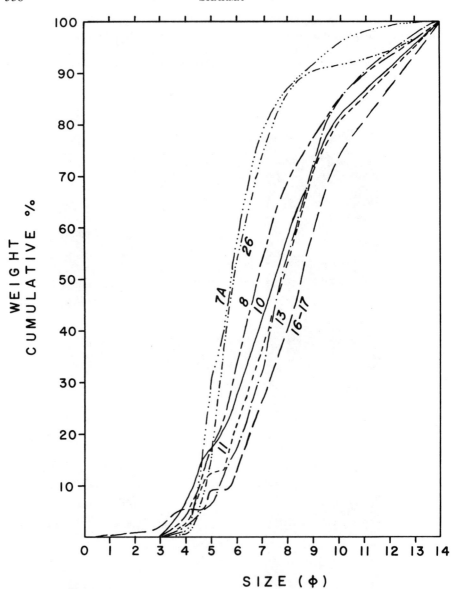

Fig. 27.16 Size distributions of ice sediments.

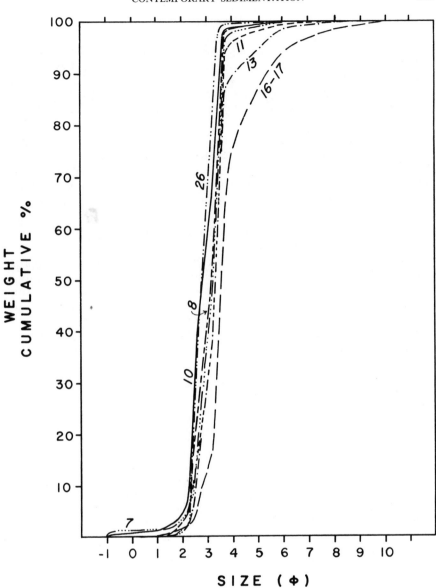

Fig. 27.17 Size distributions of bottom sediments.

Discussion

STRATY: Your interpretation of the water movement in upper Bristol Bay has already been well documented in the past few years. It was Dr. Favorite who first proposed the current that far up in the bay, and Hebard's work (1961) has followed the idea. Hydrographic research in more recent years has verified this movement through use of direct current measurements, drift card retrieval and dye studies.

SHARMA: My information is based on Hebard's work and data obtained from drift card studies. You are referring to the surface currents; however, my paper has discussed mainly the subsurface currents that affect sediment transport. My point is that these currents move along the bottom area also.

KITANO: What kinds of biogenic sediments are found in the area of your studies?

SHARMA: Biogenic material constitutes approximately 50% of the suspended sediments in the southeastern Bering shelf and only about 15% of those in the central shelf area. The bottom concentration of biogenic matter is approximately 100–200 valves per gram of sediment and is comprised mainly of diatoms which are concentrated in areas deeper than 100 m, although some specimens are found also in the shelf area. [Detailed distribution of diatoms and their identification are described in Part 2 of this chapter].

KITANO: Do you have data for the average concentrations of suspended matter and dissolved chemical species in the river waters which enter the Bering Sea?

SHARMA: No, we have never worked close enough to shore to go into the river to sample the suspended material there. This is a subject which is proposed for future study.

KITANO: Is there a possibility that sulfate is reduced to sulfide in the coastal areas of the Bering Sea?

SHARMA: Since we have not worked very close to shore, we do not have geochemical data specifically concerning reduction of sulfate; this process is a possibility, however. We were quite surprised to find a strong linear relationship between the amount of organic carbon and the clay content in sheltered areas of minimal wave action. In such areas as protected bays and the mouth of the river, the sediments remain relatively undisturbed after deposition, whereas in the open water they are continuously being recycled and churned up together with nutrients such as silicate and phosphate. For this reason, sedimentation studies should be closely tied in with biological investigations.

KASAHARA: Have you considered the disturbance caused by heavy trawl fishing in the Southeastern Bering Shelf? Data are available to calculate the areas covered by this fishing gear to assess the possible effects on the sediments.

SHARMA: I have not studied this aspect, but it would not seem to be a significant factor in view of the total forces operating in such a large area. In any case, the isolated variations of man's impact on the sediments are more than offset by the redistribution effect of frequent storm activity.

Part 2. Distribution of Recent diatoms on the eastern Bering shelf

KEI OSHITE

Hokkaido University of Education, Hakodate, Japan

GHANSHYAM D. SHARMA

Institute of Marine Science, University of Alaska, Fairbanks, Alaska

INTRODUCTION

The Bering Sea influences significantly the climate and oceanography of the Arctic and to an extent that of the entire globe. This large body of subarctic water receives significant amounts of Pacific water and also exchanges with the Arctic Ocean, thus forming a complex oceanographic circulation. It is believed that the driving force for these water movements is provided by atmospheric circulation. The climate and oceanography of the Bering Sea are intricately interrelated. The Cenozoic climatic history of the Bering Sea is complex. Glaciation caused extension and obliteration of littoral, neritic environments over an extremely smooth, gently sloping eastern Bering shelf. Distribution and migration of various environments on the shelf thus may be elucidated only if the present faunal distribution is related to the contemporary environments. This study describes the Recent subarctic diatom ecology, and it is anticipated that this information may be used to interpret ancient Bering Sea environmental and climatic conditions.

Area of study

Seventeen bottom grabs were collected from the southeastern Bering shelf during July 1960 on the *Oshoro Maru* at stations between 56°–62°N and 162°–174°W (Table 27.1; Fig. 27.18) (Faculty of Fisheries, Hokkaido University 1961). The sediments collected varied from mud near the outer shelf to

TABLE 27.1   Station locations and sediment texture on the eastern Bering shelf

Station Number	Location Latitude	Longitude	Depth (m)	Sediment Texture
1	57°40′N	171°50′W	104	Mud
2	57 43	173 40	165	Sandy gravel
3	58 21	174 00	132	Sandy mud
4	58 59	173 32	121	Sandy mud
5	61 01	173 24	77	Sandy mud
6	61 00	171 39	63	Sandy mud
7	59 58	170 00	59	Sandy mud
8	59 25	169 12	50	Muddy sand
9	58 43	167 54	45	Sandy mud
10	58 43	168 04	47	Sand
11	58 45	168 04	47	Sand
12	58 25	166 23	49	Sand
13	58 10	164 42	49	Muddy sand
14	57 46	162 01	40	Sand
15	56 50	163 04	72	Muddy sand
16	56 04	163 56	92	Muddy sand
17	55 07	164 59	105	Sandy mud

sand on the inner shelf. The sediment texture is related to water depth and gradually increases with increasing distance from shore (Sharma et al. 1970, 1973; Part 1, this chapter).

Method of study

A representative fraction (25 or 50 mg) of sediment from each station was weighed and distributed evenly on a glass slide for microscopic examination. All species present were identified, and the number of frustules of individual species was counted to obtain the distribution of each species per gram of sediments.

Results

All but one species of diatom observed were of marine origin (Table 27.2). The freshwater species *Epithemia zebra* was observed in samples from stations 4 and 6. The marine diatoms observed can be broadly classified as oceanic, neritic or littoral species. A few undefined species, however, were also observed in the sediment and are included in Table 27.2. The number of diatoms per gram of bottom sediment varies from less than 100,000 in the inner shelf to over 200,000 in the outer shelf (Fig. 27.19). The number of frustules present in sediments appears to be related to the distance from land (station 9 contained 39,000 frustules, and station 4 had 413,000 frustule per gram of sediment), water depth and sediment texture. The widely distributed and abundant diatoms observed belong to boreal or cosmopolitan species. Two arctic species (*Thalassiosira hyalina Melosira arctica*) and five south boreal species (*Asteromphalus heptactis, Coscinodiscus auguste-lineatus, Coscinodiscus divisus, Actinocyclus ehrenbergii* and *Melosira sol*) are occasionally observed in small numbers and are mostly confined to neritic and littoral zones. The number of oceanic species increases offshore, while neritic and littoral species predominate in the central and inner shelves (Figs. 27.19–27.22; Table 27.3).

Discussion

The distribution of sediments on the Bering Shelf demonstrates the influence of currents which are partly responsible for their distribution (Sharma 1970; 1973; Part 1, Chapter 27, this volume). Undoubtedly these currents will also affect the deposition and concentration of diatoms in bottom sediments. The distribution of diatoms in sediments is also significantly controlled by water depth, water mass, water turbulence, suspended load, salinity and temperature. An adequate knowledge of the above parameter is therefore essential towards an understanding of diatom distribution.

Diatoms were obtained from the continental shelf, which is extremely flat with a slope angle of less than one minute. The floor is smooth with an average depth of 50 m. The shelf can be broadly divided into three regions: inner shelf, a shallow area along the coast with a depth of less than 50 m; central shelf, with an average depth of 75 m; and outer shelf region, with water depth greater than 100 m. Two distinct water masses are present in this region: the Pacific Subarctic water and the Kuskokwim-Nushagak-Kvichak complex. The Pacific Subarctic water enters near Unimak Pass and moves eastward along the Alaska

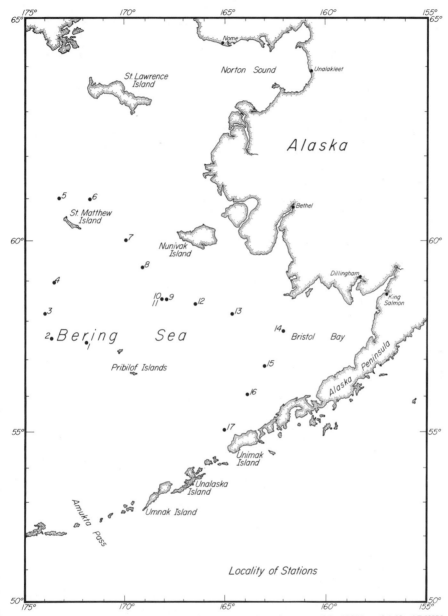

Fig. 27.18 Bottom-grab sampling locations at stations between 56–62°N and 162–174°W on the southeastern Bering shelf during July 1960, R/V *Oshoro Maru*.

TABLE 27.2 Number ($\times 10^3$) of diatom frustules found per g of sediment sample

Species	Station 1	2	3	4	5	6	7	8	9	12	13	14	15	16	17
Oceanic:															
Asteromphalus heptactis			+												
Coscinodiscus asteromphalus	9.8	1.2	1.5	10.4	0.6	0.8	0.4	0.1	0.1			0.1	0.5	0.2	
C. curvatulus			0.1									+	0.5	0.1	0.2
C. excentricus	5.0	0.1	2.1	13.4	4.2	1.2	0.2	0.1		+			0.9	1.2	5.4
C. lineatus	1.4		0.1	3.2		0.2	0.2	0.1	0.1		+			0.1	1.2
C. marginatus	10.4	0.4	1.2	7.2	2.0	0.4			+				+	0.1	1.4
C. nodulifer									+						
C. oculus-iridis	0.4	0.4	0.7	1.0		0.2	0.4	0.2	0.1			+	0.3	+	
C. radiatus	14.2	0.6	2.0	7.8	4.2	4.2	1.6	0.2		0.3	0.1	0.2	0.3	1.9	3.4
C. rothii			+												
C. stellaris				0.4		0.2	0.2			0.1	+	0.2	0.3	+	0.4
C. subtilis															0.2
C. wailesii			0.1					+	0.4	0.4	0.2		0.3		
Denticula seminae	15.2	0.2	14.8	46.8	1.8	1.4	0.2	+						0.3	9.8
Rhizosolenia hebetata	9.4	0.3	3.3	25.2	0.8	1.8	0.4					+		0.2	0.4
Neritic:															
Actinoptychus splendens												+	+		
A. undulatus	3.2	0.1	0.9	3.0	1.0	1.4	1.4	0.2	0.2	0.5	0.2	1.1	0.2	1.6	4.2
Biddulphia aurita	2.2		0.8	4.4	0.8	1.0	1.6	0.7	0.8	2.2	2.6		3.9	2.8	19.8
Coscinodiscus anguste-lineatus			0.1								+				0.4
C. divisus			0.1		0.2										
Fragilaria oceanica	3.2	0.2	0.2	7.4	6.6	5.8	7.0	2.7	4.4	12.3	13.3	12.2	9.4	3.7	2.0
Porosira glacialis	1.6	0.1	0.2												
Thalassiosira decipiens	7.2	0.4	1.9	5.2	3.6	0.4			+	+	0.2		1.5	0.7	1.8
T. hyalina	1.2	0.2	1.1	3.2	0.4			0.2	+	0.2			0.6	0.6	0.8

TABLE 27.2 (continued)

Species	Station 1	2	3	4	5	6	7	8	9	12	13	14	15	16	17
Littoral:															
Actinocyclus ehrenbergii	0.4					0.4	0.2	0.1	+	0.4	0.2	+	+	+	0.2
Arachnoidiscus ehrenbergii		0.1											+	+	0.8
Cocconeis costata															
C. scutellum															0.2
C. vitrea				0.2											0.2
Diploneis smithii	1.6														
Hyalodiscus subtilis	1.0														
Melosira arctica	0.6	0.2	0.2	1.4	1.2	2.4	1.8	0.3	0.1		0.1	0.1	0.1	0.5	0.8
M. sol		+	+	1.0	0.2	0.4		0.1	+				0.1	0.2	0.6
M. sulcata	130.8	6.7	17.0	208.4	53.2	187.4	139.6	37.2	25.8	46.1	36.5	26.7	73.8	24.0	13.2
Navicula distans	1.0		0.2	2.0	0.4	1.8	1.6	1.0	1.6	1.4	0.6	0.2	1.7	0.2	0.2
Pleurosigma normanii							0.2	0.1	0.1	0.3	0.9	0.5	0.5		
Rhabdonema arcuatum										0.1					
Rhaphoneis amphiceros	3.8		0.4	1.4	0.8	2.8	2.4	1.5	2.4	4.8	4.6	7.8	2.9	0.5	0.8
R. amphi. v. gemmifera				0.2		0.2	0.8		0.1	0.4	0.2		0.1		
Trachyneis aspera													+		
Triceratium favus													+		
Freshwater:															
Epithemia zebra				0.2		0.2									
Not defined:															
Coscinodiscus spp.	0.8	0.1													
Nitzschia spp.		+	+												
Pinnularia sp.		+		0.2	0.2								+		
Stephanopyxis spp.	9.2	0.2	0.6	0.2	3.0	2.0	1.2	+	+	0.2	0.1		1.2	1.7	0.4
Synedra spp.	0.4	0.2	1.5	4.8	1.6		0.4		0.1				0.1		
Thalassionema and Thalassiothrix spp.	13.6	0.5	2.4	48.2	11.4	9.8	4.8	0.3	1.2	1.0	0.4	0.2	1.4	2.1	10.6
Thalassiosira spp.													0.2		
TOTAL	251.0	12.2	54.3	413.0	101.4	228.6	168.9	45.9	39.1	70.8	60.5	49.6	101.6	45.3	77.1

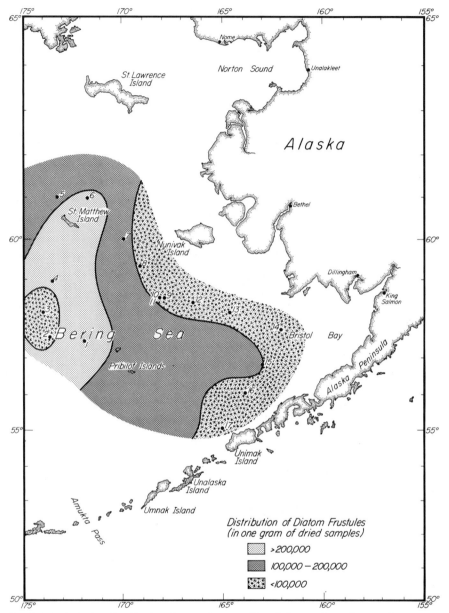

Fig. 27.19   Density distribution (per g dry sample) of diatom frustules on the eastern Bering shelf.

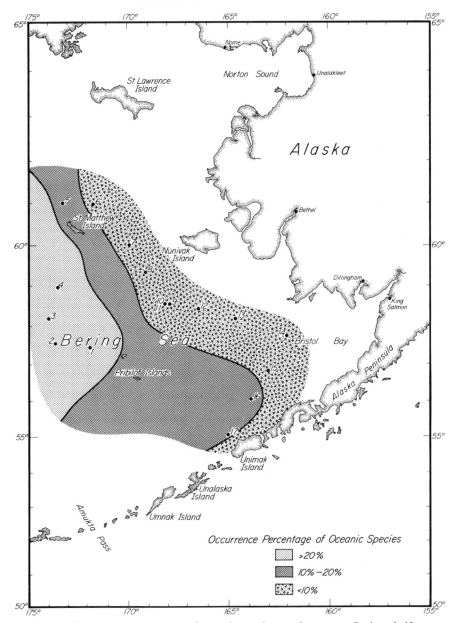

Fig. 27.20 Occurrence percentage of oceanic species on the eastern Bering shelf.

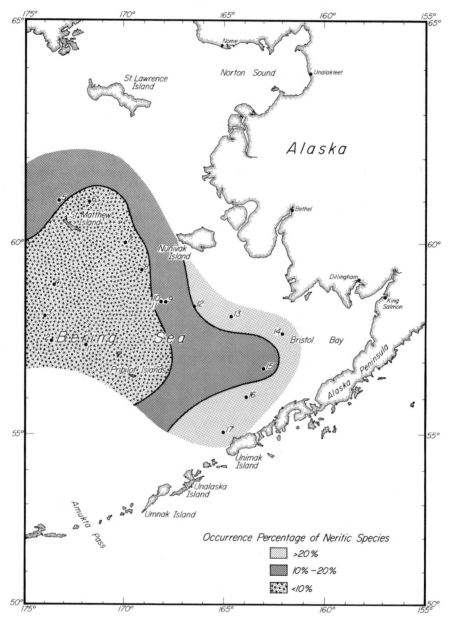

Fig. 27.21   Occurrence percentage of neritic species on the eastern Bering shelf.

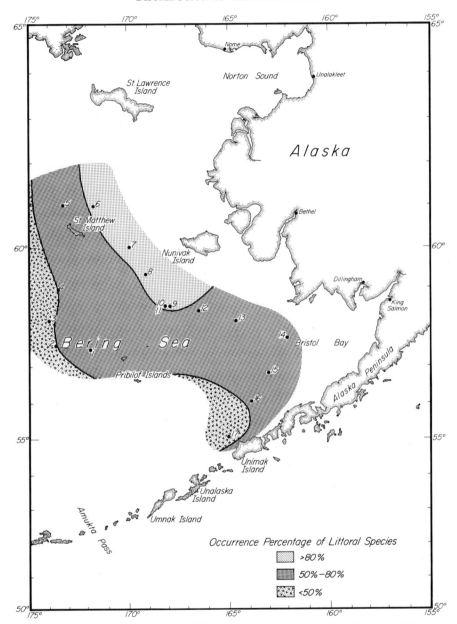

Fig. 27.22 Occurrence percentage of littoral species on the eastern Bering shelf.

TABLE 27.3 Occurrence percentage of number of diatom frustules originating from three ecological divisions

Station Number	1	2	3	4	5	6	7	8	9	12	13	14	15	16	17
Oceanic	27.5	27.8	49.3	29.2	16.5	5.6	3.1	2.9	2.9	1.6	0.7	1.5	3.6	13.7	31.2
Neritic	7.4	7.9	9.7	5.6	12.4	3.7	6.3	8.5	13.8	21.2	26.8	26.8	15.3	20.7	32.9
Littoral	55.5	57.1	32.8	52.3	55.1	85.5	86.8	87.8	80.0	75.6	71.2	71.3	78.2	57.2	21.6

Peninsula; this water is characterized by salinity of 32 to 34‰ and a temperature range of 2 to 4 C. The Kuskokwim-Nushagak-Kvichak waters are formed by mixing of freshwater and seawater. During summer months the increased surface flow from these rivers extends this water mass farther offshore, where salinity is less than 32‰ and the temperature ranges from 6 to 9 C.

The zonal distribution of oceanic, neritic, and littoral species suggests that, together with the bathymetry, the intrusion of Pacific Subarctic water near Unimak Pass and its movement along the Alaska Peninsula may also affect the diatoms in sediments. The increasing oceanic species and decreasing littoral species in the central shelf indicates the mixing of nutrient-rich oceanic water in this area. On the other hand, freshwater from the Kuskokwim, Kvichak, and Nushagak rivers along the eastern and northeastern shores of the Bering Sea provides a broad shelf area for the littoral species.

Although the data presented are not sufficient to delineate all the ecological environments, it is hoped that the comparison between the present diatom distribution and that inferred from cores representing paleo sediments may be useful for interpretation of paleoenvironments on the Bering Shelf.

COMBINED REFERENCES [Parts 1 and 2]

CREAGER, J. S., and D. A. McMANUS
    1967   Geology of the floor of the Bering and Chukchi seas; American Studies. In *The Bering land bridge*, edited by D. M. Hopkins. Stanford Univ. Press, Stanford, Calif., pp. 7–31.

FACULTY OF FISHERIES, HOKKAIDO UNIVERSITY
    1961   Data record of oceanographic observations and exploratory fishing, Vol. 5.

HEBARD, J. F.
    1961   Currents in southeastern Bering Sea. *Bull. Int. North Pac. Fish. Comm.* 5: 9–16.

LISITSYN, A. P.
    1966   *Recent sedimentation in the Bering Sea* [in Russian]. Izd. Nauka, Moscow, 584 pp. (Transl., 1969, avail. Nat. Tech. Inf. Serv., Springfield, Va., TT 68–50315).

McMANUS, D. A., and C. S. SMYTH
    1970   Turbid bottom water on the continental shelf of the northern Bering Sea. *J. Sed. Petrol.* 40(3): 869–873.

NELSON, C. H., and D. M. HOPKINS
    1972   Sedimentary processes and distribution of particulate gold in the northern Bering Sea. *U. S. Geol. Survey Prof. Paper* 689, p. 27.

SHARMA, G. D.
    1973   Graded sedimentation on Bering Shelf. In *Proceedings of 24th International Geological Congress.*, Montreal, Canada, pp. 262–271.

SHARMA, G. D., J. D. KREITNER, and D. W. HOOD

1970   Recent sedimentation in southern Bering Shelf [Abstract]. *Amer. Ass. Petrol. Geol. Bull.* 54(12): 2503–2504.

1971   Sea ice characteristics in Bering Sea. In *Proceedings of the First International Conference on Port and Ocean Engineering under Arctic Conditions,* edited by S. S. Wetteland and P. Bruun. Tech. Univ. Norway, Trondheim, pp. 211–220.

1973   Geological oceanography of the Bering Shelf. In *Arctic geology and oceanography,* edited by Y. Herman. Elsevier Publ. Co., New York.

SHARMA, G. D., A. S. NAIDU, and D. W. HOOD

1972   Bristol Bay: A model contemporary graded shelf. *Amer. Ass. Petrol. Geol. Bull.* 56(10): 2000–2012.

VENKATARATHNAM, K.

1969   Clastic sediments on the continental shelf of the northern Bering Sea. *Spec. Rep. 41, Dep. Oceanogr., Univ. Washington,* pp. 40–61.

CHAPTER 28

Lagoon contributions to sediments and water of the Bering Sea

ROBERT J. BARSDATE, MARY NEBERT *and* C. PETER McROY

Institute of Marine Science, University of Alaska, Fairbanks, Alaska

Abstract

The eastern Bering Sea is bordered by many shallow marine or estuarine lagoons. The productivity of many of these lagoons far exceeds that of equivalent areas of the adjacent Bering Sea, and the larger lagoons have tidal exchanges equivalent to the flow volume of major rivers. Izembek Lagoon is an embayment of the Bering Sea at the tip of the Alaska Peninsula and has been the principal site for sea grass community studies in the eastern Bering Sea since 1963. Our investigations of this lagoon suggest that, through biogenic modifications, the water flowing from the lagoon is as different in many important geochemical parameters as is river water entering the sea from terrestrial sources.

The vast eelgrass meadows of the lagoon annually produce 166,000 metric tons of particulate carbon, as organic matter which contains 7400 metric tons of nitrogen, 1660 metric tons of phosphorus, 3.45 metric tons of copper, and 386 metric tons of silica. Only a small fraction of the total production appears to be recycled within the lagoon. Avian herbivory accounts for no more than 3 percent of the total annual production, while decomposition and leaching of beached eelgrass is perhaps less than 0.1 percent. Detached floating eelgrass appears to gain rather than lose nitrogen, phosphorus, copper, and silica in transit from the lagoon, but the overall quantities involved are negligible. In addition to elements incorporated in detached floating eelgrass, the lagoon exports substantial quantities of dissolved carbon, nitrogen, and phosphorus. Dissolved copper and silica, on the other hand, are lost to the lagoon from the Bering Sea, at annual rates which may exceed 200 metric tons of copper and 600 metric tons of silica. Lead also is removed from the Bering Sea but in somewhat smaller quantities.

The fixed carbon and other elements from sea grasses that are eventually incorporated in the detrital organics of Bering Sea sediments may form an important component of Bering Sea foodwebs.

INTRODUCTION

The eastern margin of the Bering Sea is bordered by shallow lagoons, essentially marine in nature, which have only minor freshwater inputs from rivers or streams and freely exchange with Bering Sea or Bristol Bay water through channels between the sandspits and barrier islands that form their outer boundaries. Many of these lagoons support large stands of eelgrass (*Zostera marina* L.), a marine vascular plant which grows in shallow waters along the high latitude coasts of North America and Europe. McRoy (1968) reports eelgrass meadows in 17 coastal lagoons along the eastern Bering Sea.

Productivity and standing stock in these lagoons are very much higher than in the Bering Sea. Net daily productivities of 1 to 8 g C/m^2-day have been estimated in the lagoon eelgrass beds (McRoy 1970a), as compared with a maximum of 0.6 g C/m^2-day reported by Taniguchi (1969) for planktonic organisms in the eastern Bering Sea. Phytoplankton productivity values per m^3 in the lagoons are also higher than in the eastern Bering Sea (McRoy, Goering and Shiels 1972), but, because of the shallow depths, this accounts for only a small fraction of the total productivity of the lagoon. Benthic algae are present but not abundant.

Although the area of these lagoons is very small in comparison with the whole Bering Sea or even with the shelf area of the eastern Bering Sea, the relative magnitude of tidal exchange suggests that the possible effects of these lagoons should not be ignored in a consideration of the shelf areas of the Bering Sea. There is increasing recognition of the importance of the coastal zone in determining the composition of seawater and of the profound chemical changes undergone by both fresh and salt water entering this environment (Bien et al. 1958; Duke et al. 1966; Ketchum 1967; Pomeroy et al. 1969; Turekian 1971). Knowledge of the processes involved is essential to understanding both the workings of the natural systems and the possible effects of man's perturbations. The reduced freshwater input in the special case of the marine lagoons of western Alaska emphasizes the sediment-plant and water-plant interactions, thus tending to simplify the system and create a semi-controlled environment for process studies. Since 1963 Izembek Lagoon has been the site of intensive investigations of the lagoon ecosystem (McRoy 1966, 1970a, b; McRoy and Barsdate 1970; McRoy, Barsdate and Nebert 1972). With this considerable body of available information and the results of more recent work, we have undertaken the formulation of elemental budgets for carbon, nitrogen, phosphorus, silica, and copper in Izembek Lagoon. Included are estimates of the net flux of these elements between the lagoon and the Bering Sea, both in solution and as detached floating eelgrass, along with a comparison of these flux rates with the metabolic requirements of the lagoon eelgrass stand.

Description of study area

The shallow basin of Izembek Lagoon at the southern tip of the Alaska Peninsula is almost 41 km long and 3 to 12 km wide, with approximately 195 km of coastline (Fig. 28.1). The 218 km^2 surface area is about 78 percent tide flats and 22 percent tide channels, with eelgrass beds covering roughly 68 percent

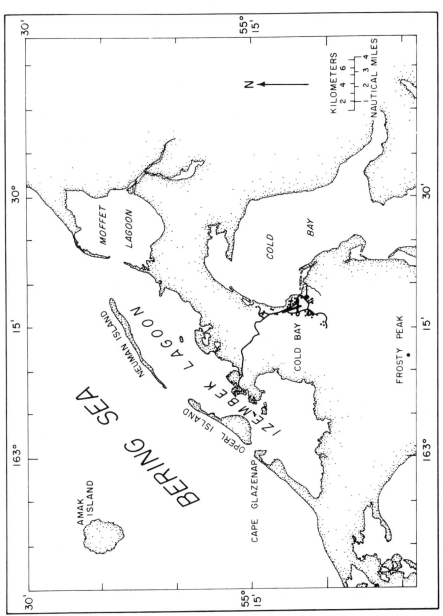

Fig. 28.1 Map of Izembek Lagoon and adjacent areas.

of the tide flats, or about 116 km^2. This is the largest known single stand of *Zostera marina* (McRoy 1966).

Four streams empty into the lagoon but are of little significance, and salinities in the Bering Sea and lagoon water differ by less than 1‰ as a result of the lack of freshwater dilution within the lagoon. Three large passes, 2 to 3 km wide and 10 to 13 m deep, between the spits and barrier islands provide the lagoon free exchange with Bering Sea water. The mean tide height of 0.98 m results in a tidal prism of 1.7 × 10^8 m^3 or 68 percent of the total lagoon volume at mean high tide. On an annual basis the tidal exchange of Izembek Lagoon is:

$$1.7 \times 10^8 \text{ m}^3 \times 2 \text{ tides per day} \times 365 \text{ days} = 1.2 \times 10^{11} \text{ m}^3$$

This is approximately 63 percent of the mean annual discharge of the Yukon River and three times that of the Kuskokwim River (river flow data from Roden, 1967).

Eelgrass productivity in Izembek Lagoon has been estimated to be 8 g C/m^2-day during the period of active growth, and eelgrass standing stock averages 1.2 kg/m^2 (McRoy 1970a). A mean phytoplankton productivity of 79 mg C/m^3-day and a standing stock of 1.2 mg chlorophyll a/m^3 have been observed in Izembek Lagoon by McRoy, Goering and Shiels (1972).

Annual elemental production of Izembek Lagoon eelgrass

The total annual production of the Izembek Lagoon eelgrass beds has been calculated from estimates of standing stock, standing stock turnover rate, and length of growing season. McRoy (1970a) has suggested a daily turnover of 2 percent of the standing stock. This rate applied to the standing stock of 1.2 kg/m^2 or 1.39 × 10^5 MT (metric tons) dry weight eelgrass for the entire lagoon gives a daily production of 24 g/m^2-day or 2.78 × 10^3 MT/day. The shallow-water environment of the eelgrass beds in Izembek Lagoon, in combination with the climatic conditions, imparts a pronounced seasonality to the eelgrass productivity; we have therefore based our calculations of annual production on an active growing season of 165 days and a 200-day dormant period during the remainder of the year. Thus the annual eelgrass production in Izembek Lagoon is equal to:

$$2 \text{ percent/day} \times 1.2 \text{ kg/m}^2 \times 116 \text{ km}^2 \times 165 \text{ days} = 4.6 \times 10^5 \text{ MT/yr}$$

This value is reasonably consistent with McRoy's (1970a) calculation of 5.1 × 10^5 MT/yr based on a somewhat different set of figures and Petersen's (1914) method of estimating annual eelgrass production by doubling the standing stock.

From total production and the concentrations of elements in eelgrass, the amounts of the elements annually produced can be assessed. Although literature values are available for the concentrations of most of the elements found in eelgrass, carbon, phosphorus, and nitrogen also have been analyzed in fresh-dried eelgrass plants collected in Izembek Lagoon. Our value of 3.6 × 10^5 ppm carbon is very similar to Vinogradov's (1953) estimate of 3.85 × 10^5 ppm. Phosphorus determinations averaged 3.6 × 10^3 ppm, which is somewhat higher than Candussio's (1960) figure of 2.86 × 10^3 ppm. Nitrogen determinations showed a concentration of 1.6 × 10^4 ppm in dried eelgrass, only slightly over 50 percent of the 3.05×10^4 ppm reported by Candussio (1960) for this element. The low nitrogen content found in Izembek Lagoon plants may be due to the time of year at which the samples were collected (September), since downward translocations of plant nitrogen may occur at the end of the growth season. It is more likely however, that variations in the concentrations of this and other elements are due to the environmental conditions under which the eelgrass grows (McRoy 1970b). Therefore, the results of analysis on Izembek eelgrass are used when available for calculations of elemental production.

The elemental composition of eelgrass compiled from both our results and the values from the literature appear in Table 28.1. Also shown is the amount of each element required for the synthesis of 4.6×10^5 MT dry weight

TABLE 28.1 Amounts of various elements annually incorporated in eelgrass in Izembek Lagoon, based on dry weight concentrations from the literature (annual eelgrass production = 4.6×10^5 MT/yr, dry wt)

Element	Concentration in eelgrass (ppm)	Annual elemental incorporation (metric tons)
Carbon	385,000[b]	177,100
	360,000[f]	166,000
Nitrogen	30,450[a]	14,010
	16,000[f]	7,400
Phosphorus	2,860[a]	1,316
	3,600[f]	1,660
Chlorine	43,680[a]	20,093
Potassium	22,640[a]	10,414
Calcium	20,010[a]	9,205
Sodium	19,590[a]	9,011
Magnesium	7,380[a]	3,395
Sulfur	7,300[a]	3,358
Manganese	1,825[a]	840
Silicon	840[a]	386
Aluminum	500[c]	230
Boron	310[a]	143
Iron	245[a]	113
Iodine	203[a]	93
Zinc	27[d]	13
Bromine	9.59[a]	4.41
Copper	7.50[a]	3.45
Barium	7.2	3.3
Fluorine	3.61[a]	1.66
Molybdenum	3.12[a]	1.44
Lead	<1[f]	<0.5
Nickel	0.4[b]	0.2
Cobalt	0.3[b]	0.1
Cadmium	0.23[b]	0.10
Rubidium	0.14	0.064
Berylium	0.12[e]	0.055

Data source:
[a] Candussio 1960
[b] Vinogradov 1953
[c] Hutchinson and Wollack 1943
[d] Burkholder and Doheny 1968
[e] Meehan and Smythe 1967
[f] This author

of eelgrass, based on the concentration given. This is the calculated annual production of eelgrass in Izembek Lagoon. Very large quantities of the major elements are incorporated annually in this eelgrass, and even for trace elements the figures are on the order of metric tons per year.

Recycling of eelgrass within the lagoon

Having established an estimate of annual eelgrass production in Izembek Lagoon and the corresponding incorporation of elements into particulate forms, we examined the processes which would either remove eelgrass or change its composition while in the lagoon. Those processes which appear to be of possible significance in recycling eelgrass nutrients within the lagoon are the decomposition and leaching of both detached floating eelgrass and beached eelgrass and herbivory by birds and marine invertebrates.

During the course of the summer, a noticeable amount of detached eelgrass leaves and plants are thrown up onto the beach, where dense windrows are seen throughout the late summer and fall. Winter storm activity removes most of this material, and by spring the beaches are relatively clear. A density of 3.4 kg dry weight beached eelgrass/m², or 83 kg/m of beach, was measured in early fall of 1969. Assuming roughly 50 percent of the beach to be covered with this density of eelgrass at any given time, an estimate of the total quantity of beached eelgrass is:

$$0.5 \times 195 \text{ km} \times 83 \text{ kg/m} = 8.1 \times 10^3 \text{ MT}$$

or about 1.75 percent of the annual eelgrass production. We then can calculate that detached eelgrass has a mean residence time of three days on the beach, as the amount of beached eelgrass is three times the daily production of 2.78×10^3 MT of eelgrass per day in the lagoon.

Most of the beached eelgrass is submerged during each high tide. As the water moves up the beach through the windrows, nutrients and other elements are released from the eelgrass, and concentrations of carbon, nitrogen, and phosphorus in the water along the beach become extremely high (Fig. 28.2). This local enrichment from the beached eelgrass is quite evident when a comparison is made between concentration levels in the incoming water just before high tide and those at slack high tide well up into the eelgrass. Increases of up to 15.4 mg dissolved organic C/liter, 57 μg dissolved inorganic N/liter, and 961 μg total P/liter were observed. Silicate was also higher in the flooded eelgrass, but the increase was relatively minor.

These changes in concentration occur within a tidal volume of about 1300 liters/m of shoreline. When calculated from the increase in dissolved organic carbon in this volume of water per tidal cycle, the approximate loss rate of eelgrass from the beach into solution is 440 MT per year, or less than 0.1 percent of the annual production. The mass wasting of beached eelgrass through the formation of small detrital particles does occur, as filters from water taken within the flooded beached eelgrass retain noticeably more particulates than found in water a few meters seaward of the beach. If our estimates of residence time for eelgrass on the beach are correct, the amounts of detritus thus produced represent a negligible loss rate from the beached eelgrass. The

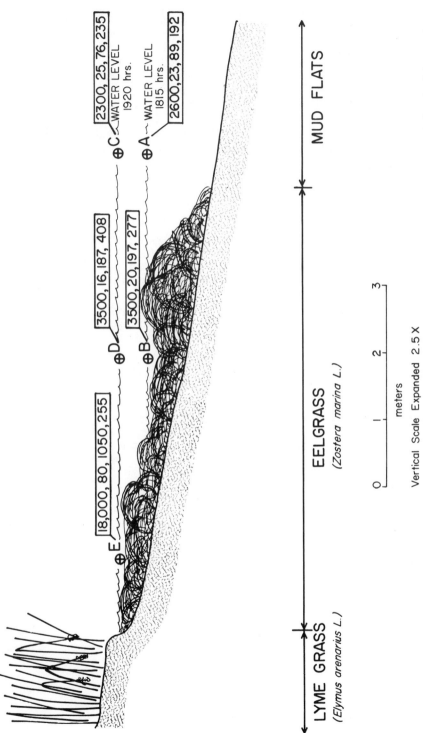

Fig. 28.2 Concentrations (μg/liter) of dissolved organic carbon, dissolved inorganic nitrogen, total phosphorus, and dissolved inorganic silica (in that order), in seawater in contact with beached eelgrass just before (1815) and at high tide (1920) on 23 September 1969.

high concentrations of dissolved nutrients and detritus, however, both may have considerable influence on the local biological environment.

In addition to the decomposition and leaching of beached eelgrass, we have considered the results of processes which act on detached floating eelgrass within the lagoon. Large mats of floating plants have been observed in the lagoon throughout the growing season, and these are moved through the passes out into the Bering Sea by tidal currents. Experimental results indicate that floating eelgrass takes up rather than loses nitrogen, phosphorus, silica, and copper in transit from the lagoon. Freshly detached eelgrass plants were incubated in 4-liter glass bottles, and concentrations of nitrogen, phosphorus, silica, copper, and lead were measured daily for up to four days. The estimated mean residence time in the lagoon for floating eelgrass is three tidal cycles, or 1.5 days, and during this period of time the decreases of N, P, Cu, and Si in the solutions surrounding experimental plants were equivalent to 3 μg N/g plant, 1.8 μg P/g plant, 0.02 μg Cu/g plant, and 52 μg Si/g plant (Table 28.2). Of the elements analysed, only lead was lost from the plants to the water, and the rate of excretion was very low—0.2 μg Pb/g plant in 1.5 days. Assuming a rate of detachment equal to the net production and using the flux rates obtained from the bottle experiments, we calculate the net gain by floating eelgrass as 0.02 percent of the annual production of nitrogen, 0.05 percent for phosphorus, 0.26 percent for copper, and 6.1 percent for silica. Although these gains may be offset to a certain extent by microbial decomposition and sloughing off of small particles of the floating eelgrass, there was little evidence of significant breakdown of plant material during the course of the 96-hour incubations.

TABLE 28.2   Uptake of dissolved nitrogen, phosphorus, copper, silica, and lead by freshly detached eelgrass plants during three tidal cycles. Units are μg/g dry wt plant—1.5 day. June values are means of duplicate experiments; single determinations were made in August. Stations I and II are shallow and deepwater habitats, respectively.

	N*	P**	Cu	Si	Pb
June, whole turions (roots and rhizomes intact)					
Station I	−0.94	7.06	−6.06	73.3	−1.14
Station II	4.04	7.54	2.24	40.0	0.42
August, whole turions (roots and rhizomes intact)					
Station I	3.30	−3.65	0.92	44.3	−0.07
Station II	3.70	−2.30	1.90	47.6	−0.26
August, without roots and rhizomes					
Station I	7.92	−14.42	3.07	22.4	0.15
Station II	3.62	5.47	1.95	72.8	−0.12
Average of June and August means determined separately					
	3.09	1.79	0.02	51.7	−0.21

*N = total dissolved inorganic nitrogen
**P = dissolved reactive phosphorus

This period is longer than either the estimated mean residence time of eelgrass on the beach or floating in the lagoon.

The dense eelgrass beds in Izembek Lagoon provide a concentrated source of food that is of particular importance to migrating waterfowl. An estimated 3×10^5 black brant consume about 2 percent of the annual eelgrass production each autumn (McRoy 1970b). Canada geese probably utilize another 0.6 percent of the annual production (McRoy 1966), while emperor geese and other species feed on an additional few tenths of a percent. Total avian herbivory appears to be somewhat less than the 3 percent of the annual eelgrass production we have used in subsequent calculations of particulate elements lost from the eelgrass beds by means of this route.

The significance of herbivory on eelgrass by marine invertebrates in Izembek Lagoon tentatively is estimated to be small. McRoy et al. (1969) describe the lagoon as developing "a somewhat restricted intertidal fauna." In the absence of any quantitative macrobenthic feeding rate information, we have omitted calculation of eelgrass loss through invertebrate herbivory.

Dissolved imports and exports

Concentrations of some elements have been found to vary considerably between Izembek Lagoon water and nearby Bering Sea water. Due to biogenic and possibly other processes within the lagoon, the water which is exchanged into the Bering Sea during each tidal cycle is either enriched or depleted in many nutrients and trace metals, and the lagoon can be considered to either import or export these elements.

The phosphorus cycle in the Izembek Lagoon eelgrass beds has been investigated in considerable detail. McRoy and Barsdate (1970) described the uptake and release of phosphorus by both the leaves and roots of eelgrass plants, suggesting that their role may be that of either a sink or a source for dissolved phosphorus. The effect of this transport on the overall phosphorus cycle of the lagoon was determined by McRoy, Barsdate and Nebert (1972). Phosphorus moved through eelgrass plants from the sediments to the water at a net rate of 62.4 mg P/m^2-day (Fig. 28.3). The lagoon exported on the order of 3 MT of phosphorus per day during the summer and fall, and the annual export has been estimated to be 495 MT of dissolved phosphorus. This flow of phosphorus can be seen in the Bering Sea as a plume of high phosphorus concentrations (>0.6 μg-atoms P/liter) west and somewhat south of Izembek Lagoon (Fig. 28.4).

Dissolved inorganic nitrogen concentrations (determined as NH_4^+, NO_2^-, and NO_3^-) have been measured in Glazenap Pass (Table 28.3) on both incoming and outgoing tides. Assuming a similar situation at each of the other two passes, for June to November of 1969 the mean net flow of 1.7 μg N/liter out of the lagoon combined with the tidal volume gives an annual export of 96 MT of dissolved inorganic nitrogen, much less than the phosphorus export. High nitrogen concentrations also are evident in the Izembek Lagoon plume, with values of 2.2 to 3.6 μg-atoms N/liter as opposed to 0.1 to 1.9 μg-atoms N/liter in nearby Bering Sea water.

Dissolved carbon data are available from only two tidal cycles in September

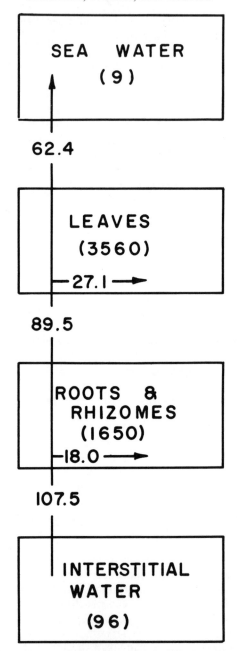

Fig. 28.3 Net daily movement of phosphorus (mg/m²) in an eelgrass stand. Amount of phosphorus (mg/m²) in each compartment is in parentheses (from McRoy, Barsdate and Nebert 1972).

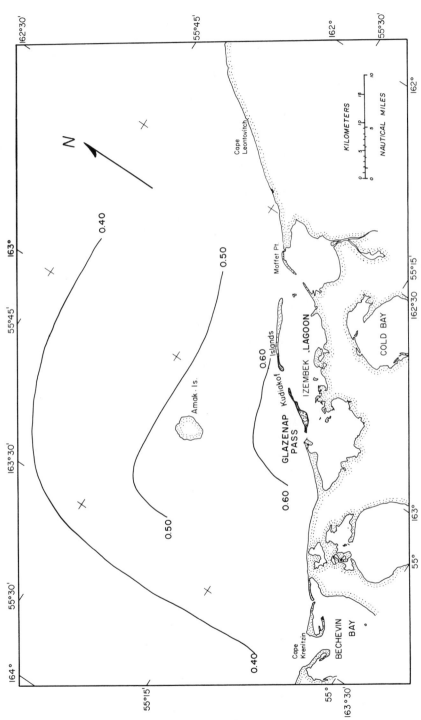

Fig. 28.4 Dissolved reactive phosphorus distribution (μg-atoms P/liter) in the Bering Sea adjacent to Izembek Lagoon, June 1968 (R/V *Acona* cruise 066).

Table 28.3 Dissolved phosphorus, nitrogen, silica, and carbon flow (µg/liter) in Glazenap Pass, 1969. The number of tidal cycles for which concentration differences were determined is given in parentheses.

	June and July		September		October		November		Mean	
	mean difference	net direction	mean difference	net direction	mean difference	net direction	mean difference	net direction	mean difference	net direction
Phosphorus	8.99 (13)	Out	9.61 (6)	Out	8.37 (4)	Out	1.31 (4)	In	8.99 (27)	Out
Nitrogen	6.45 (11)	Out	7.80 (6)	In	2.63 (4)	Out	5.42 (4)	Out	1.68 (25)	Out
Silica	4.67 (13)	Out	138.6 (6)	In	117.6 (4)	Out	55.9 (4)	Out	18.04 (27)	In
Carbon	—	—	350 (2)	Out	—	—	—	—	—	—

1969 and are suggestive, although not conclusive, of a net flow out of the lagoon. A mean of 0.35 mg dissolved carbon per liter was observed moving out of the lagoon through Glazenap Pass into the Bering Sea. Based on the assumption that this flow of carbon is the same at all three passes throughout the productive period of the eelgrass beds, an annual dissolved carbon export of 1.96×10^4 MT per year can be estimated.

Concentrations of dissolved silica were determined on a number of tidal cycles from June through November at Glazenap Pass. Making the same assumption about transport through the other passes, the annual flux of dissolved silica is on the order of 2.1×10^3 MT of silica into the lagoon from the Bering Sea. There is some uncertainty in this figure, as the silica concentrations vary widely on a short-term basis.

Copper and lead have been studied briefly but rather intensively in this

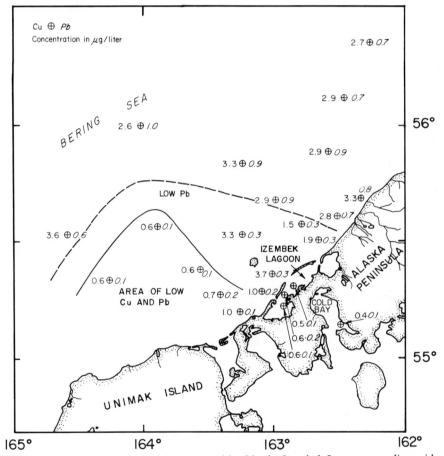

Fig. 28.5 Surface distribution of copper and lead in the Izembek Lagoon sampling grid, September, 1970 (R/V *Acona* cruise 103).

area. The values reported here are for electrochemically reactive forms of the
metals in unfiltered samples at pH less than 2. We have found, as has Matson
(1968), little difference between total dissolved copper determined after filtration
and digestion and copper determined after acidification of unfiltered samples.
Mean concentrations were 2.9 μg Cu/liter and 0.8 μg Pb/liter in the sea off
Glazenap Pass, and 0.8 μg Cu/liter and 0.2 μg Pb/liter in the plume of the lagoon
(Fig. 28.5). A series of samples taken at hourly intervals over two tide cycles
at a location about 3 km outside Glazenap Pass indicated fairly uniform copper
concentrations of the same level as found in other areas of the plume (approx.
0.8 μg Cu/liter). Occasional peaks of greater concentration can be attributed
to intrusions of normal Bering Sea water during periods of high tide (Fig. 28.6).

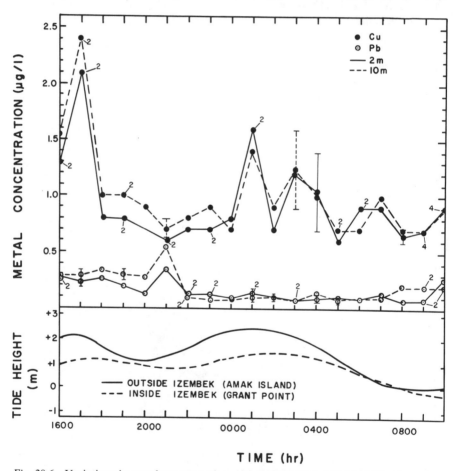

Fig. 28.6 Variations in metal concentration with time and tide height at Glazenap Pass
off Izembek Lagoon, 9–10 September 1970 (R/V *Acona* cruise 103). Brackets represent
the range for replicate analyses, and small numbers near the curves indicate coinciding
data points.

Concentrations of both metals were very low in the lagoon, 0.6 μg Cu/liter and 0.1 μg Pb/liter. Assuming constant copper flux throughout the year and a net concentration difference of 2.1 μg Cu/liter between the sea and lagoon, the annual import of dissolved ionic copper is roughly 260 MT. Simultaneous analyses for lead at the Glazenap Pass station averaged 0.2 μg/liter, the same as in the plume off the lagoon, with little suggestion of tide-related variations. Intrusions of water with open-sea concentrations of lead (0.8 μg/liter) did not occur at high tide, as the lead anomaly apparently dissipates less rapidly than does that of copper, and the lead plume extends rather far out from the lagoon. Although the maintenance of this anomaly over an extensive area requires the loss of an appreciable quantity of lead from the Bering Sea to the lagoon, the amount involved is much less than the copper flux into the lagoon, perhaps by as much as an order of magnitude.

Lagoon budget

The budget presented here (Table 28.4) for the flow of elements between the Bering Sea and Izembek Lagoon takes into account both the particulates produced in the eelgrass beds and the dissolved elements either released or taken up by the eelgrass plants, sediments or other components of the lagoon. Particulate exports from the lagoon ideally should be measured directly, but because of the large cross-sectional areas of the three passes, the rather rigorous field conditions and the variability in the rate of detached eelgrass export due to wind and weather, this has been impractical. We instead have calculated annual eelgrass production, estimated various gains and losses within the lagoon, and assumed the remainder to be exported. Avian herbivory resulted in the largest loss within the lagoon, and it amounted to less than 3 percent of the total production. Losses from beached eelgrass were very small, being less than 0.1 percent of the production. Increases in the concentration of nitrogen, phosphorus, silica, and copper occurred while eelgrass floated within the lagoon, but only for silica was this change greater than 1 percent. In the case of all elements examined, the particulate export was essentially equal to the annual production. Therefore, we conclude that the major pathway for the loss of eelgrass and its associated elements from Izembek Lagoon is exportation rather than herbivory, leaching, or decomposition within the lagoon.

Since sampling for dissolved materials was not carried out during all seasons, some assumptions were necessary for the extrapolation of the observed flux rates to an annual basis. The net flows of carbon, nitrogen, phosphorus, and silica were determined from the concentration differences in Table 28.3, assuming transport only during a 165-day eelgrass growth period. The magnitude of the copper flux into the lagoon was far too large to be attributed to the uptake and storage by eelgrass, as the standing stock of copper in eelgrass (1.0 MT) was less than twice the daily input of dissolved copper (0.71 MT Cu/day). Since the copper transport into the lagoon may not be directly related to seasonal biological phenomena, dissolved copper flux arbitrarily was assumed to continue at the observed rate throughout the year.

Rather meager information suggests that a massive amount of dissolved organic carbon—nearly 2×10^4 MT per year—moves from the lagoon into

TABLE 28.4 Annual budget for carbon, nitrogen, phosphorus, copper, and silica in Izembek Lagoon by weight and by percent of amount incorporated in the calculated annual production of eelgrass.

	Carbon		Nitrogen		Phosphorus		Copper		Silica	
	metric tons	%	metric tons	%	metric tons	%	metric tons	%	metric tons	%
Annual production	166,000	100	7,400	100	1,660	100	3.45	100	386	100
Gain while floating	—	—	1.4	0.02	0.8	0.05	0.009	0.26	23.7	6.1
Avian herbivory	−4,980	−3.0	−222	−3.0	−50	−3.0	−0.10	−3.0	−12	−3.0
Loss from beached eelgrass	−158	−0.095	−7.0	−0.094	−1.6	−0.096	−0.0033	−0.096	−0.37	−0.096
Particulate export	160,862	96.9	7,172	96.9	1,609	97.0	3.36	97.4	397	102.8
Dissolved export	19,600	11.8	96	1.3	495	29.8	−260	−7,536	−1,010	−262
Total export	180,462	108.7	7,268	98.2	2,104	126.7	−257	−7,449	−613	−159

the sea. This is equal to a substantial fraction (nearly 12 percent) of the calculated annual production of organic carbon by eelgrass. From our visual field observations, we would expect the net difference between dissolved organic carbon entering and leaving the lagoon to be much greater than the measured value of 350 μg/liter, as the water flowing from the shallower and usually warmer eelgrass beds on the outgoing tide is noticeably colored brown by organic substances. It appears likely that at least some of the organics secreted by eelgrass have such short residence times in the water that they do not exit the lagoon. Organics secreted by eelgrass confined within bottles tend to precipitate rather rapidly as an amorphous flocculent material, which in nature would enter the detritus cycle. Particulates, even those with low settling velocities, effectively are trapped in eelgrass beds (Milne and Milne 1951).

Estimates of other Bering Sea nitrogen fluxes have been made, and Hattori and Wada (Chapter 6, this volume) calculate that river flow into the eastern Bering Sea brings in 7×10^5 kg nitrogen per day, principally as nitrate, and that the net flow through the eastern Bering Sea is 1×10^7 kg NO_3-N/day. In relation to its abundance in eelgrass, a comparatively small amount of inorganic nitrogen (NH_4^+, NO_2^-, NO_3^-) is exported from the lagoon in solution, about 1 percent of the annual eelgrass nitrogen production. When combined with the particulate nitrogen, however, the sum of lagoon nitrogen exports (4.5×10^4 kg N/day) is about 6 percent of all river inputs combined and about 0.5 percent of the eastern Bering Sea nitrate-nitrogen flow. In addition to the inorganic nitrogen, there may exist a significant flux of dissolved organic nitrogen. The organic substances secreted by marine benthic algae are reported to have a nitrogen content of up to 3 percent (Smith and Young 1953). In view of the substantial dissolved organic carbon flux, there is a strong possibility that the export of dissolved organic nitrogen is much larger than that we have found for inorganic nitrogen.

As is the case for both carbon and nitrogen, phosphorus flows from the lagoon in both particulate and dissolved forms, with phosphate (reactive phosphorus) accounting for 25 percent of the total export (2.104×10^3 MT/year). The mechanism responsible for the large phosphate component has been discussed previously and will not be treated further here. Phosphorus in small particulate material (as distinguished from floating detached eelgrass) and dissolved organic phosphorus have been measured, but their contribution to the total flow of phosphorus from the lagoon is extremely small (McRoy, Barsdate and Nebert 1972).

The mean silica concentration of the water leaving the lagoon on outgoing tides is 18 μg/liter less than that coming in on flood tides, resulting in a calculated transport of 1.01×10^3 MT of silicon in solution during a 165-day growing season. The flow of particulate silicon is in the opposite direction, with 397 MT of silicon leaving the lagoon in eelgrass. Uptake by plankton apparently is the most common cause of silica depletion in coastal waters (Calvert 1966; Turekian 1971), although abiotic sorbtion of silica may be a major mechanism in some systems (Bien et al. 1958). Depletion of dissolved silica by vascular aquatic plants has not been reported previously, but the data presented here indicate that silica is taken up vigorously by detached eelgrass, even by shoots from which the roots and rhizomes have been removed (Table 28.2). This suggests that *in situ* eelgrass may remove silica from the lagoon water and in part cause the net flow of silica from the sea into Izembek Lagoon.

Copper is removed from water by detached eelgrass and possibly also by *in situ* eelgrass. The amounts of copper present in either the standing stock or annual production of eelgrass are, however, about 100-fold smaller than the extrapolated annual flux of copper into the lagoon from the sea. The mechanisms for this removal of copper from the lagoon waters have not been studied, but from the above consideration it is highly unlikely that a significant portion of the copper is stored in the biota. In shallow-water environments, cationic trace metals may be depleted from the water through interaction with the sediments (Turekian 1971; Duke et al. 1966; Pomeroy et al. 1969). Chow and Thompson (1954) reported seasonal variations in the copper concentration of surface water in the San Juan Channel, Washington; these were attributed to changes in physical conditions favoring formation or destruction of copper sulfide in the sediments.

Without a mechanistic understanding of the copper cycle and without seasonal data, the estimate of the annual copper budget of Izembek Lagoon is speculative. The processes responsible for the extraction of copper from water in the lagoon may be active for a shorter time than we have assumed, and presently unassessed pathways, such as may be involved in the suspension and transport of sediments during winter storms, may carry trace metals back into the Bering Sea from the lagoon. Despite the uncertainties involved, it appears clear that geochemical cycling of massive amounts of copper occurs in Izembek Lagoon.

DISCUSSION

Odum (1968) has generalized that ecosystems tend to retain their nutrients, but this situation is perhaps only poorly developed for temperate and high latitude systems dominated by marine vascular plants. For instance, in *Spartina* salt marshes there is a strongly developed cycle in which phosphorus, zinc, and perhaps other elements are removed from the sediments by means of vascular plant roots and translocated to upper portions of the plants to be sloughed off as detritus, which remains in part within the system (Pomeroy et al. 1969). These salt marsh estuaries are somewhat open-ended ecosystems, with rather large inputs and outputs and with no or little nutrient limitations (Pomeroy 1970). Izembek Lagoon appears to share these same features with the salt marshes. Eelgrass entering the detritus cycle, however, is much more likely to exit the system than is *Spartina*.

The abundance of particulate matter brought into the usual estuarine environment by rivers is almost totally lacking in Izembek Lagoon, and yet the lagoon annually exports immense quantities of phosphorus. Conversely, the lagoon appears to assimilate large quantities of Bering Sea copper. Although very little study has been made of the modern geology of the lagoon, it appears that the sedimentation rate there is not unusually fast, and therefore the effects of either the removal or addition of the above and other elements should be reflected clearly in the chemistry of the resulting lagoon sediments.

The large quantity of phosphate exported from Izembek Lagoon presumably has a favorable effect on the rates of plankton primary productivity and indirectly enhances productivity at many trophic levels in the eastern Bering Sea. The contributions of dissolved nitrogen and organic carbon from the lagoon, although less spectacular than phosphorus in relation to their abundance in the sea, also would tend to increase the productivity of the area. There is at present insufficient understanding of the role of metals to predict with confidence what effects, if any, result from the removal of trace metals from the sea by processes within the lagoon. Preliminary work on the sensitivity of Bering Sea phytoplankton to copper and lead, however, suggests that trace metal concentrations approach or exceed the tolerance levels of the plankton algae. Thus the trapping of metals within the lagoon also may tend to enhance Bering Sea plankton primary productivity.

The annual export of floating eelgrass from Izembek and the many other lagoons is potentially of considerable importance to the Bering Sea benthic foodwebs. As early as 1914, Boysen Jensen (1914) pointed out that microbial decomposition and zooplankton grazing are of much less importance for eelgrass than for phytoplankton. Ellson et al. (1949) report that St. Lawrence Island natives know that good cod fishing is found on detrital seagrass bottoms. Since there are no eelgrass beds on St. Lawrence Island, these detrital accumulations are the result of transport of detached plants from some other location, probably from the coastal lagoons on the Seward Peninsula. Thus it appears that the dispersion of eelgrass is great and that eelgrass may supply food for benthic faunal communities far from the original source of the eelgrass.

Discussion

KITANO: What does the distribution of copper as a trace element indicate?

BARSDATE: There seems to be a progressive increase of concentration of copper in the waters as they move from the Pacific through the Bering Sea and into the shallow areas of the shelf.

KITANO: How does the sink concentration of copper in the lagoon compare with that in the open ocean?

BARSDATE: The sink for copper in the lagoon has not been determined. On the basis of mass balance consideration alone, there is not enough eelgrass to take up the copper at the rate it seems to be entering the lagoon. One can speculate that the sediments must receive this copper, but we have not yet made any copper analyses of the sediments. High sulfide concentrations in the sediments interfere with the determination of copper by the electrochemical apparatus used for metal analyses in water.

KITANO: Does copper become complexed with the sediments?

BARSDATE: Organic and inorganic complexation are possible within the sediments and also may be involved in the transport of copper from the water to the sediments. Some of the organic materials which are secreted by eelgrass have a very short residence time in the water; when these organics flocculate, they may carry down copper which then enters the sediments. Sulfide complexation also is quite likely. In many places in the eelgrass beds there appears to be essentially no oxidized sediment zone, and sulfides may exist at or very close to the water-sediment interface.

SUGIURA: What is the concentration ratio for phosphate or nitrate inside the lagoon and out?

BARSDATE: The concentration of phosphate is perhaps threefold higher inside the lagoon than in the plume, and the plume is at least twofold higher in phosphate than the water 100 km north. Differences in nitrogen are not so clear, as on occasion the nitrate concentration in the lagoon is somewhat lower than in the waters to the north.

SUGIURA: Is the mechanism for creating a sink concentration of phosphate in the lagoon possibly a vertically occurring biological phenomenon?

BARSDATE: This seems very definitely to be the case, and it would be easy to make an experimental determination of such a condition.

HATTORI: In what form is the copper distributed vertically in the water column?

BARSDATE: This is not known. Large dissolved organic complexes, those which are too big to pass through a dialysis membrane, are almost completely absent both in the open water of the Bering Sea and in the Izembek plume, although we have found previously that they are abundant in freshwater environments. Low-molecular weight organic complexes probably are abundant, although we have inadequate techniques for their detection, and the evidence for their existence is largely by inference. Most of our analytical work has been done with unfiltered seawater acidified to pH2. This practice results in a high precision in replicate analyses, but as a result we have little information about the distribution of copper between particulate and dissolved species.

On a few occasions we have done parallel determinations for ionic copper at pH2 in filtered and unfiltered samples, for total dissolved copper, and for particulate copper. We have found the agreement between ionic copper at pH2 in unfiltered samples to agree very well with the total dissolved determinations. Even where an appreciable amount of sestonic copper is evident, only a relatively small fraction of this appears in the dissolved ionic form after acidification.

HATTORI: Have you compared the concentration of copper with that of chlorophyll or a selected biomass?

BARSDATE: On the basis of sparse data, there is a strong inverse relationship between copper concentration and both plankton primary productivity and chlorophyll in the area near the lagoon. This could be accidental rather than functional, however, because phosphate is high near the lagoon and decreases seaward.

NELSON: Two diagrams showed conflicting locations of plumes off Izembek Lagoon, where northward currents are indicated. The copper plumes seem to be severely displaced to the west, in contrast to the phosphate plumes appearing directly offshore.

BARSDATE: There appears to be a confusing tidal current there with a southwest movement at least during times of strong onshore winds. In both of the figures to which you refer, the plume appears to be displaced somewhat in the wrong direction due to this local current anomaly, but in both cases it was clearly identifiable by the conspicuous content of drifting eelgrass and other organic matter. Hebard, who reported on the currents of the area in 1961, had to make a substantial correction for wind current at one station which otherwise would have indicated clockwise circulation in the southeastern Bering Sea.

NELSON: Have you studied sedimentation rates in the lagoon area as a possible sink indicator for the trace metals?

BARSDATE: Dr. Sharma has examined the small number of core samples collected in the lagoon, but no developed sedimentilogical program exists for this area. The sedimentation rates are undoubtedly high in comparison to the open Bering Sea; however, I know of no information that suggests that the rates are exceptionally high in comparison with similar coastal environments.

NELSON: If copper shows a northward increase in Bering Sea water, then it should be noted that a corresponding increase can be expected in the sediments.

BARSDATE: Yes, that is my interpretation.

NELSON: A high copper anomaly has been associated with morainal sediments deposited into the northern Bering Sea from the Chukotka Glacier. Copper, mercury, chromium, and several other trace metals found in the sediments are being outlined in map form. A combined view of the trace-metal concentrations bath in the water and in the sediments would be illuminating.

Acknowledgments

We thank R. D. Jones, Jr., for his helpful discussions and for the use of facilities at Izembek Marine Station and M. T. Gottschalk for field and laboratory assistance. Financial support was provided by the U. S. Atomic Energy Commission Contract AT(45–1)-2229 TA No. 2 and the Oceanography Section of the National Science Foundation (Grants GB-8274 and GA-33387).

REFERENCES

BIEN, G. S., D. E. CONTOIS, and W. H. THOMAS
  1958 Removal of soluble silica from fresh water entering the sea. *Geochim. Cosmochim. Acta* 14: 35–54.

BOYSEN JENSEN, P.

1914 Studies concerning the organic matter of the sea bottom. In *Report of the Danish Biological Station to the Board of Agriculture*, edited by C. G. J. Petersen. Copenhagen, Denmark, No. 22: 1–39.

BURKHOLDER, P. R., and T. E. DOHENY

1968 The biology of eelgrass. Dept. of Conservation and Waterways Contribution No. 3, Hempstead, Long Island, 120 pp.

CALVERT, S. E.

1966 Accumulation of diatomaceous silica in the sediments of the Gulf of California. *Geol. Soc. Amer. Bull.* 77: 569–596.

CANDUSSIO, R.

1960 Chemical composition of *Zostera marina* [in Italian]. *Inst. Chimico Agrario Sperimentale Publ.* 20: 5–10.

CHOW, T. J., and T. G. THOMPSON

1954 Seasonal variation in the concentration of copper in the surface waters of San Juan Channel, Washington. *J. Mar. Res.* 3: 233–244.

DUKE, T. W., J. N. WILLIS, and T. J. PRICE

1966 Cycling of trace elements in the estuarine environment. 1. Movement and distribution of zinc 65 and stable zinc in experimental ponds. *Chesapeake Sci.* 7: 1–10.

ELLSON, J. G., B. KNAKE, and J. DASSOW

1949 Report of Alaska exploratory fishing expedition, fall of 1948, to northern Bering Sea. *U. S. Fish Wildl. Serv., Fish. Leafl.* 342.

HUTCHINSON, G. E., and A. WOLLACK

1943 Biological accumulators of aluminum. *Trans. Conn. Acad. Arts Sci.* 35: 73–128.

KETCHUM, B. H.

1967 Phytoplankton nutrients in estuaries. In *Estuaries*, edited by G. H. Lauff. Amer. Ass. Adv. Sci. Publ. No. 83, Washington, D.C., pp. 329–335.

MATSON, W. R.

1968 Trace metals, equilibrium, and kinetics of trace metal complexes in natural media. Ph.D. Thesis. MIT, Cambridge, Mass., 256 pp.

McROY, C. P.

1966 The standing stock and ecology of eelgrass (*Zostera marina* L.) in Izembek Lagoon, Alaska. M.S. thesis, Univ. Washington, Seattle. University Microfilm, Ann Arbor, Mich., 138 pp.

1968 The distribution and biogeography of *Zostera marina* (eelgrass) in Alaska. *Pacific Sci.* 22: 507–513.

1970a Standing stocks and other features of eelgrass (*Zostera marina*) populations on the coast of Alaska. *J. Fish. Res. Bd. Can.* 27: 1811–1821.

1970b On the biology of eelgrass in Alaska. Ph.D. Thesis, Univ. Alaska, Fairbanks, 156 pp.

McRoy, C. P., J. J. Goering, D. C. Burrell, M. B. Allen, and J. B. Matthews
  1969 Coastal ecosystems of Alaska. In *Coastal ecological systems of the U. S.,
        a source book for estuarine planning*, Vol. 1, edited by H. T. Odum, B. J.
        Copeland, and E. A. McMahon. Report to the FWPCA, Inst. Mar. Sci.
        Univ. North Car., pp. 125–168.

McRoy, C. P., and R. J. Barsdate
  1970 Phosphate absorption in eelgrass. *Limnol. Oceanogr.* 15: 6–13.

McRoy, C. P., R. J. Barsdate, and M. Nebert
  1972 Phosphorus cycling in an eelgrass (*Zostera marina* L.) ecosystem. *Limnol.
        Oceanogr.* 17: 58–67.

McRoy, C. P., J. J. Goering, and W. E. Shiels
  1972 Studies of primary production in the eastern Bering Sea. In *Biological ocean-
        ography of the northern North Pacific Ocean* [Motoda commemorative
        volume], edited by A. Y. Takenouti et al. Motoda Shoten-Idemitsu, Tokyo,
        pp. 199–216.

Meehan, W. R., and L. E. Smythe
  1967 Occurrence of beryllium as a trace element in environmental materials.
        Environ. Sci. Tech. 1: 839–844.

Milne, L. J., and M. J. Milne
  1951 The eelgrass catastrophe. *Sci. Amer.* 184: 52–55.

Odum, H. T.
  1968 Work circuits and systems stress. In *Symposium on Primary Production and
        Mineral Cycling in the Natural Ecosystem*, edited by H. E. Young. Univ. Maine
        Press, pp. 81–138.

Petersen, C. G. J.
  1914 Om Baendeltangens (*Zostera marina*) Aars-Production i de danske Farvande
        [in Danish, English summary]. In *Mindeskrift i Anledning af Hundredaaret for
        Japetus Steenstrups Fødsel*, edited by F. E. Jungersen and J. E. B. Warming.
        B. Lunos Bogtrykkeri, Copenhagen, No. 9, 20 pp.

Pomeroy, L. R.
  1970 The strategy of mineral cycling. *Ann. Rev. Ecol. Systematics* 1: 171–190.

Pomeroy, L. R., R. E. Johannes, E. P. Odum, and B. Roffman
  1969 The phosphorus and zinc cycles and productivity of a salt marsh. In *Pro-
        ceedings of Second National Symposium on Radioecology*, edited by D. J.
        Nelson and F. C. Evans. Ann Arbor, Mich., pp. 412–419.

Roden, G. I.
  1967 On river discharge into the northeastern Pacific Ocean and the Bering Sea.
        J. Geophys. Res. 72: 5613–5629.

Smith, D. G., and E. G. Young
  1953 On the nitrogenous constituents of *Fucus vesiculosus*. *J. Biol. Chem.* 205:
        849–858.

TANIGUCHI, A.

1969 Regional variations of surface primary productivity in the Bering Sea in summer and the vertical stability of water affecting the production. *Bull. Fac. Fish., Hokkaido Univ.* 20: 169–179.

TUREKIAN, K. K.

1971 Rivers, tributaries, and estuaries. In *Impingement of man on the oceans*, edited by D. W. Hood. Wiley-Interscience, N.Y., pp. 9–73.

VINOGRADOV, A. P.

1953 The elementary chemical composition of marine organisms. Sears Found. Mar. Res., Mem. 2, Yale Univ., New Haven, Conn., 647 pp. (Transl. from Russian).

Part 8

TECHNOLOGICAL
DEVELOPMENT

Environmental earth satellites for oceanographic-meteorological studies of the Bering Sea

E. PAUL MCCLAIN

National Oceanic and Atmospheric Administration, Washington, D.C.

Abstract

Present-day operational earth satellites already provide much information useful to the oceanographer and meteorologist in the Bering Sea area. Near-future operational and development satellites will make available more types of data and higher-resolution coverage. Because the Bering Sea is a relatively remote and inaccessible part of the world, subject moreover to incursions of polar ice, comprehensive and repetitive survey by surface vessels or aircraft would be costly, time-consuming, very difficult and to some extent hazardous. The polar-orbiting environmental earth satellite is thus in many ways an ideal sensor platform for areas such as the Bering Sea. Remote sensing does entail certain limitations, however, and it is in such areas of weakness that the data collection and relay capabilities of sun-synchronous polar satellites and earth-synchronous equatorial satellites are important.

Current operational environmental satellites

Beginning in 1970 the second generation of Improved TIROS Operational Satellites was inaugurated (Albert 1968). Environmental satellites of this ITOS series, now designated NOAA-1, NOAA-2, etc., carry vidicon camera systems (0.5–0.7 μm), two-channel (0.5–0.7 μm and 10.5–12.5 μm) scanning radiometers (SR), and solar proton monitors. These satellites are placed in near-polar, circular, sun-synchronous orbits at a nominal altitude of 1890 km. Equator-crossing times are 0300 and 1500 local time each day.

Data are received from these satellites in two modes. Global coverage on a daily basis is provided by temporary storage of information on board the spacecraft by means of tape recorders. The stored data are then read out on command to complex data-acquisition stations in Alaska and Virginia, whence they are relayed by land-line to a central facility and computer complex in Suitland, Maryland. Local coverage is provided by means of a direct-readout capability known as APT (Automatic Picture Transmission), which is readily and freely available to all nations of the world. At the present time there are over 600 APT stations operating in more than 50 countries.

By use of thermal infrared imagery (9-km resolution at nadir) on earth's night side to supplement the visual imagery (4-km resolution at nadir) on the day side, global cloud maps can be produced at 12-hourly intervals. Further-more, the infrared measurements can be used to infer additional information such as cloud-top temperatures and heights (Rao 1970) and temperatures at the sea surface, water or ice, day or night (Smith et al. 1970). The visual or infrared imagery are used directly, or after computer processing, to detect and monitor polar ice packs (McClain and Baliles 1971) and obtain information on their concentration and physical state (McClain 1973). Extensive computer processing is performed on these satellite data in order to rectify them, normalize them for variable solar illumination and camera distortions, and redisplay them on standard meteorological map projections on a regional or global basis (Bristor et al. 1966). Further manipulations of the digitized brightness or thermal values are employed to produce derived products such as average or composite charts for selected periods (Booth and Taylor 1969).

Application of NOAA satellite imagery to the Bering Sea area

Figure 29.1 is a single vidicon camera frame of the Bering Sea area taken in mid-March of 1971 when the polar ice pack was near its maximum extent. The nominal grid of latitude, longitude and coastlines is superimposed by computer during display. Figure 29.2 illustrates how earth satellite imagery can be used to monitor the changing boundaries and extent of sea ice during the ice season. It consists of a series of four vidicon camera frames, at about one-month intervals, from mid-April through mid-July in 1971. The retreat of the ice pack northward through the Bering Sea to a position well poleward of the Bering Strait in the Chukchi Sea can be noted, as well as the presence of well-defined areas of thinner ice or open water within the packs.

Lack of sufficient solar illumination limits the use of vidicon camera systems or scanning radiometers operating in the visible portion of the spectrum for ice surveillance at the highest latitudes during much of the colder half of the year. This limitation has been largely overcome by the use of thermal infrared imagery (Barnes et al. 1970); Figure 29.3 is an example of imagery obtained by direct-readout in Alaska. This image is unrectified and ungridded, and some "noise" from electrical interference is evident. The gray-scale display employed for this image results in some 32 steps of about four degrees (C) each, proceeding from black (warm) to white (cold). The coasts of Alaska and Siberia near the Bering Strait are readily seen because of the temperature contrast between the cold land and the generally warmer ice pack covering the northern

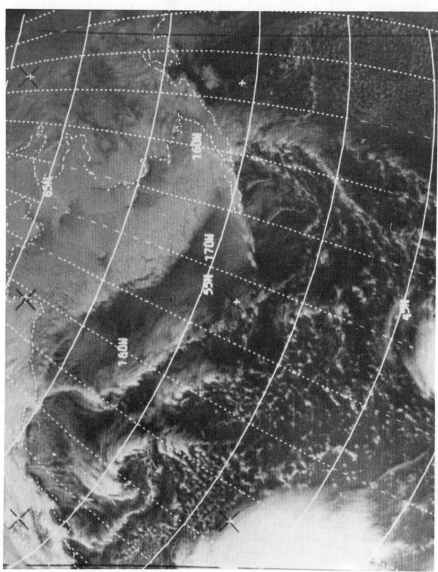

Fig. 29.1 Photograph taken by the Advanced Vidicon Camera System (AVCS) on the ITOS-1 satellite 14 March 1971.

Fig. 29.2 AVCS pictures taken by ESSA-9 and NOAA-1 satellites in the Bering Sea area. *Upper left:* ESSA-9, 16 April 1971. *Upper right:* NOAA-1, 18 May 1971. *Lower left:* ESSA-9, 24 June 1971. *Lower right:* ESSA-9, 24 July 1971.

portion of the Bering Sea. This contrast is greater than the difference in reflectivity between the snow-covered land and the ice pack seen in the vidicon pictures. This is even more evident when the infrared image is enhanced, as in Figure 29.4. The gray-scale display here proceeds in two degree steps from 274 K (black) downward to 221 K and lower (white). The southernmost boundary of the ice pack is clearly delineated by a band of relatively warm, ice-free ocean. Shallow (cool) clouds are present over much of the Bering Sea to the south of the ice, and deep (cold) clouds are found to the southeast and over the Aleutians, as well as to the west near northern Kamchatka. The distinctly warmer areas, such as are seen south of St. Lawrence Island and in the Gulf of Anadyr apparently represent regions of significantly thinner or less-concentrated ice.

A major problem in the use of either visual or infrared imagery for sea-ice surveys is proper differentiation between sea ice and clouds, whose reflectances (or temperatures) can often be comparable. Photo-interpretative methods have proved effective in this regard, but only when applied by skilled analysts who are familiar with the geography and meteorology (oceanography) of the area (Barnes et al. 1970). Characteristic differences in form, tone, texture, and particularly persistence and movement, are the principal interpretive tools.

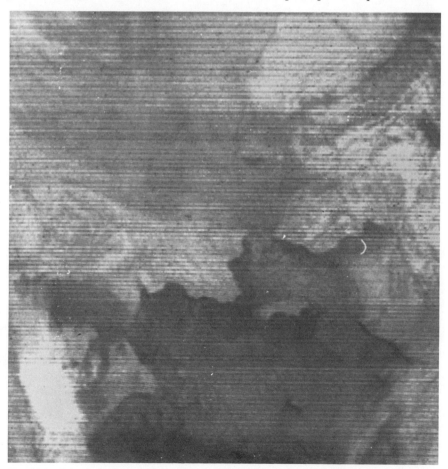

Fig. 29.3 Direct-readout thermal infrared imagery of northern Alaska from measurements taken by the ITOS-1 satellite on 15 March 1971.

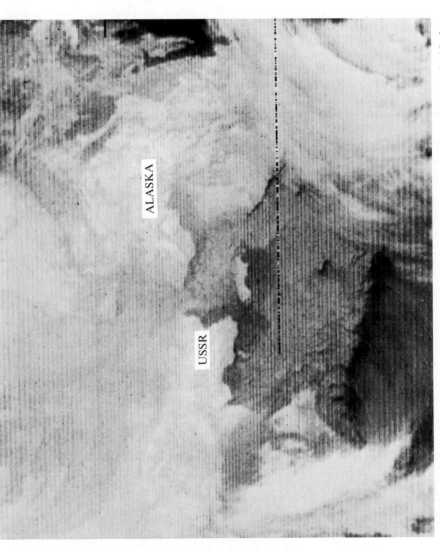

Fig. 29.4 Enhanced infrared image of ice features obtained by special display direct-readout in Alaska.

Changes in cloudiness from day to day are much greater than changes in the ice pack, and clouds obscure landmarks that would be otherwise detectable.

The transient nature of clouds in comparison with major ice and snow fields has enabled the development of an automated procedure for suppressing the cloudiness by compositing satellite data over periods of time ranging from five days upward to a month (McClain and Baker 1969). Digitized brightness values, after rectification and normalization, are composited by saving and displaying only the minimum brightness at each grid point in the array during the period in question. Such a full-resolution (4 km) Composite Minimum Brightness (CMB) chart for a 10-day period in May–June 1970 is shown in

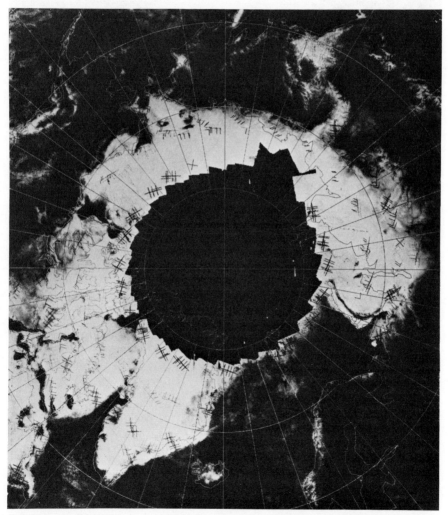

Fig. 29.5 Ten-day Composite Minimum Brightness (CMB) chart derived from ESSA-9 digitized brightness data for the period 24 May–2 June 1970.

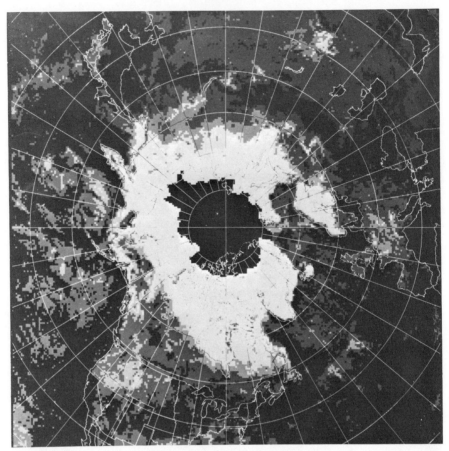

Fig. 29.6 Experimental Sea Ice chart derived from calibrated five-day CMB values for the period 24–28 April 1970. *White*: compact or very close pack, snow-covered. *Light gray*: compact or very close pack, snow free. *Medium gray*: very close or close pack, snow free, some puddling. *Dark gray*: open pack, snow free, much puddling and rotten ice. *Black*: very open pack or ice free waters. Much residual cloudiness is present and is represented in the same tone as the open pack ice.

Figure 29.5. Extremely persistent cloudiness will appear as tenuous or irregular areas of generally lower brightness than the ice pack, especially for short compositing periods.

By calibrating the CMB values externally with respect to the brightness of the Greenland ice cap (high reflectance) and cloudfree ocean areas (low reflectance), it has been found that distinct changes in brightness with time can be related to the concentration and physical condition of the ice, such as presence or absence of a snow cover or melt-water pools (McClain 1973; Wendler 1973). Redisplay of a five-day mesoscale (30-km resolution) CMB chart in five classes of ice concentration-condition is shown in Figure 29.6. In this experimental sea-ice chart, white represents that range of brightness corresponding to snow-covered, compact ice. Successively darker tones correspond to lack of a snow cover, presence of melt water, and decreasing ice concentrations (increasing amounts of open water). Residual cloudiness is particularly evident in this example because of the short compositing period, although major ice features are seen in Baffin Bay, the Bering Sea, and elsewhere.

Thermal infrared measurements from satellites have been used to qualitatively and quantitatively map sea-surface temperatures in relatively cloudfree areas (Rao et al. 1971a), and spatial and temporal compositing of these data has enabled temperature mapping even in most cloudy areas (Rao et al. 1972). Figure 29.7 is an enhanced thermal image (direct-readout) of the Gulf Stream area; quantitative isotherm analyses have been published using satellite measurements for a similar case (Rao et al. 1971b). Direct readout data have been the most useful for such studies because the stored data have been prone to tape recorder and other system noise and calibration problems.

Operational environmental satellites for the near-future

Among the next principal evolutionary changes in operational NOAA satellites is the addition of a Very High Resolution Radiometer (VHRR) on NOAA-2 in October 1972 (Schwalb 1972). The VHRR will be in addition to the present SR, and it will have the same visual and thermal infrared channels, but its ground resolution is designed to be better than 1 km at nadir. The visible channel on the SR will replace the vidicons for both global and local direct-readout cloud mapping, and it will be opened up to 0.5–1.0 μm to increase reflectances over land and facilitate detection of landmarks. The first operational atmospheric sounder, the VTPR (Vertical Temperature Profile Radiometer), will be carried on board these satellites. The VTPR scans to either side of the subpoint track and will provide essentially global coverage with a ground resolution of about 60 × 60 km at nadir.

The VHRR will be used principally in the direct readout, image mode, although limited provision is being made for remote-recording of data on tape and for digitization of selected data samples for experimental use. A special display device employing laser-exposing and dry-processing to produce a film-positive image has been developed for use with the VHRR ground station. The number of such ground stations is likely to be small in the near-future because of their cost and complexity. One station will be located in Alaska, with coverage of the Bering Sea (see Fig. 29.8), and emphasis will be upon oceanographic and hydrologic applications in this area.

The next generation of operational environmental satellites is provisionally scheduled for the latter half of this decade. Although some details remain to be settled, the payload will consist basically of the standard two-channel SR, an advanced VHRR equipped with two additional channels (probably in the near-infrared and water-vapor absorption parts of the spectrum), an advanced temperature and humidity sounder, and a solar proton monitor. They will also have a capability for data collection (from ground-based instrument platforms and balloons) and relay to central processing and dissemination facilities.

Fig. 29.7 Direct-readout thermal infrared imagery from measurements taken by the ITOS-1 satellite on 8 November 1970 off the East Coast of the United States. This is a special display used to enhance the thermal contrasts associated with the Gulf Stream and the coastal waters.

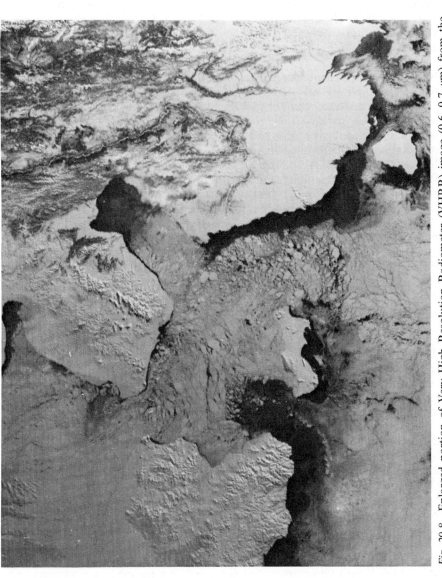

Fig. 29.8 Enlarged portion of Very High Resolution Radiometer (VHRR) image (0.6–0.7 μm) from the NOAA-2 satellite on 28 February 1973. The area shown is western Alaska, including Nunivak Island and St. Lawrence Island, the Bering Sea and extreme eastern Siberia.

Recent studies using near-infrared data from the Nimbus 3 satellite have demonstrated the capability of data in this band for the detection of thawing snow and ice surfaces (Strong et al. 1971). Earlier studies of Nimbus data in the water-vapor absorption band at 6–7 μm have shown their usefulness for the detection of cirrus clouds and water vapor in the high troposphere (Fritz and Rao 1967). These satellites will also feature, for the first time, onboard analog-to-digital conversion of data before transmission to the data acquisition station.

Near-future research and development satellites

Several research and development type satellites are scheduled in 1972 also. The first Earth Resources Technology Satellite (ERTS) is one, and the fifth in the Nimbus series is the other. An SMS (Synchronous Meteorological Satellite), the prototype of a NOAA operational model to be called GOES (Geostationary Operational Environmental Satellite), is also presently slated for launch in early 1974, but because of its location at the equator in the eastern Pacific Ocean, its usefulness for Bering Sea studies will be limited to data collection and relay from ground stations.

ERTS-A (General Electric 1971) is equipped with a 3-camera Return Beam Vidicon (RBV) and a four-channel Multispectral Scanner Subsystem (MSS), both sensitive to three bands in the visible range (0.5–0.6, 0.6–0.7, and 0.7–0.8 μm), and the latter to one band in the near-infrared (0.8–1.1 μm) spectrum. ERTS-B may add a fifth channel to the scanner, this one in the thermal infrared (10.4–12.6 μm). All bands are designed for ground resolutions of about 100 m except for the thermal infrared, which is approximately 250 m. The swath width on the ground, however, is to be only 185 km, and repeat coverage will be only every 18 days at the equator. The repeat coverage will be somewhat better at high latitudes, improving to 2–3 successive days' coverage separated by 15–16 days of no coverage in the Bering Sea area. The finer spectral and spatial resolution of the ERTS sensors in comparison with the operational satellites previously discussed should enable better and more detailed information on ice features and conditions to be obtained and better discrimination between ice and clouds (see Fig. 29.9).

The Nimbus 5 satellite will carry a Temperature-Humidity Infrared Radiometer (THIR) like the one on Nimbus 4, sensitive in the 10.5–12.5 and 6.5–7.0 μm bands and with 10 km resolution. Additionally it will carry three atmospheric sounder experiments: Infrared Temperature Profile Radiometer (ITPR), Selective Chopper Radiometer (SCR), and Nimbus E Microwave Spectrometer (NEMS). It will also employ an Electrically Scanning Microwave Radiometer (1.55 cm wave-length) with 25 km resolution. The ESMR is designed to derive the liquid water content of clouds and to detect ice cover through clouds (Wilheit et al. 1972).

A new research and development series of environmental satellites, designated EOS (Earth Observatory Satellite), is being planned as a merging and continuation of the Nimbus and ERTS programs in the period beginning about 1976. This new series of spacecraft is just in the stage of being defined, however, and it is not yet an approved program. Among the sensors presently under

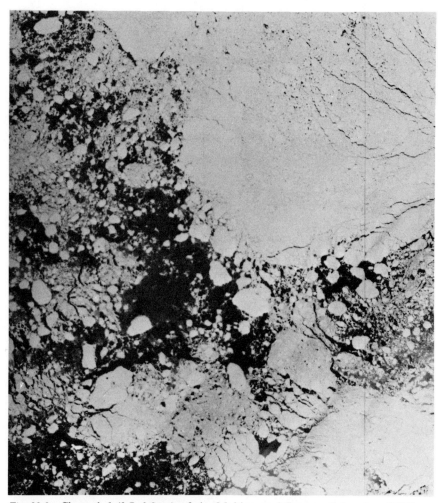

Fig. 29.9 Channel 6 (0.7–0.8μm) of the Multispectral Scanner Subsystem (MSS) on ERTS-1. A variety of ice conditions are seen in the East Siberian Sea. The darkened portions of the ice have melt water present.

consideration for the first EOS, where oceanographic capabilities are to be emphasized, are multi-channel sea-surface temperature and ocean color imaging radiometers and a microwave imaging radiometer.

Discussion

WADA: Is there any difference between the mean brightness and the mean cloudiness?

MᴄCʟᴀɪɴ: Brightness can be very highly correlated with cloudiness providing that the resolution is taken into account. This type of brightness cannot be considered strictly as an albedo, however. For example, one is not able to see extremely small cumulus clouds that are below the resolution of the measuring device, of course; likewise, there is a tendency to underestimate thin cirrus formations. As long as one is dealing mainly with large-scale cloudiness such as that associated with major storms, however, then brightness is an excellent measure of the mean cloudiness.

Nᴇsʜʏʙᴀ: Where will the satellite station be in Alaska?

MᴄCʟᴀɪɴ: The primary command and data acquisition station will be situated just outside Fairbanks; however, the data will probably be relayed (by a method not yet determined) to Anchorage, to accommodate the national weather service personnel and other major customers.

Nᴇsʜʏʙᴀ: Is there a possibility of getting an oceanographer aboard a manned orbiting observatory?

MᴄCʟᴀɪɴ: It is questionable how appropriate the participation of an ocean-ographer would be in a mission such as Skylab [which has been successfully in operation since 1973]. The orbit of this spacecraft is not a polar one, and about the only ice to be viewed by the scientist-astronauts aboard is on the Great Lakes or possibly near Labrador.

REFERENCES

Aʟʙᴇʀᴛ, E. G.
    1968   The improved TIROS operational satellite. ESSA Tech. Memo NESCTM-7, Dept. Commerce, Washington, D. C., 14 pp.

Bᴀʀɴᴇs, J. C., D. T. Cʜᴀɴɢ, and J. H. Wɪʟʟᴀɴᴅ
    1972   Improved techniques for mapping sea ice from satellite infrared data. Final Rep. to NOAA (Dept. Commerce) on Contr. E-67-70(N) from Allied Research Associates, Concord, Mass., 95 pp.

Bᴏᴏᴛʜ, A. L., and V. R. Taylor
    1969   Mesoscale archive and computer products of digitized video data from ESSA satellites. *Bull. Amer. Meteor. Soc.* 50: 431–438.

Bʀɪsᴛᴏʀ, C. L., W. M. Cᴀʟʟɪᴄᴏᴛᴛ, and R. E. Bʀᴀᴅғᴏʀᴅ
    1966   Operational processing of satellite pictures by computer. *Monthly Weather Rev.* 94: 515–527.

Fʀɪᴛᴢ, S., and P. K. Rᴀᴏ
    1967   On the infrared transmission through cirrus clouds and the estimation of relative humidity from satellites. *J. Applied Meteor.* 6: 1088–1096.

GENERAL ELECTRIC CORPORATION

1971 Earth Resources Technology Satellite data users handbook. Prep. for NASA Goddard Space Flight Center, Greenbelt, Md., under Contr. NAS 5-11320, 145 pp.

McCLAIN, E. P.

1973 Quantitative use of satellite vidicon data for delineating sea ice conditions. *Arctic* 26: 44–57.

McCLAIN, E. P., and D. R. BAKER

1969 Experimental large-scale snow and ice mapping with composite minimum brightness charts. ESSA Tech. Memo NESCTM-12, Dept. Commerce, Washington, D. C., 19 pp.

McCLAIN, E. P., and M. D. BALILES

1971 Sea ice surveillance from Earth satellites. *Mariners Weather Log* 15: 1–4.

RAO, P. K.

1970 Estimating cloud amount and height from satellite infrared radiation data. ESSA Tech. Rep. NESCTR-54, Dept. Commerce, Washington, D.C., 11 pp.

RAO, P. K., A. E. STRONG, and R. KOFFLER

1971a Gulf Stream meanders and eddies as seen in satellite infrared imagery. *J. Phys. Oceanogr.* 1: 237–239.

1971b Gulf Stream and Middle Atlantic Bight—complex thermal structure as seen from an environmental satellite. *Science* 173: 529–530 (+ cover).

RAO, P. K., W. L. SMITH, and R. KOFFLER

1972 Global sea-surface temperature distribution determined from an environmental satellite. *Monthly Weather Rev.* 100: 10–14.

SMITH, W. L., P. K. RAO, R. KOFFLER, and W. R. CURTIS

1970 Determination of sea-surface temperature from satellite high resolution infrared window radiation measurements. *Monthly Weather Rev.* 98: 604–611.

SCHWALB, A.

1972 Modified version of the Improved TIROS operational satellite (ITOS D-G). NOAA Tech. Memo NESS 35, Dept. Commerce, Washington, D. C., 48 pp.

STRONG, A. E., E. P. McCLAIN, and D. F. McGINNIS

1971 Detection of thawing snow and ice packs through combined use of visible and near-infrared measurements from Earth satellites. *Monthly Weather Rev.* 99: 828–830.

WENDLER, G.

1973 Sea ice observations by means of satellite. *J. Geophys. Res.* 78: 1427–1448.

WILHEIT, T., W. NORDBERG, J. BLINN, W. CAMPBELL, and A. EDGERTON

1972 Aircraft measurements of microwave emission from arctic sea ice. *Remote Sensing Environ.* 2: 129–139.

CHAPTER 30

Optical rectification of satellite pictures for sea-ice study

KANTARO WATANABE

Kobe Marine Observatory, Kobe, Japan

Abstract

In the United States, television pictures and infrared radiation imagery by modern meteorological satellites are automatically consolidated into rectified picture mosaics in polar stereographic and Mercator cylindrical projections by use of a high-speed computer. Although the digital rectification process has very wide application in daily weather analyses and in meteorological studies of global scope, the relatively simple method of *optical* rectification for satellite pictorial data is adequate for individual investigators who are concerned with precise analyses in certain confined localities. The principle and mechanism of an optical rectifier developed by the author for satellite pictures are presented.

INTRODUCTION

In meteorological or oceanographic applications of satellite pictorial data, all images in the television (TV) picture and the scanned infrared (IR) imagery must be first rectified into an appropriate projection map.

TV pictures by early TIROS satellites were manually rectified into regular maps by overlaying or superimposing on them radial-line, distance-circle grids (Watanabe 1961), modified Canadian grids (Hubert 1961), or latitude-longitude grids made by a graphic method (Fujita 1963) or by a computerized digital process (Frankel and Bristor 1962). After several trial optical rectifications for oblique pictures taken by TIROS satellites (K. Watanabe 1964 unpublished; T. Fujita 1964, unpublished; Senn and Davies 1966), a digital process of rectifying vertical pictures by ESSA satellites was established (Bristor et al. 1966). Since 1968, rectified picture mosaics in polar stereographic and Mercator cylindrical projections have been produced by high-speed computer at the National Environmental Satellite Services (NESS) laboratories of the

595

National Oceanic and Atmospheric Administration (NOAA) in the United States. Recently NOAA put into use a similar digital process there for rectifying IR radiation imagery scanned from ITOS and NOAA satellites.

Although these digital rectification processes can be widely utilized in daily weather analyses and meteorological studies of a global scale, local application by individual investigators is a problem. Such digital processes fail to provide adequate picture resolution in confined localities such as the Bering Sea. Facilities are limited to NESS at present for the purpose of daily weather analyses, and results are processed in the form of whole northern and southern hemispheres, respectively, in the polar stereographic projection and cover the complete equatorial belt from 30°S to 40°N in Mercator cylindrical projection. These pictures are too broad in area and offer insufficient picture resolution for precise studies in certain specific localities. For this purpose, an optical rectifier for satellite TV pictures which has been developed (K. Watanabe 1964, unpublished) is much simpler than the high-speed digital computer and can be used with ease in processing any frame of satellite pictures in localities of special interest at any time.

Relationship between the satellite picture and polar stereographic projection map
From operational meteorological satellites of the ESSA series (except ESSA-1), ITOS-1 and NOAA, TV pictures are taken in the vertical mode (Fig. 30.1). Point O represents the center of the earth, and the circle is a great circle of the earth passing through poles N and S. The lens (L) is that of a vidicon camera system located on board a satellite at altitude H when its sub-satellite point is at M_0 along the great circle of latitudinal angle ϕ. The arc A_0B_0 along the circle is a flat image of AB on the TV picture taken by the camera system. Taking the tangential plane at pole N, the arc A_0B_0 is projected as $A'B'$ onto the tangential plane from the perspective center at pole S and constitutes a part of the polar stereographic projection map. Although this illustration of the relationship between the satellite picture and polar stereographic projection map covers a specific arc (A_0B_0) along the great circle of the earth, it is applicable to any feature on the earth's surface. Thus the rectification of satellite pictures into a polar stereographic projection map is essentially a process of converting patterns in satellite pictures into projected images in the tangential plane at pole N. Such a process can be accomplished by the system of optical rectification explained in the following section.

Optical rectifier for satellite pictures
In accordance with the perspective principle illustrated in Figure 30.1, an optical rectifier can be constructed (Fig. 30.2) of three major components: a picture projector mounted with a superwide-angle lens (L); a spherical, convex screen ($A_1M_1B_1$) of opaline-coated transparent plastic; and a specially modified camera. A satellite picture (AB) in the form of a transparency is mounted in the projector and projected onto the spherical screen at the same viewing angle (image T) that the picture was taken from the satellite. By adjusting distance h between the front nodal point of the projector lens and the spherical screen, the projected image $A_1M_1B_1$ becomes geometrically similar to the actual

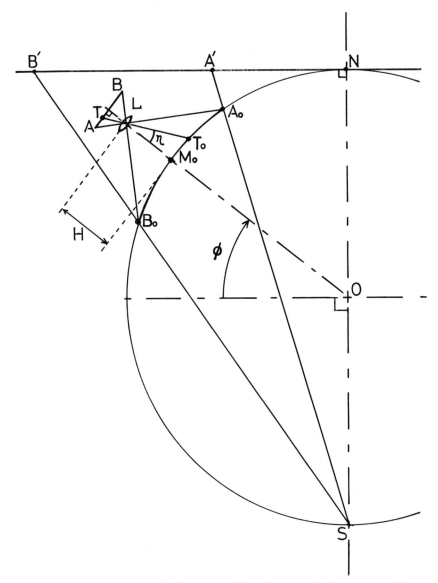

Fig. 30.1 Geometrical relationship between satellite picture and polar stereographic projection map.

feature on the earth's surface when it was taken from the satellite. The correct distance is given by the formula $h = rH/R$, where R is the radius of the earth and r the radius of the imaginary sphere of the spherical screen, and H is the altitude of the satellite. When the arc $A_1M_1B_1$ in the projected image coincides with the corresponding line of the meridian on the earth's surface passing through the object principal point by adjusting the picture orientation, the imaginary sphere containing the arc (the broken circle in Fig. 30.2) is a great circle of the imaginary sphere simulating the great circle of the earth passing through poles N and S.

The front nodal point C of the camera lens should be placed along the imaginary circle, and the position of the camera is adjusted so that the camera axis passing through the front nodal point C of the lens and the center O of the imaginary sphere deflects from the main axis LM_1O of this system by $(\frac{\pi}{2} - \phi)$, where ϕ is the latitudinal angle of the sub-satellite point when the picture was taken. The film plane must always be kept perpendicular to the axis OC. Furthermore, the plane of the camera lens passing through its front

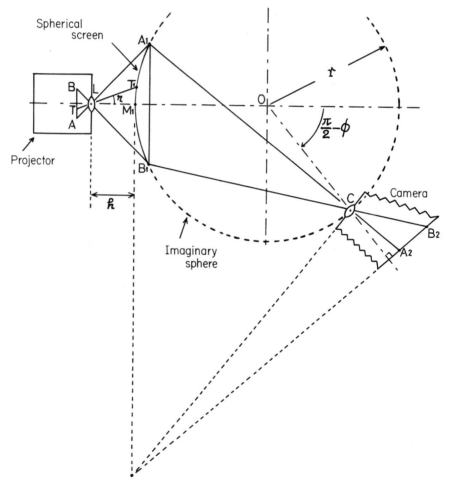

Fig. 30.2 Schematic representation of optical picture rectifier for polar stereographic projection map.

Fig. 30.3 Side view of optical picture rectifier.

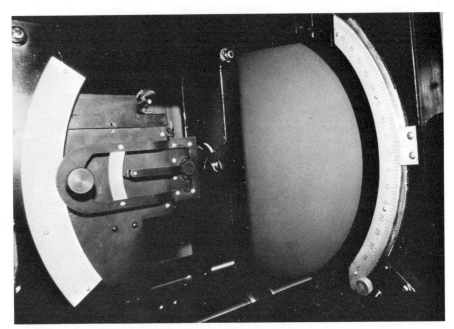

Fig. 30.4 Close-up view of projector and spherical screen.

Fig. 30.5 Camera and scale of latitudinal angle for camera position.

nodal point must be swung so that the plane intersects with the line of the tangential plane of the spherical screen at M_1 and the film plane as shown in Figure 30.2 in accordance with Scheimpflung's condition for correct focusing.

With such an optical system, the satellite picture AB is rectified into the polar stereographic projection map by photographing the projected image $A_1M_1B_1$ on the spherical screen with the camera as A_2B_2 on the film, because the image A_2B_2 becomes geometrically similar to the perspectively projected image A'B' in Figure 30.1.

The photograph in Figure 30.3 is a side view of the rectifier, where the projector is seen in the left, the camera in the right, and the angle-scale for adjusting the camera position attached to the main body in the lower right. Figure 30.4 is a close-up view of the projector and the spherical screen. The projector is movable back and forth from or to the spherical screen along the main axis of this system, LM_1O in Figure 30.2, for adjusting h using the distance scale attached to the lower part of the projector frame. A Canon superwide-angle lens for a 35-mm camera with a focal length of 19 mm is used for the projector lens. Furthermore, the projector is tilted at a proper angle around the

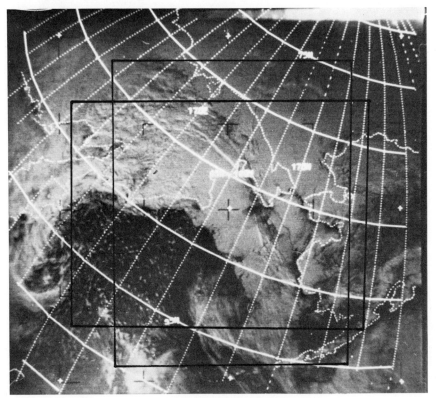

Fig. 30.6   Example of rectified ESSA-9 picture presented in Figure 30.7.

front nodal point of the projector lens using the tilt-angle scale seen in the left when oblique pictures by TIROS, ESSA-1, NIMBUS-1 and -11 are processed. The smaller angle-scale seen just right of the tilt-angle scale is for swinging the transparency mounted in the projector slightly from its proper plane when vertical pictures are processed. This swinging is necessary for getting the sharpest focus in the projected image on the spherical screen in accordance with Scheimpflung's condition. The spherical screen seen in this photograph has a disc 48 cm in diameter, made of a transparent plastic dome 78-cm radius

Fig. 30.7   ESSA-9 picture taken on 13 March 1970 (rectified in Fig. 30.6).

by coating its convex surface opaline in color. The circular angle-scale attached outside of the spherical screen is for adjusting the orientation of the satellite picture projected onto the spherical screen. The principal vertical line, the azimuth line toward the north in this case, in the projected satellite picture must be congruent with the great circle of the imaginary sphere passing through M_1 and C in Figure 30.2.

The camera in Figure 30.5 is modified from the original Topkon Horseman Press camera of 6 × 9 cm format, so that the lens can be swung around its front nodal point and the film holder can be slid 60 mm maximum. A Topkor lens 150 mm in focal length is generally used in this rectifier.

An example of products made in the first test use of the optical rectifier is presented in Figure 30.6, together with its original picture of the upper Bering Sea taken by ESSA-9 on 13 March 1970 (Fig. 30.7). Using such rectified pictures, detailed studies on the structure of ice-cover or of drift velocities of each ice-floe can be made. However, a slight distortion from the proper polar stereographic projection map is seen in areas near margins with slightly curved longitudinal lines (Fig. 30.6). It depends mostly on the incomplete adjustment of the distance h-scale in the first test use, and partly on the slight image-distortion in the original ESSA-9 picture caused by the vidicon camera system. By completing necessary adjustments and compensating the image distortion with an additional minor process, a simple and more accurate picture rectification can be produced.

Possible formats in rectification

The format of the original ESSA-9 picture in Figure 30.7 is not fully rectified in Figure 30.6. This restriction in rectification was caused by an inadequate combination of the projector lens and size of film format used in the test use for economical reasons. The left part of Figure 30.8 shows the relationship between a satellite-borne camera and the earth's surface when a picture is taken from a satellite, and the right portion illustrates the geometry of the projection of a satellite picture onto the spherical screen.

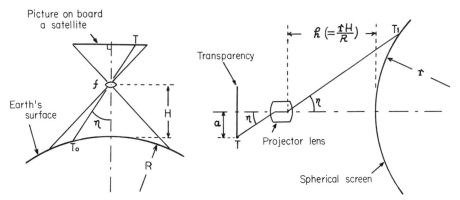

Fig. 30.8 Geometrical relationship between vertical satellite picture and its projection onto a spherical screen.

Taking the half length of the film format as a and the viewing angle, in the projection, of the margin of the format from the rear nodal point of the projector lens as η (Fig. 30.8, right), the viewing angle of the target T_0 on the earth's surface taken just on the margin of the transparency must be equal to η in the correct processing of the rectification. Then, the relation between a and η is given with the following equation,

$$\tan \eta = \frac{a}{f} - \frac{Ra}{rH},$$

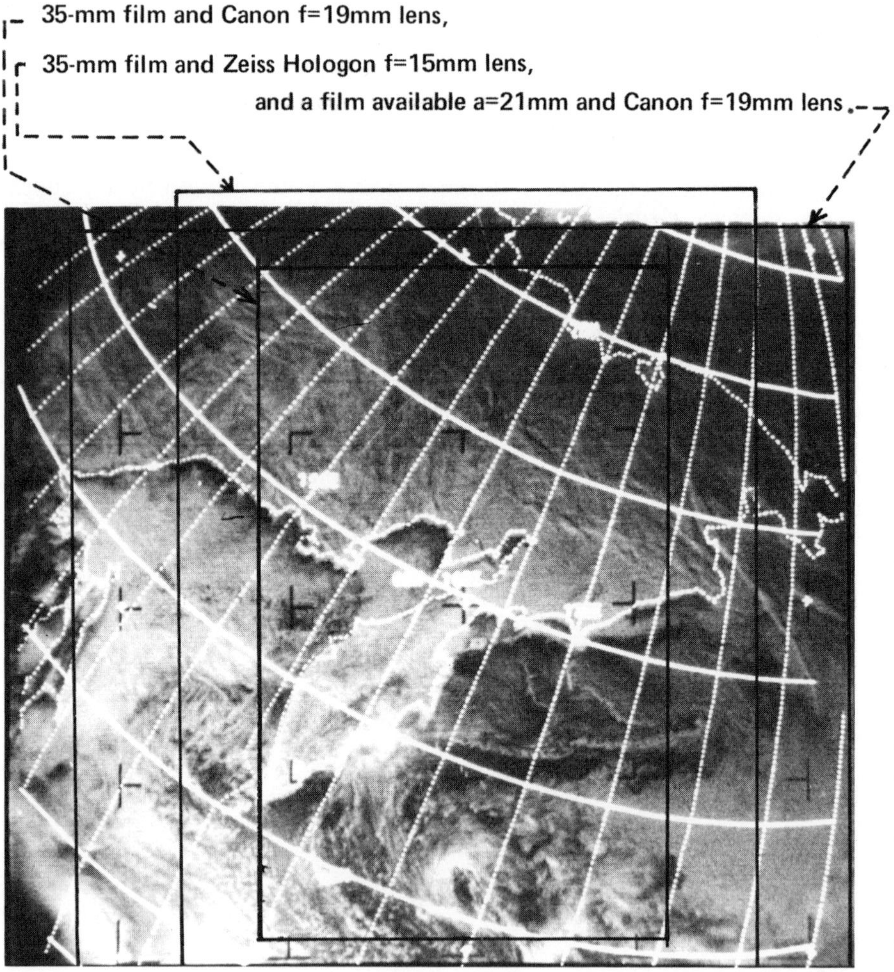

35-mm film and Canon f=19mm lens,

35-mm film and Zeiss Hologon f=15mm lens,

and a film available a=21mm and Canon f=19mm lens.

Fig. 30.9 Possible areas rectified in ESSA-9 pictures with three combinations of film format and projector lens.

where f is the focal length of the projector lens, R the radius of the earth, r the radius of the spherical screen, and H the altitude of the satellite.

Since the format of 35-mm film is 24 × 36 mm, the maximum amount of a is 18 mm. Then, the maximum amount of η comes to 39°20′ when a projector lens of $f = 19$ mm is used. The 2η means the maximum field of view in rectified pictures made with the combination of the 35-mm film and Canon superwide angle lens, $f = 19$ mm.

Similar calculations give $\eta = 47°40′$ for the combination of the 35-mm film and Zeiss Hologon ultrawide lens, $f = 15$ mm, and $\eta = 44°20′$ for the combination of a film available $a = 21$ mm and Canon superwide lens, $f = 19$ mm.

Areas in the ESSA picture which can be rectified with these three combinations of the film format and the projector lens are shown in Figure 30.9, which suggests that the combination of $a = 21$ mm and $f = 19$ mm is best and most practicable for the optical rectifier presented here.

Acknowledgments

The author would like to thank Yoshida Keiki Seisakusho Co., Ltd., and Sakai Jigu Seisakusho Co., Ltd., both in Osaka, for their cooperation in constructing the optical rectifier presented here. The contribution of the spherical screen by Minolta Camera Co., Ltd., is gratefully acknowledged. This study was conducted as a project of the Japan-U. S. Cooperative Science Program.

REFERENCES

BRISTOR, C. L., W. M. CALLICOTT, and R. E. BRADFORD
  1966 Operational processing of satellite cloud pictures by computer. *Monthly Weather Rev.* 94(8): 515–527.

FRANKEL, M., and C. L. BRISTOR
  1962 Perspective locator grids for TIROS pictures. Meteorol. Satellite Lab. Rep. No. 11, 38 pp.

FUJITA, T.
  1963 A technique for precise analysis for satellite data. Vol. 1. Photogrammetry. Meteorol. Satellite Lab. Rep. No. 14, 106 pp.
  1964 Distortion correction and optical rectification

HUBERT, L. F.
  1961 Canadian grids for TIROS-1: Additional orientation data. Supplement to Meteor. Satellite Lab. Rep. No. 5, 38 pp.

SENN, H. V., and P. J. DAVIES

    1966   Optical rectification and gridding of satellite photographs. *J. Applied Meteorol.* 5: 334–342.

WATANABE, K.

    1961   On the theory and technique of an easy method of wide-range photogrammetry for the observations of sea ice distribution. *Oceanogr. Mag.* 12(2): 77–121.

CHAPTER 31

Summary of technology development for Bering Sea study

STEVE NESHYBA

School of Oceanography, Oregon State University, Corvallis, Oregon

Each science of the real world begins with a description of that world, and this paper discusses the very real problems of data acquisition in the Bering Sea and its dissemination to users. Figure 31.1 is a reminder that logistics will play a most important role in establishing what can and what cannot be accomplished in forthcoming Bering Sea study. Logistics will not argue with the scientist as to *why* certain data may be required, nor will the scientist have a dominant role in determining *how* the data is to be acquired. The *when-where-what* of data acquisition is the principal meeting ground.

Historical data

After listening for three days to the world's most knowledgeable scientists on matters relating to the Bering Sea, I am convinced that this water is already well-measured in many ways. For example, Takenouti (Chapter 2) has shown that the southeast limit of 2 C bottom water on the Bering Sea shelf varies from year to year. Is there a cyclic pattern? Motoda and Minoda (Chapter 10) have discussed the areal coverages of certain species of zooplankton. These distribution patterns also appear to vary on a year-to-year basis. Nasu (Chapter 16) has shown the migration routes of whales in relation to hydrographic conditions. Girs (Chapter 25) has raised the question of possible coherences between the oceanographic and meteorological factors at work in the region of the Bering Sea. Cameron (Chapter 4) has advised that much value might be gained by a re-examination of existing data.

It seems important, then, to recognize that existing data, if properly ordered, may already be sufficient to yield a first-order recognition of the important interrelations between the dynamics of living resources and the dynamics of physical-chemical processes in this ocean. An early priority should be given, then, to the reacquisition of existing data, such that it may be more generally available to investigators who ought to plan their future programs around the idea that long-term oscillations are a part of their descriptions of phenomena in the Bering Sea. A possible beginning of such an ordering might be to construct an atlas. Another suggestion has been made that a data group set up under the auspices of UNESCO would be appropriate to accomplish the task.

FUTURE DATA

Data acquired in the future by a joint study of the Bering Sea should be disseminated quickly. There will be problems. The human problem—the natural inclination of a scientist to release data only after thorough verification of its merit—is a mixed asset to an international study. The data "relationship" between an international study group and existing international commissions on living resources must be defined. Agencies like the National Oceanographic Data Center, however, have already learned to treat many of the technological problems accompanying data dissemination. We should seek new, more efficient ways of displaying and conveying data; these are the principal means of exchange of ideas. For example, Dr. Tabata's film on motion of the ice pack cannot be "published" in the ordinary sense, but it certainly transmits a great deal of information to the viewer. I recommend that the groundwork be laid for an internal publication for international Bering Sea study as a vehicle for the printing of translated documents, re-printing papers from journals of limited distribution and distributing newly acquired data. A publication of this type is presently used with success by the Arctic Ice Dynamics Joint Experiment (AIDJEX).

The *how* of technological support
An international study group must acquire the data it requires from the Bering Sea in any way it can. We have already seen, at this symposium, how important will be the data support of satellites. Conventional ship support will be made available to us in direct proportion to our efforts at obtaining the support, as

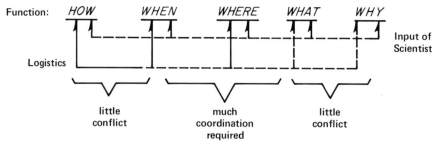

Fig. 31.1 Schematic diagram of data acquisition and dissemination for Bering Sea study.

measured by the merit of the science proposed and the efficiency with which we order priorities to the many, many tasks which are outlined here this week. Aggressiveness is important. For example, we know that new programs are underway to gain new information about the South Pacific. We collectively already know a great deal about the Bering Sea, and its economic importance is established; we should be aggressive, because international study of the Bering Sea has come into its time. It is most important to recognize the existing programs which touch in whole or in part the Bering Sea and to work out complementary activities, thereby avoiding duplication of interests and logistics. It is most important to seek new methods of data acquisition. Dr. Tabata's work on sea ice using radar is in the tradition of pioneering science technology. The technique may be extended to such applications as tracing surface-current drogues and indirect measurements of wind regimes and surface roughness.

Every major ship on an oceanographic or meteorological mission to the Bering Sea should have helicopter capability to perform STD sampling over a 100–150 km path centered on the ship course and to obtain water samples through the water column throughout a similar path for particulate matter, gas analyses and optical hydrology both *in situ* and in the laboratory. The ship should tow chemical sensors; perhaps it should perform acoustic assessment and be equipped to fish and to obtain bottom sediment samples as well. It should provide divers transportation around the Aleutian inshore regions while it makes STD profiles across Aleutian passes. It should set and retrieve short-term current meter stations on the shelf.

It is intriguing to consider possible oceanographic applications of a technique now used by modern forest fire fighters—namely, the recharging of water tanks aboard an aircraft while the craft is flying; here the technique is to skim along a local lake or other water body, refilling the tanks by scooping up water in-flight. Why not obtain surface samples from the ocean in a similar way? Expendable STD sensors deployed from aircraft are to be available soon. Can these be used profitably? Vertical, free-falling sensors of horizontal current structure are already in use as a research tool. The technique may have the same revolutionizing effect upon current measurements and studies that the STD has had upon density studies in oceanography. The sensor uses a rotating pair of electrodes to monitor passively the electric field within the water column; its ideal usage is realized when deployed synoptically over a wide area—again a potential helicopter application.

In summation, technology may provide its best input to Bering Sea study by convincing scientists that they cannot afford to *not* take advantage of the state of art in data acquisition.

The *when* function of logistical support

It should be mutually recognized by scientists on one hand and by logistical support personnel on the other hand that the statement "dynamics of . . ." implies variability at a spectrum of frequencies with respect to the Bering Sea. On the high frequency end of the spectrum, a periodicity of four to seven days covers variations due to storms. This variability is probably the highest significant dynamic feature other than the diurnal factors. Convergence and divergence of the upper layer, with accompanying upwelling, must be important. Perhaps more important than slow, continuous processes are those which accompany individual storms of a few days duration. A single storm may do more to enhance the plankton bloom than one or more weeks of relatively calm conditions. Biological rates can be closely coupled to storm frequencies, since wind stress and the local depth of the mixed layer are variable at these frequencies.

On the low frequency side, variations of the period five to seven years are already recognizable. The Nyquist sampling *rate* necessary to define these geophysical processes is one sample per 15–20 months. Again, some data at this frequency range are relatively easy to get; they can be acquired from fishing and fishery research vessels, from annual cruises taken by individual investigators (e.g., Coachman et al.) and from new programs such as MARMAP (Marine Resource Mapping, Assessment and Prediction) of the National Marine Fishery Service.

However, these programs will not yield a total picture; each is confined to specific objectives. The international Bering Sea study should attempt to complement in data coverage those programs already underway, to assure the widest coverage at this very important periodicity of five to seven years per cycle. Variations at the mid-frequency or meso-time scale of events will be most difficult to get by using ships, because of the severe winter environment of the Bering Sea. Physicists and meteorologists are fortunate in that satellite data fill this part of the spectrum quite nicely. Biologists and chemists will have less seasonal data, since their predictive models of summer conditions must be based in part on important fluctuations in winter conditions. We must give much attention to their problems. It is here that the development or use of sampling by aircraft may find its best return. Care should be exercised in the forming of predictive models (physical, chemical, or biological) that one does not overlook the importance of the so-called *rare event*. Geophysicists are even now debating how to explain *rates* of sediment deposition in relation to geologic time. It appears that some major sediment deposits cannot be accounted for by an accumulation at a steady rate over a geologic age; rather, it seems that a series of very rapid accumulations, separated in time by long periods of very low rates, is a model that best fits the facts. Long-range studies are not exempt from the rare event.

Early data support
If the end product of an extensive study of the Bering Sea is to be a series of predictive models of physical, chemical, and biological behavior of the region, then it is wise to initiate at the outset the acquisition of at least one descriptive parameter of the region on a systematic basis, with provision that the accumulation of the characteristic data be continued uninterrupted for two cycles of the lowest frequency of significant variability. The reason for this is to provide the data base against which the predictive models may be balanced. Two data fields satisfy this need: The sea-surface temperature field may be obtained either from satellite or, preferably, from aircraft equipped with radiometers. Sea-level guage stations around the Bering Basin would be

relatively easy to place into operation and are of particular importance to predictive models of circulation in that the network registers the horizontally integrated effect of wind stress. If the early work on the Bering Sea is to be descriptive of the oceanic environment of the sea, certainly circulation models will be an early result of a coordinated study.

roster of participants

International Symposium for Bering Sea Study
31 January–4 February 1973

Dr. Tadashi Asakura
Forecast Division
Japan Meteorological Agency
Ootemachi 1-3-4, Chiyoda-ku
Tokyo

Dr. Robert J. Barsdate
Institute of Marine Science
University of Alaska
Fairbanks, Alaska 99701

Dr. Carl S. Benson
Geophysical Institute
University of Alaska
Fairbanks, Alaska 99701

Dr. William M. Cameron
Pacific Environment Institute
Marine Science Branch
4160 Marine Drive
West Vancouver, B. C., Canada

Dr. Lawrence K. Coachman
Department of Oceanography
University of Washington
Seattle, Washington 98105

Dr. Nikolai C. Fadeev
Pacific Scientific Institute of
    Fisheries and Oceanography
    (TINRO)
20 Ulitsa Leninskaya
Vladivostok, USSR

Dr. Felix Favorite
Northwest Fisheries Center (NOAA)
2725 Montlake Boulevard, East
Seattle, Washington 98102

Dr. Francis H. Fay
Arctic Health Research Center
University of Alaska
Fairbanks, Alaska 99701

Prof. A. A. Girs
Arctic and Antarctic Research
    Institute
Fontanka 34
Leningrad 192104, USSR

Dr. Akihiko Hattori
Ocean Research Institute
University of Tokyo
Minamidai 1-15-1, Nakano-ku
164 Tokyo

Mr. Deane E. Holt
Office for the IDOE
National Science Foundation
1800 G Street
Washington, D. C. 20550

Dr. Donald W. Hood
Institute of Marine Science
University of Alaska
Fairbanks, Alaska 99701

Dr. Yoshio Horibe
Ocean Research Institute
University of Tokyo
Minamidai 1-15-1, Nakano-ku
164 Tokyo

Dr. Frank W. Hughes
Department of Oceanography
University of Washington
Seattle, Washington 98105

Dr. Hiroshi Kasahara
Food and Agricultural Organization
    of the United Nations
Via delle Terme di Caracalla
00100 Rome, Italy

Mrs. Eleanor J. Kelley
Institute of Marine Science
University of Alaska
Fairbanks, Alaska 99701

Dr. Kiyomitsu Kitano
Hokkaido Regional Fisheries
    Research Laboratory
Hamanakacho, Yoichi

Dr. Yasushi Kitano
Water Research Institute
Nagoya University
Furocho, Chigusa-ku
Nagoya

Dr. Kou Kusonoki
National Institute of Polar Research
Kaga 1-9-10, Itabashi-ku
Tokyo

Dr. Tatiyana G. Liubimova
All-Union Research Institute of
    Marine Fisheries and
    Oceanography (VNIRO)
17 Krasnoselskaya
Moscow B-140, USSR

Dr. E. Paul McClain
National Oceanic and Atmospheric
    Administration (NOAA), Suite 300
3737 Branch Avenue
Washington, D. C. 20031

Dr. C. Peter McRoy
Institute of Marine Science
University of Alaska
Fairbanks, Alaska 99701

Dr. Takashi Minoda
Faculty of Fisheries
Hokkaido University
Minatocho 3-1-1
Hakodate

Dr. Sigeru Motoda
Faculty of Marine Science and
    Technology
Tokai University
Orido 1000
Shimizu, Shizuoka-ken

Dr. Keiji Nasu
Far-Sea Fisheries Research
    Laboratory
Orido 1000
Shimizu, Shizuoka-ken

Miss Lidiya G. Nazarova
All-Union Research Institute of
    Marine Fisheries and
    Oceanography (VNIRO)
17 Krasnoselskaya
Moscow B-140, USSR

Dr. C. Hans Nelson
U. S. Geological Survey
345 Middlefield Road
Menlo Park, California 94025

Dr. Takahisa Nemoto
Ocean Research Institute
University of Tokyo
Minamidai 1-15-1, Nakano-ku
164 Tokyo

Dr. Steve Neshyba
School of Oceanography
Oregon State University
Corvallis, Oregon 97331

Dr. Masaharu Nishiwaki
Ocean Research Institute
University of Tokyo
Minamidai 1-15-1, Nakano-ku
164 Tokyo

Mr. Tsuneo Nishiyama
Faculty of Fisheries
Hokkaido University
Minatocho 3-1-1
Hakodate

Dr. Satoshi Nishizawa
Faculty of Agriculture
Tohoku University
Kita 6 Bancho
Sendai

Dr. Hideo Nitani
Oceanographic Data Center
Hydrographic Department, MSA
Tsukiji 5-3-1, Chuo-ku
Tokyo

Mr. Takashi Ōkawa
Sapporo District Meteorological
    Observatory
Kita-2, Nishi-18
Sapporo

Dr. Kiyotaka Ohtani
Faculty of Fisheries
Hokkaido University
Minatocho 3-1-1
Hakodate

Dr. Kei Oshite
Hokkaido University of
    Education
Hakodate

Dr. P. Kilho Park
School of Oceanography
Oregon State University
Corvallis, Oregon 97331

Mr. Mitsuru Saito
Nippon Suisan Kaisha, Ltd.
Ootemachi 2-6, Chiyoda-ku
Tokyo

Dr. Ghanshyam D. Sharma
Institute of Marine Science
University of Alaska
Fairbanks, Alaska 99701

Dr. Richard R. Straty
Auke Bay Biological Laboratory
Post Office Box 155
Auke Bay, Alaska 99821

Dr. Jiro Sugiura
Hakodate Marine Observatory
Akagawadori 181, Kameda
Hakodate

Dr. Yoshio Sugiura
Meteorological Research Institute
Koenji-kita 4-35-8, Suginami-ku
166 Tokyo

Dr. Tadashi Tabata
Institute of Low Temperature
    Science, Hokkaido University
Kita-11, Nishi-7
Sapporo

Dr. A. Yositada Takenouti
College of Marine Science and
    Technology
Tokai University
Orido 1000
Shimizu, Shizuoka-ken

Dr. Tokimi Tsujita
Faculty of Fisheries
Hokkaido University
Minatocho 3-1-1
Hakodate

Dr. Shizuo Tsunogai
Faculty of Fisheries
Hokkaido University
Minatocho 3-1-1
Hakodate

Dr. Michitaka Uda
College of Marine Science and
    Technology
Tokai University
Orido 1000
Shimizu, Shizuoka-ken

Dr. Eitaro Wada
Ocean Research Institute
University of Tokyo
Minamidai 1-15-1, Nakano-ku
164 Tokyo

Dr. Hideo Wada
Forecast Division, Japan
  Meteorological Agency
Ootemachi 1-314, Chiyoda-ku
Tokyo

Dr. Kantaro Watanabe
Kobe Marine Observatory
Nakayamatedori 7-21, Ikuta-ku
Kobe

Dr. Norman J. Wilimovsky
Institute of Animal Resource Ecology
University of British Columbia
Vancouver (8), B. C., Canada

Dr. Tatsuo Yusa
Tohoku Regional Fisheries
  Research Laboratory
Niihama 3-119
Shiogama, Miyagi-ken

author index

617